T0228353

BRIDGE ENGINEERING

HANDBOOK

BRIDGE
ENGINEERING
HANDBOOK

VOLUME III

EDITED BY
WAI-FAH CHEN and LIAN DUAN

CRC Press
Taylor & Francis Group
Boca Raton London New York

CRC Press is an imprint of the
Taylor & Francis Group, an **informa** business

First published 1999 by CRC Press
Taylor & Francis Group
6000 Broken Sound Parkway NW, Suite 300
Boca Raton, FL 33487-2742

Reissued 2019 by CRC Press

© 1999 by Taylor & Francis Group.
CRC Press is an imprint of Taylor & Francis Group, an Informa business

No claim to original U.S. Government works

This book contains information obtained from authentic and highly regarded sources. Reasonable efforts have been made to publish reliable data and information, but the author and publisher cannot assume responsibility for the validity of all materials or the consequences of their use. The authors and publishers have attempted to trace the copyright holders of all material reproduced in this publication and apologize to copyright holders if permission to publish in this form has not been obtained. If any copyright material has not been acknowledged please write and let us know so we may rectify in any future reprint.

Except as permitted under U.S. Copyright Law, no part of this book may be reprinted, reproduced, transmitted, or utilized in any form by any electronic, mechanical, or other means, now known or hereafter invented, including photocopying, microfilming, and recording, or in any information storage or retrieval system, without written permission from the publishers.

For permission to photocopy or use material electronically from this work, please access www.copyright.com (http://www.copyright.com/) or contact the Copyright Clearance Center, Inc. (CCC), 222 Rosewood Drive, Danvers, MA 01923, 978-750-8400. CCC is a not-for-profit organization that provides licenses and registration for a variety of users. For organizations that have been granted a photocopy license by the CCC, a separate system of payment has been arranged.

Trademark Notice: Product or corporate names may be trademarks or registered trademarks, and are used only for identification and explanation without intent to infringe.

A Library of Congress record exists under LC control number:

Publisher's Note
The publisher has gone to great lengths to ensure the quality of this reprint but points out that some imperfections in the original copies may be apparent.

Disclaimer
The publisher has made every effort to trace copyright holders and welcomes correspondence from those they have been unable to contact.

ISBN 13: 978-0-367-26344-7 (set)
ISBN 13: 978-0-367-25335-6 (hbk)
ISBN 13: 978-0-367-25341-7 (pbk)
ISBN 13: 978-0-429-28728-2 (ebk)

Visit the Taylor & Francis Web site at http://www.taylorandfrancis.com and the CRC Press Web site at http://www.crcpress.com

Foreword

Among all engineering subjects, bridge engineering is probably the most difficult on which to compose a handbook because it encompasses various fields of arts and sciences. It not only requires knowledge and experience in bridge design and construction, but often involves social, economic, and political activities. Hence, I wish to congratulate the editors and authors for having conceived this thick volume and devoted the time and energy to complete it in such short order. Not only is it the first handbook of bridge engineering as far as I know, but it contains a wealth of information not previously available to bridge engineers. It embraces almost all facets of bridge engineering except the rudimentary analyses and actual field construction of bridge structures, members, and foundations. Of course, bridge engineering is such an immense subject that engineers will always have to go beyond a handbook for additional information and guidance.

I may be somewhat biased in commenting on the background of the two editors, who both came from China, a country rich in the pioneering and design of ancient bridges and just beginning to catch up with the modern world in the science and technology of bridge engineering. It is particularly to the editors' credit to have convinced and gathered so many internationally recognized bridge engineers to contribute chapters. At the same time, younger engineers have introduced new design and construction techniques into the treatise.

This Handbook is divided into seven sections, namely:

- Fundamentals
- Superstructure Design
- Substructure Design
- Seismic Design
- Construction and Maintenance
- Special Topics
- Worldwide Practice

There are 67 chapters, beginning with bridge concepts and aesthestics, two areas only recently emphasized by bridge engineers. Some unusual features, such as rehabilitation, retrofit, and maintenance of bridges, are presented in great detail. The section devoted to seismic design includes soil-foundation-structure interaction. Another section describes and compares bridge engineering practices around the world. I am sure that these special areas will be brought up to date as the future of bridge engineering develops.

May I advise each bridge engineer to have a desk copy of this volume with which to survey and examine both the breadth and depth of bridge engineering.

T. Y. Lin
Professor Emeritus, University of California at Berkeley
Chairman, Lin Tung-Yen China, Inc.

Preface

The *Bridge Engineering Handbook* is a unique, comprehensive, and state-of-the-art reference work and resource book covering the major areas of bridge engineering with the theme "bridge to the 21st century." It has been written with practicing bridge and structural engineers in mind. The ideal readers will be M.S.-level structural and bridge engineers with a need for a single reference source to keep abreast of new developments and the state-of-the-practice, as well as to review standard practices.

The areas of bridge engineering include planning, analysis and design, construction, maintenance, and rehabilitation. To provide engineers a well-organized, user-friendly, and easy-to-follow resource, the Handbook is divided into seven sections. *Section I, Fundamentals*, presents conceptual design, aesthetics, planning, design philosophies, bridge loads, structural analysis, and modeling. *Section II, Superstructure Design*, reviews how to design various bridges made of concrete, steel, steel-concrete composites, and timbers; horizontally curved, truss, arch, cable-stayed, suspension, floating, movable, and railroad bridges; and expansion joints, deck systems, and approach slabs. *Section III, Substructure Design*, addresses the various substructure components: bearings, piers and columns, towers, abutments and retaining structures, geotechnical considerations, footings, and foundations. *Section IV, Seismic Design*, provides earthquake geotechnical and damage considerations, seismic analysis and design, seismic isolation and energy dissipation, soil–structure–foundation interactions, and seismic retrofit technology and practice. *Section V, Construction and Maintenance*, includes construction of steel and concrete bridges, substructures of major overwater bridges, construction inspections, maintenance inspection and rating, strengthening, and rehabilitation. *Section VI, Special Topics*, addresses in-depth treatments of some important topics and their recent developments in bridge engineering. *Section VII, Worldwide Practice*, provides the global picture of bridge engineering history and practice from China, Europe, Japan, and Russia to the U.S.

The Handbook stresses professional applications and practical solutions. Emphasis has been placed on ready-to-use materials, and special attention is given to rehabilitation, retrofit, and maintenance. The Handbook contains many formulas and tables that give immediate answers to questions arising from practical works. It describes the basic concepts and assumptions, omitting the derivations of formulas and theories, and covers both traditional and new, innovative practices. An overview of the structure, organization, and contents of the book can be seen by examining the table of contents presented at the beginning, while an in-depth view of a particular subject can be seen by examining the individual table of contents preceding each chapter. References at the end of each chapter can be consulted for more-detailed studies.

The chapters have been written by many internationally known authors from different countries covering bridge engineering practices, research, and development in North America, Europe, and the Pacific Rim. This Handbook may provide a glimpse of a rapidly growing trend in global economy in recent years toward international outsourcing of practice and competition in all dimensions of engineering. In general, the Handbook is aimed toward the needs of practicing engineers, but materials may be reorganized to accommodate undergraduate and graduate level bridge courses. The book may also be used as a survey of the practice of bridge engineering around the world.

The authors acknowledge with thanks the comments, suggestions, and recommendations during the development of the Handbook by Fritz Leonhardt, Professor Emeritus, Stuttgart University, Germany; Shouji Toma, Professor, Horrai-Gakuen University, Japan; Gerard F. Fox, Consulting Engineer; Jackson L. Durkee, Consulting Engineer; Michael J. Abrahams, Senior Vice President, Parsons, Brinckerhoff, Quade & Douglas, Inc.; Ben C. Gerwick, Jr., Professor Emeritus, University of California at Berkeley; Gregory F. Fenves, Professor, University of California at Berkeley; John M. Kulicki, President and Chief Engineer, Modjeski and Masters; James Chai, Senior Materials and Research Engineer, California Department of Transportation; Jinrong Wang, Senior Bridge Engineer, URS Greiner; and David W. Liu, Principal, Imbsen & Associates, Inc.

We wish to thank all the authors for their contributions and also to acknowledge at CRC Press Nora Konopka, Acquiring Editor, and Carol Whitehead and Sylvia Wood, Project Editors.

Wai-Fah Chen
Lian Duan

Editors

Wai-Fah Chen is a George E. Goodwin Distinguished Professor of Civil Engineering and Head of the Department of Structural Engineering, School of Civil Engineering at Purdue University. He received his B.S. in civil engineering from the National Cheng-Kung University, Taiwan, in 1959, M.S. in structural engineering from Lehigh University, Bethlehem, Pennsylvania in 1963, and Ph.D. in solid mechanics from Brown University, Providence, Rhode Island in 1966.

Dr. Chen's research interests cover several areas, including constitutive modeling of engineering materials, soil and concrete plasticity, structural connections, and structural stability. He is the recipient of numerous engineering awards, including the AISC T.R. Higgins Lectureship Award, the ASCE Raymond C. Reese Research Prize, and the ASCE Shortridge Hardesty Award. He was elected to the National Academy of Engineering in 1995, and was awarded an Honorary Membership in the American Society of Civil Engineers in 1997. He was most recently elected to the Academia Sinica in Taiwan.

Dr. Chen is a member of the Executive Committee of the Structural Stability Research Council, the Specification Committee of the American Institute of Steel Construction, and the editorial board of six technical journals. He has worked as a consultant for Exxon's Production and Research Division on offshore structures, for Skidmore, Owings and Merril on tall steel buildings, and for World Bank on the Chinese University Development Projects.

A widely respected author, Dr. Chen's works include *Limit Analysis and Soil Plasticity* (Elsevier, 1975), the two-volume *Theory of Beam-Columns* (McGraw-Hill, 1976–77), *Plasticity in Reinforced Concrete* (McGraw-Hill, 1982), *Plasticity for Structural Engineers* (Springer-Verlag, 1988), and *Stability Design of Steel Frames* (CRC Press, 1991). He is the editor of two book series, one in structural engineering and the other in civil engineering. He has authored or coauthored more than 500 papers in journals and conference proceedings. He is the author or coauthor of 18 books, has edited 12 books, and has contributed chapters to 28 other books. His more recent books are *Plastic Design and Second-Order Analysis of Steel Frames* (Springer-Verlag, 1994), the two-volume *Constitutive Equations for Engineering Materials* (Elsevier, 1994), *Stability Design of Semi-Rigid Frames* (Wiley-Interscience, 1995), and *LRFD Steel Design Using Advanced Analysis* (CRC Press, 1997). He is editor-in-chief of *The Civil Engineering Handbook* (CRC Press, 1995, winner of the Choice Outstanding Academic Book Award for 1996, *Choice Magazine*), and the *Handbook of Structural Engineering* (CRC Press, 1997).

Lian Duan is a Senior Bridge Engineer with the California Department of Transportation, U.S., and Professor of Structural Engineering at Taiyuan University of Technology, China.

He received his B.S. in civil engineering in 1975, M.S. in structural engineering in 1981 from Taiyuan University of Technology, and Ph.D. in structural engineering from Purdue University, West Lafayette, Indiana in 1990. Dr. Duan worked at the Northeastern China Power Design Institute from 1975 to 1978.

Dr. Duan's research interests cover areas including inelastic behavior of reinforced concrete and steel structures, structural stability and seismic bridge analysis and design. He has authored or coauthored more than 60 papers, chapters, and reports, and his research has focused on the development of unified interaction equations for steel beam-columns, flexural stiffness of reinforced concrete members, effective length factors of compression members, and design of bridge structures.

Dr. Duan is also an esteemed practicing engineer. He has designed numerous building and bridge structures. Most recently, he has been involved in the seismic retrofit design of the San Francisco-Oakland Bay Bridge West spans and made significant contributions to the project. He is coeditor of the *Structural Engineering Handbook* CRCnetBase 2000 (CRC Press, 2000).

Contributors

Michael I. Abrahams
Parsons, Brinckerhoff, Quade &
    Douglas, Inc.
New York, New York

Mohamed Akkari
California Department of
    Transportation
Sacramento, California

Fadel Alameddine
California Department of
    Transportation
Sacramento, California

Masoud Alemi
California Department of
    Transportation
Sacramento, California

S. Altman
California Department of
    Transportation
Sacramento, California

Rambabu Bavirisetty
California Department of
    Transportation
Sacramento, California

David P. Billington
Department of Civil Engineering
    and Operations Research
Princeton University
Princeton, New Jersey

Michael Blank
U.S.Army Corps of Engineers
Philadelphia, Pennsylvania

Simon A. Blank
California Department of
    Transportation
Walnut Creek, California

Michel Bruneau
Department of Civil Engineering
State University of New York
Buffalo, New York

Chun S. Cai
Florida Department of
    Transportation
Tallahassee, Florida

James Chai
California Department of
    Transportation
Sacramento, California

Hong Chen
J. Muller International, Inc.
Sacramento, California

Kang Chen
MG Engineering, Inc.
San Francisco, California

Wai-Fah Chen
School of Civil Engineering
Purdue University
West Lafayette, Indiana

Nan Deng
Bechtel Corporation
San Francisco, California

Robert J. Dexter
Department of Civil Engineering
University of Minnesota
Minneapolis, Minnesota

Ralph J. Dornsife
Washington State Department of
    Transportation
Olympia, Washington

Lian Duan
California Department of
    Transportation
Sacramento, California

Mingzhu Duan
Quincy Engineering, Inc.
Sacramento, California

Jackson Durkee
Consulting Structural Engineer
Bethlehem, Pennsylvania

Marc O. Eberhard
Department of Civil and
    Environmental Engineering
University of Washington
Seattle, Washington

Johnny Feng
J. Muller International, Inc.
Sacramento, California

Gerard F. Fox
HNTB (Ret.)
Garden City, New York

John W. Fisher
Department of Civil Engineering
Lehigh University
Bethlehem, Pennsylvania

Kenneth J. Fridley
Washington State University
Pullman, Washington

John H. Fujimoto
California Department of
    Transportation.
Sacramento, California

Mahmoud Fustok
California Department of
    Transportation
Sacramento, California

Ben C. Gerwick, Jr.
Ben C. Gerwick, Inc.
Consulting Engineers
San Francisco, California

Mahmoud Fustok
California Department of
   Transportation
Sacramento, California

Ben C. Gerwick, Jr.
Ben C. Gerwick, Inc.
Consulting Engineers
San Francisco, California

Chao Gong
ICF Kaiser Engineers
Oakland, California

Frederick Gottemoeller
Rosales Gottemoeller & Associates,
   Inc.
Columbia, Maryland

Fuat S. Guzaltan
Parsons, Brickerhoff, Quade &
   Douglas, Inc.
Princeton, New Jersey

Danjian Han
Department of Civil Engineering
South China University of
   Technology
Guangzhou, China

Ikuo Harazaki
Honshu–Shikoku Bridge Authority
Tokyo, Japan

Lars Hauge
COWI
Consulting Engineers and Planners
Lyngby, Denmark

Oscar Henriquez
Department of Civil Engineering
California State University
Long Beach, California

Susan E. Hida
California Department of
   Transportation
Sacramento, California

Dietrich L. Hommel
COWI
Consulting Engineers and Planners
Lyngby, Denmark

Ahmad M. Itani
University of Nevada
Reno, Nevada

Kevin I. Keady
California Department of
   Transportation
Sacramento, California

Michael D. Keever
California Department of
   Transportation
Sacramento, California

Sangjin Kim
Kyungpook National University
Taeg, South Korea

F. Wayne Klaiber
Department of Civil Engineering
Iowa State University
Ames, Iowa

Michael Knott
Moffatt & Nichol Engineers
Richmond, Virginia

Steven Kramer
University of Washington
Seattle, Washington

Alexander Krimotat
SC Solutions, Inc.
Santa Clara, California

John M. Kulicki
Modjeski and Masters, Inc.
Harrisburg, Pennsylvania

John Kung
California Department of
   Transportation
Sacramento, California

Farzin Lackpour
Parsons, Brickerhoff, Quade &
   Douglas, Inc.
Princeton, New Jersey

Don Lee
California Department of
   Transportation
Sacramento, California

Fritz Leonhardt
Stuttgart University
Stuttgart, Germany

Fang Li
California Department of
   Transportation
Sacramento, California

Guohao Li
Department of Bridge Engineering
Tongji University
Shanghai, People's Republic of
   China

Xila Liu
Department of Civil Engineering
Tsinghua University
Beijing, China

Luis R. Luberas
U.S.Army Corps of Engineers
Philadelphia, Pennsylvania

M. Myint Lwin
Washington State Department of
   Transportation
Olympia, Washington

Jyouru Lyang
California Department of
   Transportation
Sacramento, California

Youzhi Ma
Geomatrix Consultants, Inc.
Oakland, California

Alfred R. Mangus
California Department of
   Transportation
Sacramento, California

W. N. Marianos, Jr.
Modjeski and Masters, Inc.
Edwardsville, Illinois

Brian Maroney
California Department of
   Transportation
Sacramento, California

Serge Montens
Jean Muller International
St.-Quentin-en-Yvelines
France

Jean M. Muller
Jean M. Muller International
St.-Quentin-en-Yvelines
France

Masatsugu Nagai
Department of Civil and
   Environmental Engineering
Nagaoka University of Technology
Nagaoka, Japan

Andrzej S. Nowak
Department of Civil and
   Environmental Engineering
University of Michigan
Ann Arbor, Michigan

Atsushi Okukawa
Honshu–Shikoku Bridge Authority
Kobe, Japan

Dan Olsen
COWI
Consulting Engineers and Planners
Lyngby, Denmark

Klaus H. Ostenfeld
COWI
Consulting Engineers and Planners
Lyngby, Denmark

Joseph Penzien
International Civil Engineering
   Consultants, Inc.
Berkeley, California

Philip C. Perdikaris
Department of Civil Engineering
Case Western Reserve University
Cleveland, Ohio

Joseph M. Plecnik
Department of Civil Engineering
California State University
Long Beach, California

Oleg A. Popov
Joint Stock Company
   Giprotransmost (Tramos)
Moscow, Russia

Zolan Prucz
Modjeski and Masters, Inc.
New Orleans, Louisiana

Mark L. Reno
California Department of
   Transportation
Sacramento, California

James Roberts
California Department of
   Transportation
Sacramento, California

Norman F. Root
California Department of
   Transportation
Sacramento, California

Yusuf Saleh
California Department of
   Transportation
Sacramento, California

Thomas E. Sardo
California Department of
   Transportation
Sacramento, California

Gerard Sauvageot
J. Muller International
San Diego, California

Charles Scawthorn
EQE International
Oakland, California

Charles Seim
T. Y. Lin International
San Francisco, California

Vadim A. Seliverstov
Joint Stock Company
   Giprotransmost (Tramos)
Moscow, Russia

Li-Hong Sheng
California Department of
   Transportation
Sacramento, California

Donald F. Sorgenfrei
Modjeski and Masters, Inc.
New Orleans, Louisiana

Jim Springer
California Department of
   Transportation
Sacramento, California

Shawn Sun
California Department of
   Transportation
Sacramento, California

Shuichi Suzuki
Honshu-Shikoku Bridge Authority
Tokyo, Japan

Andrew Tan
Everest International Consultants,
   Inc.
Long Beach, California

Man-Chung Tang
T. Y. Lin International
San Francisco, California

Shouji Toma
Department of Civil Engineering
Hokkai-Gakuen University
Sapporo, Japan

M. S. Troitsky
Department of Civil Engineering
Concordia University
Montreal, Quebec
Canada

Keh-Chyuan Tsai
Department of Civil Engineering
National Taiwan University
Taipei, Taiwan
Republic of China

Keh-Chyuan Tsai
Department of Civil Engineering
National Taiwan University
Taipei, Taiwan
Republic of China

Wen-Shou Tseng
International Civil Engineering
  Consultants, Inc.
Berkeley, California

Chia-Ming Uang
Department of Civil Engineering
University of California
La Jolla, California

Shigeki Unjoh
Public Works Research Institute
Tsukuba Science City, Japan

**Murugesu
Vinayagamoorthy**
California Department of
  Transportation
Sacramento, California

Jinrong Wang
URS Greiner
Roseville, California

Linan Wang
California Department of
  Transportation
Sacramento, California

Terry J. Wipf
Department of Civil Engineering
Iowa State University
Ames, Iowa

Zaiguang Wu
California Department of
  Transportation
Sacramento, California

Rucheng Xiao
Department of Bridge Engineering
Tongji University
Shanghai, China

Yan Xiao
Department of Civil Engineering
University of Southern California
Los Angeles, California

Tetsuya Yabuki
Department of Civil Engineering
and Architecture
University of Ryukyu
Okinawa, Japan

Quansheng Yan
College of Traffic and
  Communication
South China University of
  Technology
Guangzhou, China

Leiming Zhang
Department of Civil Engineering
Tsinghua University
Beijing, China

Rihui Zhang
California Department of
  Transportation
Sacramento, California

Ke Zhou
California Department of
  Transportation
Sacramento, California

Contents

SECTION V Construction and Maintenance

SECTION VI Special Topics

SECTION VII Worldwide Practice

Section V

Construction and Maintenance

45

Steel Bridge Construction

Jackson Durkee

Consulting Structural Engineer,
Bethlehem, Pa.

0-8493-7434-0/00/$0.00+$.50
© 2000 by CRC Press LLC

45.1 Introduction

This chapter addresses some of the principles and practices applicable to the construction of medium- and long-span steel bridges — structures of such size and complexity that construction engineering becomes an important or even the governing factor in the successful fabrication and erection of the superstructure steelwork.

We begin with an explanation of the fundamental nature of construction engineering, then go on to explain some of the challenges and obstacles involved. The basic considerations of cambering are explained. Two general approaches to the fabrication and erection of bridge steelwork are described, with examples from experience with arch bridges, suspension bridges, and cable-stayed bridges.

The problem of erection-strength adequacy of trusswork under erection is considered, and a method of appraisal offered that is believed to be superior to the standard working-stress procedure.

Typical problems with respect to construction procedure drawings, specifications, and practices are reviewed, and methods for improvement suggested. The need for comprehensive bridge erection-engineering specifications, and for standard conditions for contracting, is set forth, and the design-and-construct contracting procedure is described.

Finally, we take a view ahead, to the future prospects for effective construction engineering in the U.S.

The chapter also contains a large number of illustrations showing a variety of erection methods for several types of major steel bridges.

45.2 Construction Engineering in Relation to Design Engineering

With respect to bridge steelwork the differences between construction engineering and design engineering should be kept firmly in mind. Design engineering is of course a concept and process well known to structural engineers; it involves preparing a set of plans and specifications — known as the contract documents — that define the structure in its completed configuration, referred to as the geometric outline. Thus, the design drawings describe to the contractor the steel bridge superstructure that the owner wants to see in place when the project is completed. A considerable design engineering effort is required to prepare a good set of contract documents.

Construction engineering, however, is not so well known. It involves governing and guiding the fabrication and erection operations needed to produce the structural steel members to the proper cambered or "no-load" shape, and get them safely and efficiently "up in the air" in place in the structure, so that the completed structure under the deadload conditions and at normal temperature will meet the geometric and stress requirements stipulated on the design drawings.

Four key considerations may be noted: (1) design engineering is widely practiced and reasonably well understood, and is the subject of a steady stream of technical papers; (2) construction engineering is practiced on only a limited basis, is not as well understood, and is hardly ever discussed; (3) for medium- and long-span bridges, the construction engineering aspects are likely to be no less important than design engineering aspects; and (4) adequately staffed and experienced construction-engineering offices are a rarity.

45.3 Construction Engineering Can Be Critical

The construction phase of the total life of a major steel bridge will probably be much more hazardous than the service-use phase. Experience shows that a large bridge is more likely to suffer failure during erection than after completion. Many decades ago, steel bridge design engineering had progressed to the stage where the chance of structural failure under service loadings became altogether remote. However, the erection phase for a large bridge is inherently less secure, primarily because of the prospect of inadequacies in construction engineering and its implementation at the job site. The hazards associated with the erection of large steel bridges will be readily apparent from a review of the illustrations in this chapter.

Figure 45.1   Failure of a steel girder bridge during erection, 1995. Steel bridge failures such as this one invite suspicion that the construction engineering aspects were not properly attended to.

For significant steel bridges the key to construction integrity lies in the proper planning and engineering of steelwork fabrication and erection. Conversely, failure to attend properly to construction engineering constitutes an invitation to disaster. In fact, this thesis is so compelling that whenever a steel bridge failure occurs during construction (see for example Figure 45.1), it is reasonable to assume that the construction engineering investigation was either inadequate, not properly implemented, or both.

45.4 Premises and Objectives of Construction Engineering

During the erection sequences the various components of steel bridges may be subjected to stresses that are quite different from those which will occur under the service loadings and which have been provided for by the designer. For example, during construction there may be a derrick moving and working on the partially erected structure, and the structure may be cantilevered out some distance causing tension-designed members to be in compression and vice versa. Thus, the steelwork contractor needs to engineer the bridge members through their various construction loadings, and strengthen and stabilize them as may be necessary. Further, the contractor may need to provide temporary members to support and stabilize the structure as it passes through its successive erection configurations.

In addition to strength problems there are also geometric considerations. The steelwork contractor must engineer the construction sequences step by step to ensure that the structure will fit properly together as erection progresses, and that the final or closing members can be moved into position and connected. Finally, of course, the steelwork contractor must carry out the engineering studies needed to ensure that the geometry and stressing of the completed structure under normal temperature will be in accordance with the requirements of the design plans and specifications.

45.5 Fabrication and Erection Information Shown on Design Plans

Regrettably, the level of engineering effort required to accomplish safe and efficient fabrication and erection of steelwork superstructures is not widely understood or appreciated in bridge design offices, nor indeed by many steelwork contractors. It is only infrequently that we find a proper level of capability and effort in the engineering of construction.

The design drawings for an important bridge will sometimes display an erection scheme, even though most designers are not experienced in the practice of erection engineering and usually expend only a minimum or even superficial effort on erection studies. The scheme portrayed may not be practical, or may not be suitable in respect to the bidder or contractor's equipment and experience. Accordingly, the bidder or contractor may be making a serious mistake if he relies on an erection scheme portrayed on the design plans.

As an example of misplaced erection effort on the part of the designer, there have been cases where the design plans show cantilever erection by deck travelers, with the permanent members strengthened correspondingly to accommodate the erection loadings; but the successful bidder elected to use water-borne erection derricks with long booms, thereby obviating the necessity for most or all of the erection strengthening provided on the design plans. Further, even in those cases where the contractor would decide to erect by cantilevering as anticipated on the plans, there is hardly any way for the design engineer to know what will be the weight and dimensions of the contractor's erection travelers.

45.6 Erection Feasibility

Of course, the bridge designer does have a certain responsibility to his client and to the public in respect to the erection of the bridge steelwork. This responsibility includes: (1) making certain, during the design stage, that there is a feasible and economical method to erect the steelwork; (2) setting forth in the contract documents any necessary erection guidelines and restrictions; and (3) reviewing the contractor's erection scheme, including any strengthening that may be needed, to verify its suitability. It may be noted that this latter review does not relieve the contractor from responsibility for the adequacy and safety of the field operations.

Bridge annals include a number of cases where the design engineer failed to consider erection feasibility. In one notable instance the design plans showed the 1200 ft (366 m) main span for a long crossing over a wide river as an esthetically pleasing steel tied-arch. However, erection of such a span in the middle of the river was impractical; one bidder found that the tonnage of falsework required was about the same as the weight of the permanent arch-span steelwork. Following opening of the bids, the owner found the prices quoted to be well beyond the resources available, and the tied-arch main span was discarded in favor of a through-cantilever structure, for which erection falsework needs were minimal and practical.

It may be noted that design engineers can stand clear of serious mistakes such as this one, by the simple expedient of conferring with prospective bidders during the preliminary design stage of a major bridge.

45.7 Illustrations of Challenges in Construction Engineering

Space does not permit comprehensive coverage of the numerous and difficult technical challenges that can confront the construction engineer in the course of the erection of various types of major steel bridges. However, some conception of the kinds of steelwork erection problems, the methods available to resolve them, and the hazards involved can be conveyed by views of bridges in various stages of erection; refer to the illustrations in the text.

45.8 Obstacles to Effective Construction Engineering

There is an unfortunate tendency among design engineers to view construction engineering as relatively unimportant. This view may be augmented by the fact that few designers have had any significant experience in the engineering of construction.

Further, managers in the construction industry must look critically at costs, and they can readily develop the attitude that their engineers are doing unnecessary theoretical studies and calculations, detached from the practical world. (And indeed, this may sometimes be the case.) Such management apprehension can constitute a serious obstacle to staff engineers who see the need to have enough money in the bridge tender to cover a proper construction engineering effort for the project. There is the tendency for steelwork construction company management to cut back the construction engineering allowance, partly because of this apprehension and partly because of the concern that other tenderers will not be allotting adequate money for construction engineering. This effort is often thought of by company management as "a necessary evil" at best — something they would prefer not to be bothered with or burdened with.

Accordingly, construction engineering tends to be a difficult area of endeavor. The way for staff engineers to gain the confidence of management is obvious — they need to conduct their investigations to a level of technical proficiency that will command management respect and support, and they must keep management informed as to what they are doing and why it is necessary. As for management's concern that other bridge tenderers will not be putting into their packages much money for construction engineering, this concern is no doubt often justified, and it is difficult to see how responsible steelwork contractors can cope with this problem.

45.9 Examples of Inadequate Construction Engineering Allowances and Effort

Even with the best of intentions, the bidder's allocation of money to construction engineering can be inadequate. A case in point involved a very heavy, long-span cantilever truss bridge crossing a major river. The bridge superstructure carried a contract price of some $30 million, including an allowance of $150,000, or about one-half of 1%, for construction engineering of the permanent steelwork (i.e., not including such matters as design of erection equipment). As fabrication and erection progressed, many unanticipated technical problems came forward, including brittle-fracture aspects of certain grades of the high-strength structural steel, and aerodynamic instability of H-shaped vertical and diagonal truss members. In the end the contractor's construction engineering effort mounted to about $1.3 million, almost nine times the estimated cost.

Another significant example — this one in the domain of buildings — involved a design-and-construct project for airplane maintenance hangars at a prominent international airport. There were two large and complicated buildings, each 100 × 150 m (328 × 492 ft) in plan and 37 m (121 ft) high with a 10 m (33 ft) deep space-frame roof. Each building contained about 2450 tons of structural steelwork. The design-and-construct steelwork contractor had submitted a bid of about $30 million, and included therein was the magnificent sum of $5,000 for construction engineering, under the expectation that this work could be done on an incidental basis by the project engineer in his "spare time."

As the steelwork contract went forward it quickly became obvious that the construction engineering effort had been grossly underestimated. The contractor proceeded to staff-up appropriately and carried out in-depth studies, leading to a detailed erection procedure manual of some 270 pages showing such matters as erection equipment and its positioning and clearances; falsework requirements; lifting tackle and jacking facilities; stress, stability, and geometric studies for gravity and wind loads; step-by-step instructions for raising, entering, and connecting the steelwork components; closing and swinging the roof structure and portal frame; and welding guidelines and procedures. This erection procedure manual turned out to be a key factor in the success of the fieldwork. The cost of this construction engineering effort amounted to about ten times the estimate, but still came to a mere one-fifth of 1% of the total contract cost.

In yet another example, a major steelwork general contractor was induced to sublet the erection of a long-span cantilever truss bridge to a reputable erection contractor, whose quoted *price* for the work was less than the general contractor's estimated *cost*. During the erection cycle the general contractor's engineers made some visits to the job site to observe progress, and were surprised and disconcerted to observe how little erection engineering and planning had been accomplished. For example, the erector had made no provision for installing jacks in the bottom-chord jacking points for closure of the main span; it was left up to the field forces to provide the jack bearing components inside the bottom-chord joints and to find the required jacks in the local market. When the job-built installations were tested it was discovered that they would not lift the cantilevered weight, and the job had to be shut down while the field engineer scouted around to find larger-capacity jacks. Further, certain compression members did not appear to be properly braced to carry the erection loadings; the erector had not engineered those members, but just assumed they were adequate. It became obvious that the erector had not appraised the bridge members for erection adequacy and had done little or no planning and engineering of the critical evolutions to be carried out in the field.

Many further examples of inadequate attention to construction engineering could be presented. Experience shows that the amounts of money and time allocated by steelwork contractors for the engineering of construction are frequently far less than desirable or necessary. Clearly, effort spent on construction engineering is worthwhile; it is obviously more efficient and cheaper, and certainly much safer, to plan and engineer steelwork construction in the office in advance of the work, rather than to leave these important matters for the field forces to work out. Just a few bad moves on site, with the corresponding waste of labor and equipment hours, will quickly use up sums of money much greater than those required for a proper construction engineering effort — not to mention the costs of any job accidents that might occur.

The obvious question is "Why is construction engineering not properly attended to?" Do not contractors learn, after a bad experience or two, that it is both necessary and cost effective to do a thorough job of planning and engineering the construction of important bridge projects? Experience and observation would seem to indicate that some steelwork contractors learn this lesson, while many do not. There is always pressure to reduce bid prices to the absolute minimum, and to add even a modest sum for construction engineering must inevitably reduce the prospect of being the low bidder.

45.10 Considerations Governing Construction Engineering Practices

There are no textbooks or manuals that define how to accomplish a proper job of construction engineering. In bridge construction (and no doubt in building construction as well) the engineering of construction tends to be a matter of each firm's experience, expertise, policies, and practices. Usually there is more than one way to build the structure, depending on the contractor's ingenuity and engineering skill, his risk appraisal and inclination to assume risk, the experience of his fabrication and erection work forces, his available equipment, and his personal preferences. Experience shows that each project is different; and although there will be similarities from one bridge of a given type to another, the construction engineering must be accomplished on an individual project basis. Many aspects of the project at hand will turn out to be different from those of previous similar jobs, and also there may be new engineering considerations and requirements for a given project that did not come forward on previous similar work.

During the estimating and bidding phase of the project the prudent, experienced bridge steelwork contractor will "start from scratch" and perform his own fabrication and erection studies, irrespective of any erection schemes and information that may be shown on the design plans. These studies can involve a considerable expenditure of both time and money, and thereby place that contractor at a disadvantage in respect to those bidders who are willing to rely on hasty, superficial studies, or — where the design engineer has shown an erection scheme — to simply assume that it has been engineered correctly and proceed to use it. The responsible contractor, on the other hand, will appraise the feasible construction methods and evaluate their costs and risks, and then make his selection.

After the contract has been executed the contractor will set forth how he intends to fabricate and erect, in detailed plans that could involve a large number of calculation sheets and drawings along with construction procedure documents. It is appropriate for the design engineer on behalf of his client to review the contractor's plans carefully, perform a check of construction considerations, and raise appropriate questions. Where the contractor does not agree with the designer's comments the two parties get together for review and discussion, and in the end they concur on essential factors such as fabrication and erection procedures and sequences, the weight and positioning of erection equipment, the design of falsework and other temporary components, erection stressing and strengthening of the permanent steelwork, erection stability and bracing of critical components, any erection check measurements that may be needed, and span closing and swinging operations.

The design engineer's approval is needed for certain fabrication plans, such as the cambering of individual members; however, in most cases the designer should stand clear of actual *approval* of the contractor's construction plans since he is not in a position to accept construction responsibility, and too many things can happen during the field evolutions over which the designer has no control.

It should be emphasized that even though the design engineer has usually had no significant experience in steelwork construction, the contractor should welcome his comments and evaluate them carefully and respectfully. In major bridge projects many construction matters can be improved upon or get out of control, and the contractor should take advantage of every opportunity to augment his prospects and performance. The experienced contractor will make sure that he works constructively with the design engineer, standing well clear of antagonistic or confrontational posturing.

45.11 Camber Considerations

One of the first construction engineering problems to be resolved by the steel bridge contractor is the cambering of individual bridge components. The design plans will show the "geometric outline" of the bridge, which is its shape under the designated load condition — commonly full dead load — at normal temperature. The contractor, however, fabricates the bridge members under the no-load condition, and at the "shop temperature" — the temperature at which the shop measuring tapes have been standardized and will have the correct length. The difference between the shape of a member under full dead load and normal temperature, and its shape at the no-load condition and shop temperature, is defined as member camber.

While camber is inherently a simple concept, it is frequently misunderstood; indeed, it is often not correctly defined in design specifications and contract documents. For example, beam and girder camber has been defined in specifications as "the convexity induced into a member to provide for vertical curvature of grade and to offset the anticipated deflections indicated on the plans when the member is in its erected position in the structure. Cambers shall be measured in this erected position..." This definition is not correct, and reflects a common misunderstanding of a key structural engineering term. Camber of bending members is not convexity, nor does it have anything to do with grade vertical curvature, nor is it measured with the member in the erected position. Camber — of a bending member, or any other member — is the *difference in shape* of the member under its no-load fabrication outline as compared with its geometric outline; and it is "measured" — i.e., the cambered dimensions are applied to the member — not when it is in the *erected* position (whatever that might be), but rather, when it is in the *no-load* condition.

In summary, camber is a *difference* in shape and not the shape itself. Beams and girders are commonly cambered to compensate for deadload bending, and truss members to compensate for deadload axial force. However, further refinements can be introduced as may be needed; for example, the arch-rib box members of the Lewiston-Queenston bridge (Fig. 45.4) were cambered to compensate for deadload axial force, bending, and shear.

A further common misunderstanding regarding cambering of bridge members involves the effect of the erection scheme on cambers. The erection scheme may require certain members to be strengthened, and this in turn will affect the cambers of those members (and possibly of others as well, in the case of statically indeterminate structures). However, the fabricator should address the matter of cambering only after the final sizes of all bridge members have been determined. Camber is a function of member properties, and there is no merit to calculating camber for members whose cross-sectional areas may subsequently be increased because of erection forces.

Thus, the erection scheme may affect the required member properties, and these in turn will affect member cambering; but the erection scheme does not *of itself* have any effect on camber. Obviously, the temporary stress-and-strain maneuvers to which a member will be subjected, between its no-load condition in the shop and its full-deadload condition in the completed structure, can have no bearing on the camber calculations for the member.

To illustrate the general principles that govern the cambering procedure, consider the main trusses of a truss bridge. The first step is to determine the erection procedure to be used, and to augment the strength of the truss members as may be necessary to sustain the erection forces. Next, the bridge deadload weights are determined, and the member deadload forces and effective cross-sectional areas are calculated.

Consider now a truss chord member having a geometric length of 49.1921 ft panel-point-to-panel-point and an effective cross-sectional area of 344.5 in.^{2}, carrying a deadload compressive force of 4230 kips. The bridge normal temperature is 45F and the shop temperature is 68F. We proceed as follows:

1. Assume that the chord member is in place in the bridge, at the full dead load of -4230 kips and the normal temperature of 45F.

2. Remove the member from the bridge, allowing its compressive force to fall to zero. The member will increase in length by an amount ΔL_s:

$$\Delta L_s = \frac{SL}{AE} = \frac{4230 \text{ kips} \times 49.1921}{344.5 \text{ in.}^2 \times 29000 \text{ kips}}$$

$$= 0.0208 \text{ft}$$

3. Now raise the member temperature from 45F to 68F. The member will increase in length by an additional amount ΔL_t:

$$\Delta L_t = L\omega t = (49.1921 + 0.0208) \text{ } ft \times$$

$$0.0000065 / \text{deg} \times (68 - 45) \text{deg}$$

$$= 0.0074 \text{ } ft$$

4. The total increase in member length will be:

$$\Delta L = \Delta L_s + \Delta L_t = 0.0208 + 0.0074$$

$$= 0.0282 \text{ } ft$$

5. The theoretical cambered member length — the no-load length at 68F — will be:

$$L_{tc} = 49.1921 + 0.0282 = 49.2203 \text{ } ft$$

6. Rounding L_{tc} to the nearest 1/32 in., we obtain the cambered member length for fabrication as:

$$L_{fc} = 49 \text{ ft } 2\frac{21}{32} \text{i}$$

Accordingly, the general procedure for cambering a bridge member of any type can be summarized as follows:

1. Strengthen the structure to accommodate erection forces, as may be needed.

2. Determine the bridge deadload weights, and the corresponding member deadload forces and effective cross-sectional areas.
3. Starting with the structure in its geometric outline, remove the member to be cambered.
4. Allow the deadload force in the member to fall to zero, thereby changing its shape to that corresponding to the no-load condition.
5. Further change the shape of the member to correspond to that at the shop temperature.
6. Accomplish any rounding of member dimensions that may be needed for practical purposes.
7. The total change of shape of the member — from geometric (at normal temperature) to no-load at shop temperature — constitutes the member camber.

It should be noted that the gusset plates for bridge-truss joints are always fabricated with the connecting-member axes coming in at their *geometric* angles. As the members are erected and the joints fitted-up, secondary bending moments will be induced at the truss joints under the steel-load-only condition; but these secondary moments will disappear when the bridge reaches its full-deadload condition.

45.12 Two General Approaches to Fabrication and Erection of Bridge Steelwork

As has been stated previously, the objective in steel bridge construction is to fabricate and erect the structure so that it will have the geometry and stressing designated on the design plans, under full dead-load at normal temperature. This geometry is known as the geometric outline. In the case of steel bridges there have been, over the decades, two general procedures for achieving this objective:

1. The "field adjustment" procedure — Carry out a continuing program of steelwork surveys and measurements in the field as erection progresses, in an attempt to discover fabrication and erection deficiencies; and perform continuing steelwork adjustments in an effort to compensate for such deficiencies and for errors in span baselines and pier elevations.
2. The "shop control" procedure — Place total reliance on first-order surveying of span baselines and pier elevations, and on accurate steelwork fabrication and erection augmented by meticulous construction engineering; and proceed with erection without any field adjustments, on the basis that the resulting bridge deadload geometry and stressing will be as good as can possibly be achieved.

Bridge designers have a strong tendency to overestimate the capability of field forces to accomplish accurate measurements and effective adjustments of the partially erected structure, and at the same time they tend to underestimate the positive effects of precise steel bridgework fabrication and erection. As a result, we continue to find contract drawings for major steel bridges that call for field evolutions such as the following:

1. **Continuous trusses and girders** — At the designated stages, measure or "weigh" the reactions on each pier, compare them with calculated theoretical values, and add or remove bearing-shoe shims to bring measured values into agreement with calculated values.
2. **Arch bridges** — With the arch ribs erected to midspan and only the short, closing "crown sections" not yet in place, measure thrust and moment at the crown, compare them with calculated theoretical values, and then adjust the shape of the closing sections to correct for errors in span-length measurements and in bearing-surface angles at skewback supports, along with accumulated fabrication and erection errors.
3. **Suspension bridges** — Following erection of the first cable wire or strand across the spans from anchorage to anchorage, survey its sag in each span and adjust these sags to agree with calculated theoretical values.
4. **Arch bridges and suspension bridges** — Carry out a deck-profile survey along each side of the bridge under the steel-load-only condition, compare survey results with the theoretical profile,

Figure 45.2   Erection of arch ribs, Rainbow Bridge, Niagara Falls, New York, 1941. Bridge span is 950 ft (290 m), with rise of 150 ft 46 m); box ribs are 3 × 12 ft (0.91 × 3.66 m). Tiebacks were attached starting at the end of the third tier and jumped forward as erection progressed (see Figure 45.3). Much permanent steelwork was used in tieback bents. Derricks on approaches load steelwork onto material cars that travel up arch ribs. Travelers are shown erecting last full-length arch-rib sections, leaving only the short, closing crown sections to be erected. Canada is at right, the U.S. at left. (Courtesy of Bethlehem Steel Corporation.)

and shim the suspender sockets so as to render the bridge floorbeams level in the completed structure.

5. **Cable-stayed bridges** — At each deck-steelwork erection stage, adjust tensions in the newly erected cable stays so as to bring the surveyed deck profile and measured stay tensions into agreement with calculated theoretical data.

There are two prime obstacles to the success of "field adjustment" procedures of whatever type: (1) field determination of the actual geometric and stress conditions of the partially erected structure and its components will not necessarily be definitive, and (2) calculation of the corresponding "proper" or "target" theoretical geometric and stress conditions will most likely prove to be less than authoritative.

45.13 Example of Arch Bridge Construction

In the case of the arch bridge closing sections referred to heretofore, experience on the construction of two major fixed-arch bridges crossing the Niagara River gorge from the U.S. to Canada — the Rainbow and the Lewiston-Queenston arch bridges (see Figures 45.2 through 45.5) — has demonstrated the difficulty, and indeed the futility, of attempts to make field-measured geometric and stress conditions agree with calculated theoretical values. The broad intent for both structures was to make such adjustments in the shape of the arch-rib closing sections at the crown (which were nominally about 1ft [0.3m] long) as would bring the arch-rib actual crown moments and thrusts into agreement with the calculated theoretical values, thereby correcting for errors in span-length measurements, errors in bearing-surface angles at the skewback supports, and errors in fabrication and erection of the arch-rib sections.

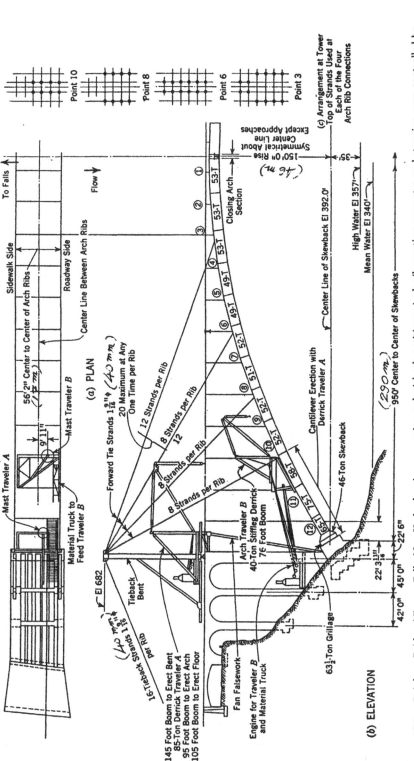

Figure 45.3 Rainbow Bridge, Niagara Falls, New York, showing successive arch tieback positions. Arch-rib erection geometry and stressing were controlled by means of measured tieback tensions in combination with surveyed arch-rib elevations.

Figure 45.4 Lewiston-Queenston arch bridge, near Niagara Falls, New York, 1962. The longest fixed-arch span in the U.S. at 1000 ft (305 m); rise is 159 ft (48 m). Box arch-rib sections are typically about 3 × 13-1/2 ft (0.9 × 4.1 m) in cross-section and about 44-1/2 ft (13.6 m) long. Job was estimated using erection tiebacks (same as shown in Figure 45.3), but subsequent studies showed the long, sloping falsework bents to be more economical (even if less secure looking). Much permanent steelwork was used in the falsework bents. Derricks on approaches load steelwork onto material cars that travel up arch ribs. The 115-ton-capacity travelers are shown erecting the last full-length arch-rib sections, leaving only the short, closing crown sections to be erected. Canada is at left, the U.S. at right. (Courtesy of Bethlehem Steel Corporation.)

Following extensive theoretical investigations and on-site measurements the steelwork contractor found, in the case of each Niagara arch bridge, that there were large percentage differences between the field-measured and the calculated theoretical values of arch-rib thrust, moment, and line-of-thrust position, and that the measurements could not be interpreted so as to indicate what corrections to the theoretical closing crown sections, if any, should be made. Accordingly, the contractor concluded that the best solution in each case was to abandon any attempts at correction and simply install the theoretical-shape closing crown sections. In each case, the contractor's recommendation was accepted by the design engineer.

Points to be noted in respect to these field-closure evolutions for the two long-span arch bridges are that accurate jack-load closure measurements at the crown are difficult to obtain under field conditions; and calculation of corresponding theoretical crown thrusts and moments are likely to be questionable because of uncertainties in the dead loading, in the weights of erection equipment, and in the steelwork temperature. Therefore, attempts to adjust the shape of the closing crown sections so as to bring the actual stress condition of the arch ribs closer to the presumed theoretical condition are not likely to be either practical or successful.

It was concluded that for long, flexible arch ribs, the best construction philosophy and practice is (1) to achieve overall geometric control of the structure by performing all field survey work and steelwork fabrication and erection operations to a meticulous degree of accuracy, and then (2) to rely on that overall geometric control to produce a finished structure having the desired stressing and geometry. For the Rainbow arch bridge, these practical construction considerations were set forth definitively by the contractor in [2]. The contractor's experience for the Lewiston-Queenston arch bridge was similar to that on Rainbow, and was reported — although in considerably less detail — in [10].

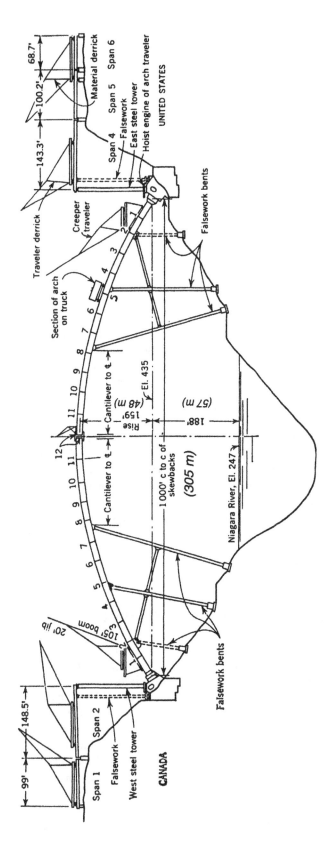

Figure 45.5 Lewiston-Queenston arch bridge, near Niagara Falls, New York. Crawler cranes erect steelwork for spans 1 and 6 and erect material derricks theron. These derricks erect traveler derricks, which move forward and erect supporting falsework and spans 2, 5, and 4. Traveler derricks erect arch-rib sections 1 and 2 and supporting falsework at each skewback, then set up creeper derricks, which erect arches to midspan.

45.14 Which Construction Procedure Is To Be Preferred?

The contractor's experience on the construction of the two long-span fixed-arch bridges is set forth at length since it illustrates a key construction theorem that is broadly applicable to the fabrication and erection of steel bridges of all types. This theorem holds that the contractor's best procedure for achieving, in the completed structure, the deadload geometry and stressing stipulated on the design plans, is generally as follows:

1. Determine deadload stress data for the structure at its geometric outline (under normal temperature), based on accurately calculated weights for all components.
2. Determine the cambered (i.e., "no-load") dimensions of each component. This involves determining the change of shape of each component from the deadload geometry, as its deadload stressing is removed and its temperature is changed from normal to the shop temperature. (Refer to Section 45.11).
3. Fabricate, with all due precision, each structural component to its proper no-load dimensions — except for certain flexible components such as wire rope and strand members, which may require special treatment.
4. Accomplish shop assembly of members and "reaming assembled" of holes in joints, as needed.
5. Carry out comprehensive engineering studies of the structure under erection at each key erection stage, determining corresponding stress and geometric data, and prepare a step-by-step erection procedure plan, incorporating any check measurements that may be necessary or desirable.
6. During the erection program, bring all members and joints to the designated alignment prior to bolting or welding.
7. Enter and connect the final or closing structural components, following the closing procedure plan, without attempting any field measurements thereof or adjustments thereto.

In summary, the key to construction success is to accomplish the field surveys of critical baselines and support elevations with all due precision, perform construction engineering studies comprehensively and shop fabrication accurately, and then carry the erection evolutions through in the field without any second guessing and ill-advised attempts at measurement and adjustment.

It may be noted that no special treatment is accorded to statically indeterminate members; they are fabricated and erected under the same governing considerations applicable to statically determinate members, as set forth above. It may be noted further that this general steel bridge construction philosophy does not rule out check measurements altogether, as erection goes forward; under certain special conditions, measurements of stressing and/or geometry at critical erection stages may be necessary or desirable in order to confirm structural integrity. However, before the erector calls for any such measurements he should make certain that they will prove to be practical and meaningful.

45.15 Example of Suspension Bridge Cable Construction

In order to illustrate the "shop control" construction philosophy further, its application to the main cables of the first Wm. Preston Lane, Jr., Memorial Bridge, crossing the Chesapeake Bay in Maryland, completed in 1952 (Figure 45.6), will be described. Suspension bridge cables constitute one of the most difficult bridge erection challenges. Up until "first Chesapeake" the cables of major suspension bridges had been adjusted to the correct position in each span by means of a sag survey of the first-erected cable wires or strands, using surveying instruments and target rods. However, on first Chesapeake, with its 1600 ft (488 m) main span, 661 ft (201 m) side spans, and 450 ft (137 m) back spans, the steelwork contractor recommended abandoning the standard cable-sag survey and adopting the "setting-to-mark" procedure for positioning the guide strands — a significant new concept in suspension bridge cable construction.

The steelwork contractor's rationale for "setting to marks" was spelled out in a letter to the design engineer (see Figure 45.7). (The complete letter is reproduced because it spells out significant construction

Figure 45.6   Suspension spans of first Chesapeake Bay Bridge, Maryland, 1952. Deck steelwork is under erection and is about 50% complete. A typical four-panel through-truss deck section, weighing about 100 tons, is being picked in west side span, and also in east side span in distance. Main span is 1600 ft (488 m) and side spans are 661 ft (201 m); towers are 324 ft (99 m) high. Cables are 14 in. (356 mm) in diameter and are made up of 61 helical bridge strands each (see Figure 45.8).

philosophies.) This innovation was accepted by the design engineer. It should be noted that the contractor's major argument was that setting to marks would lead to more accurate cable placement than would a sag survey. The minor arguments, alluded to in the letter, were the resulting savings in preparatory office engineering work and in the field engineering effort, and most likely in construction time as well.

Each cable consisted of 61 standard helical-type bridge strands, as shown in Figure 45.8. To implement the setting-to-mark procedure each of three bottom-layer "guide strands" of each cable (i.e., strands 1, 2, and 3) was accurately measured in the manufacturing shop under the simulated full-deadload tension, and circumferential marks were placed at the four center-of-saddle positions of each strand. Then, in the field, the guide strands (each about 3955 ft [1205 m] long) were erected and positioned according to the following procedure:

1. Place the three guide strands for each cable "on the mark" at each of the four saddles and set normal shims at each of the two anchorages.
2. Under conditions of uniform temperature and no wind, measure the sag differences among the three guide strands of each cable, at the center of each of the five spans.
3. Calculate the "center-of-gravity" position for each guide-strand group in each span.
4. Adjust the sag of each strand to bring it to the center-of gravity position in each span. This position was considered to represent the correct theoretical guide-strand sag in each span.

The maximum "spread" from the highest to the lowest strand at the span center, prior to adjustment, was found to be 1-3/4 in. (44 mm) in the main span, 3-1/2 in. (89 mm) in the side spans, and 3-3/4 in (95 mm) in the back spans. Further, the maximum change of perpendicular sag needed to bring the guide strands to the center-of-gravity position in each span was found to be 15/16 in. (24 mm) for the main span, 2-1/16 in. (52 mm) for the side spans, and 2-1/16 in. (52 mm) for the back spans. These small adjustments testify to the accuracy of strand fabrication and to the validity of the setting-to-mark strand adjustment procedure, which was declared to be a success by all parties concerned. It seems doubtful that such accuracy in cable positioning could have been achieved using the standard sag-survey procedure.

With the first-layer strands in proper position in each cable, the strands in the second and subsequent layers were positioned to hang correctly in relation to the first layer, as is customary and proper for suspension bridge cable construction.

This example provides good illustration that the construction-engineering philosophy referred to as the shop-control procedure can be applied advantageously not only to typical rigid-type steel structures, such as continuous trusses and arches, but also to flexible-type structures, such as suspension bridges.

July 6th, 1951
JJ:MM
C-1756

[To the design engineer]

Gentlemen: Attention of Mr. _____
                  Re: <u>Chesapeake Bay Bridge — Suspension Span Cables</u>

In our studies of the method of cable erection, we have arrived at the conclusion that setting of the guide strands to measured marks, instead of to surveyed sag, is a more satisfactory and more accurate method. Since such a procedure is not in accordance with the specifications, we wish to present for your consideration the reasoning which has led us to this conclusion, and to describe in outline form our proposed method of setting to marks.

On previous major suspension bridges, most of which have been built with parallel-wire instead of helical-strand cables, the thought has evidently been that setting the guide wire or guide strand to a computed sag, varying with the temperature, would be the most accurate method. This is associated with the fact that guide wires were never measured and marked to length. These established methods were carried over when strand-type cables came into use. An added reason may have been the knowledge that a small error in length results in a relatively large error in sag; and on the present structure the length-error to sag-error ratios are 1:2.4 and 1:1.5 for the main span and side spans, respectively.

However, the reading of the sag in the field is a very difficult operation because of the distances involved, the slopes of the side spans and backstays, the fact that even slight wind causes considerable motion to the guide strand, and for other practical reasons. We also believe that even though readings are made on cloudy days or at night, the actual temperature of all portions of the structure which will affect the sag cannot be accurately known. We are convinced that setting the guide strands according to the length marks thereon, which are placed under what amount to laboratory or ideal conditions at the manufacturing plant, will produce more accurate results than would field measurement of the sag.

To be specific, consider the case of field determination of sag in the main span, where it is necessary to establish accessible platforms, and an H.I. and a foresight somewhat below the desired sag elevation; and then to sight on the foresight and bring a target, hung from the guide strand, down to the line-of-sight. In the present case it is 1600 ft (488 m) to the foresight and 800 ft (244 m) to the target. Even if the line-of-sight were established just right, it would be only under perfect conditions of temperature and air — if indeed then — that such a survey would be precise. The difficulties are still greater in the side spans and back spans, where inclined lines-of-sight must be established by a series of offset measurements from distant bench marks. There is always the danger, particularly in the present location and at the time now scheduled, that days may be lost in waiting for the right conditions of weather to make an instrument survey feasible.

There is a second factor of doubt involved. The strand is measured under a known stress and at a known modulus, with "mechanical stretch" taken out. It is then reeled to a relatively small diameter and unreeled at the bridge site. Under its own weight, and until the full dead load has been applied, there is an indeterminable loss in mechanical set, or loss of modulus. A strand set to proper sag for the final modulus will accordingly be set too low, and the final cable will be below plan elevation. This possible error can only be on the side that is less desirable. Evidently, also, it could be on the order of 1-1/2 in. (40 mm) of sag increase for 1% of temporary reduction in modulus. If the strand were set to sag based on the assumed smaller modulus than will exist for the fully loaded condition, we doubt whether this smaller modulus could be chosen closely enough to ensure that the final sag would be correct. We are assured, however, by our manufacturing plant, that even though the modulus under bare-cable weight may be subject to unknown variation, the modulus which existed at the pre-stressing bed under the measuring tension will be duplicated when this same tension is

Figure 45.7   Setting cable guide strands to marks.

reached under dead load. Therefore, if the guide stand is set to measured marks, the doubt as to modulus is eliminated.

A third source of error is temperature. In past practice the sag has been adjusted, by reference to a chart, in accordance with the existing temperature. Granted that the adjustment is made in the early morning (the fog having risen but the sun not), it is hard to conceive that the actual average temperature in 3955 ft (1205 m) of strand will be that recorded by any thermometer. The mainspan sag error is about 0.7 in. (18 mm) per deg C of temperature.

These conditions are all greatly improved at the strand pre-stressing bed. There seems to be no reason to doubt that the guide strands can be measured and marked to an insignificant degree of error, at a stipulated stress and under a well-soaked and determinable temperature. Any errors in sag level must result from something other than the measured length of the guide strand.

There is one indispensable condition which, however, holds for either method of setting. That is, that the total distance from anchorage to anchorage, and the total calculated length of strand under its own-weight stress, must agree within the limits of shimming provided in the anchorages. Therefore, this distance in the field must be checked to close agreement. While the measured length of strand will be calculated with precision, it is interesting to note that in this calculation, it is not essential that the modulus be known with exactness. The important factor is that the strand length under the final deadload stress will be calculated exactly; and since that length is measured under the corresponding average strand stress, knowledge of the modulus is not a consideration. If the modulus at deadload stress is not as assumed, the only effect will be a change of deflection under live load, and this is minor. We emphasize again that the stand length under dead load, and the length as measured in the prestressing bed, will be identical regardless of the modulus.

The calculations for the bare-cable position result in pulled-back positions for the tops of the towers and cable bents, in order to control the unbalanced forces tending to slip the strands in the saddles. These pullback distances may be slightly in error without the slipping forces overcoming friction and thereby becoming apparent. Such errors would affect the final sags of strands set to sag. However, they would have no effect on the final sags of strands set-to-mark at the saddles; these errors change the temporary strand sags only, and under final stress the sags and the shaft leans will be as called for by the design plans.

It sometimes has happened that a tower which at its base is square to the bridge axis, acquires a slight skew as it rises. The amount of this skew has never, so far as we know, been important. If it is disregarded and the guide strands are attached without any compensating change, then the final loading will, with virtual certainty, pull the tower square. All sources of possible maladjustment have now been discussed except one — the errors in the several span lengths at the base of the towers and bents. The intention is to recognize and accept these, by performing the appropriate check measurements; and to correct for them by slipping the guide strands designated amounts through the saddles such that the center-of-saddle mark on the strand will be offset by that same amount from the centerline of the saddle.

If we have left unexplained herein any factor that seems to you to render our procedure questionable, we are anxious to know of it and discuss it with you in the near future; and we will be glad to come to your offices for this purpose. The detailed preparations for observing strand sags would require considerable time, and we are not now doing any work along those lines.

Yours very truly,
Chief Engineer

Figure 45.7 *(Continued)* Setting cable guide strands to marks.

One important caveat: the steelwork contractor must be a firm of suitable caliber and experience.

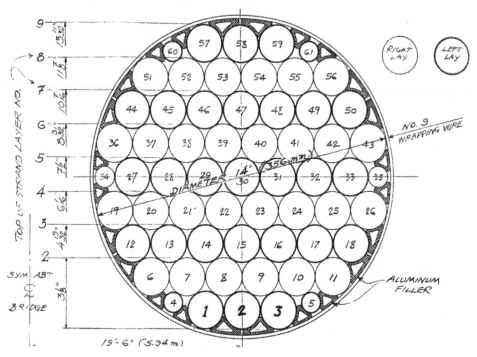

Figure 45.8   Main cable of first Chesapeake Bay suspension bridge, Maryland. Each cable consists of 61 helical-type bridge strands, 55 of 1-11/16 in. (43 mm) and 6 of 29/32 in. (23 mm) diameter. Strands 1, 2, and 3 were designated "guide strands" and were set to mark at each saddle and to normal shims at anchorages.

45.16 Example of Cable-Stayed Bridge Construction

In the case cable-stayed bridges, the first of which were built in the 1950s, it appears that the governing construction-engineering philosophy calls for field measurement and adjustment as the means for control of stay-cable and deck-structure geometry and stressing. For example, we have seen specifications calling for the completed bridge to meet the following geometric and stress requirements:

1. The deck elevation at midspan shall be within 12 in. (305 mm) of theoretical.
2. The deck profile at each cable attachment point shall be within 2 in. (50mm) of a parabola passing through the actual (i.e., field-measured) midspan point.
3. Cable-stay tensions shall be within 5% of the "corrected theoretical" values..

Such specification requirements introduce a number of problems of interpretation, field measurement, calculation, and field correction procedure, such as the following:

1. Interpretation:

   - The specifications are silent with respect to transverse elevation differentials. Therefore, two deck-profile control parabolas are presumably needed, one for each side of the bridge.

2. Field measurement of actual deck profile:

   - The temperature will be neither constant nor uniform throughout the structure during the survey work.

Figure 45.9 Cable-stayed orthotropic-steel-deck bridge over Mississippi River at Luling, La., 1982; view looking northeast. The main span is 1222 ft (372 m); the A-frame towers are 350 ft (107 m) high. A barge-mounted ringer derrick erected the main steelwork, using a 340 ft (104 m) boom with a 120 ft (37 m) jib to erect tower components weighing up to 183 tons, and using a shorter boom for deck components. Cable stays at the ends of projecting cross girders are permanent; others are temporary erection stays. Girder section 16-west of north portion of bridge, erected a few days previously, is projecting at left; companion girder section 16-east is on barge ready for erection (see Figure 45.10).

- The survey procedure itself will introduce some inherent error.

3. Field measurement of cable-stay tensions:

- Hydraulic jacks, if used, are not likely to be accurate within 2%, perhaps even 5%; further, the exact point of "lift off" will be uncertain.
- Other procedures for measuring cable tension, such as vibration or strain gaging, do not appear to define tensions within about 5%.
- All cable tensions cannot be measured simultaneously; an extended period will be needed, during which conditions will vary and introduce additional errors.

4. Calculation of "actual" bridge profile and cable tensions:

- Field-measured data must be transformed by calculation into "corrected actual" bridge profiles and cable tensions, at normal temperature and without erection loads.
- Actual dead weights of structural components can differ by perhaps 2% from nominal weights, while temporary erection loads probably cannot be known within about 5%.
- The actual temperature of structural components will be uncertain and not uniform.
- The mathematical model itself will introduce additional error.

5. "Target condition" of bridge:

Figure 45.10 Luling Bridge deck steelwork erection, 1982; view looking northeast (refer to Figure 45.9). Twin box girders are 14 ft (4.3 m) deep; deck plate is 7/16 in. (11 mm) thick. Girder section 16-east is being raised into position (lower right) and will be secured by large-pin hinge bars prior to fairing-up of joint holes and permanent bolting. Temporary erection stays are jumped forward as girder erection progresses.

- The "target condition" to be achieved by field adjustment will differ from the geometric condition, because of the absence of the deck wearing surface and other such components; it must therefore be calculated, introducing additional error.

6. Determining field corrections to be carried out by erector, to transform "corrected actual" bridge into "target condition" bridge:
   - The bridge structure is highly redundant, and changing any one cable tension will send geometric and cable-tension changes throughout the structure. Thus, an iterative correction procedure will be needed.

It seems likely that the total effect of all these practical factors could easily be sufficient to render ineffective the contractor's attempts to fine tune the geometry and stressing of the as-erected structure in order to bring it into agreement with the calculated bridge target condition. Further, there can be no assurance that the specifications requirements for the deck-profile geometry and cable-stay tensions are even compatible; it seems likely that *either* the deck geometry *or* the cable tensions may be achieved, but not *both*.

Specifications clauses of the type cited seem clearly to constitute unwarranted and unnecessary field-adjustment requirements. Such clauses are typically set forth by bridge designers who have great confidence in computer-generated calculation, but do not have a sufficient background in and understanding of the practical factors associated with steel bridge construction. Experience has shown that field procedures for major bridges developed unilaterally by design engineers should be reviewed carefully to determine whether they are practical and desirable and will in fact achieve the desired objectives.

In view of all these considerations, the question comes forward as to what design and construction principles should be followed to ensure that the deadload geometry and stressing of steel cable-stayed bridges will fall within acceptable limits. Consistent with the general construction-engineering procedures recommended for other types of bridges, we should abandon reliance on field measurements followed by adjustments of geometry and stressing, and instead place prime reliance on proper geometric control of bridge components during fabrication, followed by accurate erection evolutions as the work goes forward in the field.

Accordingly, the proper construction procedure for cable-stayed steel bridges can be summarized as follows:

1. Determine the actual bridge baseline lengths and pier-top elevations to a high degree of accuracy.
2. Fabricate the bridge towers, cables, and girders to a high degree of geometric precision.
3. Determine, in the fabricating shop, the final residual errors in critical fabricated dimensions, including cable-stay lengths after socketing, and positions of socket bearing surfaces or pinholes.
4. Determine "corrected theoretical" positioning for each individual cable stay.
5. During erection, bring all tower and girder structural joints into shop-fabricated alignment, with fair holes, etc.
6. At the appropriate erection stages, achieve "corrected theoretical" positioning for each cable stay.
7. With the structure in the all-steel-erected condition (or other appropriate designated condition), check it over carefully to determine whether any significant geometric or other discrepancies are in evidence. If there are none, declare conditions acceptable and continue with erection.

This construction engineering philosophy can be summarized by stating that if the steelwork fabrication and erection are properly engineered and carried out, the geometry and stressing of the completed structure will fall within acceptable limits; whereas, if the fabrication and erection are not properly done, corrective measurements and adjustments attempted in the field are not likely to improve the structure, or even to prove satisfactory. Accordingly, in constructing steel cable-stayed bridges we should place full reliance on accurate shop fabrication and on controlled field erection, just as is done on other types of steel bridges, rather than attempting to make measurements and adjustments in the field to compensate for inadequate fabrication and erection.

45.17 Field Checking at Critical Erection Stages

As has been stated previously, the best governing procedure for steel bridge construction is generally the shop control procedure, wherein full reliance is placed on accurate fabrication of the bridge components as the basis for the integrity of the completed structure. However, this philosophy does not rule out the desirability of certain checks in the field as erection goes forward, with the objective of providing assurance that the work is on target and no significant errors have been introduced.

It would be impossible to catalog those cases during steel bridge construction where a field check might be desirable; such cases will generally suggest themselves as the construction engineering studies progress. We will only comment that these field-check cases, and the procedures to be used, should be looked at carefully, and even skeptically, to make certain that the measurements will be both desirable and practical, producing meaningful information that can be used to augment job integrity.

45.18 Determination of Erection Strength Adequacy

Quite commonly, bridge member forces during the erection stages will be altogether different from those that will prevail in the completed structure. At each critical erection stage the bridge members must be reviewed for strength and stability, to ensure structural integrity as the work goes forward. Such a construction engineering review is typically the responsibility of the steelwork erector, who carries out thorough erection studies of the structure and calls for strengthening or stabilizing of members as needed. The erector submits the studies and recommendations to the design engineer for review and comment, but normally the full responsibility for steelwork structural integrity during erection rests with the erector.

Figure 45.11 First Quebec railway cantilever bridge, 23 August 1907. Cantilever erection of south main span, six days before collapse. The tower traveler erected the anchor span (on falsework) and then the cantilever arm; then erected the top-chord traveler, which is shown erecting suspended span at end of cantilever arm. The main span of 1800 ft(549 m) was the world's longest of any type. The sidespan bottom chords second from pier (arrow) failed in compression because latticing connecting chord corner angles was deficient under secondary bending conditions.

In the U.S., bridgework design specifications commonly require that stresses in steel structures under erection shall not exceed certain multiples of design allowable stresses. Although this type of erection stress limitation is probably safe for most steel structures under ordinary conditions, it is not necessarily adequate for the control of the erection stressing of large monumental-type bridges. The key point to be understood here is that fundamentally, there is no logical fixed relationship between design allowable stresses, which are based upon somewhat uncertain long-term service loading requirements along with some degree of assumed structural deterioration, and stresses that are safe and economical during the bridge erection stages, where loads and their locations are normally well defined and the structural material is in new condition. Clearly, the basic premises of the two situations are significantly different, and "factored design stresses" must therefore be considered unreliable as a basis for evaluating erection safety.

There is yet a further problem with factored design stresses. Large truss-type bridges in various erection stages may undergo deflections and distortions that are substantial compared with those occurring under service conditions, thereby introducing apprehension regarding the effect of the secondary bending stresses that result from joint rigidity.

Recognizing these basic considerations, the engineering department of a major U.S. steelwork contractor went forward in the early 1970s to develop a logical philosophy for erection strength appraisal of large structural steel frameworks, with particular reference to long-span bridges, and implemented this philosophy with a stress analysis procedure. The effort was successful and the results were reported in a paper published by the American Society of Civil Engineers in 1977[6]. This stress analysis procedure, designated the erection rating factor (ERF) procedure, is founded directly upon basic structural principles, rather than on bridge-member design specifications, which are essentially irrelevant to the problem of erection stressing.

It may be noted that a significant inducement toward development of the ERF procedure was the failure of the first Quebec cantilever bridge in 1907 (see Figures 45.11 and 45.12). It was quite obvious that evaluation of the structural safety of the Quebec bridge at advanced cantilever erection stages such as that portrayed in Figure 45.11, by means of the factored-design-stress procedure, would inspire no confidence and would not be justifiable.

The erection rating factor (ERF) procedure for a truss bridge can be summarized as follows:

1. Assume either (a) pin-ended members (no secondary bending), (b) plane-frame action (rigid truss joints, secondary bending in one plane), or (c) space-frame action (bracing-member joints also rigid, secondary bending in two planes), as engineering judgement dictates.
2. Determine, for each designated erection stage, the member primary forces (axial) and secondary forces (bending) attributable to gravity loads and wind loads.
3. Compute the member stresses induced by the combined erection axial forces and bending moments.
4. Compute the ERF for each member at three or five locations: at the middle of the member; at each joint, inside the gusset plates (usually at the first row of bolts); and, where upset member plates or gusset plates are used, at the stepped-down cross-section outside each joint.
5. Determine the minimum computed ERF for each member and compare it with the stipulated minimum value.
6. Where the computed minimum ERF equals or exceeds the stipulated minimum value, the member is considered satisfactory. Where it is less, the member may be inadequate; reevaluate the critical part of it in greater detail and recalculate the ERF for further comparison with the stipulated minimum. (Initially calculated values can often be increased significantly.)
7. Where the computed minimum ERF remains less than the stipulated minimum value, strengthen the member as required.

Note that member forces attributable to wind are treated the same as those attributable to gravity loads. The old concept of "increased allowable stresses" for wind is not considered to be valid for erection conditions and is not used in the ERF procedure. Maximum acceptable ℓ/r and b/t values are included in the criteria. ERFs for members subjected to secondary bending moments are calculated using interaction equations.

45.19 Philosophy of the Erection Rating Factor

In order that the structural integrity and reliability of a steel framework can be maintained throughout the erection program, the minimum probable (or "minimum characteristic") strength value of each member must necessarily be no less than the maximum probable (or "maximum characteristic") force value, under the most adverse erection condition. In other words, the following relationship is required:

$$S - \Delta S \geq F + \Delta F \qquad (45.1)$$

where
S = computed or nominal strength value for the member
ΔS = maximum probable member strength underrun from the computed or nominal value
F = computed or nominal force value for the member
ΔF = maximum probable member force overrun from the computed or nominal value

Equation 45.1 states that in the event the actual strength of the structural member is less than the nominal strength, S, by an amount ΔS, while at same time the actual force in the member is greater than the nominal force, F, by an amount ΔF, the member strength will still be no less than the member force, and so the member will not fail during erection. This equation provides a direct appraisal of erection realities, in contrast to the allowable-stress approach based on factored design stresses.

Proceeding now to rearrange the terms in Equation 45.1, we find that

Figure 45.12 Wreckage of south anchor span of first Quebec railway cantilever bridge, 1907. View looking north from south shore a few days after the collapse of 29 August 1907, the worst disaster in the history of bridge construction. About 20,000 tons of steelwork fell into the St. Lawrence River, and 75 workmen lost their lives.

$$S\left(1 - \frac{\Delta S}{S}\right) \geq F\left(1 + \frac{\Delta F}{F}\right) ; \quad \frac{S}{F} \geq \qquad\qquad (45.2)$$

The ERF is now defined as

$$\mathrm{ERF} \equiv \qquad\qquad (45.3)$$

that is, the nominal strength value, S, of the member divided by its nominal force value, F. Thus, for erection structural integrity and reliability to be maintained, it is necessary that

$$\mathrm{ERF} \geq \frac{1 + \frac{\iota}{\cdot}}{1 - \frac{\iota}{\cdot}} \qquad\qquad (45.4)$$

45.20 Minimum Erection Rating Factors

In view of possible errors in (1) the assumed weight of permanent structural components, (2) the assumed weight and positioning of erection equipment, and (3) the mathematical models assumed for purposes of erection structural analysis, it is reasonable to assume that the actual member force for a given erection condition may exceed the computed force value by as much as 10%; that is, it is reasonable to take $\Delta F/F$ as equal to 0.10.

For tension members, uncertainties in (1) the area of the cross-section, (2) the strength of the material, and (3) the member workmanship, indicate that the actual member strength may be up to 15% less than the computed value; that is, $\Delta S/S$ can reasonably be taken as equal to 0.15. The additional uncertainties associated with compression member strength suggest that $\Delta S/S$ be taken as 0.25 for those members. Placing these values into Equation 45.4, we obtain the following minimum ERFs:

$$\text{Tension members:} \quad \text{ERF}_{t,min} \ = \ (1+0.10)/(1-0.15)$$

$$= \ 1.294, \text{say } 1.30$$

$$\text{Compression members:} \quad \text{ERF}_{c,min} \ = \ (1+0.10)/(1-0.25)$$

$$= \ 1.467, \text{say } 1.45$$

The proper interpretation of these expressions is that if, for a given tension (compression) member, the ERF is calculated as 1.30 (1.45) or more, the member can be declared safe for the particular erection condition. Note that higher, or lower, values of erection rating factors may be selected if conditions warrant.

The minimum ERFs determined as indicated are based on experience and judgment, guided by analysis and test results. They do not reflect any specific probabilities of failure and thus are not based on the concept of an acceptable risk of failure, which might be considered the key to a totally rational approach to structural safety. This possible shortcoming in the ERF procedure might be at least partially overcome by evaluating the parameters $\Delta F/F$ and $\Delta S/S$ on a statistical basis; however, this would involve a considerable effort, and it might not even produce significant results.

It is important to recognize that the ERF procedure for determining erection strength adequacy is based directly on fundamental strength and stability criteria, rather than being only indirectly related to such criteria through the medium of a design specification. Thus, the procedure gives uniform results for the erection rating of framed structural members irrespective of the specification that was used to design the members. Obviously, the end use of the completed structure is irrelevant to its strength adequacy during the erection configurations, and therefore the design specification should not be brought into the picture as the basis for erection appraisal.

Experience with application of the ERF procedure to long-span truss bridges has shown that it places the erection engineer in much better contact with the physical significance of the analysis than can be obtained by using the factored-design-stress procedure. Further, the ERF procedure takes account of secondary stresses, which have generally been neglected in erection stress analysis.

Although the ERF procedure was prepared for application to truss bridge members, the simple governing structural principle set forth by Equation 45.1 could readily be applied to bridge members and components of any type.

45.21 Deficiencies of Typical Construction Procedure Drawings and Instructions

At this stage of the review it is appropriate to bring forward a key problem in the realm of bridge construction engineering: the strong tendency for construction procedure drawings to be insufficiently clear, and for step-by-step instructions to be either lacking or less than definitive. As a result of these deficiencies it is not uncommon to find the contractor's shop and field evolutions to be going along under something less than suitable control.

Figure 45.13 Visiting the work site. It is of first-order importance for bridge construction engineers to visit the site regularly and confer with the job superintendent and his foremen regarding practical considerations. Construction engineers have much to learn from the work forces in shop and field, and vice versa. (Courtesy of Bethlehem Steel Corporation.)

Shop and field operations personnel who are in a position to speak frankly to construction engineers will sometimes let them know that procedure drawings and instructions often need to be clarified and upgraded. This is a pervasive problem, and it results from two prime causes: (1) the fabrication and erection engineers responsible for drawings and instructions do not have adequate on-the-job experience, and (2) they are not sufficiently skilled in the art of setting forth on the documents, clearly and concisely, exactly what is to be done by the operations forces — and, sometimes of equal importance, what *is not* to be done.

This matter of clear and concise construction procedure drawings and instructions may appear to be a pedestrian matter, but it is decidedly not. *It is a key issue of utmost importance to the success of steel bridge construction.*

45.22 Shop and Field Liaison by Construction Engineers

In addition to the need for well-prepared construction procedure drawings and instructions, it is essential for the staff engineers carrying out construction engineering to set up good working relations with the shop and field production forces, and to visit the work sites and establish effective communication with the personnel responsible for accomplishing what is shown on the documents.

Construction engineers should review each projected operation in detail with the work forces, and upgrade the procedure drawings and instructions as necessary, as the work goes forward. Further, engineers should be present at the work sites during critical stages of fabrication and erection. As a component of these site visits, the engineers should organize special meetings of key production personnel to go over critical operations in detail — complete with slides and blackboard as needed — thereby providing the work forces with opportunities to ask questions and discuss procedures and potential problems, and providing engineers the opportunity to determine how well the work forces understand the operations to be carried out.

This matter of liaison between the office and the work sites — like the preceding issue of clear construction procedure documents — may appear to be somewhat prosaic; again, however, it *is a matter of paramount importance*. Failure to attend to these two key issues constitutes a serious problem in steel bridge construction, and opens the door to high costs and delays, and even to erection accidents.

45.23 Comprehensive Bridge Erection-Engineering Specifications

The erection rating factor (ERF) procedure for determination of erection strength adequacy, as set forth heretofore for bridge trusswork, could readily be extended to cover bridge members and components of any type under erection loading conditions. Bridge construction engineers should work toward this objective, in order to release erection strength appraisal from the limitations of the commonly used factored-design-stress procedure.

Looking still further ahead, it is apparent that there is need in the bridge engineering profession for comprehensive erection engineering specifications for steel bridge construction. Such specifications should include guidelines for such matters as devising and evaluating erection schemes, determining erection loads, evaluating erection strength adequacy of all types of bridge members and components, designing erection equipment, and designing temporary erection members such as falsework, tiedowns, tiebacks, and jacking struts. The specifications might also cover contractual considerations associated with construction engineering.

The key point to be recognized here is that the use of bridge *design* specifications as the basis for erection engineering studies, as is currently the custom, is not appropriate. Erection engineering is a related but different discipline, and should have its own specifications. However, given the current fragmented state of construction engineering in the U.S. (refer to Section 45.26), it is difficult to envision how such erection engineering specifications could be prepared. Proprietary considerations associated with each erection firm's experience and procedures could constitute an additional obstacle.

45.24 Standard Conditions for Contracting

A further basic problem in respect to the future of steel bridge construction in the U.S. lies in the absence of standard conditions for contracting.

On through the 19th century both the design and the construction of a major bridge in the U.S. were frequently the responsibility of a single prominent engineer, who could readily direct and coordinate the work and resolve problems equitably. Then, over, the first thirty years or so of the 20th century this system was progressively displaced by the practice of competitive bidding on plans and specifications prepared by a design engineer retained by the owner. As a result the responsibility for the structure previously carried by the designer-builder became divided, with the designer taking responsibility for service integrity of the completed structure while prime responsibility for structural adequacy and safety during construction was assumed by the contractor. Full control over the preparation of the plans and specifications — the contract documents — was retained by the design engineer.

This divided responsibility has resulted in contract documents that may not be altogether equitable, since the designer is inevitably under pressure to look after the immediate financial interests of his client, the owner. Documents prepared by only one party to a contract can hardly be expected to reflect the appropriate interest of the other party. However, until about mid-20th-century design and construction responsibilities for major bridgework, although divided between the design engineer and the construction engineer, were nonetheless usually under the control of leading members of the bridge engineering profession who were able to command the level of communication and cooperation needed for resolution of inevitable differences of opinion within a framework of equity and good will.

Since the 1970s there has been a trend away from this traditional system of control. The business and management aspects of design firms have become increasingly important, while at the same time steelwork construction firms have become more oriented toward commercial and legal considerations. Professional design and construction engineers have lost stature correspondingly. As a result of these adverse trends, bridgework specifications are being ever more stringently drawn, bidding practices are becoming increasingly aggressive, claims for extra reimbursement are proliferating, insurance costs for all concerned are rising, and control of bridge engineering and construction is being influenced to an increasing extent by administrators and attorneys. These developments have not benefited the bridge owners, the design engineering profession, the steelwork construction industry, or the public — which must ultimately pay all of the costs of bridge construction.

It seems clear that in order to move forward out of this unsatisfactory state of affairs, a comprehensive set of standard conditions for contracting should be developed to serve as a core document for civil engineering construction — a document that would require only the addition of special provisions in order to constitute the basic specifications for any major bridge construction project. Such standard conditions would have to be prepared "off line" by a group of high-level engineering delegates having well established engineering credentials.

A core contract document such as the one proposed has been in general use in Great Britain since 1945, when the first edition of the *Conditions of Contract and Forms of Tender, Agreement and Bond for Use in Connection with Works of Civil Engineering Construction* was published by The Institution of Civil Engineers (ICE). This document, known informally as *The ICE Conditions of Contract*, is now in its 6th edition [1]. It is kept under review and revised as necessary by a permanent Joint Contracts Committee consisting of delegates from The Institution of Civil Engineers, The Association of Consulting Engineers, and The Federation of Civil Engineering Contractors. This document is used as the basis for the majority of works of civil engineering construction that are contracted in Great Britain, including steel bridges.

Further comments on the perceived need for U.S. standard conditions for contracting can be found in [7].

45.25 Design-and-Construct

As has been mentioned, design-and-construct was common practice in the U.S. during the 19th century. Probably the most notable example was the Brooklyn Bridge, where the designer-builders were John A. Roebling and his son Washington A. Roebling. Construction of the Brooklyn Bridge was begun in 1869 and completed in 1883. Design-and-construct continued in use through the early years of the 20th century; the most prominent example from that era may be the Ambassador suspension bridge between Detroit, Michigan and Windsor, Ontario, Canada, completed in 1929. The Ambassador Bridge was designed and built by the McClintic-Marshall Construction Co., Jonathan Jones, chief engineer; it has an 1850 ft main span, at that time the world's record single span.

Design-and-construct has not been used for a major steel bridge in the U.S. since the Ambassador Bridge. However, the procedure has seen significant use throughout the 20th century for bridges in other countries and particularly in Europe; and most recently design-construct-operate-maintain has come into the picture. Whether these procedures will find significant application in the U.S. remains to be seen.

The advantages of design-and-construct are readily apparent:

1. More prospective designs are likely to come forward, than when designs are obtained from only a single organization.
2. Competitive designs are submitted at a preliminary level, making it possible for the owner to provide some input to the selected design between the preliminary stage and design completion.
3. The owner knows the price of the project at the time the preliminary design is selected, as compared with design-bid where the price is not known until the design is completed and bids are received.
4, As the project goes forward the owner deals with only a single entity, thereby reducing and simplifying his administrative effort.
5. The design-and-construct team members must work effectively together, eliminating the antagonisms and confrontations that can occur on a design-bid project.

A key requirement in the design-and-construct system for a project is the meticulous preparation of the request-for-proposals (RFP). The following essentials should be covered in suitable detail and clarity in the RFP:

1. Description of project to be constructed.
2. Scope of work.
3. Structural component types and characteristics: which are required, which are acceptable, and which are not acceptable.
4. Minimum percentages of design and construction work that must be performed by the team's own forces.
5. Work schedules; time incentives and disincentives.
6. Procedure to be followed when actual conditions are found to differ from those assumed.
7. Quality control and quality assurance factors.
8. Owner's approval prerogatives during final-design stage and construction stage.
9. Applicable local, state and federal regulations.
10. Performance and payment bonding requirements.
11. Warranty requirements.
12. Owner's procedure for final approval of completed project.

In preparing the RFP, the owner should muster all necessary resources from both inside and outside his organization. Political considerations should be given due attention. Document drafts should receive the appropriate reviews, and an RFP brought forward that is in near-final condition. Then, at the start of the contracting process, the owner will typically proceed as follows:

1. Announce the project and invite prospective teams to submit qualifications.
2. Prequalify a small number of teams, perhaps three to five, and send the draft RFP to each.
3. Hold a meeting with the prequalified teams for informal exchange of information and to discuss questions.
4. Prepare the final RFP and issue it to each prequalified team, and announce the date on which proposals will be due.

The owner will customarily call for the proposals to be submitted in two separate components: the design component, showing the preliminary design carried to about the 25% level; and the monetary component, stating the lump-sum bid. Before the bids are opened the owner will typically carry out a scoring process for the preliminary designs, not identifying the teams with their designs, using a 10 point or 100 point grading scale and giving consideration to the following factors:

1. Quality of the design,
2. Bridge aesthetics.
3. Fabrication and erection feasibility and reliability.
4. Construction safety aspects.
5. Warranty and long-term maintenance considerations.
6. User costs.

Using these and other such scoring factors (which can be assigned weights if desired), a final overall design score is assigned to each preliminary design. Then the lump-sum bids are opened. A typical procedure is to divide each team's bid price by its design score, yielding and overall price rating, and to award the contract to the design-and-construct team having the lowest price rating.

Following the contract award the successful team will proceed to bring its preliminary design up to the final-design level, with no site work permitted during this interval. It is customary for the owner to award each unsuccessful submitting team a stipend to partially offset the costs of proposal preparation.

45.26   Construction Engineering Procedures and Practices — The Future

The many existing differences of opinion and procedures in respect to proper governance of steelwork fabrication and erection for major steel bridges raises the question: How do proper bridge construction guidelines come into existence and find their way into practice and into bridge specifications? Looking back over the period roughly from 1900 to 1975, we find that the major steelwork construction companies in the U.S. developed and maintained competent engineering departments that planned and engineered large bridges (and smaller ones as well) through the fabrication and erection processes with a high degree of proficiency. Traditionally, the steelwork contractor's engineers worked in cooperation with design-office engineers to develop the full range of bridgework technical factors, including construction procedure and practices.

However, times have changed; since the 1970s, major steel bridge contractors have all but disappeared in the U.S., and further, very few bridge design offices have on their staffs engineers experienced in fabrication and erection engineering. As a result, construction engineering often receives less attention and effort than it needs and deserves, and this is not a good omen for the future of the design and construction of large bridges in the U.S.

Bridge construction engineering is not a subject that is or can be taught in the classroom; it must be learned on the job with major steelwork contractors. The best route for an aspiring young construction engineer is to spend significant amounts of time in the fabricating shop and at the bridge site, interspersed with time doing construction-engineering technical work in the office. It has been pointed out previously that although construction engineering and design engineering are related, they constitute different practices and require diverse backgrounds and experience. Design engineering can essentially be learned in the design office; construction engineering, however, cannot — it requires a background of experience at work sites. Such experience, it may be noted, is valuable also for design engineers; however, it is not as necessary for them as it is for construction engineers.

The training of future steelwork construction engineers in the U.S. will be handicapped by the demise of the "Big Two" steelwork contractors in the 1970s. Regrettably, it appears that surviving steelwork contractors in the U.S. generally do not have the resources for supporting strong engineering departments, and so there is some question as to where the next generation of steel bridge construction engineers in the U.S. will be coming from.

45.27 Concluding Comments

In closing this review of steel bridge construction it is appropriate to quote from the work of an illustrious British engineer, teacher, and author, the late Sir Alfred Pugsley [15]:

> A further crop of [bridge] accidents arose last century from overloading by traffic of various kinds, but as we have seen, engineers today concentrate much of their effort to ensure that a margin of strength is provided against this eventuality. But there is one type of collapse that occurs almost as frequently today as it has over the centuries: collapse at a late stage of erection.
>
> The erection of a bridge has always presented its special perils and, in spite of ever-increasing care over the centuries, few great bridges have been built without loss of life. Quite apart from the vagaries of human error, with nearly all bridges there comes a critical time near completion when the success of the bridge hinges on some special operation. Among such are ... the fitting of a last section ... in a steel arch, the insertion of the closing central [members] in a cantilever bridge, and the lifting of the roadway deck [structure] into

position on a suspension bridge. And there have been major accidents in many such cases. It may be wondered why, if such critical circumstances are well known to arise, adequate care is not taken to prevent an accident. Special care is of course taken, but there are often reasons why there may still be "a slip bewixt cup and lip". Such operations commonly involve unusually close cooperation between constructors and designers, and between every grade of staff, from the laborers to the designers and directors concerned; and this may put a strain on the design skill, on detailed inspection, and on practical leadership that is enough to exhaust even a Brunel.

In such circumstances it does well to ... recall [the] dictum ... that "it is essential not to have faith in human nature. Such faith is a recent heresy and a very disastrous one." One must rely heavily on the lessons of past experience in the profession. Some of this experience is embodied in professional papers describing erection processes, often (and particularly to young engineers) superficially uninteresting. Some is crystallized in organizational habits, such as the appointment of resident engineers from both the contracting and [design] sides. And some in precautions I have myself endeavored to list ...

It is an easy matter to list such precautions and warnings, but quite another for the senior engineers responsible for the completion of a bridge to stand their ground in real life. This is an area of our subject that depends in a very real sense on the personal qualities of bridge engineers ... At bottom, the safety of our bridges depends heavily upon the integrity of our engineers, particularly the leading ones.

45.28 Further Illustrations of Bridges Under Construction, Showing Erection Methods

Figure 45.14   Royal Albert Bridge across River Tamar, Saltash, England, 1857. The two 455 ft (139 m) main spans, each weighing 1060 tons, were constructed on shore, floated out on pairs of barges, and hoisted about 100 ft (30 m) to their final position using hydraulic jacks. Pier masonry was built up after each 3 ft (1 m) lift.

Figure 45.15 Eads Bridge across the Mississippi River, St. Louis, Mo., 1873. The first important metal arch bridge in the U.S., supported by four planes of hingeless trussed arches having chrome-steel tubular chords. Spans are 502-520-502 ft (153-158-153 m). During erection, arch trusses were tied back by cables passing over temporary towers built on the piers. Arch ribs were packed in ice to effect closure.

Figure 45.16 Glasgow (Missouri) railway truss bridge, 1879. Erection on full supporting falsework was commonplace in the 19th century. The world's first all-steel bridge, with five 315 ft (96 m) through-truss simple spans, crossing the Missouri River.

Figure 45.17 Niagara River railway cantilever truss bridge, near Niagara Falls, New York, 1883. Massive wood erection traveler constructed side span on falsework, then cantilevered half of main span to midspan. Erection of other half of bridge was similar. First modern-type cantilever bridge, with 470 ft (143 m) clear main span having a 120 ft (37 m) center suspended span.

The massive cantilevers of the Forth bridge, shown under erection, were conceived in the shadow of the Tay bridge disaster.

Figure 45.18 Construction of monumental Forth Bridge, Scotland, 1888. Numerous small movable booms were used, along with erection travelers for cantilevering the two 1710 ft (521 m) main spans. The main compression members are tubes 12 ft (3.65 m) in diameter; many other members are also tubular. Total steelwork weight is 51,000 tons. Records are not clear regarding such essentials as cambering and field fitting of individual members in this heavily redundant railway bridge. The Forth is arguably the world's greatest steel structure.

Figure 45.19 Pecos River railway viaduct, Texas, 1892. Erection by massive steam-powered wood traveler having many sets of falls and very long reach. Cantilever-truss main span has 185 ft (56 m) clear opening.

Figure 45.20   Raising of suspended span, Carquinez Strait Bridge, California, 1927. The 433 ft (132 m) suspended span, weighting 650 tons, was raised into position in 35 min., driven by four counterweight boxes having a total weight of 740 tons.

Figure 45.21   First Cooper River cantilever bridge, Charleston, S.C., 1929. Erection travelers constructed 450 ft (137 m) side spans on falsework, then went on to erect 1050 ft (320 m) main span (including 437.5 ft [133 m] suspended span) by cantilevering to midspan.

Figure 45.22 Erecting south tower of Golden Gate Bridge, San Francisco, 1935. A creeper traveler with two 90 ft (27 m) booms erects a tier of tower cells for each leg, then is jumped to the top of that tier and proceeds to erect the next tier. The tower legs are 90 ft (27 m) center-to-center and 690 ft (210 m) high. When the traveler completed the north tower (in background) it erected a Chicago boom on the west tower leg, which dismantled the creeper, erected tower-top bracing, and erected two small derricks (one shown) to service cable erection. Each tower contains 22,200 tons of steelwork.

Figure 45.23 Balanced-cantilever erection, Governor O.K. Allen railway/highway cantilever bridge, Baton Rouge, La., 1939. First use of long balanced-cantilever erection procedure in the U.S. On each pier 650 ft (198 m) of steelwork, about 4000 tons, was balanced on the 40 ft (12 m) base formed by a sloping falsework bent. The compression load at the top of the falsework bent was measured at frequent intervals and adjusted by positioning a counterweight car running at bottom-chord level. The main spans are 848-650-848 ft (258-198-258 m); 650 ft span shown. (Courtesy of Bethlehem Steel Corporation.)

Figure 45.24 Tower erection, second Tacoma Narrows Bridge, Washington, 1949. This bridge replaced first Tacoma Narrows bridge, which blew down in a 40 mph (18 m/sec) wind in 1940. The tower legs are 60 ft (18 m) on centers and 462 ft (141 m) high. Creeper traveler is shown erecting west tower, in background. On east tower, creeper erected a Chicago boom at top of south leg; this boom dismantled creeper, then erected tower-top bracing and a stiffleg derrick, which proceeded to dismantle Chicago boom. Tower manhoist can be seen at second-from-topmost landing platform. Riveting cages are approaching top of tower. Note tower-base erection kneebraces, required to ensure tower stability in free-standing condition (see Figure 45.27).

Figure 45.25 Aerial spinning of parallel-wire main cables, second Tacoma Narrows suspension bridge, Washington, 1949. Each main cable consists of 8702 parallel galvanized high-strength wires of 0.196 in (4.98 mm) diameter, laid up as 19 strands of mostly 460 wires each. Following compaction the cable became a solid round mass of wires with a diameter of 20-1/4 in. (514 mm).

Figure 45.25a Tramway starts across from east anchorage carrying two wire loops. Three 460-wire strands have been spun, with two more under construction. Tramway spinning wheels pull wire loops across the three spans from east anchorage to west anchorage. Suspended footbridges provide access to cables. Spinning goes on 24 hours per day.

Figure 45.25b   Tramway arrives at west anchorage. Wire loops shown in Figure 45.25a are removed from spinning wheels and placed around strand shoes at west anchorage. This tramway then returns empty to east anchorage, while tramway for other "leg" of endless hauling rope brings two wire loops across for second strand that is under construction for this cable.

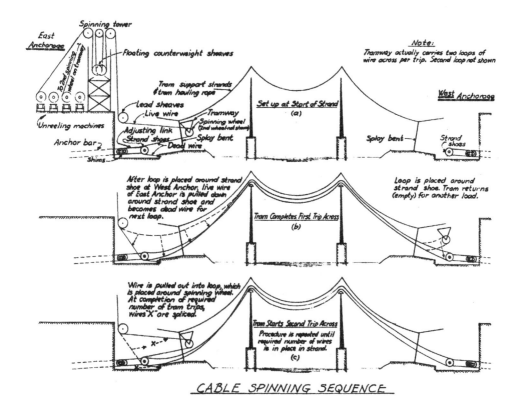

Figure 45.26a   Erection of individual wire loops.

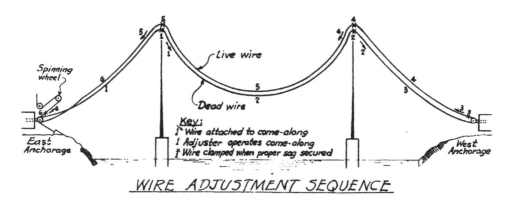

Figure 45.26b   Adjustment of individual wire loops.

Figure 45.26   Cable-spinning procedure for constructing suspension bridge parallel-wire main cables, showing details of aerial spinning method for forming individual 5 mm wires into strands containing 400 to 500 wires. Each wire loop is erected as shown in Figure 45.26a (refer to Figure 45.25), then adjusted to the correct sag as shown in Figure 45.26b. Each completed strand is banded with tape, then adjusted to the correct sag in each span. With all strands in place, they are compacted to form a solid round homogeneous mass of cable wires. The aerial spinning method was developed by John Roebling in the mid-19th century.

Figure 45.27 Erection of suspended deck steelwork, second Tacoma Narrows Bridge, Washington, 1950. Chicago boom on tower raises deck steelwork components to deck level, where they are transported to deck travelers by material cars. Each truss double panel is connected at top-chord level to previously erected trusses, and left open at bottom-chord level to permit temporary upward deck curvature, which results from partial loading condition of main suspension cables. Main span (at right) is 2800 ft (853 m), and side spans are 1100 ft (335 m). Stiffening trusses are 33 ft (10 m) deep and 60 ft (18 m) on centers. Tower-base kneebraces (see Figure 45.24) show clearly here.

| (a) | (b) | (c) |

Figure 45.28 Moving deck traveler forward, second Tacoma Narrows Bridge, Washington, 1950. Traveler pulling-falls leadline passes around sheave beams at forward end of stringers, and is attached to front of material car (at left). Material car is pulled back toward tower, advancing traveler two panels to its new position at end of deck steelwork. Arrows show successive positions of material car. (a) Traveler at start of move, (b) traveler advanced one panel, and (c) traveler at end of move.

Figure 45.29 Erecting closing girder sections of Passaic River Bridge, New Jersey Turnpike, 1951. Huge double-boom travelers, each weighing 270 tons, erect closing plate girders of the 375 ft (114 m) main span. Closing girders are 14 ft (4.3 m) deep and 115 ft (35 m) long and weigh 146 tons each. Sidewise entry was required (as shown) because of long projecting splice material. Longitudinal motion was provided at one pier, where girders were jacked to effect closure. Closing girders were laterally stable without floor steel fill-in, such that derrick falls could be released immediately. (Courtesy of Bethlehem Steel Corporation.)

(a)   (b)

Figure 45.30 Floating-in erection of a truss span, first Chesapeake Bay Bridge, Maryland, 1951. Erected 300 ft (91 m) deck-truss spans form erection dock, providing a work platform for two derrick travelers. A permanent deck-truss span serves as a falsework truss supported on barges and is shown carrying the 470 ft (143 m) anchor arm of the through-cantilever truss. This span is floated to its permanent position, then landed onto its piers by ballasting the barges. (a) Float leaves erection dock, and (b) float arrives at permanent position. (Courtesy of Bethlehem Steel Corporation.)

Figure 45.31 Floating-in erection of a truss span, first Chesapeake Bay Bridge, Maryland, 1952. A 480 ft (146 m) truss span, weighting 850 tons, supported on falsework consisting of a permanent deck-truss span along with temporary members, is being floated-in for landing onto its piers. Suspension bridge cables are under construction in background. (Courtesy of Bethlehem Steel Corporation.)

Figure 45.32 Erection of a truss span by hoisting, first Chesapeake Bay Bridge, Maryland, 1952. A 360 ft (110 m) truss span is floated into position on barges and picked clear using four sets of lifting falls. Suspension bridge deck is under construction at right. (Courtesy of Bethlehem Steel Corporation.)

Figure 45.33 Erection of suspension bridge deck structure, first Chesapeake Bay Bridge, Maryland, 1952. A typical four-panel through-truss deck section, weighting 99 tons, has been picked from the barge and is being raised into position using four sets of lifting falls attached to main suspension cables. Closing deck section is on barge, ready to go up next. (Courtesy of Bethlehem Steel Corporation.)

Figure 45.34 Greater New Orleans cantilever bridge, Louisiana, 1957. Tall double-boom deck travelers started at ends of main bridge and erected anchor spans on falsework, then the 1575 ft (480 m) main span by cantilevering to midspan. (Courtesy of Bethlehem Steel Corporation.)

Figure 45.35 Tower erection, second Delaware Memorial Bridge, Wilmington, Del., 1966. Tower erection traveler has reached topmost erecting position and swings into place 23-ton closing top-strut section. Tower legs were jacked apart about 2 in. (50 mm) to provide entering clearance. Traveler jumping beams are in topmost working position, above cable saddles. Tower steelwork is about 418 ft (127 m) high. Cable anchorage pier is under construction at right. First Delaware Memorial Bridge (1951) is at left. Main span of both bridges is 2150 ft (655 m). (Courtesy of Bethlehem Steel Corporation.)

Figure 45.36 Erecting orthotropic-plate decking panel, Poplar Street Bridge, St. Louis, Mo., 1967. A five-span, 2165 ft (660 m) continuous box-girder bridge, main span 600 ft (183 m). Projecting box ribs are 5-1/2 × 17 ft (1.7 × 5.2 m) in cross-section, and decking section is 27 × 50 ft (8.2 × 15.2 m). Decking sections were field welded, while all other connections were field bolted. Box girders are cantilevered to falsework bents using overhead "positioning travelers" (triangular structure just visible above deck at left) for intermediate support. (Courtesy of Bethlehem Steel Corporation.)

Figure 45.37 Erection of parallel-wire-stand (PWS) cables, Newport Bridge suspension spans, Narragansett Bay, R.I., 1968. Bridge engineering history was made at Newport with the development and application of shop-fabricated parallel-wire socketed strands for suspension bridge cables. Each Newport cable was formed of seventy-six 61-wire PWS, each 4512 ft (1375 m) long and weighing 15 tons. Individual wires are 0.202 in. (5.13 mm) in diameter and are zinc coated. Parallel-wire cables can be constructed of PWS faster and at lower cost than by traditional air spinning of individual wires (see Figures 45.25 and 45.26). (Courtesy of Bethlehem Steel Corporation.)

Figure 45.37a Aerial tramway tows PWS from west anchorage up side span, then on across other spans to east anchorage. Strands are about 1-3/4 in. (44 mm) in diameter.

Figure 45.37b   Cable formers maintain strand alignment in cables prior to cable compaction. Each finished cable is about 15-1/4 in.(387 mm) in diameter. (Courtesy of Bethlehem Steel Corporation.)

Figure 45.38   Pipe-type anchorage for parallel-wire-strand (PWS) cables, Newport Bridge suspension spans, Narragansett Bay, R.I., 1967. Pipe anchorages shown will be embedded in anchorage concrete. Socketed end of each PWS is pulled down its pipe from upper end, then seated and shim-adjusted against heavy bearing plate at lower end. Pipe-type anchorage is much simpler and less costly than standard anchor-bar type used with aerial-spun parallel-wire cables (see Figure 45.25b). (Courtesy of Bethlehem Steel Corporation.)

Sept. 1, 1970   J. L. DURKEE ET AL   3,526,570
                        PARALLEL WIRE STRAND

Filed Aug. 25, 1966   4 Sheets-Sheet 1

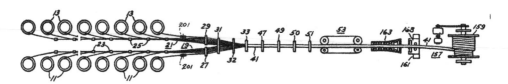

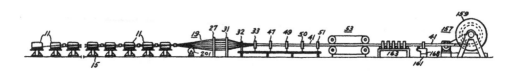

Figure 45.39 Manufacturing facility for production of shop-fabricated parallel-wire strands (PWS). Prior to 1966, parallel-wire suspension bridge cables had to be constructed wire-by-wire in the field using the aerial spinning procedure developed by John Roebling in the mid-19th century (refer to Figures 45.25 and 45.26). In the early 1960s a major U.S. steelwork contractor originated and developed a procedure for manufacturing and reeling parallel-wire strands, as shown in these patent drawings. A PWS can contain up to 127 wires (see Figures 45.45 and 45.46). (top) Plan view of PWS facility. Turntables 11 contain "left-hand" coils of wire and turntables 13 contain "right-hand" coils, such that wire cast is balanced in the formed strand. Fairleads 23 and 25 guide the wires into half-layplates 27 and 29, followed by full layplates 31 and 32 whose guide holes delineate the hexagonal shape of final strand 41. (bottom) Elevation view of PWS facility. Hexagonal die 33 contains six spring-actuated rollers that form the wires into regular-hexagon shape; and similar roller dies 47, 49, 50, and 51 maintain the wires in this shape as PWS 41 is pulled along by hexagonal dynamic clamp 53. The PWS is bound manually with plastic tape at about 3 ft (1 m) intervals as it passes along between roller dies. The PWS passes across roller table 163, then across traverse carriage 168, which is operated by traverse mechanism 161 to direct the PWS properly onto reel 159. Finally, the reeled PWS is moved off-line for socketing. Note that wire measuring wheels 201 can be installed and used for control of strand length.

Figure 45.40 Suspended deck steelwork erection, Newport Bridge suspension spans, Narragansett Bay, R.I., 1968. Closing mainspan deck section is being raised into position by two cable travelers, each made up of a pair of 36 in. (0.91 m) wide-flange rolled beams that ride cables on wooden wheels. Closing section is 40-1/2 ft (12 m) long at top-chord level, 66 ft (20 m) wide and 16 ft (5 m) deep, and weighs about 140 tons. (Courtesy of Bethlehem Steel Corporation.)

Figure 45.41 Erection of Kansas City Southern Railway box-girder bridge, near Redland, Okla., by "launching," 1970. This nine-span continuous box-girder bridge is 2110 ft (643 m) long, with a main span of 330 ft (101 m). Box cross section is 11 × 14.9 ft (3.35 × 4.54 m). Girders were launched in two "trains," one from north end and one from south end. A "launching nose" was used to carry leading end of each girder train up onto skidway supports as train was pushed out onto successive piers. Closure was accomplished at center of main span. (Courtesy of Bethlehem Steel Corporation.)

Figure 45.41a Leading end of north girder train moves across 250 ft (76 m) span 4, approaching pier 5. Main span, 330 ft (101 mm), is to right of pier 5.

Figure 45.41b Launching nose rides up onto pier 5 skidway units, removing girder-train leading-end sag.

Figure 45.41c Leading end of north girder train is now supported on pier 5.

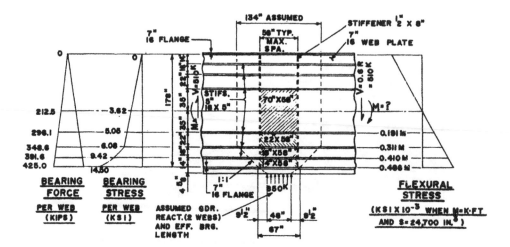

Figure 45.42a Typical assumed erection loading of box-girder web panels in combined moment, shear, and transverse compression.

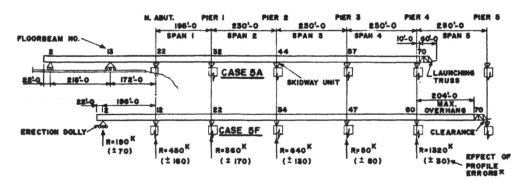

Figure 45.42b Launch of north girder train from pier 4 to pier 5.

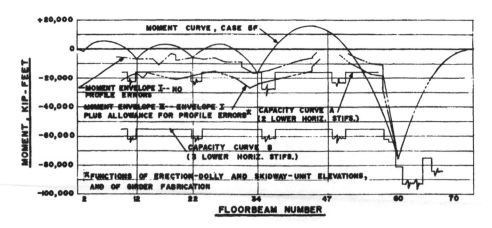

Figure 45.42c Negative-moment envelopes occurring simultaneously with reaction, for launch of north girder train to pier 5.

Figure 45.42 Erection strengthening to withstand launching, Kansas City Southern Railway box-girder bridge, near Redland, Okla. (see Figure 45.41).

Figure 45.43 Erection of west arch span of twin-arch Hernando de Soto Bridge, Memphis, Tenn., 1972. The two 900 ft (274 m) continuous-truss tied-arch spans were erected by a high-tower derrick boat incorporating a pair of barges. West-arch steelwork (shown) was cantilevered to midspan over two pile-supported falsework bents. Projecting east-arch steelwork (at right) was then cantilevered to midspan (without falsework) and closed with falsework-supported other half-arch. (Courtesy of Bethlehem Steel Corporation.)

Figure 45.44 Closure of east side span, Commodore John Barry cantilever truss bridge, Chester, Pa., 1973. High-tower derrick boat (in background) started erection of trusses at both main piers, supported on falsework; then erected top-chord travelers for main and side spans. Sidespan traveler carried steelwork erection to closure, as shown, and falsework bent was then removed. East-mainspan traveler then cantilevered steelwork (without falsework) to midspan, concurrently with cantilever erection by west-half mainspan traveler, and trusses were closed at midspan. Commodore Barry has 1644-ft (501 m) main span, the longest cantilever span in the U.S., and 822-ft (251 m) side spans. (Courtesy of Bethlehem Steel Corporation.)

Figure 45.45 Reel of parallei-wire strand (PWS), Akashi Kaikyo suspension bridge, Kobe, Japan, 1994. Each sock-eted PWS is made up of 127 0.206 in. (5.23 mm) wires, is 13,360 ft (4073 m) long, and weighs 96 tons. Plastic-tape bindings secure the strand wires at 1 m intervals. Sockets can be seen on right side of reel. These PWS are the longest and heaviest ever manufactured. (Courtesy of Nippon Steel — Kobe Steel.)

Figure 45.46 Parallel-wire-strand main cable, Akashi Kaikyo suspension bridge, Kobe, Japan, 1994. Main span is 6532 ft (1991 m), by far the world's longest. The PWS at right is being towed across spans, supported on rollers. Completed cable is made up of 290 PWS, making a total of 36,830 wires, and has a diameter of 44.2 in. (1122 mm) following compaction — largest bridge cables built to date. Each 127-wire PWS is about 2-3/8 in. (60 mm) in diameter. (Courtesy of Nippon Steel — Kobe Steel.)

Figure 45.47 Artist's rendering of proposed Messina Strait suspension bridge connecting Sicily with mainland Italy. The Messina Strait crossing has been under discussion since about 1870, under investigation since about 1955, and under active design since about 1975. The first realistic proposals for a crossing were made in 1969 in response to an international competition sponsored by the Italian government. There were 158 submissions — eight American, three British, three French, one German, one Swedish, and the remaining Italian. Forty of the submissions showed a single-span or multi-span suspension bridge. The enormous bridge shown has a single span of 10827 ft (3300 m) and towers 1250 ft (380 m) high. The bridge construction problems for such a span would be tremendously challenging. (Courtesy of Stretto di Messina, S.p.A.)

References

1. Conditions of Contract and Forms of Tender, Agreement and Bond for Use in Connection with Works of Civil Engineering Construction, 6th ed. (commonly known as "The ICE Conditions of Contract"), Inst. Civil Engrs. (U.K.), 1991.
2. Copp, J.I., de Vries, K., Jameson, W.H., and Jones, J. 1945. Fabrication and Erection Controls, Rainbow Arch Bridge Over Niagara Gorge — a Symposium, *Transactions ASCE*, vol. 110.
3. Durkee, E.L., 1945. Erection of Steel Superstructure, Rainbow Arch Bridge Over Niagara Gorge — A Symposium, *Transactions ASCE*, vol. 110.
4. Durkee, J.L., 1966. Advancements in Suspension Bridge Cable Construction, *Proceedings, International Symposium on Suspension Bridges*, Laboratorio Nacional de Engenharia Civil, Lisbon.
5. Durkee, J.L., 1972. Railway Box-Girder Bridge Erected by Launching, *J. Struct. Div., ASCE*, July.
6. Durkee, J.L., and Thomaides, S.S., 1977. Erection Strength Adequacy of Long Truss Cantilevers, *J. Struct. Div., ASCE*, January.
7. Durkee, J.L., 1977. Needed: U.S. Standard Conditions for Contracting, *J. Struct. Div., ASCE*, June.
8. Durkee, J.L., 1982. Bridge Structural Innovation: A Firsthand Report, *J. Prof. Act., ASCE*, July.
9. Enquiry into the Basis of Design and Methods of Erection of Steel Box Girder Bridges. Final Report of Committee, 4 vols. (commonly known as "The Merrison Report"), HMSO (London), 1973/4.
10. Feidler, L.L., Jr., 1962. Erection of the Lewiston-Queenston Bridge, *Civil Engrg., ASCE*, November.
11. Freudenthal, A.M., Ed., 1972. The Engineering Climatology of Structural Accidents, *Proceedings of the International Conference on Structural Safety and Reliability*, Pergamon Press, Elmsford, N.Y.
12. Holgate, H., Kerry, J.G.G., and Galbraith, J., 1908. Royal Commission Quebec Bridge Inquiry Report, Sessional Paper No. 154, vols. I and II, S.E. Dawson, Ottawa, Canada.
13. Leto, I.V., 1994. Preliminary design of the Messina Strait Bridge, *Proc. Inst. Civil Engrs.* (U.K.), vol. 102(3), August.
14. Petroski, H., 1993. Predicting Disaster, *American Scientist*, vol. 81, March.
15. Pugsley, A., 1968. The Safety of Bridges, *The Structural Engineer*, U.K., July.
16. Ratay, R.T., Ed., 1996. *Handbook of Temporary Structures in Construction*, 2nd ed., McGraw-Hill, New York.
17. Schneider, C.C., 1905. The Evolution of the Practice of American Bridge Building, *Transactions ASCE*, vol. 54.
18. Sibly, P.G. and Walker, A.C., 1977. Structural Accidents and Their Causes, *Proc. Inst. Civil Engrs.* (U.K.), vol. 62(1), May.
19. Smith, D.W., 1976. Bridge Failures, *Proc. Inst. Civil Engrs.* (U.K.), vol. 60(1), August.

46

Concrete Bridge Construction

Simon A. Blank
*California Department
of Transportation*

Michael M. Blank
U.S. Army Corps of Engineers

Luis R. Luberas
U.S. Army Corps of Engineers

46.1 Introduction

This chapter will focus on the principles and practices related to construction of concrete bridges in which construction engineering contributes greatly to the successful completion of the projects. We will first present the fundamentals of construction engineering and analyze the challenges and obstacles involved in such processes and then introduce the problems in relation to design, construction practices, project planning, scheduling and control, which are the ground of future factorial improvements in effective construction engineering in the United States. Finally, we will discuss prestressed concrete, high-performance concrete, and falsework in some detail.

46.2 Effective Construction Engineering

The construction industry is a very competitive business and many companies who engage in this marketplace develop proprietary technology in their field. In reality, most practical day-to-day issues are very common to the whole industry. Construction engineering is a combination of art and

0-8493-7434-0/00/$0.00+$.50
© 2000 by CRC Press LLC

science and has a tendency to become more the art of applying science (engineering principles) and approaches to the construction operations. Construction engineering includes design, construction operation, and project management. The final product of the design team effort is to produce drawings, specifications, and special provisions for various types of bridges. A fundamental part of construction engineering is construction project management (project design, planning, scheduling, controlling, etc.).

Planning starts with analysis of the type and scope of the work to be accomplished and selection of techniques, equipment, and labor force. Scheduling includes the sequence of operations and the interrelation of operations both at a job site and with external aspects, as well as allocation of manpower and equipment. Controlling consists of supervision, engineering inspection, detailed procedural instructions, record maintenance, and cost control. Good construction engineering analysis will produce more valuable, effective, and applicable instructions, charts, schedules, etc.

The objective is to plan, schedule, and control the construction process such that every construction worker and every activity contributes to accomplishing tasks with minimum waste of time and money and without interference. All construction engineering documents (charts, instructions, and drawings) must be clear, concise, definitive, and understandable by those who actually perform the work. As mentioned before, the bridge is the final product of design team efforts. When all phases of construction engineering are completed, this product — the bridge — is ready for to take service loading. In all aspects of construction engineering, especially in prestressed concrete, design must be integrated for the most effective results. The historical artificial separation of the disciplines — design and construction engineering — was set forth to take advantage of the concentration of different skills in the workplace. In today's world, the design team and construction team must be members of one team, partners with one common goal. That is the reason partnering represents a new and powerful team-building process, designed to ensure that projects become positive, ethical, and win–win experiences for all parties involved.

The highly technical nature of a prestressing operation makes it essential to perform preconstruction planning in considerable detail. Most problems associated with prestressed concrete could have been prevented by properly planning before the actual construction begins. Preconstruction planning at the beginning of projects will ensure that the structure is constructed in accordance with the plans, specifications, special provisions, and will also help detect problems that might arise during construction. It includes (1) discussions and conferences with the contractor, (2) review of the responsibilities of other parties, and (3) familiarization with the plans, specifications, and special provisions that relate to the planned work, especially if there are any unusual conditions. The preconstruction conference might include such items as scheduling, value of engineering, grade control, safety and environmental issues, access and operational considerations, falsework requirements, sequence of concrete placement, and concrete quality control and strength requirements. Pre-construction planning has been very profitable and in many has cases resulted in substantial reduction of labor costs. More often in prestressed concrete construction, the details of tendon layout, selection of prestressing system, mild-steel details, etc. are left up to general contractors or their specialized subcontractors, with the designer showing only the final prestress and its profile and setting forth criteria. And contractors must understand the design consideration fully to select the most efficient and economical system. Such knowledge may in many cases provide a competitive edge, and construction engineering can play a very important role in it.

46.3   Construction Project Management

46.3.1 General Principles

Construction project management is a fundamental part of construction engineering. It is a feat that few, if any, individuals can accomplish alone. It may involve a highly specialized technical field or science, but it always includes human interactions, attitudes and aspects of leadership, common sense, and resourcefulness. Although no one element in construction project management will create success, failure in one of the foregoing elements will certainly be enough to promote failure and to escalate costs. Today's construction environment requires serious consultation and management of the following life-cycle elements: design (including specifications, contract clauses, and drawings), estimating, budgeting, scheduling, procurement, biddability–constructibility–operability (BCO) review, permits and licenses, site survey, assessment and layout, preconstruction and mutual understanding conference, safety, regulatory requirements, quality control (QC), construction acceptance, coordination of technical and special support, construction changes and modifications, maintenance of progress drawings (redlines), creating as-built drawings, project records, among other elements.

Many construction corporations are becoming more involved in environmental restoration either under the Resource Conservation and Recovery Act (RCRA) or under the Comprehensive Environmental Response, Compensation and Liability Act (CERCLA, otherwise commonly known as the Superfund). This new involvement requires additional methodology and considerations by managers. Some elements that would otherwise be briefly covered or completely ignored under normal considerations may be addressed and required in a site Specific Health and Environmental Response Plan (SHERP). Some elements of the SHERP may include site health and safety staff, site hazard analysis, chemical and analytical protocol, personal protective equipment requirements and activities, instrumentation for hazard detection, medical surveillance of personnel, evacuation plans, special layout of zones (exclusion, reduction and support), and emergency procedures.

Federal government contracting places additional demands on construction project management in terms of added requirements in the area of submittals and transmittals, contracted labor and labor standards, small disadvantaged subcontracting plans, and many other contractual certification issues, among others. Many of these government demands are recurring elements throughout the life cycle of the project which may require adequate resource allocation (manpower) not necessary under the previous scenarios.

The intricacies of construction project management require the leadership and management skills of a unique individual who is not necessarily a specialist in any one of the aforementioned elements but who has the capacity to converse and interface with specialists in the various fields (i.e., chemists, geologists, surveyors, mechanics, etc.). An individual with a combination of an engineering undergraduate degree and a graduate business management degree is most likely to succeed in this environment. Field management experience can substitute for an advanced management degree.

It is the purpose of this section to discuss and elaborate elements of construction project management and to relate some field experiences and considerations. The information presented here will only promote further discussion and is not intended to be all-inclusive.

46.3.2 Contract Administration

Contract administration focuses on the relationships between the involved parties during the contract performance or project duration. Due to the nature of business, contract administration embraces numerous postaward and preaward functions. The basic goals of contract administration are to assure that the owner is satisfied and all involved parties are compensated on time for their efforts. The degree and intensity of contract administration will vary from contact to contract depending upon the size and complexity of the effort to be performed. Since money is of the essence, too many resources can add costs and expenditures to the project, while insufficient resources may also cost in loss of time, in inefficiencies, and in delays. A successful construction project management program is one that has the vision and flexibility to allocate contract administrative personnel and resources wisely and that maintains a delicate balance in resources necessary to sustain required efficiencies throughout the project life cycle.

46.3.3 Project Design

Project design is the cornerstone of construction project management. In this phase, concepts are drawn, formulated, and created to satisfy a need or request. The design is normally supported by sound engineering calculations, estimates, and assumptions. Extensive reviews are performed to minimize unforeseen circumstances, avoiding construction changes or modifications to the maximum extent possible in addition to verifying facts, refining or clarifying concepts, and dismissing assumptions. This phase may be the ideal time for identification and selection of the management team.

Normally, 33, 65, 95, and 100% design reviews are standard practice. The final design review follows the 95% design review which is intended for the purpose of assuring that review comments have been either incorporated into the design or dismissed from consideration. Reviews include design analysis reviews and BCO reviews. It can be clearly understood from the nomenclature that a BCO encompasses all facets of a project. Biddability relates to how the contact requirements are worded to assure clarity of purpose or intent and understanding by potential construction contractors. Constructibility concentrates on how components of the work or features of the work are assembled and how they relate to the intended final product. The main purpose of the constructibility review is to answer questions, such as whether it can be built in the manner represented in the contact drawings and specifications. Interaction between mechanical, civil, electrical, and other related fields is also considered here. Operability includes aspects of maintenance and operation, warranties, services, manpower, and resource allocation during the life of the finished work.

The finished product of the design phase should include construction drawings illustrating dimensions, locations, and details of components; contract clauses and special clauses outlining specific needs of the construction contractor; specifications for mechanical, civil, and electrical or special equipment; a bidding and payment schedule with details on how parties will be compensated for work performed or equipment produced and delivered; responsibilities; and operation and maintenance (O&M) requirements. In many instances, the designer is involved throughout the construction phase for design clarification or interpretation, incorporation of construction changes or modifications to the project, and possible O&M reviews and actions. It is not uncommon to have the designer perform contract management services for the owner.

There are a number of computer software packages readily available to assist members of the management team in writing, recording, transmitting, tracking, safekeeping, and incorporating BCO comments. Accuracy of records and safekeeping of documentation regarding this process has proved to be valuable when a dispute, claim, design deficiency, or liability issue are encountered later during the project life cycle.

46.3.4 Planning and Scheduling

Planning and schedulings are ongoing tasks throughout the project until completion and occupancy by a certain date occur. Once the design is completed and the contractor selected to perform the work, the next logical step may be to schedule and conduct a preconstruction conference. Personnel representing the owner, designer, construction contractor, regulatory agencies, and any management/oversight agency should attend this conference. Among several key topics to discuss and understand, construction planning and scheduling is most likely to be the main subject of discussion. It is during this conference that the construction contractor may present how the work will be executed. The document here is considered the "baseline schedule." Thereafter, the baseline schedule becomes a living document by which progress is recorded and measured. Consequently, the baseline schedule can be updated and reviewed in a timely manner and becomes the construction progress schedule. As stated previously, the construction progress schedule is the means by which the construction contractor records progress of work, anticipates or forecasts requirements so proper procurement and allocation of resources can be achieved, and reports the construction status of work upwardly to the owner or other interested parties. In addition, the construction contractor may use progress schedule information to assist in increasing efficiencies or to formulate the basis of payment for services provided or rendered and to anticipate cash flow requirements. The construction progress schedule can be updated as needed, or mutually agreed to by the parties, but for prolonged projects it is normally produced monthly.

A dedicated scheduler, proper staffing, and adequate computer and software packages are important to accomplish this task properly. On complex projects, planning and scheduling is a full-time requirement.

46.3.5 Safety and Environmental Considerations

Construction of any bridge is a hazardous activity by nature. No person may be required to work in surroundings or under conditions that are unsafe or dangerous to his or her health. The construction project management team must initiate and maintain a safety and health program and perform a job hazard analysis with the purpose of eliminating or reducing risks of accidents, incidents, and injuries during the performance of the work. All features of work must be evaluated and assessed in order to identify potential hazards and implement necessary precautions or engineer controls to prevent accidents, incidents, and injuries.

Frequent safety inspections and continued assessment are instrumental in maintaining the safety aspects and preventive measures and considerations relating to the proposed features of work. In the safety area, it is important for the manager to be able to distinguish between accidents/incidents and injuries. Lack of recorded work-related injuries is not necessarily a measure of how safe the work environment is on the project site. The goal of every manager is to complete the job in an accident/incident- and injury-free manner, as every occurrence costs time and money.

Today's construction operational speed, government involvement, and community awareness are placing more emphasis, responsibilities, and demands on the designer and construction contractor to protect the environment and human health. Environmental impact statements, storm water management, soil erosion control plans, dust control plan, odor control measures, analytical and disposal requirements, Department of Transportation (DOT) requirements for overland shipment, activity hazard analysis, and recycling are some of the many aspects that the construction project management team can no longer ignore or set aside. As with project scheduling and planning, environmental and safety aspects of construction may require significant attention from a member of the construction management team. When not properly coordinated and executed, environmental considerations and safety requirements can delay the execution of the project and cost significant amounts of money.

46.3.6 Implementation and Operations

Construction implementation and operations is the process by which the construction project manager balances all construction and contract activities and requirements in order to accomplish the tasks. The bulk of construction implementation and operations occurs during the construction phase of the project. The construction project management team must operate in synchronization and maintain good communication channels in order to succeed in this intense and demanding phase. Many individuals in this field may contend that the implementation and operation phase of the construction starts with the site mobilization. Although it may be an indicator of actual physical activity taking place on site, construction implementation and operations may include actions and activities prior to the mobilization to the project site.

Here, a delicate balance is attempted to be maintained between all activities taking place and those activities being projected. Current activities are performed and accomplished by field personnel with close monitoring by the construction management staff. Near (approximately 1 week ahead), intermediate (approximately 2 to 4 weeks), and distant future (over 4 weeks) requirements are identified, planned, and scheduled in order to procure equipment and supplies, schedule work crews, and maintain efficiencies and progress. Coordinating progress and other meetings and conferences may take place during the implementation and operation phase.

46.3.7 Value Engineering

Some contracts include an opportunity for contractors to submit a value engineering (VE) recommendation. This recommendation is provided to either the owner or designer. The purpose of the VE is to promote or increase the value of the finished product while reducing the dollars spent or invested; in other words, to provide the desired function for minimum cost(s). VE is not intended to reduce performance, reliability, maintainability, or life expectancy below the level required to perform the basic function. Important VE evaluation criteria performed are in terms of "collateral savings" — the measurable net reductions in the owner's/agency's overall costs of construction, operations, maintenance, and/or logistics support. In most cases, collateral savings are shared between the owner/agency and the proponent of the VE by reducing the contract price or estimated cost in the amount of the instant contract savings and by providing the proponents of the VE a share of the savings by adding the amount calculated to the contract price or fee.

46.3.8 Quality Management

During the construction of a bridge, construction quality management (CQM) play a major role in quality control and assurance. CQM refers to all control measures and assurance activities instituted by the parties to achieve the quality established by the contract requirements and specifications. It encompasses all phases of the work, such as approval of submittals, procurements, storage of materials and equipment, coordination of subcontractor activities, and the inspections and the tests required to ensure that the specified materials are used and that installations are acceptable to produce the required product. The key elements of the CQM are the contractor quality control (CQC) and quality assurance (QA). To be effective, there must be a planned program of actions, and lines of authority and responsibilities must be established. CQC is primarily the construction contractor's responsibily while QA is primarily performed by an independent agency (or other than the construction contractor) on behalf of the designer or owner. In some instances, QA may be performed by the designer. In this manner, a system of checks and balances is achieved minimizing the conflicts between quality and efficiency normally developed during construction. Consequently, CQM is a combined responsibility.

In the CQC, the construction contractor is primarily responsible for (1) producing the quality product on time and in compliance with the terms of the contract; (2) verifying and checking the adequacy of the construction contractor's quality control program of the scope and character necessary to achieve the quality of construction outlined in the contract; and (3) producing and maintaining acceptable records of its QC activities. In the QA, the designated agency is primarily responsible for (1) establishing standards and QC requirements; (2) verifying and checking adequacy of the construction contractor's QC (QA for acceptance), performing special tests and inspections as required in the contract, and determining that reported deficiencies have been corrected; and (3) assuring timely completion.

46.3.9 Partnership and Teamwork

A great deal of construction contract success, as discussed before, is attributable to partnering. Partnering should be undertaken and initiated at the earliest stage during the construction project management cycle. Some contracts may have a special clause which is intended to encourage the construction contractor to establish clear channels of communication and effective working relationships. The best approach to partnering is for the parties to volunteer to participate.

Partnering differs from the team-building concept. Team building may encourage establishing open communications and relationships when all parties share liabilities, risk, and money exposure, but not necessarily share costs of risks. The immediate goal of partnering is to establish mutual agreement(s) at the initial phases of the project on the following areas: identification of common goals; identification of common interests; establishment of lines of communication; establishment of lines of authority and decision making; commitment to cooperative problem solving, among others.

Partnering takes the elements of luck, hope, and personality out of determining project success. It facilitates workshops in which stakeholders in a specific project or program come together as a team that results in breakthrough success for all parties involved. For example, the Office of Structure Construction (OSC) of the California Department of Transportation (Caltrans) has a vision of delivery of structure construction products of the highest possible quality in partnership with their clients. And this work is not only of high quality, but is delivered in the safest, most cost-effective, and fastest manner possible. In partnership with the districts or other clients, the Office of Structure Construction (OSC) does the following to fulfill its purpose:

- Administers and inspects the construction of the Caltrans transportation structures and related facilities in a safe and efficient manner;
- Provides specialized equipment and training, standards, guidelines, and procedural manuals to ensure consistency of inspection and administration by statewide OSC staff;
- Provides consultations on safety for OSC staff and district staff performing structure construction inspection work;
- Conducts reviews and provides technical consultation and assistance for trenching and shoring temporary support and falsework construction reviews;
- Provides technical recommendations on the preparations of structure claims and the contract change orders (CCOs);
- Provides construction engineering oversight on structure work on non-state-administrated projects;
- Conducts BCO review.

46.3.10 Project Completion and Turnover of Facility

Success in construction project management may be greatly impacted during project completion and turnover of the facilities to the user or owner. The beginning of the project completion and turnover phase may be identified by one of the following: punch list developed, prefinal inspections scheduled, support areas demobilized, site restoration initiated, just to mention a few. Many of the problems encountered during this last phase may be avoided or prevented with proper user or owner participation and involvement during the previous phases, particularly during the construction where changes and modifications may have altered the original design. A good practice in preventing conflicts during the completion and turnover of the facilities is to invite the owner or user to all construction progress meetings and acceptance inspections. In that manner, the user or owner is completely integrated during the construction with ample opportunity to provide feedback and be part of the decision-making process. In addition, by active participation, the owner or user is being informed and made aware of changes, modifications, and/or problems associated with the project.

46.4 Major Construction Considerations

Concrete bridge construction involves site investigation; structure design; selection of materials — steel, concrete, aggregates, and mix design; workmanship of placment and curing of concrete; handling and maintenance of the structure throughout its life. Actually, site investigations are made of any structure, regardless of how insignificant it may be. The site investigation is very important for intelligent design of the bridge structures and has a significant influence on selection of the material and mix. A milestone is to investigate the fitness of the location to satisfy the requirements of the bridge structure. Thus, investigation of the competence of the foundation to carry the service load safely and an investigation of the existence of forces or substances that may attack the concrete structure can proceed. Of course, the distress or failure may have several contributing causal factors: unsuitable materials, construction methods, loading conditions; faulty mix design; design mistakes; conditions of exposure; curing condition, or environmental factors.

46.5 Structural Materials

46.5.1 Normal Concrete

Important Properties

Concrete is the only material that can be made on site, and is practically the most dependable and versatile construction material used in bridge construction. Good durable concrete is quality concrete that meets all structural and aesthetic requirements for a period of structure life at minimum cost. We are looking for such properties as workability in the fresh condition; strength in accordance with design, specifications, and special provisions; durability; volume stability; freedom from blemishes (scaling, rock pockets, etc.); impermeability; economy; and aesthetic appearance. Concrete when properly designed and fabricated can actually be crack-free not only under normal service loads, but also under moderate overload, which is very attractive for bridges that are exposed to an especially corrosive atmosphere.

The codes and specifications usually specify the minimum required strength for various parts of a bridge structure. The required concrete strength is determined by design engineers. For cast-in-place concrete bridges, a compressive strength of 3250 to 5000 psi (22 to 33 MPa) is usual. For precast structure compressive strength of 4000 to 6000 psi (27 to 40 MPa) is often used. For special precast, prestressed structures compressive strength of 6000 to 8000 psi (40 to 56 MPa) is used. Other properties of concrete are related to the strength, although not necessarily dependent on the strength.

Workability is the most important property of fresh concrete and depends on the properties and proportioning of the materials: fine and coarse aggregates, cement, water, and admixtures. Consistency, cohesiveness, and plasticity are elements of workability. Consistency is related to the fluidity of mix. Just adding water to a batch of concrete will make the concrete more fluid or "wetter," but the quality of the concrete will diminish. Consistency increases when water is added and an average of 3% in total water per batch will change the slump about 1 in. (2.54 cm). The research and practice show that workability is a maximum in concrete of medium consistency, between 3 in. (7.62 cm) and 6 in. (15.24 cm) slump. Very dry or wet mixes produce less-workable concrete. Use of relatively harsh and dry mixes is allowed in structures with large cross sections, but congested areas containing much reinforcement steel and embedded items require mixes with a high degree of workability.

A good and plastic mixture is neither harsh nor sticky and will not segregate easily. Cohesiveness is not a function of slump, as very wet (high-slump) concrete lacks plasticity. On the other hand, a low-slump mix can have a high degree of plasticity. A harsh concrete lacks plasticity and cohesiveness and segregates easily.

Workability has a great effect on the cost of placing concrete. Unworkable concrete, not only requires more labor and effort in placing, but also produces rock pockets and sand streaks, especially in small congested forms. It is a misconception that compaction or consolidation of concrete in the form can be done with minimum effort if concrete is fluid or liquid to flow into place. It is obvious that such concrete will flow in place but segregate badly, so that large aggregate will settle out of the mortar and excess water will rise to the top surface. And unfortunately, this error in workmanship will become apparent after days, even months later, showing up as cracks, low strength, and general inferiority of concrete. The use of high-range water-reducing admixtures (superplasticizers) allows placing of high-slump, self-leveling concrete. They increase strength of concrete and provide great workability without adding an excessive amount of water. As an example of such products used in the Caltrans is PolyHeed 997 which meets the requirements for a Type A, water-reducing admixture specified in ASTM C 494-92, Corps of Engineers CRD-C 87-93, and AASHTO M 194-87, the Standard Specifications for chemical admixtures for concrete.

Special Consideration for Cold-Weather Construction

Cold weather can damage a concrete structure by freezing of fresh concrete before the cement has achieved final set and by repeated cycles of freezing of consequent expansion of water in pores and openings in hardened concrete. Causes of poor frost resistance include poor design of construction joints, segregation of concrete during placement; leaky formwork; poor workmanship, resulting in honeycomb and sand streaks; insufficient or absent drainage, permitting water to accumulate against concrete. In order to provide resistance against frost adequate drainage should be designed. If horizontal construction joints are necessary, they should be located below the low-water or above the high-water line about 2 to 3 ft (0.6 to 1 m). Previously placed concrete must be cleaned up completely. Concrete mix should have a 7% (max) air for ½ in. (12.7 mm) or ¾ in. (19 mm) (max) aggregate, ranging down to 3 to 4% for cobble mixes. It is essential to use structurally sound aggregates with low porosity. The objective of frost-resistant concrete mix is to produce good concrete with smooth, dense, and impermeable surface. This can be implemented by good construction techniques used in careful placement of concrete as near as possible to its final resting place, avoiding segregation, sand streaks, and honeycomb under proper supervision, quality control, and assurance.

Sudden changes in temperature can stress concrete and cause cracking or crazing. A similar condition exists when cold water is applied to freshly stripped warm concrete, particularly during hot weather. For the best results, the temperature difference should not exceed 25°F between concrete and curing water. In cases when anchor bolt holes were left exposed to weather, filled with water, freezing of water exerted sufficient force to crack concrete. This may happen on the bridge pier cap under construction.

Concrete Reinforcement and Placement

The optimum conditions for structural use is a medium slump of concrete and compaction by vibrators. A good concrete with low slump for the placing conditions can be ruined by insufficient or improper consolidation. Even workable concrete may not satisfy the needs of the bridge structure if it is not properly consolidated, preferably by vibration. An abrupt change in size, and congestion of reinforcement not only makes proper placing of concrete difficult but also causes cracks to develop. Misplacement of reinforcement within concrete will greatly contribute to development of structural cracks. The distress and failure of concrete are mostly caused by ignorance, carelessness, wrong assumptions, etc.

Concrete Mix and Trial Batches

The objective of concrete mix designs and trial batches is to produce cost-effective concrete with sufficient workability, strength, durability, and impermeability to meet the conditions of placing, finishing characteristics, exposure, loading, and other requirements of bridge structures. A complete discussion of concrete mixes and materials can be found in many texts such as *Concrete Manual* by Waddel [1]. The purpose of trial batches is to determine strength, water–cement ratio, combined grading of aggregates, slump, type and proportioning of cement, aggregates, entrained air, and admixtures as well as scheduling of trial batches and uniformity. Trial batches should always be made for bridge structures, especially for large and important ones. They should also be made in cases where there is no adequate information available for existing materials used in concrete mixes, and they are subjected to revision in the field as conditions require.

Consideration to Exposure Condition

Protection of waterfront structures should be considered when they are being designed. Designers often carefully consider structural and aesthetic aspects without consideration of exposure conditions. Chemical attack is aggravated in the presence of water, especially in transporting the chemiclas into the concrete through cracks, honeycombs, or pores in surfaces. Use of chamfers and fillers is good construction practice. Chamfering helps prevent spalling and chipping from moving objects. Fillets in reentrant corners eliminate possible scours or cracking. Reinforcement should be well covered with sound concrete and in most cases the 3 in. (7.62 cm) coverage is specified. First-class nonreactive and well-graded aggregates in accordance with the UBC standard should be used. Cement Type II or Type Y with a low of C_3 should be used. Careful consideration should be given to the use of an approved pozzolan with a record of successfully usage in a similar exposure. Mix design should contain an adequate amount of entrained air and other parameters in accordance with specifications or a special provision for a particular project. The concrete should be workable with slump and water–cement ratio as low as possible and containing at least 560 pcy (332 kg/m³). To reduce mixing water for the same workability and, by the same token, to enhance strength and durability, a water-reducing admixture is preferred. The use of calcium chloride and Type III cement for acceleration of hardening and strength development is precluded. Concrete should be handled and placed with special care to avoid segregation and prevent honeycomb and sand streaks. The proper cure should be taken for at least seven days before exposure.

46.5.2   High-Performance Concrete

High-performance concrete (HPC) is composed of the same materials used in normal concrete, but proportioned and mixed to yield a stronger, more durable product. HPC structures last much longer and suffer less damage from heavy traffic and climatic condition than those made with conventional concrete. To promote the use of HPC in highway structures in the United States, a group of concrete experts representing the state DOTs, academia, the highway industry, and the Federal Highway Administration (FHWA) has developed a working definition of HPC, which includes performance criteria and the standard tests to evaluate performance when specifying an HPC mixture. The designer determines what level of strength, creep, shrinkage, elasticity, freeze/thaw durability, abrasion resistance, scaling resistance, and chloride permeability are needed. The definition specifies what tests grade of HPC satisfies those requirements and what tests to perform to confirm that the concrete meets that grade.

An example of the mix design for the 12,000-psi high-strength concrete used in the Orange County courthouse in Florida follows:

Gradient	Weight (pounds)
Cement, Type 1	900
Fly ash, Class F	72
Silica fume	62
Natural sand	980
No. 8 granite aggregate	1,780
Water	250
Water reducer	2 oz per cubic hundredweight
Superplasticizer	35 oz per cubic hundredweight

The Virginia and Texas DOTs have already started using HPC that is ultra-high-strength concrete 12,000 to 15,000 psi (80 to 100 MPa) in bridge construction and rehabilitation of the existing bridges [2].

46.5.3   Steel

All reinforcing steel for bridges is required to conform to specifications of ASTM Designation A615, Grade 60 or low-alloy steel deformed bars conforming to ASTM Designation A706. Prestressing steel: high-tensile wire conforming to ASTM Designations: A421, including Supplement I, High-tensile wire strand A416, Uncoated high-strength steel bars: A722, are usually used. All prestressing steel needs to be protected against physical damage and rust or other results of corrosion at all times from manufacture to grouting or encasing in concrete. Prestressing steel that has physical damage at any time needs to be rejected. Prestressing steel for post-tensioning that is installed in members prior to placing and curing of the concrete needs to be continuously protected against rust or other corrosion until grouted, by means of a corrosion inhibitor placed in the ducts or applied to the steel in the duct.

The corrosion inhibitor should conform to the specified requirements. When steam curing is used, prestressing steel for post-tensioning should not be installed until the stem curing is completed. All water used for flushing ducts should contain either quick lime (calcium oxide) or slaked lime (calcium hydroxide) in the amount of 0.01 kg/l. All compressed air used to blow out ducts should be oil free.

46.6 Construction Operations

46.6.1 Prestressing Methods

If steel reinforcement in reinforced concrete structures is tensioned against the concrete, the structure becomes a prestressed concrete structure. This can be accomplished by using pretensioning and post-tensioning methods.

Pretensioning

Pretensioning is accomplished by stressing tendons, steel wires, or strands to a predetermined amount. While stress is maintained in the tendons, concrete is placed in the structure. After the concrete in the structure has hardened, the tendons are released and the concrete bonded to the tendons becomes prestressed.

Widely used in pretensioning techniques are hydraulic jacks and strands composed of several wires twisted around a straight center wire. Pretensioning is a major method used in the manufacture of prestressed concrete in the United States. The basic principles and some of the methods currently used in the United States were imported from Europe, but much has been done in the United States to develop and adapt manufacturing procedures. One such adaptation employs pretensioned tendons which do not pass straight through the concrete member, but are deflected or draped into a trajectory that approximates a curve. This method is very widely practiced in the fabrication of precast bridge girders in the United States.

Post-Tensioning

A member is called as posttensioned when the tendons are tensioned after the concrete has hardened and attained sufficient strength (usually 70% final strength) to withstand the prestressing force, and each end of the tendons are anchored. Figure 46.1 shows a typical post-tensioning system. A common method used in the United States to prevent tendons from bonding to the concrete during placing and curing of the concrete is to encase the tendon in a mortar-tight metal tube or flexible metal hose before placing it in the forms. The metal hose or tube is referred to as a sheath or duct and remains in the structure. After the tendons have been stressed, the void between the tendons and the duct is filled with grout. The tendons become bonded to the structural concrete and protected from corrosion [3].Construction engineers can utilize prestressing very effectively to overcome excessive temporary stresses or deflections during construction, for example, using cantilevering techniques in lieu of falsework.

Prestressing is not a fixed state of stress and deformation, but is time dependent. Both concrete and steel may be deformed inelastically under continued stress. After being precompressed, concrete continues to shorten with time (creep). Loss of moisture with time also contribute to a shortening (shrinkage). In order to reduce prestress losses due to creep and shrinkage and to increase the level of precompression, use of higher-strength not only steel but also higher-strength concrete, that has low creep, shrinkage, and thermal response is recommended. New chemical admixtures such as high-range water-reducing admixtures (superplasticizers) and slag used for producing high-performance concrete and for ultra-high-strength concrete. The new developments are targeted to producing high-strength steel that is "stabilized" against stress relaxation which leads to a reduction of stress in tendons, thus reducing the prestress in concrete.

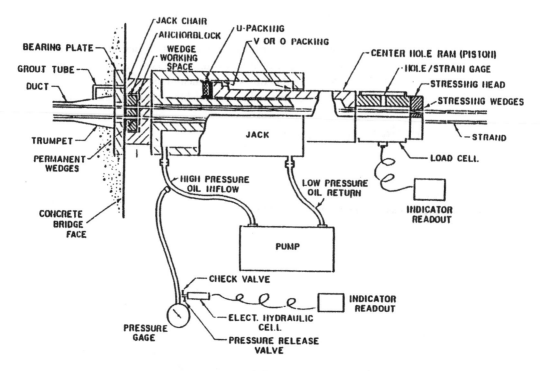

FIGURE 46.1   Typical post-tensioning system.

46.6.2 Fabrication and Erection Stages

During construction, not all elements of a bridge have the same stresses they were designed for. That is the reason it is a very important part of construction engineering to be aware of this and to make sure that appropriate steps have been taken. For example, additional reinforcement will be added to the members in the fabrication stage and delivered to the job site for erection.

In the case of cast-in-place box-girder bridge construction the sequences of prestressing tendons have to be engineered step-by-step to ensure that the structure will have all parameters for future service load after completion of this stage.

The sequence of the erection itself may produce additional stresses that structures or portions of the structures were not designed for. These stresses and the stability of structures during erection are a big concern that is often overlooked by designers and contractors — construction sequences play a very important role in the erection of a segmental type of bridge. It seems that we have to give more attention to analysis of the role of the construction engineering implementation of such erections. And, yes, sometimes the importance of construction engineering to accomplish safe and efficient fabrication and erection of bridge structures (precast, prestressed girders, cast-in-pile) is not sufficiently emphasized by design engineers and/or fabrication, erection contractors.

Unfortunately, we have to admit that the design set of drawings even for an important bridge does not include the erection scheme. And, of course, we can show many examples of misplaced erection efforts on the part of the designer, but our goal is to show why it happened and to make efforts to pay more attention to the fabrication and erection stages. Even if such an erection scheme is included in the design drawings, contractors are not supposed to rely solely on what is provided by the designer's erection plan.

Sometimes a design can be impractical, or it may not be suitable in terms of the erection contractor's equipment and experience. Because the erection plans usually are very generalized and because not enough emphasis is given to the importance of this stage, it is important that the

FIGURE 46.2   Pine Valley Creek Bridge — construction at Pier 4.

designer understand the contractor's proposed method so that the designer can determine if these methods are compatible with the plans, specifications, and requirements of the contract. This is the time that any differences should be resolved. The designer should also discuss any contingency plan in case the contractor has problems. In many instances, the designer is involved throughout the construction phase for design and specification clarification or interpretation, incorporation of construction changes or modifications to the project, and possible O&M reviews/action.

46.6.3 Construction of Segmental Bridges

The first precast segmental box-girder bridge was built by Jean Muller, the Choisy-le-Roi-Bridge crossing the Seine River in 1962. In North America (Canada), a cast-in-place segmental bridge on the Laurentian Autoroute, near Ste. Adele, Quebec, in 1964 and a precast segmental bridge crossing the Lievre River near Notre Dame du Laus also in Quebec in 1967 were constructed. In the United States, the first precast segmental bridge was completed in 1973 in Corpus Christi (Texas). The Pine Valley Creek Bridge with five spans (270 + 340 + 450 + 380 + 270 ft) supported by 340-ft-high pier as shown in Figures 46.2 to 46.5 is the first cast-in-place segmental bridge constructed in the United States in 1974 using the cantilever method. The ends of the bridge are skewed to fit the bridge into the canyon. The superstructure consists of two parallel box structures each providing a roadway width of 40 ft between railings. The superstructures are separated by a 38-ft median.

Segmental cantilever construction is a fairly recent development, and the concept has been improved and used successfully to build bridges throughout the world. Its unique characteristic of needing no ground-supported falsework makes the method attractive for use over congested streets, waterways, deep gorges, or ocean inlets. It has been used for spans of less than 100 ft, all the way to the current record span of 755 ft over the Urato River in Japan. Another advantage of the method lies in its economy and efficiency of material use. Construction of segmental bridges can be classified by three methods: balanced cantilever, span-by-span, and progressive placement or incremental launching. For detailed discussion see Chapter 11.

FIGURE 46.3   Pine Valley Creek Bridge — pier construction

FIGURE 46.4   Pine Valley Creek Bridge — girder construction

FIGURE 46.5 Pine Valley Creek Bridge — construction completion.

46.6.4 Construction of HPC Bridges

The first U.S. bridge was built with HPC under the Strategic Highway Research Program (SHRP) in Texas in 1996. The FHWA and the Texas DOT in cooperation with the Center for Transportation Research (CTR) at the University of Texas at Austin sponsored a workshop to showcase HPC for bridges in Houston in 1996. The purpose of the event was to introduce the new guidelines to construction professionals and design engineers, and to show how HPC was being used to build more durable structures. It was also focused on the pros and cons of using HPC, mix proportioning, structural design, HPC in precast prestressed and cast-in-place members, long-term performance, and HPC projects in Nebraska, New Hampshire, and Virginia. The showcase had a distinctly regional emphasis because local differences in cements, aggregates, and prestressing fabricators have a considerable impact on the design and construction of concrete structures. In Texas, concrete can be produced with compressive strength of 13,000 to 15,000 psi (900 to 1000 MPa).

The Louetta Road Overpass using HPC is expected to have a useful life of 75 to 100 years, roughly double the average life of a standard bridge. A longer life span means not only lower user cost, but motorists will encounter fewer lane closures and other delays caused by maintenance work. At the present time 15 HPC bridge have been built in the United States.

The first bridge to utilize HPC fully in all aspects of design and construction is the Louetta Road Overpass on State Highway 249 in Houston. The project consists of two U-beam bridges carrying two adjacent lanes of traffic. The spans range from 121.5 to 135.5 ft (37 to 41.3 m) long. The HPC is about twice as strong as conventional concrete. It costs an average of $260/m^2 ($24/ft^2) of deck area, a price compatible with the 12 conventional concrete bridges on the same project. The second Texas HPC bridge located in San Angelo carries the eastbound lanes of U.S. Route 67 over the North Concho River, U.S. 87, and the South Orient railroad. The 954-ft (291-m) HPC I-beam bridge runs parallel to a conventional concrete bridge. The HPC was chosen for the east-bound lanes because the span crossing the North Concho River was 157 ft (48 m) long. This distance exceeds the capacity of Texas conventional prestressed concrete U-beam simple-span construction. The San Angelo Bridge presents an ideal opportunity for comparing HPC and conventional concrete. The first spans

FIGURE 46.6 Falsework at I-80 HOV construction, Richmond, CA.

of two bridges are the same length and width making it easy to compare the cost and performance between HPC and conventional concrete. The comparison indicated that conventional concrete lanes of the first span required seven beams with 5.6 ft (1.7 m) spacing, while the HPC span required only four beams with 11 ft (3.4 m) spacing.

46.7 Falsework

Falsework may be defined as a temporary framework on which the permanent structure is supported during its construction. The term *falsework* is universally associated with the construction of cast-in-place concrete structures, particularly bridge superstructures. The falsework provides a stable platform upon which the forms may be built and furnish support for the bridge superstructure.

Falsework is used in both building and bridge construction. The temporary supports used in building work are commonly referred to as "shoring." It is also important to note the difference between "formwork" and "falsework." Formwork is used to retain plastic concrete in its desired shape until it has hardened. It is designed to resist the fluid pressure of plastic concrete and additional pressure generated by vibrators. Because formwork does not carry dead load of concrete, it can be removed as soon as the concrete hardens. Falsework does carry the dead load of concrete, and therefore it has to remain in place until the concrete becomes self-supporting. Plywood panels on the underside of a concrete slab serve both as a formwork and as a falsework member. For design, however, such panels are considered to be forms in order to meet all design and specification requirements applied to them.

Bridge falsework can be classified in two types: (1) conventional systems (Figure 46.6) in which the various components (beams, posts, caps, bracings, etc.) are erected individually to form the completed system and (2) proprietary systems in which metal components are assembled into modular units that can be stacked, one above the other, to form a series of towers that compose the vertical load-carrying members of the system.

The contractor is responsible for designing and constructing safe and adequate falsework that provides all necessary rigidity, supports all load composed, and produces the final product (structure) according to the design plans, specifications, and special provisions. It is very important also to keep in mind that approval by the owner of falsework working drawings or falsework inspection will in no way relieve the contractor of full responsibility for the falsework. In the state of California, any falsework height that exceeds 13 ft (4 m) or any individual falsework clear span that exceeds 17 ft (5 m) or where provision for vehicular, pedestrian, or railroad traffic through the falsework is made, the drawings have to be signed by the registered civil engineer in the state of California. The design drawings should include details of the falsework removal operations, methods and sequences of removal, and equipment to be used. The drawings must show the size of all load-supporting members, connections and joints, and bracing systems. For box-girder structures, the drawings must show members supporting sloping exterior girders, deck overhangs, and any attached construction walkway. All design-controlling dimensions, including beam length and spacing, post locations and spacing, overall height of falsework bents, vertical distance between connectors in diagonal bracing must be shown.

It is important that falsework construction substantially conform to the falsework drawings. As a policy consideration, minor deviations to suit field conditions or the substitution of materials will be permitted if it is evident by inspection that the change does not increase the stresses or deflections of any falsework members beyond the allowable values, nor reduce the load-carrying capacity of the overall falsework system. If revision is required, the approval of revised drawings by the state engineer is also required. Any change in the approved falsework design, however minor it may appear to be, has the potential to affect adversely the structural integrity of the falsework system. Therefore, before approving any changes, the engineer has to be sure that such changes will not affect the falsework system as a whole.

References

1. Waddel, J. J., *Concrete Manual*, 1994.
2. *Focus Mag.*, May 1996.
3. Libby, J., *Modern Prestressed Concrete*, 1994.
4. Gerwick, B.C., Jr., *Construction of Prestressed Concrete Structures*, 1994.
5. Fisk, E. R., *Construction Project Administration*, John Wiley & Sons, New York, 1994.
6. Blank, M. M., *Selected Published Articles from 1986 to 1995*, Naval Air Warfare Center, Warmminster, PA, 1996.
7. Blank, M. M. and Blank, S. A., *Effective Construction Management Tools*, Management Division, Naval Air Warfare Center, Warmminster, PA, 1995.
8. Godfrey, K. A., *Partnering in Design and Construction*, McGraw-Hill, New York, 1995.
9. Blank, M. M. et al., Partnering: the key to success, *Found. Drilling*, March/April, 1997.
10. Kubai, M. T. *Engineering Quality in Construction Partnering and TQM*, McGraw-Hill, New York, 1994.
11. Rubin, D.K., A burning sensation in Texas, *Eng. News Rec.*, July 12, 1993.
12. Partnering Guide for Environmental Missions of the Air Force, Army and Navy, Prepared by a Tri-Service Committee, Air Force, Army, and Navy, July 1996.
13. Post, R. G., Effective partnering, *Construction*, June 1996.
14. Schriener, J., Partnering, TQM, ADF, low insurance cost, *Eng. News Rec.*, January 15, 1995.
15. Caltrans, *Standard Specifications*, California Department of Transportation, Sacramento, 1997.
16. Caltrans, *A Guide for Field Inspection of Cast-in-Place Post-Tensioned Structures*, California Department of Transportation, Sacramento, 1992.

47

Substructures of Major Overwater Bridges

Ben C. Gerwick, Jr.
*Ben C. Gerwick Inc. and University
of California, Berkeley*

47.1 Introduction

The design and construction of the piers for overwater bridges present a series of demanding criteria. In service, the pier must be able to support the dead and live loads successfully, while resisting environmental forces such as current, wind, wave, sea ice, and unbalanced soil loads, sometimes even including downslope rock fall. Earthquake loadings present a major challenge to design, with cyclic reversing motions propagated up through the soil and the pier to excite the superstructure. Accidental forces must also be resisted. Collision by barges and ships is becoming an increasingly serious hazard for bridge piers in waterways, both those piers flanking the channel and those of approaches wherever the water depth is sufficient.

0-8493-7434-0/00/$0.00+$.50
© 2000 by CRC Press LLC

Soil–structure foundation interaction controls the design for dynamic and impact forces. The interaction with the superstructure is determined by the flexibility of the entire structural system and its surrounding soil.

Rigid systems attract very high forces: under earthquake, the design forces may reach 1.0 g, whereas flexible structures, developing much less force at longer periods, are subject to greater deflection drift. The design must endeavor to obtain an optimal balance between these two responses. The potential for scour due to currents, amplified by vortices, must be considered and preventive measures instituted.

Constructibility is of great importance, in many cases determining the feasibility. During construction, the temporary and permanent structures are subject to the same environmental and accidental loadings as the permanent pier, although for a shorter period of exposure and, in most cases, limited to a favorable time of the year, the so-called weather window. The construction processes employed must therefore be practicable of attainment and completion. Tolerances must be a suitable compromise between practicability and future performance. Methods adopted must not diminish the future interactive behavior of the soil–structure system.

The design loadings for overwater piers are generally divided into two limit states, one being the limit state for those loadings of high probability of occurrence, for which the response should be essentially elastic. Durability needs to be considered in this limit state, primarily with respect to corrosion of exposed and embedded steel. Fatigue is not normally a factor for the pier concepts usually considered, although it does enter into the considerations for supplementary elements such as fender systems and temporary structures such as dolphins if they will be utilized under conditions of cyclic loading such as waves. In seismic areas, moderate-level earthquakes, e.g., those with a return period of 300 to 500 years, also need to be considered.

The second limit state is that of low-probability events, often termed the "safety" or "extreme" limit state. This should include the earthquake of long return period (1000 to 3000 years) and ship collision by a major vessel. For these, a ductile response is generally acceptable, extending the behavior of the structural elements into the plastic range. Deformability is essential to absorb these high-energy loads, so some damage may be suffered, with the provision that collapse and loss of life are prevented and, usually, that the bridge can be restored to service within a reasonable time.

Plastic hinging has been adopted as a principle for this limit state on many modern structures, designed so that the plastic hinging will occur at a known location where it can be most easily inspected and repaired. Redundant load paths are desirable: these are usually only practicable by the use of multiple piles.

Bridge piers for overwater bridges typically represent 30 to 40% of the overall cost of the bridge. In cases of deep water, they may even reach above 50%. Therefore, they deserve a thorough design effort to attain the optimum concept and details.

Construction of overwater bridge piers has an unfortunate history of delays, accidents, and even catastrophes. Many construction claims and overruns in cost and time relate to the construction of the piers. Constructibility is thus a primary consideration.

The most common types of piers and their construction are described in the following sections.

47.2 Large-Diameter Tubular Piles

47.2.1 Description

Construction of steel platforms for offshore petroleum production as well as deep-water terminals for very large vessels carrying crude oil, iron, and coal, required the development of piling with high axial and lateral capacities, which could be installed in a wide variety of soils, from soft sediments to rock. Lateral forces from waves, currents, floating ice, and earthquake as well as from berthing dominated the design. Only large-diameter steel tubular piles have proved able to meet these criteria (Figures 47.1 and 47.2).

FIGURE 47.1 Large-diameter steel tubular pile, Jamuna River Bridge, Bangladesh

FIGURE 47.2 Driving large-diameter steel tubular pile.

FIGURE 47.3 Steel tubular pile being installed from jack-up barge. Socket will be drilled into rock and entire pile filled with tremie concrete.

Such large piling, ranging from 1 to 3 m in diameter and up to over 100 m in length required the concurrent development of very high energy pile-driving hammers, an order of magnitude higher than those previously available. Drilling equipment, powerful enough to drill large-diameter sockets in bedrock, was also developed (Figure 47.3).

Thus when bridge piers were required in deeper water, with deep sediments of varying degrees or, alternatively, bare rock, and where ductile response to the lateral forces associated with earthquake, ice, and ship impact became of equal or greater importance than support of axial loads, it was only natural that technology from the offshore platform industry moved to the bridge field.

The results of this "lateral" transfer exceeded expectations in that it made it practicable and economical to build piers in deep waters and deep sediments, where previously only highly expensive and time-consuming solutions were available.

47.2.2 Offshore Structure Practice

The design and construction practices generally follow the Recommended Practice for Planning, Designing and Constructing Fixed Offshore Platforms published by the American Petroleum Institute, API-RP2A [1]. This recommended practice is revised frequently, so the latest edition should always be used. Reference [2] presents the design and construction from the construction contractor's point of view.

There are many variables that affect the designs of steel tubular piles: diameter, wall thickness (which may vary over the length), penetration, tip details, pile head details, spacing, number of piles, geometry, and steel properties. There must be consideration of the installation method and its effect on the soil–pile interaction. In special cases, the tubular piles may be inclined, i.e., "raked" on an angle from vertical.

In offshore practice, the piles are almost never filled with concrete, whereas for bridge piers, the designer's unwillingness to rely solely on skin friction for support over a 100-year life as well as concern for corrosion has led to the practice of cleaning out and filling with reinforced concrete. A recent advance has been to utilize the steel shell along with the concrete infill in composite action to increase strength and stiffness. The concrete infill is also utilized to resist local buckling under overload and extreme conditions. Recent practice is to fill concrete in zones of high moment.

Tubular piles are used to transfer the superimposed axial and lateral loads and moments to the soil. Under earthquake, the soil imparts dynamic motions to the pile and hence to the structure. These interactions are highly nonlinear. To make matters even more complex, the soils are typically nonuniform throughout their depth and have different values of strength and modulus.

In design, axial loads control the penetration while lateral load transfer to the soil determines the pile diameter. Combined pile stresses and installation stresses determine the wall thickness. The interaction of the pile with the soil is determined by the pile stiffness and diameter. These latter lead to the development of a *P–y* curve, *P* being the lateral shear at the head of the pile and *y* being the deflection along the pile. Although the actual behavior is very complex and can only be adequately solved by a computerized final design, an initial approximation of three diameters can give an assumed "point of fixity" about which the top of the pile bends.

Experience and laboratory tests show that the deflection profile of a typical pile in soft sediments has a first point of zero deflection about three diameters below the mudline, followed by deflection in reverse bending and finally a second point of zero displacement. Piles driven to a tip elevation at or below this second point have been generally found to develop a stable behavior in lateral displacement even under multiple cycles of high loading.

If the deflection under extreme load is significant, *P*–Δ effects must also be considered. Bridge piers must not only have adequate ultimate strength to resist extreme lateral loads but must limit the displacement to acceptable values. If the displacement is too great, the *P*–Δ effect will cause large additional bending moments in the pile and consequently additional deflection.

The axial compressive behavior of piles in bridge piers is of dominant importance. Settlement of the pile under service and extreme loads must be limited. The compressive axial load is resisted by skin friction along the periphery of the pile, by end bearing under the steel pile tip, and by the end bearing of the soil plug in the pile tip. This latter must not exceed the skin friction of the soil on the inside of the pile, since otherwise the plug will slide upward. The actual characteristics of the soil plug are greatly affected by the installation procedures, and will be discussed in detail later.

Axial tension due to uplift under extreme loads such as earthquakes is resisted by skin friction on the periphery and the deadweight of the pile and footing block.

Pile group action usually differs from the summation of individual piles and is influenced by the stiffness of the footing block as well as by the applied bending moments and shears. This group action and its interaction with the soil are important in the final design, especially for dynamic loading such as earthquakes.

API-RP2A Section G gives a design procedure for driven steel tubular piles as well as for drilled and grouted piles.

Corrosion and abrasion must be considered in determining the pile wall thickness. Corrosion typically is most severe from just below the waterline to just above the wave splash level at high tide, although another vulnerable location is at the mudline due to the oxygen gradient. Abrasion typically is most severe at the mudline because of moving sands, although suspended silt may cause abrasion throughout the water column.

Considering a design lifetime of a major bridge of 100 years or more, coatings are appropriate in the splash zone and above, while sacrificial anodes may be used in the water column and at the mudline. Additional pile wall thickness may serve as sacrificial steel: for seawater environment, 10 to 12 mm is often added.

47.2.3 Steel Pile Design and Fabrication

Tubular steel piles are typically fabricated from steel plate, rolled into "cans" with the longitudinal seam being automatically welded. These cans are then joined by circumferential welds. Obviously, these welds are critical to the successful performance of the piles. During installation by pile hammer, the welds are often stressed very highly under repeated blows: defective welds may crack in the weld or the heat-affected zone (HAZ). Welds should achieve as full joint penetration as practicable, and the external weld profile should merge smoothly with the base metal on either side.

API-RP2A, section L, gives guidance on fabrication and welding. The fabricated piles should meet the specified tolerances for both pile straightness and for cross section dimensions at the ends. These latter control average diameter and out-of-roundness. Out-of-roundness is of especial concern as it affects the ability to match adjacent sections for welding.

Inspection recommendations are given in API-RP2A, section N. Table N.4-1, with reference to structural tubulars, calls for 10% of the longitudinal seams to be verified by either ultrasonic (UT) or radiography (RT). For the circumferential weld seams and the critical intersection of the longitudinal and circumferential seams, 100% UT or RT is required.

Because of the typically high stresses to which piles supporting bridge piers are subjected, both under extreme loads and during installation, as well as the need for weldability of relatively thick plates, it is common to use a fine-grained steel of 290 to 350 MPa yield strength for the tubular piles.

Pile wall thickness is determined by a number of factors. The thickness may be varied along the length, being controlled at any specific location by the loading conditions during service and during installation.

The typical pile used for a bridge pier is fixed at the head. Hence, the maxima combined bending and axial loads will occur within the 1½ diameters immediately below the bottom of the footing. Local buckling may occur. Repeated reversals of bending under earthquake may even lead to fracture. This area is therefore generally made of thicker steel plate. Filling with concrete will prevent local buckling. General column buckling also needs to be checked and will usually be a maximum at a short distance below the mudline.

Installation may control the minimum wall thickness. The hammer blows develop high compressive waves which travel down the pile, reflecting from the tip in amplified compression when high tip resistance is encountered. When sustained hard driving with large hammers is anticipated, the minimum pile wall thickness should be $t = 6.35 + D/100$ where t and D are in millimeters. The drivability of a tubular pile is enhanced by increasing the wall thickness. This reduces the time of driving and enables greater penetration to be achieved.

During installation, the weight of the hammer and appurtenances may cause excessive bending if the pile is being installed on a batter. Hydraulic hammers usually are fully supported on the pile, whereas steam hammers and diesel hammers are partially supported by the crane.

If the pile is cleaned out during driving in order to enable the desired penetration to be achieved, external soil pressures may develop high circumferential compression stresses. These interact with the axial driving stresses and may lead to local buckling.

The tip of the pile is subject to very high stresses, especially if the pile encounters boulders or must be seated in rock. This may lead to distortion of the tip, which is then amplified during successive blows. In extreme cases, the tip may "tear" or may "accordion" in a series of short local axial buckles. Cast steel driving shoes may be employed in such cases; they are usually made of steels of high toughness as well as high yield strength. The pile head also must be thick enough to withstand both the local buckling and the bursting stresses due to Poisson's effect.

The transition between sections of different pile wall thickness must be carefully detailed. In general, the change in thickness should not be more than 12 mm at a splice and the thicker section should be beveled on a 1:4 slope.

FIGURE 47.4   Large-diameter tubular steel pile being positioned.

47.2.4   Transportation and Upending of Piles

Tubular piles may be transported by barge. For loading, they are often simply rolled onto the barge, then blocked and chained down. They may also be transported by self-flotation. The ends are bulkheaded during deployment. The removal of the bulkheads can impose serious risks if not carefully planned. One end should be lifted above water for removal of that bulkhead, then the other. If one bulkhead is to be removed underwater by a diver, the water inside must first be equalized with the outside water; otherwise the rush of water will suck the diver into the pipe. Upending will produce high bending moments which limit the length of the sections of a long pile (Figure 47.4). Otherwise the pile may be buckled.

47.2.5   Driving of Piles

The driving of large-diameter tubular piles [2] is usually done by a very large pile hammer. The required size can be determined by both experience and the use of a drivability analysis, which incorporates the soil parameters.

Frequently, the tubular pile for a bridge pier is too long or too heavy to install as a single section. Hence, piles must be spliced during driving. To assist in splicing, stabbing guides may be preattached to the tip of the upper segment, along with a backup plate. The tip of the upper segment should be prebeveled for welding.

FIGURE 47.5   Arrangement of internal jet piping and "spider" struts in large-diameter tubular pile.

Splicing is time-consuming. Fortunately, on a large-diameter pile of 2 to 4 m diameter, there is usually space to work two to three crews concurrently. Weld times of 4 to 8 h may be required. Then the pile must cool down (typically 2 h) and NDT performed. Following this, the hammer must be repositioned on top of the pile. Thus a total elapsed time may be 9 to 12 h, during which the skin friction on the pile sides "sets up," increasing the driving resistance and typically requiring a number of blows to break the pile loose and resume penetration.

When very high resistance is encountered, various methods may be employed to reduce the resistance so that the design pile tip may be reached. Care must be taken that these aids do not lessen the capacity of the pile to resist its design loads.

High resistance of the tubular pile is primarily due to plugging of the tip; the soil in the tip becomes compacted and the pile behaves as a displacement pile instead of cutting through the soil. The following steps may be employed.

1. *Jetting internally to break up the plug, but not below the tip.* The water level inside must be controlled, i.e., not allowed to build up much above the outside water level, in order to prevent piping underneath. Although a free jet or arrangement of jets may be employed, a very effective method is to manifold a series of jets around the circumference and weld the down-going pipes to the shell (Figure 47.5). Note that these pipes will pick up parasitic stresses under the pile hammer blows.

2. *Clean out by airlift.* This is common practice when using large-diameter tubular piles for bridge piers but has serious risks associated with it. The danger arises from the fact that an airlift can remove water very rapidly from the pile, creating an unbalanced head at the tip, and allowing run-in of soil. Such a run-in can result in major loss of resistance, not only under the tip in end bearing but also along the sides in skin friction.

   Unfortunately, this problem has occurred on a number of projects! The prevention is to have a pump operating to refill the pile at the same rate as the airlift empties it — a very difficult matter to control. If structural considerations allow, a hole can be cut in the pile wall so that the water always automatically balances. This, of course, will only be effective when the hole is below water. The stress concentrations around such a hole need to be carefully evaluated. Because of the risks and the service consequences of errors in field control, the use of an airlift is often prohibited. The alternative method, one that is much safer, is the use of a grab bucket (orange peel bucket) to remove the soil mechanically. Then, the water level can be controlled with relative ease.

3. *Drilling ahead a pilot hole, using slurry.* If the pile is kept full of slurry to the same level as the external water surface, then a pilot hole, not to exceed 75% of the diameter, may be drilled ahead from one to two diameters. Centralizers should be used to keep the drilled hole properly

aligned. Either bentonite or a polymer synthetic slurry may be used. In soils such as stiff clay or where a binder prevents sloughing, seawater may be used. Reverse circulation is important to prevent erosion of the soils due to high-velocity flow. Drilling ahead is typically alternated with driving. The final seating should be by driving beyond the tip of the drilled hole to remobilize the plug resistance.

4. *External jetting.* External jetting relieves the skin friction during driving but sometimes permanently reduces both the lateral and axial capacity. Further, it is of only secondary benefit as compared with internal jetting to break up the plug. In special cases, it may still be employed. The only practicable method to use with long and large tubular piles is to weld the piping on the outside or inside with holes through the pile wall. Thus, the external jetting resembles that used on the much larger open caissons. As with them, low-pressure, high-volume water flow is most effective in reducing the skin friction. After penetration to the tip, grout may be injected to partially restore the lateral and axial capacity.

47.2.6 Utilization of Piles in Bridge Piers

There are several possible arrangement for tubular piles when used for bridge piers. These differ in some cases from those used in offshore platforms.

1. The pile may be driven to the required penetration and left with the natural soil inside. The upper portion may then be left with water fill or, in some cases, be purposely left empty in order to reduce mass and weight; in this case it must be sealed by a tremie concrete plug. To ensure full bond with the inside wall, that zone must be thoroughly cleaned by wire brush on a drill stem or by jet.

For piles fixed at their head, at least 2 diameters below the footing are filled with concrete to resist local buckling. Studs are installed in this zone to ensure shear transfer.

2. The pile, after driving to final penetration, is cleaned out to within one diameter of the tip. The inside walls are cleaned by wire brush or jet. A cage of reinforcing steel may be placed to augment the bending strength of the tubular shell. Centralizers should be used to ensure accurate positioning. The pile is then filled with tremie concrete. Alternatively, an insert steel tubular with plugged tip may be installed with centralizers, and the annular space filled with tremie grout. The insert tubular may need to be temporarily weighted and/or held down to prevent flotation in the grout.

Complete filling of a tubular pile with concrete is not always warranted. The heat of hydration is a potential problem, requiring special concrete mix design and perhaps precooling.

The reasons for carrying out this practice, so often adopted for bridge piers although seldom used in offshore structures, are

a. Concern over corrosion loss of the steel shell over the 100-year lifetime;
b. A need to ensure positively the ability of the permanent plug to sustain end bearing;
c. Prevention of local buckling near the mudline and at the pile head;
d. To obtain the benefits of composite behavior in stiffness and bending capacity.

If no internal supplemental reinforcement is required, then the benefits of (b), (c), and (d) may be achieved by simple filling with tremie concrete. To offset the heat of hydration, the core may be placed as precast concrete blocks, subsequently grouted into monolithic behavior. Alternatively, an insert pile may be full length. In this case, only the annulus is completely filled. The insert pile is left empty except at the head and tip.

The act of cleaning out the pile close to the tip inevitably causes stress relaxation in the soil plug below the clean-out. This will mean that under extreme axial compression, the pile will undergo a small settlement before it restores its full resistance. To prevent this, after the concrete plug has hardened, grout may be injected just beneath the plug, at a pressure that will restore the compactness of the soil but not so great as to pipe under the tip or fracture the foundation, or the pile may be re-seated by driving.

3. The tubular pile, after being installed to design penetration, may be filled with sand up to two diameters below the head, then with tremie concrete to the head. Reinforcing steel may be placed in the concrete to transfer part of the moment and tension into the footing block. Studs may be pre-installed on that zone of the pile to ensure full shear transfer. The soil and sand plug will act to limit local buckling at the mudline under extreme loads.

4. A socket may be drilled into rock or hard material beyond the tip of the driven pile, and then filled with concrete. Slurry is used to prevent degradation of the surface of the hole and sloughing. Seawater may be used in some rocks but may cause slaking in others such as shales and siltstone. Bentonite slurry coats the surface of the hole; the hole should be flushed with seawater just before concreting. Synthetic slurries are best, since they react in the presence of the calcium ion from the concrete to improve the bond. Synthetic polymer slurries biodegrade and thus may be environmentally acceptable for discharge into the water.

When a tubular pile is seated on rock and the socket is then drilled below the tip of the pile, it often is difficult to prevent run-in of sands from around the tip and to maintain proper circulation. Therefore, after landing, a hole may be drilled a short distance, for example, with a churn drill or down-the-hole drill and then the pile reseated by the pile hammer.

Either insert tubulars or reinforcing steel cages are placed in the socket, extending well up into the pile. Tremie concrete is then placed to transfer the load in shear. In the case where a tubular insert pile is used, its tip may be plugged. Then grout may be injected into the annular space to transfer the shear.

Grout should not be used to fill sockets of large-diameter tubulars. The heat of hydration will damage the grout, reducing its strength. Tremie concrete should be used instead, employing small-size coarse aggregate, e.g., 15 mm, to ensure workability and flowability.

Although most sockets for offshore bridge piers have been cylindrical extensions of the tubular pile, in some offshore oil platforms belled footings have been constructed to transfer the load in end bearing. Hydraulically operated belling tools are attached to the drill string. Whenever transfer in end bearing is the primary mechanism, the bottom of the hole must be cleaned of silt just prior to the placement of concrete.

47.2.7 Prestressed Concrete Cylinder Piles

As an alternative to steel tubular piling, prestressed concrete cylinder piles have been used for a number of major overwater bridges, from the San Diego–Coronado and Dunbarton Bridges in California to bridges across Chesapeake Bay and the Yokohama cable-stayed bridge (Figures 47.6 and 47.7). Diameters have ranged from 1.5 to 6 m and more. They offer the advantage of durability and high axial compressive capacity. To counter several factors producing circumferential strains, especially thermal strains, spiral reinforcement of adequate cross-sectional area is required. This spiral reinforcement should be closely spaced in the 2-m zone just below the pile cap, where sharp reverse bending occurs under lateral loading.

Pile installation methods vary from driving and jetting of the smaller-diameter piles to drilling in the large-diameter piling (Figure 47.8).

47.2.8 Footing Blocks

The footing block constructed at the top of large-diameter tubular piles serves the purpose of transmitting the forces from the pier shaft to the piles. Hence, it is subjected to large shears and significant moments. The shears require extensive vertical reinforcement, for both global shear (from the pier shaft) and local shear (punching shear from the piles). Large concentrations of reinforcement are required to distribute the moments. Post-tensioned tendons may be effectively utilized.

FIGURE 47.6 Large-diameter prestressed concrete pile, Napa River Bridge, California.

FIGURE 47.7 Prestressed concrete cylinder pile for Oosterschelde Bridge, the Netherlands.

Although the primary forces typically produce compression in the upper surface of the footing block, secondary forces and particularly high temporary stresses caused by the heat of hydration produce tension in the top surface. Thus, adequate horizontal steel must be provided in the top and bottom in both directions.

FIGURE 47.8 Installing concrete cylinder pile by internal excavation, jetting, and pull-down force from barge.

The heat of hydration of the cemetitious materials in a large footing block develops over a period of several days. Due to the mass of the block, the heat in the core may not dissipate and return to ambient for several weeks.

The outside surface meantime has cooled and contracted, producing tension which often leads to cracking. Where inadequate reinforcement is provided, the steel may stretch beyond yield, so that the cracks become permanent. If proper amounts of reinforcement are provided, then the cracking that develops will be well distributed, individual cracks will remain small, and the elastic stress in the reinforcement will tend to close the cracks as the core cools.

Internal laminar cracking may also occur, so vertical reinforcement and middepth reinforcement should also be considered.

Footing blocks may be constructed in place, just above water, with precast concrete skirts extending down below low water in order to prevent small boats and debris from being trapped below. In this case, the top of the piles may be exposed at low water, requiring special attention to the prevention of corrosion.

Footing blocks may be constructed below water. Although cofferdams may be employed, the most efficient and economical way is usually to prefabricate the shell of the footing block. This is then floated into place. Corner piles are then inserted through the structure and driven to grade. The prefabricated box is then lowered down by ballasting, supported and guided by the corner piles. Then the remaining piles are threaded through holes in the box and driven. Final connections are made by tremie concrete.

Obviously, there are variations of the above procedure. In some cases, portions of the box have been kept permanently empty, utilizing their buoyancy to offset part of the deadweight.

Transfer of forces into the footing block requires careful detailing. It is usually quite difficult to transfer full moment by means of reinforcing inside the pile shell. If the pile head can be dewatered,

FIGURE 47.9 Large steel sheet pile cofferdam for Second Delaware Memorial Bridge, showing bracing frames.

reinforcing steel bars can be welded to the inside of the shell. Cages set in the concrete plug at the head may employ bundled bars with mechanical heads at their top. Alternatively the pile may be extended up through the footing block. Shear keys can be used to transfer shear. Post-tensioning tendons may run through and around the pile head.

47.3 Cofferdams for Bridge Piers

47.3.1 Description

The word *cofferdam* is a very broad term to describe a construction that enables an underwater site to be dewatered. As such, cofferdams can be large or small. Medium-sized cofferdams of horizontal dimensions from 10 to 50 m have been widely used to construct the foundations of bridge piers in water and soft sediments up to 20 m in depth; a few have been larger and deeper (Figure 47.9). Typical bridge pier cofferdams are constructed of steel sheet piles supported against the external pressures by internal bracing.

A few very large bridge piers, such as anchorages for suspension bridges, have utilized a ring of self-supporting sheet pile cells. The interior is then dewatered and excavated to the required depth. A recent such development has been the building of a circular ring wall of concrete constructed by the slurry trench method (Figures 47.10 and 47.11). Concrete cofferdams have also used a ring wall of precast concrete sheet piles or even cribs.

47.3.2 Design Requirements

Cofferdams must be designed to resist the external pressures of water and soil [3]. If, as is usual, a portion of the external pressures is designed to be resisted by the internal passive pressure of the soil, the depth of penetration must be selected conservatively, taking into account a potential sudden reduction in passive pressure due to water flow beneath the tip as a result of unbalanced water pressures or jetting of piles. The cofferdam structure itself must have adequate vertical support for self-load and equipment under all conditions.

In addition to the primary design loads, other loading conditions and scenarios include current and waves, debris and ice, overtopping by high tides, flood, or storm surge. While earthquake-induced loads, acting on the hydrodynamic mass, have generally been neglected in the past, they

FIGURE 47.10   Slurry wall cofferdam for Kawasaki Island ventilation shaft, Trans-Tokyo Bay Tunnels and Bridge.

FIGURE 47.11   Concrete ring wall cofferdam constructed by slurry trench methods.

are now often being considered on major cofferdams, taking into account the lower input acceler-ations appropriate for the reduced time of exposure and, where appropriate, the reduced conse-quences.

Operating loads due to the mooring of barges and other floating equipment alongside need to be considered. The potential for scour must be evaluated, along with appropriate measures to reduce the scour. When the cofferdam is located on a sloping bank, the unbalanced soil loads need to be properly resisted. Accidental loads include impact from boats and barges, especially those working around the site.

FIGURE 47.12 Dewatering the cofferdam for the main tower pier, Second Delaware Memorial Bridge.

The cofferdam as a whole must be adequately supported against the lateral forces of current waves, ice, and moored equipment, as well as unbalanced soil loads. While a large deep-water cofferdam appears to be a rugged structure, when fully excavated and prior to placement of the tremie concrete seal, it may be too weak to resist global lateral forces. Large tubular piles, acting as spuds in conjunction with the space-frame or batter piles may be needed to provide stability.

The cofferdam design must be such as to integrate the piling and footing block properly. For example, sheet piles may prevent the installation of batter piles around the periphery. To achieve adequate penetration of the sheet piles and to accommodate the batter piles, the cofferdam may need to be enlarged. The arrangement of the bracing should facilitate any subsequent pile installation.

To enable dewatering of the cofferdam (Figure 47.12), a concrete seal is constructed, usually by the tremie method. This seal is designed to resist the hydrostatic pressure by its own buoyant weight and by uplift resistance provided by the piling, this latter being transferred to the concrete seal course by shear (Figure 47.13).

In shallow cofferdams, a filter layer of coarse sand and rock may permit pumping without a seal. However, in most cases, a concrete seal is required. In some recent construction, a reinforced concrete footing block is designed to be constructed underwater, to eliminate the need for a separate concrete seal. In a few cases, a drainage course of stone is placed below the concrete seal; it is then kept dewatered to reduce the uplift pressure. Emergency relief pipes through the seal course will prevent structural failure of the seal in case the dewatering system fails.

The underwater lateral pressure of the fresh concrete in the seal course and footing block must be resisted by external backfill against the sheet piles or by internal ties.

47.3.3 Internally Braced Cofferdams

These are the predominant type of cofferdams. They are usually rectangular in shape, to accommodate a regular pattern of cross-lot bracing.

FIGURE 47.13 Pumped-out cofferdam showing tremie concrete seal and predriven steel H-piles.

The external wall is composed of steel sheet piles of appropriate section modulus to develop bending resistance. The loading is then distributed by horizontal wales to cross-lot struts. These struts should be laid out on a plan which will permit excavation between them, to facilitate the driving of piling and to eliminate, as far as practicable, penetration of bracing through the permanent structure.

Wales are continuous beams, loaded by the uniform bearing of sheet piles against them. They are also loaded axially in compression when they serve as a strut to resist the lateral loads acting on them end-wise. Wales in turn deliver their normal loads to the struts, developing concentrated local bearing loads superimposed upon the high bending moments, tending to produce local buckling. Stiffeners are generally required.

While stiffeners are readily installed on the upperside, they are difficult to install on the underside and difficult to inspect. Hence, these stiffeners should be pre-installed during fabrication of the members.

The wales are restrained from global buckling in the horizontal plane by the struts. In the vertical plane they are restrained by the friction of the sheet piles, which may need to be supplemented by direct fixation. Blocking of timber or steel shims is installed between the wales and sheet piles to fit the irregularities in sheet pile installation and to fill in the needed physical clearances.

Struts are horizontal columns, subject to high axial loading, as well as vertical loads from self-weight and any equipment that is supported by them. Their critical concern is stability against buckling. This is countered in the horizontal plane by intersecting struts but usually needs additional support in the vertical plane, either by piling or by trussing two or more levels of bracing.

The orthogonal horizontal bracing may be all at one elevation, in which case the intersections of the struts have to be accommodated, or they may be vertically offset, one level resting on top of the other. This last is normally easier since, otherwise, the intersections must be detailed to transmit the full loads across the joint. This is particularly difficult if struts are made of tubular pipe sections. If struts are made of wide-flanged or H-section members, then it will usually be found preferable to construct them with the weak axis in the vertical plane, facilitating the detailing of strut-to-strut intersections as well as strut-to-wale intersections. In any event, stiffeners are required to prevent buckling of the flanges.

For deep-water piers, the cofferdam bracing is best constructed as a space-frame, with two or more levels joined together by posts and diagonals in the vertical plane. This space-frame may be completely prefabricated and set as a unit, supported by vertical piles. These supporting piles are typically of large-diameter tubular members, driven through sleeves in the bracing frame and connected to it by blocking and welding.

The setting of such a space-frame requires a very large crane barge or equivalent, with both adequate hoisting capacity and reach. Sometimes, therefore, the bracing frame is made buoyant, to be partially or wholly self-floating. Tubular struts can be kept empty and supplemental buoyancy can be provided by pontoons.

Another way to construct the bracing frame is to erect one level at a time, supported by large tubular piles in sleeves. The lower level is first erected, then the posts and diagonal bracing in the vertical plane. The lower level is then lowered by hoists or jacks so that the second level can be constructed just above water and connections made in the dry.

A third way is to float in the prefabricated bracing frame on a barge, drive spud piles through sleeves at the four corners, and hang the bracing frame from the piles. Then the barge is floated out at low tide and the bracing frame lowered to position.

47.3.4 Circular Cofferdams

Circular cofferdams are also employed, with ring wales to resist the lateral forces in compression. The dimensions are large, and the ring compression is high. Unequal loading is frequently due to differential soil pressures. Bending moments are very critical, since they add to the compression on one side. Thus the ring bracing must have substantial strength against buckling in the horizontal plane.

47.3.5 Excavation

Excavation should be carried out in advance of setting the bracing frame or sheet piles, whenever practicable. Although due to side slopes the total volume of excavation will be substantially increased, the work can be carried out more efficiently and rapidly than excavation within a bracing system.

When open-cut excavation is not practicable, then it must be carried out by working through the bracing with a clamshell bucket. Struts should be spaced as widely as possible so as to permit use of a large bucket. Care must be taken to prevent impact with the bracing while the bucket is being lowered and from snagging the bracing from underneath while the bucket is being hoisted. These accidental loads may be largely prevented by temporarily standing up sheet piles against the bracing in the well being excavated, to act as guides for the bucket.

Except when the footing course will be constructed directly on a hard stratum or rock, overexcavation by 1 m or so will usually be found beneficial. Then the overexcavation can be backfilled to grade by crushed rock.

47.3.6 Driving of Piles in Cofferdams

Pilings can be driven before the bracing frame and sheet piles are set. They can be driven by underwater hammers or followers. To ensure proper location, the pile driver be equipped with telescopic leads, or a template be set on the excavated river bottom or seafloor.

Piling may alternatively be driven after the cofferdam has been installed, using the bracing frame as a template. In this case, an underwater hammer presents problems of clearance due to its large size, especially for batter piles. Followers may be used, or, often, more efficiently, the piles may be lengthened by splicing to temporarily extend all the way to above water. They are then cut off to grade after the cofferdam has been dewatered. This procedure obviates the problems occasioned if a pile fails to develop proper bearing since underwater splices are not needed. It also eliminates cutoff waste. The long sections of piling cutoff after dewatering can be taken back to the fabrication yard and re-spliced for use on a subsequent pier.

All the above assumes driven steel piling, which is the prevalent type. However, on several recent projects, drilled shafts have been constructed after the cofferdam has been excavated. In the latter case, a casing must be provided, seated sufficiently deep into the bottom soil to prevent run-in or blowout [see Section 47.2.6, item (4)].

Driven timber or concrete piles may also be employed, typically using a follower to drive them below water.

47.3.7 Tremie Concrete Seal

The tremie concrete seal course functions to resist the hydrostatic uplift forces to permit dewatering. As described earlier, it usually is locked to the foundation piling to anchor the slab. It may be reinforced in order to enable it to distribute the pile loads and to resist cracking due to heat of hydration.

Tremie concrete is a term derived from the French to designate concrete placed through a pipe. The term has subsequently evolved to incorporate both a concrete mix and a placement procedure. Underwater concreting has had both significant successes and significant failures. Yet the system is inherently reliable and concrete equal or better than concrete placed in the dry has been produced at depths up to 250 m. The failures have led to large cost overruns due to required corrective action. They have largely been due to inadvertently allowing the concrete to flow through or be mixed with the water, which has caused washout of the cement and segregation of the aggregate.

Partial washout of cement leads to the formation of a surface layer of laitance which is a weak paste. This may harden after a period of time into a brittle chalklike substance.

The tremie concrete mix must have an adequate quantity of cementitious materials. These can be a mixture of portland cement with either fly ash or blast furnace slag (BFS). These are typically proportioned so as to reduce the heat of hydration and to promote cohesiveness. A total content of cementitious materials of 400 kg/m^3 (~700 lb/cy) is appropriate for most cases.

Aggregates are preferably rounded gravel so they flow more readily. However, crushed coarse aggregate may be used if an adequate content of sand is provided. The gradation of the combined aggregates should be heavy toward the sand portion — a 45% sand content appears optimum for proper flow. The maximum size of coarse aggregate should be kept small enough to flow smoothly through the tremie pipe and any restrictions such as those caused by reinforcement. Use of 20 mm maximum size of coarse aggregate appears optimum for most bridge piers.

A conventional water-reducing agent should be employed to keep the water/cementitious material ratio below 0.45. Superplasticizers should not normally be employed for the typical cofferdam, since the workability and flowability may be lost prematurely due to the heat generated in the mass concrete. Retarders are essential to prolong the workable life of the fresh mix if superplasticizers are used.

Other admixtures are often employed. Air entrainment improves flowability at shallow water depths but the beneficial effects are reduced at greater depths due to the increased external pressure. Weight to reduce uplift is also lost.

Microsilica may be included in amounts up to 6% of the cement to increase the cohesiveness of the mix, thus minimizing segregation. It also reduces bleed. Antiwashout admixtures (AWA) are also employed to minimize washout of cementitious materials and segregation. They tend to promote self-leveling and flowability. Both microsilica and AWA may require the use of superplasticizers in which case retarders are essential. However, a combination of silica fume and AWA should be avoided as it typically is too sticky and does not flow well.

Heat of hydration is a significant problem with the concrete seal course, as well as with the footing block, due to the mass of concrete. Therefore, the concrete mix is often precooled, e.g., by chilling of the water or the use of ice. Liquid nitrogen is sometimes employed to reduce the temperature of the concrete mix to as low as 5°C. Heat of hydration may be reduced by incorporating substantial amounts of fly ash to replace an equal portion of cement. BFS–cement can also be used to reduce

FIGURE 47.14 Placing underwater concrete through hopper and tremie pipe, Verrazano Narrows Bridge, New York.

heat, provided the BFS is not ground too fine, i.e., not finer than 2500 cm²/g and the proportion of slag is at least 70% of the total.

The tremie concrete mix may be delivered to the placement pipe by any of several means. Pumping and conveyor belts are best because of their relatively continuous flow. The pipe for pumping should be precooled and insulated or shielded from the sun; conveyor belts should be shielded. Another means of delivery is by bucket. This should be air-operated to feed the concrete gradually to the hopper at the upper end of the tremie pipe. *Placement down the tremie pipe should be by gravity feed only* (Figure 47.14).

Although many placements of tremie concrete have been carried out by pumping, there have been serious problems in large placements such as cofferdam seals. The reasons include:

1. Segregation in the long down-leading pipe, partly due to formation of a partial vacuum and partly due to the high velocity;
2. The high pressures at discharge;
3. The surges of pumping.

Since the discharge is into fresh concrete, these phenomena lead to turbulence and promote intermixing with water at the surface, forming excessive laitance.

These discharge effects can be contrasted with the smooth flow from a gravity-fed pipe in which the height of the concrete inside the tremie pipe automatically adjusts to match the external pressure of water vs. the previously placed concrete. For piers at considerable depths, this balance point will be about half-way down. The pipe should have an adequate diameter in relation to the maximum size of coarse aggregate to permit remixing: a ratio of 8 to 1 is the minimum. A slight inclination of the tremie pipe from the vertical will slow the feed of new concrete and facilitate the escape of entrapped air.

For starting the tremie concrete placement, the pipe must first be filled slightly above middepth. This is most easily done by plugging the end and placing the empty pipe on the bottom. The empty pipe must be negatively buoyant. It also must be able to withstand the external hydrostatic pressure as well as the internal pressure of the underwater concrete. Joints in the tremie pipe should be gasketed and bolted to prevent water being sucked into the mix by venturi action. To commence placement, with the tremie pipe slightly more than half full, it is raised 150 mm off the bottom. The temporary plug then comes off and the concrete flows out. The above procedure can be used both for starting and for resuming a placement, as, for example, when the tremie is relocated, or after a seal has been inadvertently lost.

The tremie pipe should be kept embedded in the fresh concrete mix a sufficient distance to provide backpressure on the flow (typically 1 m minimum), but not so deep as to become stuck in the concrete due to its initial set. This requires adjustment of the retarding admixture to match the rate of concrete placement and the area of the cofferdam against the time of set, keeping in mind the acceleration of set due to heat as the concrete hydrates.

Another means for initial start of a tremie concrete placement is to use a pig which is forced down the pipe by the weight of the concrete, expelling the water below. This pig should be round or cylindrical, preferably the latter, equipped with wipers to prevent leakage of grout and jamming by a piece of aggregate. An inflated ball, such as an athletic ball (volleyball or basketball) must never be used; these collapse at about 8 m water depth! A pig should not be used to restart a placement, since it would force a column of water into the fresh concrete previously placed.

Mixes of the tremie concrete described will flow outward on a slope of about 1 on 8 to 1 on 10. With AWAs, an even flatter surface can be obtained.

A trial batch with underwater placement in a shallow pit or tank should always be done before the actual placement of the concrete seal. This is to verify the cohesiveness and flowability of the mix. Laboratory tests are often inadequate and misleading, so a large-scale test is important. A trial batch of 2 to 3 m^3 has often been used.

The tremie concrete placement will exert outward pressure on the sheet piles, causing them to deflect. This may in turn allow new grout to run down past the already set concrete, increasing the external pressure. To offset this, the cofferdam can be partially backfilled before starting the tremie concreting and tied across the top. Alternatively, dowels can be welded on the sheets to tie into the concrete as it sets; the sheet piles then have to be left in place.

Due to the heat of hydration, the concrete seal will expand. Maximum temperature may not be achieved for several days. Cooling of the mass is gradual, starting from the outside, and ambient temperature may not be achieved for several weeks. Thus an external shell is cooling and shrinking while the interior is still hot. This can produce severe cracking, which, if not constrained, will create permanent fractures in the seal or footing. Therefore, in the best practice, reinforcing steel is placed in the seal to both provide a restraint against cracking and to help pull the cracks closed as the mass cools.

After a relatively few days, the concrete seal will usually have developed sufficient strength to permit dewatering. Once exposed to the air, especially in winter, the surface concrete will cool too fast and may crack. Placing insulation blankets will keep the temperature more uniform. They will, of course, have to be temporarily moved to permit the subsequent work to be performed.

47.3.8 Pier Footing Block

The pier footing block is next constructed. Reinforcement on all faces is required, not only for structural response but also to counteract thermal strains. Reference is made to Section 31.2.8 in which reinforcement for the footing block is discussed in more detail.

The concrete expands as it is placed in the footing block due to the heat of hydration. At this stage it is either still fresh or, if set, has a very low modulus. Then it hardens and bonds to the tremie concrete below. The lock between the two concrete masses is made even more rigid if piling protrudes through the top of the tremie seal, which is common practice. Now the footing block cools and tries to shrink but is restrained by the previously placed concrete seal. Vertical cracks typically form. Only if there is sufficient bottom reinforcement in both directions can this shrinkage and cracking be adequately controlled. Note that these tensile stresses are permanently locked into the bottom of the footing block and the cracks will not close with time, although creep will be advantageous in reducing the residual stresses.

After the footing block has hardened, blocking may be placed between it and the sheet piles. This, in turn, may permit removal of the lower level of bracing. As an alternative to bracing, the footing block may be extended all the way to the sheet piles, using a sheet of plywood to prevent adhesion.

47.3.9 Pier Shaft

The pier shaft is then constructed. Block-outs may be required to allow the bracing to pass through. The internal bracing is removed in stages, taking care to ensure that this does not result in over-loading a brace above. Each stage of removal should be evaluated.

Backfill is then placed outside the cofferdam to bring it up to the original seabed. The sheet piles can then be removed. The first sheets are typically difficult to break loose and may require driving or jacking in addition to vibration. Keeping in mind the advantage of steel sheet piles in preventing undermining of the pier due to scour, as well as the fact that removal of the sheets always loosens the surrounding soil, hence reducing the passive lateral resistance, it is often desirable to leave the sheet piles in place below the top of the footing. They may be cut off underwater by divers; then the tops are pulled by vibratory hammers.

Antiscour stone protection is now placed, with an adequate filter course or fabric sheet in the case of fine sediments.

47.4 Open Caissons

47.4.1 Description

Open caissons have been employed for some of the largest and deepest bridge piers [4]. These are an extension of the "wells" which have been used for some 2000 years in India. The caisson may be constructed above its final site, supported on a temporary sand island, and then sunk by dredging out within the open wells of the caisson, the deadweight acting to force the caisson down through the overlying soils (Figure 47.15). Alternatively, especially in sites overlain by deep water, the caisson may be prefabricated in a construction basin, floated to the site by self-buoyancy, augmented as necessary by temporary floats or lifts, and then progressively lowered into the soils while building up the top.

Open caissons are effective but costly, due to the large quantity of material required and the labor for working at the overwater site. Historically, they have been the only means of penetrating thorough deep overlying soils onto a hard stratum or bedrock. However, their greatest problem is maintaining stability during the early phases of sinking, when they are neither afloat nor firmly embedded and supported. Long and narrow rectangular caissons are especially susceptible to tipping, whereas square and circular caissons of substantial dimensions relative to the water depth are inherently more stable. Once the caisson tips, it tends to drift off position. It is very difficult to bring it back to the vertical without overcorrecting.

When the caisson finally reaches its founding elevation, the surface of rock or hard stratum is cleaned and a thick tremie concrete base is placed. Then the top of the caisson is completed by casting a large capping block on which to build the pier shaft.

FIGURE 47.15 Open-caisson positioned within steel jackets on "pens."

47.4.2 Installation

The sinking of the cofferdam through the soil is resisted by skin friction along the outside and by bearing on the cutting edges. Approximate values of resistance may be obtained by multiplying the friction factor of sand on concrete or steel by the at-rest lateral force at that particular stage, $f = ØK_0wh^2$ where f is the unit frictional resistance, $Ø$ the coefficient of friction, w the underwater unit weight of sand, K_0 the at-rest coefficient of lateral pressure, and h the depth of sand at that level. f is then summed up over the embedded depth. In clay, the cohesive shear controls the "skin friction." The bearing value of the cutting edges is generally the "shallow bearing value," i.e., five times the shear strength at that elevation.

These resistances must be overcome by deadweight of the caisson structure, reduced by the buoyancy acting on the submerged portions. This deadweight may be augmented by jacking forces on ground anchors.

The skin friction is usually reduced by lubricating jets causing upward flow of water along the sides. Compressed air may be alternated with water through the jets; bentonite slurry may be used to provide additional lubrication. The bearing on the cutting edges may be reduced by cutting jets built into the walls of the caisson or by free jets operating through holes formed in the walls. Finally, vibration of the soils near and around the caisson may help to reduce the frictional resistance.

When a prefabricated caisson is floated to the site, it must be moored and held in position while it is sunk to and into the seafloor. The moorings must resist current and wave forces and must assist in maintaining the caisson stable and in a vertical attitude. This latter is complicated by the need to build up the caisson walls progressively to give adequate freeboard, which, of course, raises the center of gravity.

Current force can be approximately determined by the formula

$$F = CA\rho\frac{V^2}{2g}$$

where C varies from 0.8 for smooth circular caissons to 1.3 for rectangular caissons, A is the area, ρ is the density of water, and V is the average current over the depth of flotation. Steel sheet piles develop high drag, raising the value of C by 20–30%.

As with all prismatic floating structures, stability requires that a positive metacentric height be maintained. The formula for metacentric height, $\overline{GM}$ is

$$\overline{GM} = \overline{KB} - \overline{KG} + \overline{BM}$$

where $\overline{KB}$ is the distance from the base to the center of buoyancy, $\overline{KG}$ the distance to the center of gravity, and $\overline{BM} = I/V$.

I is the moment of inertia on the narrowest (most sensitive) axis, while V is the displaced volume of water. For typical caissons, a $\overline{GM}$ of +1 m or more should be maintained.

The forces from mooring lines and the friction forces from any dolphins affect the actual attitude that the structure assumes, often tending to tip it from vertical. When using mooring lines, the lines should be led through fairleads attached near the center of rotation of the structure. However, this location is constantly changing, so the fairlead attachment points may have to be shifted upward from time to time.

Dolphins and "pens" are used on many river caissons, since navigation considerations often preclude mooring lines. These are clusters of piles or small jackets with pin piles and are fitted with vertical rubbing strips on which the caisson slides.

Once the caisson has been properly moored on location, it is ballasted down. As it nears the existing river or harbor bottom, the current flow underneath increases dramatically. When the bottom consists of soft sediments, these may rapidly scour away in the current. To prevent this, a mattress should be first installed.

Fascine mattresses of willow, bamboo, or wood with filter fabric attached are ballasted down with rock. Alternatively, a layer of graded sand and gravel, similar to the combined mix for concrete aggregate, can be placed. The sand on top will scour away, but the final result will be a reverse filter.

In order to float a prefabricated caisson to the site initially, false bottoms are fitted over the bottom of the dredging wells. These false bottoms are today made of steel, although timber was used on many of the famous open caissons from the 19th and the early part of the 20th centuries. They are designed to resist the hydrostatic pressure plus the additional force of the soils during the early phases of penetration. Once the caisson is embedded sufficiently to ensure stability, the false bottoms are progressively removed so that excavation can be carried out through the open wells. This removal is a very critical and dangerous stage, hazardous both to the caisson and to personnel. The water level inside at this stage should be slightly higher than that outside. Even then, when the false bottom under a particular well is loosened, the soil may suddenly surge up, trapping a diver. The caisson, experiencing a sudden release of bearing under one well, may plunge or tip.

Despite many innovative schemes for remote removal of false bottoms, accidents have occurred. Today's caissons employ a method for gradually reducing the pressure underneath and excavating some of the soil before the false bottom is released and removed. For such constructions, the false bottom is of heavily braced steel, with a tube through it, typically extending to the water surface. The tube is kept full of water and capped, with a relief valve in the cap. After the caisson has penetrated under its own weight and come to a stop, the relief valve is opened, reducing the pressure to the hydrostatic head only. Then the cap is removed. This is done for several (typically, four) wells in a balanced pattern. Then jets and airlifts may be operated through the tube to remove the soil under those wells. When the caisson has penetrated sufficiently far for safety against tipping, the wells are filled with water; the false bottoms are removed and dredging can be commenced.

47.4.3 Penetration of Soils

The penetration is primarily accomplished by the net deadweight, that is, the total weight of concrete steel and ballast less the buoyancy. Excavation within the wells is carried down in a balanced pattern until the bearing stratum is reached. Then tremie concrete is placed, of sufficient depth to transfer the design bearing pressures to the walls.

The term *cutting edge* is applied to the tips of the caisson walls. The external cutting edges are shaped as a wedge while the interior ones may be either double-wedge or square. In the past, concern over concentrated local bearing forces led to the practice of making the cutting edges of heavy and expensive fabricated steel. Today, high-strength reinforced concrete is employed, although if obstructions such as boulders, cobbles, or buried logs are anticipated or if the caisson must penetrate rock, steel armor should be attached to prevent local spalling.

The upper part of the caisson may be replaced by a temporary cofferdam, allowing the pier shaft dimensions to be reduced through the water column. This reduces the effective driving force on the caisson but maintains and increases its inherent stability.

The penetration requires the progressive failure of the soil in bearing under the cutting edges and in shear along the sides. Frictional shear on the inside walls is reduced by dredging while that on the outside walls is reduced by lubrication, using jets as previously described.

Controlling the penetration is an essentially delicate balancing of these forces, attempting to obtain a slight preponderance of sinking force. Too great an excess may result in plunging of the caisson and tipping or sliding sidewise out of position. That is why pumping down the water within the caisson, thus reducing buoyancy, is dangerous; it often leads to sudden inflow of water and soil under one edge, with potentially catastrophic consequences.

Lubricating jets may be operated in groups to limit the total volume of water required at any one time to a practicable pump capacity. In addition to water, bentonite may be injected through the lubricating jets, reducing the skin friction. Compressed air may be alternated with water jetting.

Other methods of aiding sinking are employed. Vibration may be useful in sinking the caisson through sands, especially when it is accompanied by jetting. This vibration may be imparted by intense vibration of a steel pile located inside the caisson or even by driving on it with an impact hammer to liquefy the sands locally.

Ground anchors inserted through preformed holes in the caisson walls may be jacked against the caisson to increase the downward force. They have the advantage that the actual penetration may be readily controlled, both regarding force exerted and displacement.

Since all the parameters of resistance and of driving force vary as the caisson penetrates the soil, and because the imbalance is very sensitive to relatively minor changes in these parameters, it is essential to plan the sinking process in closely spaced stages, typically each 2 to 3 m. Values can be precalculated for each such stage, using the values of the soil parameters, the changes in contact areas between soil and structure, the weights of concrete and steel and the displaced volume. These need not be exact calculations; the soil parameters are estimates only since they are being constantly modified by the jetting. However, they are valuable guides to engineering control of the operations.

There are many warnings from the writings of engineers in the past, often based on near-failures or actual catastrophes.

1. Verify structural strength during the stages of floating and initial penetration, with consideration for potentially high resistance under one corner or edge.
2. In removing false bottoms, be sure the excess pressure underneath has first been relieved.
3. Do not excavate below cutting edges.
4. Check outside soundings continually for evidence of scour and take corrective steps promptly.
5. Blasting underneath the cutting edges may blow out the caisson walls. Blasting may also cause liquefaction of the soils leading to loss of frictional resistance and sudden plunging. If blasting is needed, do it before starting penetration or, at least, well before the cutting edge reaches the hard strata so that a deep cushion of soil remains over the charges.
6. If the caisson tips, avoid drastic corrections. Instead, plan the correction to ensure a gradual return to vertical and to prevent the possibility of tipping over more seriously on the other side. Thus steps such as digging deeper on the high side and overballasting on the high side should be last resorts and, then, some means mobilized to arrest the rotation as the caisson nears vertical. Jacking against an external dolphin is a safer and more efficient method for correcting the tipping (Figure 47.16).

FIGURE 47.16 Open caisson for Sunshine Bridge across the Mississippi River. This caisson tipped during removal of false bottoms and is shown being righted by jacking against dolphins.

Sinking the caisson should be a continuous process, since once it stops the soil friction and/or shear increase significantly and it may be difficult to restart the caisson's descent.

47.4.4 Founding on Rock

Caissons founded on bare rock present special difficulties. The rock may be leveled by drilling, blasting, and excavation, although the blasting introduces the probability of fractures in the underlying rock. Mechanical excavation may therefore be specified for the last meter or two. Rotary drills and underwater road-headers can be used but the process is long and costly. In some cases, hydraulic rock breakers can be employed; in other cases a hammer grab or star chisel may be used. For the Prince Edward Island Bridge, the soft rock was excavated and leveled by a very heavy clamshell bucket. Hydraulic backhoes and dipper dredges have been used elsewhere. A powerful cutter-head dredge has been planned for use at the Strait of Gibraltar.

47.4.5 Contingencies

The planning should include methods for dealing with contingencies. The resistance of the soil and especially of hard strata may be greater than anticipated. Obstructions include sunken logs and even sunken buried barges and small vessels, as well as cobbles and boulders. The founding rock or stratum may be irregular, requiring special means of excavating underneath the cutting edge at high spots or filling in with concrete in the low spots. One contingency that should always be addressed is what steps to take if the caisson unexpectedly tips.

Several innovative solutions have been used to construct caissons at sites with especially soft sediments. One is a double-walled self-floating concept, without the need for false bottoms. Double-walled caissons of steel were used for the Mackinac Strait Bridge in Michigan. Ballast is progressively filled into the double-wall space while dredging is carried out in the open wells.

In the case of extremely soft bottom sediments, the bottom may be initially stabilized by ground improvement, for example, with surcharge by dumped sand, or by stone columns, so that the caisson may initially land on and penetrate stable soil. Great care must, of course, be exercised to maintain control when the cutting edge breaks through to the native soils below, preventing erratic plunging.

FIGURE 47.17 Excavating within pressurized working chamber of pneumatic caisson for Rainbow Suspension Bridge, Tokyo. (Photo courtesy of Shiraishi Corporation.)

This same principle holds true for construction on sand islands, where the cutting edge and initial lifts of the caisson may be constructed on a stratum of gravel or other stable material, then the caisson sunk through to softer strata below. Guides or ground anchors will be of benefit in controlling the sinking operation.

47.5 Pneumatic Caissons

47.5.1 Description

These caissons differ from the open caisson in that excavation is carried out beneath the base in a chamber under air pressure. The air pressure is sufficient to offset some portion of the ambient hydrostatic head at that depth, thus restricting the inflow of water and soil.

Access through the deck for workers and equipment and for the removal of the excavated soil is through an airlock. Personnel working under air pressure have to follow rigid regimes regarding duration and must undergo decompression upon exit. The maximum pressures and time of exposure under which personnel can work is limited by regulations. Many of the piers for the historic bridges in the United States, e.g., the Brooklyn Bridge, were constructed by this method.

47.5.2 Robotic Excavation

To overcome the problems associated with working under air pressure, the health hazards of "caisson disease," and the high costs involved, robotic cutters [5] have been developed to excavate and remove the soil within the chamber without human intervention. These were recently implemented on the piers of the Rainbow Suspension Bridge in Tokyo (Figures 47.17 and 47.18).

The advantage of the pneumatic caisson is that it makes it possible to excavate beneath the cutting edges, which is of special value if obstructions are encountered. The great risk is of a "blowout" in which the air escapes under one edge, causing a rapid reduction of pressure, followed by an inflow of water and soil, endangering personnel and leading to sudden tilting of the caisson. Thus, the use of pneumatic caissons is limited to very special circumstances.

FIGURE 47.18   Excavating beneath cutting edge of pneumatic caisson. (Photo courtesy of Shiraishi Corporation.)

47.6 Box Caissons

47.6.1 Description

One of the most important developments of recent years has been the use of box caissons, either floated in or set in place by heavy-lift crane barges [4]. These box caissons, ranging in size from a few hundred tons to many thousands of tons, enable prefabrication at a shore site, followed by transport and installation during favorable weather "windows" and with minimum requirements for overwater labor. The development of these has been largely responsible for the rapid completion of many long overwater bridges, cutting the overall time by a factor of as much as three and thus making many of these large projects economically viable.

The box caisson is essentially a structural shell that is placed on a prepared underwater foundation. It is then filled with concrete, placed by the tremie method previously described for cofferdam seals. Alternatively, sand fill or just ballast water may be used.

Although many box caissons are prismatic in shape, i.e., a large rectangular base supporting a smaller rectangular column, others are complex shells such as cones and bells. When the box caisson is seated on a firm foundation, it may be underlain by a meter or two of stone bed, consisting of densified crushed rock or gravel that has been leveled by screeding. After the box has been set, underbase grout is often injected to ensure uniform bearing.

47.6.2 Construction

The box caisson shell is usually the principal structural element although it may be supplemented by reinforcing steel cages embedded in tremie concrete. This latter system is often employed when joining a prefabricated pier shaft on top of a previously set box caisson.

Box caissons may be prefabricated of steel; these were extensively used on the Honshu–Shikoku Bridges in Japan (Figures 47.19 and 47.20). After setting, they were filled with underwater concrete, in earlier cases using grout-intruded aggregate, but in more recent cases tremie concrete.

For reasons of economy and durability, most box caissons are made of reinforced concrete. Although they are therefore heavier, the concurrent development of very large capacity crane barges and equipment has made their use fully practicable. The weight is advantageous in providing stability in high currents and waves.

FIGURE 47.19 Steel box caisson being positioned for Akashi Strait Bridge, Japan, despite strong currents.

FIGURE 47.20 Akashi Strait Suspension Bridge is founded on steel box caissons filled with tremie concrete.

47.6.3 Prefabrication Concepts

Prefabrication of the box caissons may be carried out in a number of interesting ways. The caissons(s) may be constructed on the deck of a large submersible barge. In the case of the two concrete caissons for the Tsing Ma Bridge in Hong Kong, the barge then moved to a site where it could submerge to launch the caissons. They were floated to the site and ballasted down onto the predredged rock base. After sealing the perimeter of the cutting edge, they were filled with tremie concrete.

In the case of the 66 piers for the Great Belt Western Bridge in Denmark, the box caissons were prefabricated on shore in an assembly-line process. (Figures 47.21 and 47.22). They were progressively moved out onto a pier from which they could be lifted off and carried by a very large crane barge to their site. They were then set onto the prepared base. Finally, they were filled with sand and antiscour stone was placed around their base.

FIGURE 47.21   Prefabrication of box caisson piers for Great Belt Western Bridge, Denmark.

FIGURE 47.22   Prefabricated concrete box caissons are moved by jacks onto pier for load-out.

A similar procedure has been followed for the approach piers on the Oresund Bridge between Sweden and Denmark and on the Second Severn Bridge in Southwest England. For the Great Belt Eastern Bridge, many of the concrete box caissons were prefabricated in a construction basin (Figure 47.23). Others were fabricated on a quay wall.

For the Prince Edward Island Bridge, bell-shaped piers, with open bottom, weighing up to 8000 tons, were similarly prefabricated on land and transported to the load-out pier and onto a barge, using transporters running on pile-supported concrete beams (Figures 47.24 through 47.26). Meanwhile, a shallow trench had been excavated in the rock seafloor, in order to receive the lower end of the bell. The bell-shaped shell was then lowered into place by the large crane barge. Tremie concrete was placed to fill the peripheral gap between bell and rock.

47.6.4 Installation of Box Caissons by Flotation

Large concrete box caissons have been floated into location, moored, and ballasted down onto the prepared base (Figure 47.27). During this submergence, they are, of course, subject to current, wave, and wind forces. The moorings must be sufficient to control the location; "taut moorings" are therefore used for close positioning.

FIGURE 47.23 Large concrete box caissons fabricated in construction basin for subsequent deployment to site by self-flotation, Great Belt Eastern Bridge, Denmark.

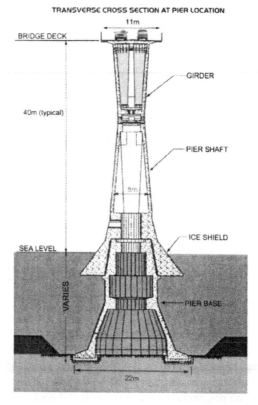

FIGURE 47.24 Schematic representation of substructure for Prince Edward Island Bridge, Canada. Note ice shield, designed to reduce forces from floating ice in Northumberland Strait.

The taut moorings should be led through fairleads on the sides of the caisson, in order to permit lateral adjustment of position without causing tilt. In some cases where navigation requirements prevent the use of taut moorings, dolphins may be used instead. These can be faced with a vertical rubbing strip or master pile. Tolerances must be provided in order to prevent binding.

FIGURE 47.25   Prefabrication of pier bases, Prince Edward Island Bridge, Canada.

FIGURE 47.26   Prefabricated pier shaft and icebreaker, Prince Edward Island Bridge, Canada.

Stability is of critical importance for box caissons which are configured such that the water plane diminishes as they are submerged. It is necessary to calculate the metacentric height, $\overline{GM}$, at every change in horizontal cross section as it crosses the water plane, just as previously described for open caissons.

During landing, as during the similar operation with open caissons, the current under the caisson increases, and scour must be considered. Fortunately, in the case of box caissons, they are being

FIGURE 47.27 Prefabricated box caisson is floated into position, Great Belt Eastern Bridge, Denmark. Note temporary cofferdam above concrete caisson.

landed either directly on a leveled hard stratum or on a prepared bed of densified stone, for which scour is less likely.

As the base of the caisson approaches contact, the prism of water trapped underneath has to escape. This will typically occur in a random direction. The reaction thrust of the massive water jet will push the caisson to one side. This phenomenon can be minimized by lowering the last meter slowly.

Corrections for the two phenomena of current scour and water-jet thrust are in opposition to one another, since lowering slowly increases the duration of exposure to scour. Thus it is essential to size and compact the stone of the stone bed properly and also to pick a time of low current, e.g., slack tide for installation.

47.6.5 Installing Box Caissons by Direct Lift

In recent years, very heavy lift equipment has become available. Jack-up, floating crane barges, and catamaran barges, have all been utilized (Figures 48.28 through 48.31). Lifts up to 8000 tons have been made by crane barge on the Great Belt and Prince Edward Island Bridges.

The box caissons are then set on the prepared bed. Where it is impracticable to screed a stone bed accurately, landing seats may be preset to exact grade under water and the caisson landed on them, and tremie concrete filled in underneath.

Heavy segments, such as box caissons, are little affected by current — hence can be accurately set to near-exact location in plan. Tolerances of the order of 20 to 30 mm are attainable.

47.6.6 Positioning

Electronic distance finders (EDF), theodolites, lasers, and GPS are among the devices utilized to control the location and grade. Seabed and stone bed surveys may be by narrow-beam high-frequency sonar and side-scan sonar. At greater depths, the sonar devices may be incorporated in an ROV to get the best definition.

47.6.7 Grouting

Grouting or concreting underneath is commonly employed to ensure full bearing. It is desirable to use low-strength, low-modulus grout to avoid hard spots. The edges of the caisson have to be sealed by penetrating skirts or by flexible curtains which can be lowered after the caisson is set in place,

FIGURE 47.28 Installing precast concrete box caissons for Second Severn Crossing, Bristol, England. Extreme tidal range of 10 m and high tidal current imposed severe demands on installation procedures and equipment.

FIGURE 47.29 Lifting box caisson from quay wall on which it was prefabricated and transporting it to site while suspended from crane barge.

since otherwise the tremie concrete will escape, especially if there is a current. Heavy canvas or submerged nylon, weighted with anchor chain and tucked into folds, can be secured to the caisson during prefabrication. When the caisson is finally seated, the curtains can be cut loose; they will restrain concrete or grout at low flow pressures. Backfill of stone around the edges can also be used to retain the concrete or grout.

Heat of hydration is also of concern, so the mix should not contain excessive cement. The offshore industry has developed a number of low-heat, low modulus, thixotropic mixes suitable for this use. Some of them employ seawater, along with cement, fly ash, and foaming agents. BFS cement has also been employed.

Box caissons may be constituted of two or more large segments, set one on top of the other and joined by overlapping reinforcement encased in tremie concrete. The segments often are match-cast to ensure perfect fit.

FIGURE 47.30 Setting prefabricated box caisson on which is mounted a temporary cofferdam, Great Belt Eastern Bridge, Denmark.

FIGURE 47.31 Setting 7000-ton prefabricated box caisson, Great Belt Western Bridge, Denmark.

47.7 Present and Future Trends

47.7.1 Present Practice

There is a strong incentive today to use large prefabricated units, either steel or concrete, that can be rapidly installed with large equipment, involving minimal on-site labor to complete. On-site operations, where required, should be simple and suitable for continuous operation. Filling prefabricated shells with tremie concrete is one such example.

Two of the concepts previously described satisfy these current needs. The first, a box caisson — a large prefabricated concrete or steel section — can be floated in or lifted into position on a hard seafloor. The second, large-diameter steel tubular piles, can be driven through soft and variable soils to be founded in a competent stratum, either rock or dense soils. These tubular piles are especially suitable for areas of high seismicity, where their flexibility and ductility can be exploited to reduce the acceleration transmitted to the superstructure (Figure 47.32). In very deep water, steel-framed

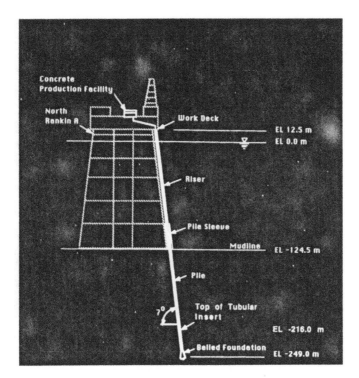

FIGURE 47.32 Conceptual design for deep-water bridge pier, utilizing prefabricated steel jacket and steel tubular pin piles.

jackets may be employed to support the piles through the water column (Figure 47.33). The box caisson, conversely, is most suitable to resist the impact forces from ship collision. The expanding use of these two concepts is leading to further incremental improvements and adaptations which will increase their efficiency and economy.

Meanwhile cofferdams and open caissons will continue to play an important but diminishing role. Conventional steel sheet pile cofferdams are well suited to shallow water with weak sediments, but involve substantial overwater construction operations.

47.7.2 Deep Water Concepts

The Japanese have had a study group investigating concepts for bridge piers in very deep water, and soft soils. One initial concept that has been pursued is that of the circular cofferdam constructed of concrete by the slurry wall process [3]. This was employed on the Kobe anchorage for the Akashi Strait Bridge and on the Kawasaki Island Ventilation Structure for the Trans-Tokyo Bay tunnels, the latter with a pumped-out head of 80 m, in extremely soft soils in a zone of high seismicity (see Section 47.3.1).

Floating piers have been proposed for very deep water, some employing semisubmersible and tension "leg-platform" concepts from the offshore industry. While technically feasible, the entire range of potential adverse loadings, including accidental flooding, ship impact, and long-period swells, need to be thoroughly considered. Tethered pontoons of prestressed concrete have been successfully used to support a low-level bridge across a fjord in Norway.

Most spectacular of all proposed bridge piers are those designed in preliminary feasibility studies for the crossing of the Strait of Gibraltar. Water depths range from 305 m for a western crossing to 470 m for a shorter eastern crossing. Seafloor soils are highly irregular and consist of relatively weak sandstone locally known as flysch. Currents are strong and variable. Wave and swell exposure is significant. For these depths, only offshore platform technology seems appropriate.

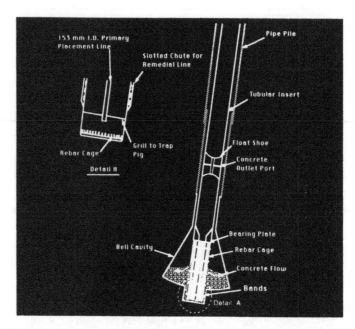

FIGURE 47.33 Belled footing provides greater bearing area for driven-and-drilled steel tubular pile.

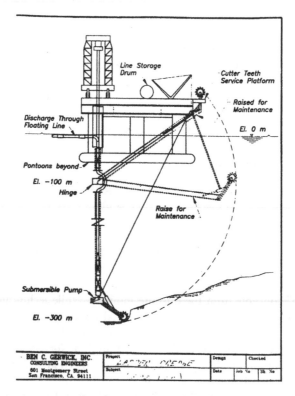

FIGURE 47.34 Concept for preparation of seabed for seating of prefabricated box piers in 300 m water depth, Strait of Gibraltar.

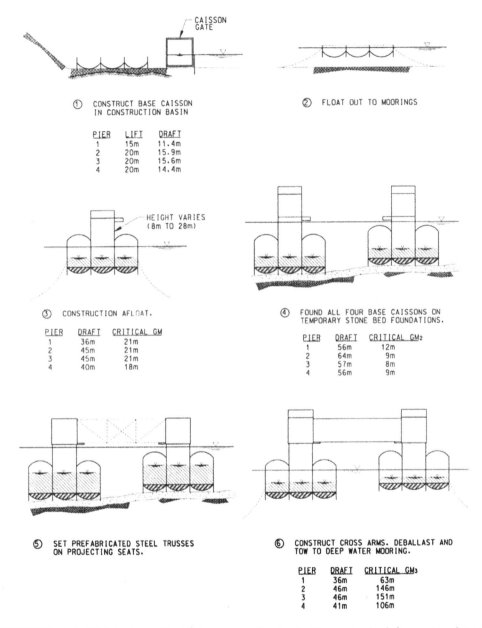

FIGURE 47.35 (a,b,c) Fabrication and installation concept for piers in 300 m water depth for crossing of Strait of Gibraltar.

Both steel jackets with pin piles and concrete offshore structures were investigated. Among the other criteria that proved extremely demanding were potential collision by large crude oil tankers and, below water, by nuclear submarines.

These studies concluded that the concrete offshore platform concept was a reasonable and practicable extension of current offshore platform technology. Leveling and preparing a suitable foundation is the greatest challenge and requires the integration and extension of present systems of dredging well beyond the current state of the art (Figure 47.34). Conceptual systems for these structures have been developed which indicate that the planned piers are feasible by employing an extension of the concepts successfully employed for the offshore concrete platforms in the North Sea (Figure 47.35).

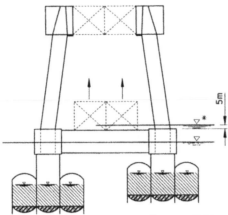

⑦ BALLAST DOWN TO MID-HEIGHT OF CROSS-ARMS.
CONSTRUCT ALL 4 SHAFTS AND CONICAL TOPS.
RAISE PLATFORM FABRICATION TRUSSES AND
SECURE TO SHAFT TOP CONES AND UPPER
FALSEWORK (TRUSS WEIGTHS ~ 2000T)

PIER	DRAFT*	CRITICAL GM₄
1	96m	22m
2	106m	34m
3	95m	36m
4	97m	27m

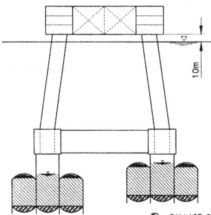

⑧ BALLAST DOWN TO DEEP DRAFT. COMPLETE
UPPER PLATFORM, VERTICAL CROSS WALLS
FORMED WITH REBAR TIED BUT NO CONCRETE
PLACED.

PIER	DRAFT*	CRITICAL GM₅
1	103m	5m
2	281m	21m
3	246m	16m
4	196m	6m

FIGURE 47.35   (continued)

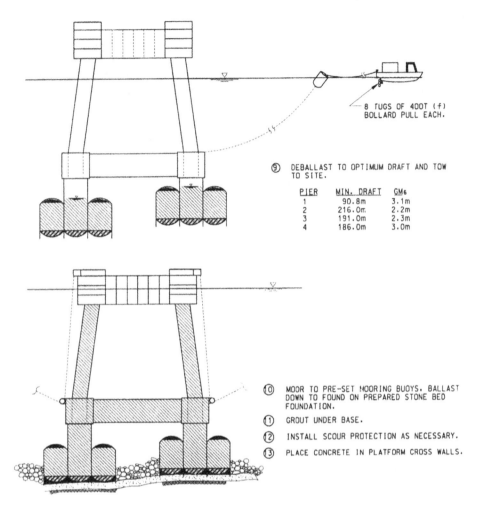

⑨ DEBALLAST TO OPTIMUM DRAFT AND TOW
    TO SITE.

PIER	MIN. DRAFT	GM_6
1	90.8m	3.1m
2	216.0m	2.2m
3	191.0m	2.3m
4	186.0m	3.0m

8 TUGS OF 400T (f)
BOLLARD PULL EACH.

⑩ MOOR TO PRE-SET MOORING BUOYS. BALLAST
    DOWN TO FOUND ON PREPARED STONE BED
    FOUNDATION.

⑪ GROUT UNDER BASE.

⑫ INSTALL SCOUR PROTECTION AS NECESSARY.

⑬ PLACE CONCRETE IN PLATFORM CROSS WALLS.

FIGURE 47.35   (continued)

Reference

1. API-RP2A, Recommended Practice for Planning, Designing and Constructing Fixed Offshore Platforms, 1993.
2. Gerwick, Ben C., Jr., *Construction of Offshore Structures*, John Wiley & Sons, New York, 1986.
3. Ratay, R. T., Ed., Cofferdams, in *Handbook of Temporary Structures in Construction*, 2nd ed., McGraw-Hill, New York, 1996, chap. 7.
4. O'Brien, J. J., Havers, J. A., and Stubbs, F. W., *Standard Handbook of Heavy Construction*, 3rd ed., McGraw-Hill, New York, 1996, chap. B-11 (Marine Equipment), chapter D-4, (Cofferdams and Caissons).
5. Shiraishi, S., Unmanned excavation systems in pneumatic caissons, in *Developments in Geotechnical Engineering*, A.A. Balkema, Rotterdam, 1994.

48

Bridge Construction Inspection

Mahmoud Fustok
*California Department
of Transportation*

Masoud Alemi
*California Department
of Transportation*

48.1 Introduction

Bridge construction inspection provides quality assurance for building bridges and plays a very important role in the bridge industry. Bridge construction involves two types of structures: permanent and temporary structures. Permanent structures, including foundations, abutments, piers, columns, wingwalls, superstructures, and approach slabs, are those that perform the structural functions of a bridge during its service life. Temporary structures, such as shoring systems, guying systems, forms and falsework, are those that support the permanent structure during its erection and construction.

This chapter discusses inspection principles, followed by the guidelines for inspecting materials, construction operations, component construction, and temporary structures. It also touches on safety considerations and documentation.

0-8493-7434-0/00/$0.00+$.50
© 2000 by CRC Press LLC

48.2   Inspection Objectives and Responsibilities

48.2.1   Objectives

The objective of construction inspection is to ensure that the work is being performed according to the project plans, specifications [1,2], and the appropriate codes including AASHTO, AWS [3,4] as necessitated by the project. The specifications describe the expected quality of materials; standard methods of work; methods and frequency of testing; and the variation or tolerance allowed.. Design and construction of temporary structures should also meet the requirements of construction manuals [5–8].

48.2.2   Responsibilities of the Inspector

The inspector's primary responsibility is to make sure that permanent structures are constructed in accordance with project plans and specifications and to ensure that the operations and/or products meet the quality standard. The inspector is also responsible to determine the design adequacy of temporary structures proposed for use by the contractor. The qualified inspector should have a thorough knowledge of specifications and should exercise good judgment. The inspector should keep a detailed diary of daily observations, noting particularly all warnings and instructions given to the contractor. The inspector should maintain continual communication with the contractor and resolve issues before they become problems.

48.3   Material Inspection

48.3.1   Concrete

At the beginning of the project, a set of concrete mix designs should be proposed by the contractor for use in the project. These mix designs are based on the specification requirements, the desired workability of the mix, and availability of local resources. The proposed concrete mix designs should be reviewed and approved by the inspector.

Method and frequency of sampling and testing of concrete are covered in the specifications. Concrete cylinders are sampled, cured, and tested to determine their compressive strength. The following tests are conducted to check some other concrete properties including:

- Cement content;
- Cleanness value of the coarse aggregate;
- Sand equivalent of the fine aggregate;
- Fine, coarse, and combined aggregate grading;
- Uniformity of concrete.

48.3.2   Reinforcement

Reinforcing steel properties and fabrication should conform to the specifications. A Certificate of Compliance and a copy of the mill test report for each heat and size of reinforcing steel should be furnished to the inspector. These reports should show the physical and chemical analysis of the reinforcing bars.

Bars should not be bent in a manner that damages the material. Check for cracking on Grade 60 reinforcing steel where radii of hooks have been bent too tight. Bars with kinks or improper bends must not be used in the project. Hooks and bends must conform to the specifications.

For epoxy-coated reinforcing bars, a Certificate of Compliance conforming to the specifications must be furnished for each shipment. Epoxy coating material must be tested for specification compliance. Bars are tested for coating thickness and for adhesion requirements. Any damage to the coating caused by shipment and/or installation must be repaired with patching material compatible with the coating material.

48.3.3 Structural Steel

The contractor furnishes to the inspector a copy of all mill orders, certified mill test reports, and a Certificate of Compliance for all fabricated structural steel to be used in the work [9]. In addition, the test reports for steels with specified impact values include the results of Charpy V-notch impact tests. Structural steel which used to fabricate fracture critical members should meet the more stringent Charpy V-notch requirements [4].

48.4 Operation Inspection

48.4.1 Layout and Grades

Bridge inspectors spend a major part of their time making sure that the structure is being built at the correct locations and elevations as shown on the project plans. Lines and grades are provided at reference points by surveyors. Based on these reference points the contractor establishes the lines and grades for building the structure. Horizontal alignment usually consists of a series of circular curves connected with tangents. Vertical alignment usually consists of a series of parabolic curves that also are connected with tangents. When a bridge is on a horizontal curve, its cross section is sloped to counteract the centrifugal forces. This slope is called the superelevation. Therefore, the inspector uses the reference points and basic geometric principles for horizontal and vertical curves to check the bridge geometry. The following layout and grades are to be checked in the field:

1. Pile locations and cutoff elevations;
2. Footings location and grades;
3. Column pour grades;
4. Abutments and wing walls pour grades;
5. Falsework grade points;
6. Lost deck grade points;
7. Overhang jacks and edge of deck.

The deck contour sheet (DCS) is a scaled topographic plan that shows the top of the bridge deck elevations [10]. With some manipulation, the DCS provides elevations for various components and construction stages of a bridge, including abutments, columns, falsework, lost deck, and edge of deck. Keeping tight control on alignment and grades with the DCS will produce smooth vehicle ridability and an aesthetically pleasing bridge.

Falsework (FW) grades are adjusted to meet the soffit elevations. FW points should be placed at locations where the FW will be adjusted, and may be shot at each stage. For example, FW points should not be set on girder centerlines of a cast-in-place prestressed box-girder bridge; otherwise they will be covered by the prestress ducts.

Lost deck dowels (LDD) are the control points to construct the top deck of the bridge. Accurate field layout and DCS layout of the LDD points are essential. Before the soffit and stem pour, the contractor places LDD at predetermined points. After the girder pour, the inspector measures the elevations of the LDD and compares them to those picked from the DCS; then the amount of adjustment needed to the deck grades is calculated.

Inspection of the edge of deck (EOD) grades is very critical for controlling and smooth operation of deck-finishing tools. Additionally, EOD controls the thickness of the concrete slab. It is recommended to locate EOD grade points at points of form adjustment.

FW bents are erected at elevations higher than the theoretical grades to offset the anticipated FW settlements. For FW used to construct cast-in-place concrete bridges, camber strips are usually placed on top of FW stringers to offset (1) deflection of the FW beam under its own weight and the actual load imposed, (2) difference between beam profile and bridge profile grade, and (3) difference between beam profile and any permanent camber.

Steel plate girders are fabricated with built-in upward camber to offset vertical deflections due to the girder weight and other dead loads such as deck and barrier, deflection caused by concrete creep, and to provide any vertical curve required by the profile. The total values are usually called the web camber. Header camber is the web camber minus the girder deflection due to its own weight. If an FW bent is needed to erect a field splice for a steel girder, the grades of the FW bent should be set to no-load elevations. No-load elevation is equal to the plan elevation plus the sum of deflections minus the depth of the superstructure.

48.4.2 Concrete Pour

Ready-mixed concrete is usually delivered to the field by concrete trucks. Each load of ready-mixed concrete should be accompanied by a ticket showing volume of concrete, mix number, time of batch, and reading of the revolution counter. The inspector should ensure the concrete meets the specification requirements regarding:

- Elapse time of batch;
- Number of drum revolutions;
- Concrete temperature;
- Concrete slump (penetration).

Addition of water to the delivered concrete, at the job site, should be approved by the inspector and it should follow the specifications.

The inspector should ensure that concrete is consolidated using vibrators by methods that will not cause segregation of aggregates and will result in a dense homogeneous concrete free of voids and rock pockets. The vibrator should not be dragged horizontally over the top of the concrete surface. Special care must be taken in vibrating areas where there is a high concentration of reinforcing steel. To prevent concrete from segregation caused by excessive free fall, a double-belting hopper or an elephant trunk should be used to guide concrete placing for piles, foundations, and walls.

While placing concrete into a footing, pouring concrete at one location and using vibrators to spread it should not be allowed since it causes aggregate segregation. Care should be taken when topping off column pour to the proper grade. For a cast-in-place box-girder bridge, columns are usually poured 30 mm higher than the theoretical grade to help butt soffit plywood and hide the joint.

While placing concrete, soffit thickness and height of concrete in stems should be checked. Concrete in bent caps should be placed at the proper grade to allow room for minimum concrete clearance over the main cap top reinforcement as shown on the project plans. The appropriate rebar concrete cover is essential in preventing steel rusting.

During bridge deck placement, the pour front should not exceed 3 to 4 m ahead of the finishing machine. Application of curing compound should be performed by power-operated equipment, and it should follow the finishing machine closely.

48.4.3 Reinforcement Placing

Reinforcement should be properly placed as shown on the plans in terms of grade, size, quantity, and location of steel rebar. Reinforcement should be firmly and securely held in position by wiring at intersections and splices with wire ties. The "pigtails" on wire ties should be flattened so that they maintain the minimum concrete coverage from formed surfaces. Rust bleeding can occur from pigtails which extend to the concrete surface.

Positioning of reinforcing steel in the forms is usually accomplished by the use of precast mortar blocks. The use of metal, plastic, and wooden support chairs is normally not permitted. For soffit and deck, blocks should be sufficient to keep mats off plywood a distance equal to the specified rebar clearance. For short- to medium-height columns, reinforcement clearance can be checked from the top of the form with a mirror or flashlight. For foundations, adequate blocking should be provided to hold the bottom mat in proper position. If column steel rests on the bottom mat, extra blocking may be required. Reinforcing steel should be protected from bond breaker substances.

Splicing of reinforcing bars can be done by lapping, butt welding, mechanical butt splice, or mechanical lap splicing.

For lap splicing A 615 steel rebar, the following splicing length is recommended [1]:

For Grade 300:

- Bar No. 25 and smaller is $30D$
- Bar Nos. 30 and 35 is $45D$

For Grade 400:

- Bar No. 25 or smaller is $45D$
- Bar Nos. 30 and 35 is $60D$

where D is diameter of the smaller bar joined.

Reinforcing bars larger than No. 35 should not be spliced by lapping.

For welded rebar, welds should be the right size and be free of cracks, lack of fusion, undercutting, and porosity. Butt welds should be made with multiple passes while flare welds may be made in one pass. For quality assurance, radiographic examinations might be performed on full-penetration butt-welded splices. Radiographs can be made by either X ray or gamma ray.

For mechanical butt splices, splicing should be performed in accordance with the manufacturer's recommendations. Tests of sample splices should be done for quality assurance.

Reinforcement for abutment and wingwalls should be checked before forms are buttoned up.

Proper reinforcing steel placement in deck, especially truss bars, is very important, since the moment-carrying capacity of a bridge deck is greatly sensitive to the effective depth of the section. String lines are used between grade points to check steel clearance and deck thickness.

48.4.4 Welding of Structural Steel

The inspector is responsible for the quality assurance inspection (QAI) of all welding. The inspector should ascertain that equipment, procedures, and techniques are in accordance with specifications. Contract specifications usually refer to welding codes such as AWS D1.5 [4]. All welding should be performed in accordance with an approved welding procedure and by a certified welder. All welding materials such as electrodes, fluxes, and shielding gases must be properly packaged, stored, and dried. Quality of the welds is largely dependent on the welding equipment. The equipment should be checked to ensure that it is in good working condition. Travel speed and rate of flow of shielding gases should be monitored. The actual heat input should not exceed the maximum heat input that was tested and approved. Preheat and interpass temperature specifications should be adhered to as they affect cooling rate and heat input. When postheat is specified, the temperature and duration must be monitored.

Base metal at the welding area (root face, groove face) must be within allowable roughness tolerances. All mill scale should be removed from surfaces where girder web to flange welds are made.

Inspectors must verify that all nondestructive testing (NDT) has been performed and passed the specified requirements [11,12]. The inspector should maintain a record of all locations of inspected areas with the NDT report and findings, together with the method of repairs and NDT test results of weld repairs.

Fillet weld profile should be within final dimensional requirements for leg and throat size and surface contour. Bumps and craters due to starts and stops, weld rollover, and insufficient leg and throat must be ground and repaired to acceptable finish.

48.4.5 High-Strength Bolts

The contact surfaces of all high-strength bolted connections should be thoroughly cleaned of rust, mill scale, dirt, grease, paint, lacquer, or other material. Before the installation of fasteners, the inspector should check the marking, surface condition of bolts, nuts, and washers for compliance with the specifications. Nuts and bolts that are not galvanized should be clean and dry or lightly lubricated. Nuts for high-strength galvanized bolts should be overtapped after galvanizing, and then treated with a lubricant [13].

High-strength bolts may be tensioned by use of calibrated power wrench, a manual torque wrench, the turn-of-nut method, or by tightening and using a direct tension indicator. The inspector should observe calibration and testing procedures required to confirm that the selected procedure is properly applied and the required tensions are provided [14]. The inspector should monitor the installation in the work to ensure that the selected procedure, as demonstrated in the testing, is routinely properly applied.

To inspect completed joints, the following procedure is recommended [15]. A representative sample of bolts from the diameter, length, and grade of the bolts used in the work should be tightened in the tension-measuring device by any convenient means to an initial condition equal to approximately 15% of the required fastener tension and then to the minimum tension specified. Tightening beyond the initial condition must not produce greater nut rotation than 1.5 times that permitted in the specifications. The inspection wrench should be applied to the tightened bolts and the torque necessary to turn the nut or head 5° should be determined.

The Coronet load indicator (CLI) is a simple and accurate aid for tightening and inspecting high-strength bolts. The CLI is a hardened round washer with bumps on one face. As the bolt and nut are tightened, the clamping force flattens the bumps, placed against the underside of the bolt head. The nut should be tightened until the gap is reduced to 0.38 mm. This requirement applies to both A325 and A490 bolts. Once the gap is reduced to the required dimension, the bolt and nut are properly tightened. Visual gap inspection is usually adequate by comparing them against gaps which were checked with the feeler gauge. All CLI for A325 are round. CLI for A490 has three ears or the letter V stamped at three places.

Reuse of ASTM A490 and galvanized ASTM A325 bolts is not allowed. Reuse of ASTM A325 bolts may be allowed if it is approved by the inspector. Reuse does not include retightening bolts which may have been loosened by the tightening of adjacent bolts.

48.5 Component Inspection

48.5.1 Foundation

Conventional bridge foundations can be classified in three types: (1) spread footing foundation, (2) pile foundation, (3) special-case foundations such as pile shafts, tiebacks, soil nails, and tiedowns. The first two types of foundations are more common, and therefore, their inspections will be discussed. Problems that may be encountered during foundation construction should be discussed with the designer and the engineering geologist who performed the foundation studies.

FIGURE 48.1   Foundation forms.

48.5.1.1   Spread Footing Foundations

Spread footing foundations support the load by bearing directly on the foundation stratum. The conformity of the foundation material with the log of the test boring should be checked. Additionally, the bearing surface should be free of disturbed material and be compacted if it is necessary.

Foundations are poured using forms (Figure 48.1) or using the soil as a form in neat-line excavation method. Inspection of the forms is covered in Section 48.6.4. The "neat-line" excavation method is usually done for column and retaining wall footings. The toe of retaining wall should be placed against undisturbed material. Depth and dimensions of footing need to be checked. Standing water and all sloughed material in the excavation must be removed prior to placement of concrete into the foundation. The foundation material should be wet down but not saturated. Footings more than 760 mm vertical dimension, and with a top layer of reinforcement, should be reconsolidated by a vibrator for a depth of 300 mm. Reconsolidation should be done not less than 15 min after the initial leveling of the top of the foundation has been completed. A curing compound will usually be used on top of the footing. If the construction joint is sandblasted before the completion of the curing period, the exposed area should be cured using an alternative method for the remaining time.

48.5.1.2   Pile Foundations

Pile foundations transmit design loads into adjacent soil through pile friction, end bearing, or both. Tops of piles are never exact. Determine the pile with highest elevation and block up bottom-mat steel reinforcement to horizontal accordingly. The highest pile may have to be cut off if the grade is unreasonable.

There are two major types of piles: cast-in-drilled-hole piles and driven piles.

Cast-in-Drilled-Hole Concrete Piles
For cast-in-drilled-hole piles, the inspector should check the following:

- Diameter, depth, and straightness of drilled holes;
- Cleanness of the bottom of holes from water and loose materials.

Material encountered during drilling should be compared with that shown on the log of test borings. If there is a significant difference, the designer should be informed.

For 400-mm-diameter piles, the top 4500 mm of concrete should be vibrated; for larger-diameter piles the full length should be vibrated [16].

Using steel casing is one method to prevent soil cave-ins and intrusion of groundwater. Casing is pulled when placing concrete keeping its bottom below the concrete surface. Waiting too long to pull the casing may cause the concrete to set up and may lead to the following problems:

- The concrete comes up with the casing.
- The casing cannot be removed.
- The concrete may not fill the voids left by the casing.

Use of a concrete mix with fluidity at the high end of the allowable range will help to mitigate these problems.

A slurry displacement method can be used to prevent cave-in of unstable soil and intrusion of groundwater into the drilled hole. The drilling slurry remains in the drilled hole until it is displaced by concrete. Concrete is placed using a delivery system with rigid tremie tube or a rigid pump tube, starting at the bottom of the drilled hole. Sampling and testing of drilling slurry is an important quality control requirement. The following properties of drilling slurries should be monitored: density, sand content, pH value, and viscosity [16].

Inspection tubes are installed inside the spiral or hoop reinforcement in a straight alignment in order to facilitate pile testing. Inspection tubes permit the insertion of a testing probe that measures the density of the pile concrete. A radiographic technique, commonly called gamma ray scattering, is used to measure the density. If the pile is accepted, the inspection tubes are cleaned and filled with grout.

Driven Piles

Prior to start-up of a pile-driving operation, the inspector should check the hammer type and the pile size. Piles should be marked for logging. During the pile-driving operation, the inspector should monitor plumbness or batter of the pile, and log the pile penetration.

Charts of calibration curves are developed for different pile capacity and hammer. By using the energy theory, a penetration-per-blow chart which corresponds to the specified capacity of the piles can be developed. One of the commonly used formulas to determine the bearing capacity of a pile is the ENR formula:

$$p = \frac{E}{6(s + 2.54)} \quad q \tag{48.1}$$

where
p = safe load in kilonewtons
E = manufacturer's rating for energy developed by the hammer in joules
s = average penetration per blow in millimeters

On a large structure which necessitates long piles to be driven in order to satisfy the design load-bearing requirements, monitoring tests may be performed to determine the pile-driving chart.

For pipe steel piles, piles should have the specified diameter, length, and wall thickness as shown on the project plans. If the piles are to be spliced, welding should be performed by a certified welder and the quality of the welded joints should be checked. The method of pipe splicing should be in accordance with project plans and specification requirements. Steel shells may be driven open or closed ended. For shells that are driven open ended, soil should be augured out, and the pile should be cleaned before it is filled with concrete.

For precast concrete piles, the following should be checked:

FIGURE 48.2 Crane used to hold the column cage; column is being cured with plastic wrap.

- Pile is free of damage or cracks;
- Size and length of piles;
- Age of pile, minimum 14 days before it is allowed to be driven [1].

If piles are driven through new embankment greater than 1500 mm, predrilling is required [1]. Problems with the driving operation can be categorized into three types:

1. Hard driving occurs if the soil is too dense or the hammer does not have enough energy to drive the pile. This problem can be solved by predrilling, jetting, or using a larger low-velocity hammer. If a pile is undergoing hard driving and suddenly experiences a large movement, this could indicate a fracture of the pile belowground. In this case the pile should be extracted and replaced or a replacement pile should be driven next to it.
2. Soft pile occurs when the pile is driven to the specified tip elevation but has not attained the specified bearing capacity. This pile is set for a minimum of 12 h, then retapped. If the retapped pile will not attain the bearing capacity, then the contractor has to furnish longer piles.
3. Alignment of piles: if the pile begins to move out of plumbness, correction should be made. The pile may have to be pulled and redriven.

48.5.2 Concrete Columns and Pile Shaft

For medium to long columns, column cages should be held at the top with a crane until footing concrete is set as insurance for the guying system (Figure 48.2). For extremely tall columns, a crane holds the cage until the column is poured.

For fixed columns, make sure ties to the bottom mat are placed in accordance with the project plans to prevent the cage moving during placement of concrete into the footing. For pinned columns, check key details and verify that adequate blocking is provided to support the steel cage at the proper height above a key.

Check the sequence of attaching and removing the guying system to ensure it is done in accordance with the approved plans. Location of utilities should be checked prior to forming a column.

FIGURE 48.3 Wingwall and abutment construction.

Column forms are usually removed a few days after the concrete pour. Columns should be covered with plastic wrap until they are cured (Figure 48.2).

Pile shafts are encountered in bedrock material usually close to canyons or hillside areas with limited room for footing foundations. They are primarily a cast-in-drilled-hole footing with neat-line excavation. The column has the same size extension as the pile shaft or a slightly smaller section. Shoring is required in areas that are not solid rock and any excavation out of the neat area should be filled with concrete. Since blasting is the most common method of excavation, extreme caution is necessary to protect workers and the public.

48.5.3 Abutment and Wingwalls

Abutment and wingwalls are normally formed and poured at the same time for seat-type abutments (Figure 48.3). For the diaphragm abutments of cast-in-place, prestressed-concrete box girders, wingwalls should be placed after stressing. Utility openings in wingwalls and/or abutments should be checked in accordance with the project plans. Bearing pad and internal key layout are usually checked after pour strips are in place.

48.5.4 Superstructure

The following is a summary of the construction inspection for concrete box and steel plate girders which are commonly used for building short- to medium-size bridges.

48.5.4.1 Cast-in-Place Concrete Box Girder

Concrete box girders are constructed in two stages. First, FW is erected and stem and soffit are formed and poured (Figure 48.4). Second, lost deck is built and then the concrete for the bridge deck is poured (Figure 48.5). For prestressed concrete, once the bridge deck is cured, the frame is prestressed, and then the FW is removed. Erection of stem and soffit will be discussed here, and bridge prestressing and deck erection will be discussed in following sections.

The following items need to be checked during construction of soffit and stems:

- Size of camber strips placed on top of FW stringers;
- Location of utility and/or soffit access openings and the corresponding reinforcing details;

FIGURE 48.4 Bridge stem and soffit forms for concrete box girder.

FIGURE 48.5 Bridge deck reinforcement with bidwell finishing machine.

- Location and elevation of block-outs for deck drain pipes;
- Size and profile of the prestressing ducts;
- Smoothness of duct-to-flare connection at bearing plate for alignment of line profile (Figure 48.6);
- Tattletales readings; readings exceeding those anticipated should be investigated to determine if they are due to FW failure or due to excessive soil settlement; corrective measure should be taken accordingly.

FIGURE 48.6 Bridge soffit and girder reinforcements with prestressing duct profile.

Inspectors should make sure that

- Trumpets are properly secured to the bearing plates;
- Ducts and intermediate vents are secured in place;
- Proper gap is maintained between snap ties and ducts to prevent any damage to prestressing ducts;
- Tendon openings are sealed to prevent water or debris from entering;
- Size and location of prestress bearing plates at abutment or hinge diaphragms are in accordance with the approved prestressing working drawings;
- Elastomeric bearing pads are installed properly at abutments and the remainder of the abutment seat area is covered with expanded polystyrene of the same thickness as the pads and that joints are sealed with tape to prevent grout leakage;
- Curing compound is applied appropriately and on time.

Due to the tremendous amount of force involved in prestressing a concrete bridge, careful consideration should be given to this operation. Caution should be exercised around the prestressing jack during the stressing operation and around ducts while they are not grouted (Figure 48.7).

Prior to placement of tendons in prestressing ducts, the following should be verified:

- Ducts are unobstructed and free of water and debris;
- Strands are free of rust.

Prior to stressing, the contractor should submit the required calibration curves for specific jack/gauge combinations. It is the inspector's responsibilitu to determine

- That tendons are installed in accordance with plans
- Whether stressing should be done from one end or two ends
- That stressing sequence is performed in accordance with plans

During stressing, strands are painted on both ends to check slippage. Prestressing calibration curves are plotted, and measured elongation is compared to calculated elongation using the actual area and modulus of elasticity [17].

FIGURE 48.7 Prestressing strands and anchor set.

During the grouting of tendons, the following needs to be checked:

- Water/cement ratio;
- Efflux time;
- That grout is continuously agitated.

48.5.4.2 Steel Girder

Structural steel is usually inspected at the fabrication site. In the field, girders should be checked for damage that may have occurred during transportation to the job. All bearing assemblies should be set level and to the elevations shown on the plans [18]. Full bearing on concrete should be obtained under bearing assemblies (Figure 48.8).

Fracture-critical members (FCM) are tension members or tension components of bending members, the failure of which may result in collapse of the bridge. FMCs are usually identified on the plans or described in the contract documents. FCMs are subject to the additional provisions of Section 12 of AASHTO/AWS D1.5. All welds to FCMs are considered fracture critical and should conform to the requirements of the fracture control plan.

All surfaces of structural steel that are to be painted should be blast-cleaned to produce a dense, uniform surface with an angular anchor pattern [19]. On the same day that blast cleaning is done, structures should be painted with undercoats prior to their erection. Steel surfaces that are inaccessible for painting after erection should be fully painted before they are erected. Steel areas where paint has been damaged due to erection should be cleaned and painted with undercoats before the application of any subsequent paint. Subsequent painting should not be performed until the cleaned surfaces are dry. Succeeding applications of paint should be of such shade to provide contrast with the paint being covered. Paint thickness is measured with a thickness gauge that is calibrated by a magnetic film.

48.5.4.3 Precast Prestressed Girders

Like steel beams, precast prestressed girders are inspected at the fabricator site. Precast concrete members are usually cured by the water or steam method. Upon arrival to the project site, girders should be inspected for damage that may have occurred while being transported to the project site.

FIGURE 48.8   Bearing assembly of steel plate girder.

Precast girders should be handled, transported, and erected using extreme care to avoid twisting, racking, or other distortion that would result in cracking or damage. Girders are placed on elastomeric pads at certain locations shown in the plans. Girders are braced and held together by temporary wooden blocking.

The top of the girder elevation is determined by profiling each girder. The profile grade will determine the location of the finished grade of the top slab and the location of the slab forms. After girders have been profiled and finished grades have been determined, placing forms for the top slab can start. Prestressed concrete panels are a type of slab form that is left in place and becomes the bottom part of the concrete slab. They are normally 100 mm thick [20], rectangular shaped, and vary in width and length.

48.5.4.4 Concrete Deck

The bridge top deck is probably the most critical part in terms of smooth vehicle ridability and aesthetically pleasing bridge. Smoothness of the bridge surface and the approach slabs should be tested by a bridge profilograph. Two profiles will be obtained in each lane. Surfaces that fail to conform to the smoothness tolerances should be ground until these tolerances are achieved. Grinding should not reduce the concrete cover on reinforcing steel to less than 40 mm [1]. The following items need to be checked during the construction of bridge deck:

1. Adequacy of sandblasting on top of girders of cast-in-place concrete bridges;
2. Tops of girders are free of dust;
3. Block-outs for joint seal assemblies;
4. Overhang chamfer and screed pipes have smooth line;
5. Clearance of finishing machine roller to the steel mat (see Figure 48.5), and height of deck drain inlets in relation to finishing machine roller;
6. Tattletales for additional FW settlement;
7. The curing rugs periodically for dampness after completion of deck pour.

Inspectors should ensure that:

1. Soffit vents, access openings, drains, and their support systems are clear from steel rebar and located in accordance with the project plans [21];

FIGURE 48.9 Bridge deck is being cured with moist rugs.

2. Application of rugs or mats is begun within 4 h after completion of deck finishing, and no later than the following morning (Figure 48.9);
3. Prior to prestressing or release of FW, deck surface is inspected for crack intensity to ensure that it meets the allowable tolerance set forth in the specification.

48.6 Temporary Structures

48.6.1 Falsework

FW is used to provide temporary support to the superstructure during construction. To construct a cast-in-place concrete bridge, a complete FW system is needed. Such a system consists of FW foundation, bents, stringers, joists, and forms (Figure 48.10). To erect a steel girder, a simple FW system might be needed. Such a system consists of FW bent and foundation.

FW is designed by the contractor and approved by the inspector. FW is designed to resist vertical loads, as well as longitudinal and transverse horizontal loads. The inspector needs to have a basic understanding of design of timber members, steel members, cables, and to be familiar with application of the FW manual to check the submitted calculations and plans accordingly.

Temporary bracing or other means are needed to hold FW in stable condition during erection and removal.

Typical FW includes timber pads, corbels, steel cells, timber or steel posts, steel or timber FW bent caps, steel stringers, and timber joists. Cables are usually used to resist lateral load in the longitudinal direction. Timber, cable, or steel bar bracing is used to resist lateral loads in the transverse direction. Forms are placed on top of the joist.

When FW pad foundations are used, soil-bearing capacity may be approximated based on observed soil classification or by performing soil load tests. FW pads must be placed on a level and firm material. Pad foundations should be protected from flooding and from undermining by surface runoff. Continuous pads should be inspected to ensure the pad joints are located according to the approved FW plans.

FIGURE 48.10   FW system for cast-in-place concrete bridge.

Timber FW materials should be inspected for defects. Bolts and/or nails can be used to connect timber framing. Inspectors should check edge distance, end distance, and minimum spacing as required by the FW manual [5]. For nailed or spiked connections, ensure that adequate penetrations are provided. If the grades of FW bent are adjusted at the bottom of the post, check diagonal bracing and its connection for any distortion caused by differential movement. Wedging might be needed to ensure full bearing at contact surfaces.

The use of worn or kinked cable should not be permitted. Cables should be looped around a thimble with a minimum diameter corresponding to the cable diameter. Proper clip installation is critical and should be inspected carefully. Check preloading of cables used for internal bracing. Preloading of cables in a frame should be done simultaneously to prevent distortion.

FW with traffic openings should be in accordance with approved FW plans and specifications. In addition, adequacy of vertical and horizontal clearance as shown on the FW plans and as specified in the project specifications should be checked. Inspect the FW lighting system to ensure that the system is detailed according to the approved FW lighting plans.

During concrete pour, inspect FW for the following: excessive settlement, crushing of wedges, and deflection of bracing or distortion of its connection.

48.6.2 Shoring, Sloping, and Benching

Shoring, sloping, and benching are methods to prevent excavation cave-ins. Application of these methods depends on soil type, depth, and size of excavation.

Shoring systems are used to support the sides of excavations from cave-in. Steel members, timber members, or a combination of both are used to construct shoring systems. Trenching is similar to shoring, but the excavation is narrow relative to its length; the width at bottom is less than 5 m. During excavation, verify that the soil properties are the same as was anticipated in the design. Make sure that shoring members have the size and spacing as shown in the approved plans. If a tieback system is used, cables should be preloaded. Inspections should be made after a rain storm or other hazardous conditions. The inspector should ensure that ladders, ramps, or other safe means are provided in excavated areas for providing safe access.

FIGURE 48.11   Temporary deck form for steel bridge.

48.6.3   Guying Systems

Guying systems are temporary structures used to stabilize column cages during construction. Guying systems usually consist of a set of cables connecting the cage or the form to heavy loads such as deadmen or K-rail. This system is designed to resist an assumed wind load applied to the cage. The inspector should check the guying system calculations submitted by the contractor for the adequacy of the system and also to ensure that the erection sequence and the removing of guys to place forms is performed in accordance with the approved plans.

48.6.4   Concrete Forms

Concrete forms should be mortar-tight and with sufficient strength to prevent excessive deflection during placement of concrete. Forms are used to hold the concrete in its plastic state until it is hardened. Forms should be cleaned of all dirt, mortar, debris, nails, and wires. Forms that will later be removed should be thoroughly coated with form oil prior to use (Figure 48.11). Forms should be wet down before placing concrete.

For foundations, attention should be given to bracing of forms to prevent any movement during concrete placement. The bottom of forms should be checked for gaps that may cause excessive leakage of concrete.

48.7   Safety

The primary responsibility of the inspector is to ensure that a safe working environment and practices are maintained at the project site. They should set an example by following the code of safe practice and also by using personal safety equipment including hard hats, gloves, and protective clothing. In addition, they must enforce the safety issues as specified in the contract specifications. This may involve monitoring the operation of equipment and other construction equipment including barricades, warning lights, and reflectors to ensure that they are installed in accordance with the plans and specifications [22].

Prior to entering elevated or excavated areas, the inspector should ascertain that safe access is provided and proper worker protection is in place.

48.8 Record Keeping and As-Built Plans

In addition to construction inspection, the inspector is also responsible for maintaining an accurate and complete record of work that is being performed by contractors. Project records and reports are necessary to determine that contract requirements have been met so that payments can be made to the contractor. Project records should be kept current, complete, accurate, and should be submitted on time.

It is critical that the inspector keep a written diary of the activities that take place in the field. The diary should contain information concerning the work being inspected, including unusual incidents and important conversations. This information may become very critical in case of legal action for litigation involving construction claims or job failure.

As-built plans should reflect any deviation that may exist between the project plans and what was built in the field. Accurate and complete as-built plans are very important and useful for maintaining the bridge, and any future work on the bridge. The as-built plans also provide input and information for future seismic retrofit of the bridge.

48.9 Summary

This chapter discusses the construction inspection of new bridges. It provided guidelines for inspecting the main materials commonly used in building bridges and the major construction operations. Although more emphasis is placed on typical short- to medium-length bridges, the same principles are applied to other type of bridges.

References

1. Caltrans, *Standard Specifications*, California Department of Transportation, Sacramento, 1995.
2. Caltrans, *Standard Plans*, California Department of Transportation, Sacramento, 1995.
3. AASHTO, *LRFD Bridge Design Specifications*, American Association of State Highway and Transportation Officials, Washington, D.C., 1994.
4. AWS, *Bridge Welding Code*, ANSI/AASHTO/AWS D1.5-95, American Welding Society, Miami, FL, 1995.
5. Caltrans, *Falsework Manual*, Office of Structure Construction, California Department of Transportation, Sacramento, 1988.
6. Caltrans, *Trenching & Shoring Manual*, Office of Structure Construction, California Department of Transportation, Sacramento, 1990.
7. FHWA, Lateral Support Systems and Underpinning, Vol. 1: Design and Construction, Report No. FHWA-RD-75-128, U.S. Federal Highway Administration, Washington, D.C., 1976.
8. Mehtlan, J., *Outline of Field Construction Procedure*, Office of Structure Construction, California Department of Transportation, Sacramento, 1988.
9. Barsom, J. M., Properties of bridge steels, Vol. I, Chap. 3, in *Highway Structures Design Handbook*, American Institute of Steel Construction, 1994.
10. Caltrans, *Bridge Construction Survey Guide*, Office of Structure Construction, California Department of Transportation, Sacramento, 1991.
11. AWS, *Welding Inspection*, 2nd ed., American Welding Society, Miami, FL, 1980.
12. AWS, *Guide for the Visual Inspection of Welds*, ANSI/AWS B1.11-88, American Welding Society, Miami, FL, 1997.
13. FHWA, High-Strength Bolts for Bridges, FHWA-SA-91-031, U.S. Department of Transportation, Washington, D.C., 1991.

14. AISC, Mechanical fasteners for steel bridges, Vol. I, Chap. 4A, in *Highway Structures Design Handbook,* American Institute of Steel Construction, Chicago, 1996.

15. Caltrans, *Bridge Construction Record & Procedures,* California Department of Transportation, Sacramento, 1994.

16. Caltrans, *Foundation Manual,* Office of Structure Construction California Department of Transportation, Sacramento, 1996.

17. Caltrans, *Prestress Manual,* California Department of Transportation, Sacramento, 1992.

18. AISC, Steel erection for highway, railroad and other bridge structures, Vol. I, Chap. 14, in *Highway Structures Design Handbook,* American Institute of Steel Construction, Chicago, 1994.

19. Caltrans, *Source Inspection Manual,* Office of Materials Engineering and Testing Services, California Department of Transportation, Sacramento, 1995.

20. TxDOT, *Bridge Construction Inspection,* Texas Department of Transportation, Austin, TX, 1997.

21. Caltrans, *Bridge Deck Construction Manual,* Office of Structure Construction, California Department of Transportation, Sacramento, 1991.

22. FHWA, Bridge Inspector's Training Manual/90, FHWA-PD-91-015, U.S. Department of Transportation, Washington, D.C., 1991.

49

Maintenance Inspection and Rating

Murugesu Vinayagamoorthy
*California Department
  of Transportation*

49.1 Introduction

Before the 1960s, little emphasis was given to inspection and maintenance of bridges in the United States. After the 1967 tragic collapse of the Silver Bridge at Point Pleasant in West Virginia, national interest in the inspection and maintenance rose considerably. The U.S. Congress passed the Federal Highway Act of 1968 which resulted in the establishment of the National Bridge Inspection Standard (NBIS). The NBIS sets the national policy regarding bridge inspection procedure, inspection frequency, inspector qualifications, reporting format, and rating procedures. In addition to the establishment of NBIS, three manuals — FHWA Bridge Inspector's Training Manual 70 [1], AASHO Manual for Maintenance Inspection of Bridges [2], and FHWA Recording and Coding Guide for the Structure Inventory and Appraisal of the Nation's Bridges [3] — have been developed and

0-8493-7434-0/00/$0.00+$.50
© 2000 by CRC Press LLC

updated [4–10] since the 1970s. These manuals along with the NBIS provide definitive guidelines for bridge inspection. Over the past three decades, the bridge inspection program evolved into one of the most-sophisticated bridge management systems. This chapter will focus only on the basic, fundamental requirements for maintenance inspection and rating.

49.2 Maintenance Documentation

Each bridge document needs to have items such as structure information, structural data and history, description on and below the structure, traffic information, load rating, condition and appraisal ratings, and inspection findings. The inspection findings should have the signature of the inspection team leader.

All states in the United States are encouraged, but not mandated, to use the codes and instructions given in the Recording and Coding Guide [8,9] while documenting the bridge inventory. In order to maintain the nation's bridge inventory, FHWA requests all state agencies to submit data on the Structure Inventory and Appraisal (SI&A) Sheet. The SI&A sheet is a tabulation of pertinent information about an individual bridge. The information on SI&A sheet is a valuable aid to establish maintenance and replacement priorities and to determine the maintenance cost of the nation's bridges.

49.3 Fundamentals of Bridge Inspection

49.3.1 Qualifications and Responsibilities of Bridge Inspectors

The primary purpose of bridge inspection is to maintain the public safety, confidence, and investment in bridges. Ensuring public safety and investment decision requires a comprehensive bridge inspection. To this end, a bridge inspector should be knowledgeable in material and structural behavior, bridge design, and typical construction practices. In addition, inspectors should be physically strong because the inspection sometimes requires climbing on rough, steep, and slippery terrain, working at heights, or working for days.

Some of the major responsibilities of a bridge inspector are as follows:

- Identifying minor problems that can be corrected before they develop into major repairs;
- Identifying bridge components that require repairs in order to avoid total replacement;
- Identifying unsafe conditions;
- Preparing accurate inspection records, documents, and recommendation of corrective actions; and
- Providing bridge inspection program support.

In the United States, NBIS requires a field leader for highway bridge inspection teams. The field team leader should be either a professional engineer or a state certified bridge inspector, or a Level III bridge inspector certified through the National Institute for Certification of Engineering Technologies. It is the responsibility of the inspection team leader to decide the capability of individual team members and delegate their responsibilities accordingly. In addition, the team leader is responsible for the safety of the inspection team and establishing the frequency of bridge inspections.

49.3.2 Frequency of Inspection

NBIS requires that each bridge that is opened to public be inspected at regular intervals not exceeding 2 years. The underwater components that cannot be visually evaluated during periods of low flow or examined by feel for their physical conditions should be inspected at an interval not exceeding 5 years.

The frequency, scope, and depth of the inspection of bridges generally depend on several parameters such as age, traffic characteristics, state of maintenance, fatigue-prone details, weight limit posting level, and known deficiencies. Bridge owners may establish the specific frequency of inspection based on the above factors.

49.3.3 Tools for Inspection

In order to perform an accurate and comprehensive inspection, proper tools must be available. As a minimum, an inspector needs to have a 2-m (6-ft) pocket tape, a 30-m (100-ft) tape, a chipping hammer, scrapers, flat-bladed screwdriver, pocketknife, wire brush, field marking crayon, flashlight, plumb bob, binoculars, thermometer, tool belt with tool pouch, and a carrying bag. Other useful tools are a shovel, vernier or jaw-type calipers, lighted magnifying glass, inspection mirrors, dye penetrant, 1-m (4-ft) carpenter's level, optical crack gauge, paint film gauge, and first-aid kits. Additional special inspection tools are survey, nondestructive testing, and underwater inspection equipment.

Inspection of a bridge prompts several unique challenges to bridge inspectors. One of the challenges to inspectors is the accessibility of bridge components. Most smaller bridges can be accessed from below without great effort, but larger bridges need the assistance of accessing equipment and vehicles. Common access equipment are ladders, rigging, boats or barges, floats, and scaffolds. Common access vehicles are manlifts, snoopers, aerial buckets, and traffic protection devices. Whenever possible, it is recommended to access the bridge from below since this eliminates the need for traffic control on the bridge. Setting up traffic control may create several problems, such as inconvenience to the public, inspection cost, and safety of the public and inspectors.

49.3.4 Safety during Inspection

During the bridge inspection, the safety of inspectors and of the public using the bridge or passing beneath the bridge should be given utmost importance. Any accident can cause pain, suffering, permanent disability, family hardship, and even death. Thus, during the inspection, inspectors are encouraged to follow the standard safety guidelines strictly.

The inspection team leader is responsible for creating a safe environment for inspectors and the public. Inspectors are always encouraged to work in pairs. As a minimum, inspectors must wear safety vests, hard hats, work gloves, steel-toed boots, long-sleeved shirts, and long pants to ensure their personal safety. Other safety equipment are safety goggles, life jackets, respirator, gloves, and safety belt. A few other miscellaneous safety items include walkie-talkies, carbon monoxide detectors, and handheld radios.

Field clothes should be appropriate for the climate and the surroundings of the inspection location. When working in a wooded area, appropriate clothing should be worn to protect against poisonous plants, snakes, and disease-carrying ticks. Inspectors should also keep a watchful eye for potential hazardous environments around the inspection location. When entering a closed bridge box cells, air needs to be checked for the presence of oxygen and toxic or explosive gases. In addition, care should be taken when using existing access ladders and walkways since the ladder rungs may be rusted or broken. When access vehicles such as snoopers, booms, or rigging are used, the safe use of this equipment should be reviewed before the start of work.

49.3.5 Reports of Inspection

Inspection reports are required to establish and maintain a bridge history file. These reports are useful in identifying and assessing the repair requirements and maintenance needs of bridges. NBIS requires that the findings and results of a bridge inspection be recorded on standard inspection forms. Actual field notes and numerical conditions and appraisal ratings should be included in the standard inspection form. It is also important to recognize that these inspection reports are legal documents and could be used in future litigation.

Descriptions in the inspection reports should be specific, detailed, quantitative, and complete. Narrative descriptions of all signs of distress, failure, or defects with sufficient accuracy should be noted so that another inspector can make a comparison of condition or rate of disintegration in the future. One example of a poor description is, "Deck is in poor condition." A better description would be, "Deck is in poor condition with several medium to large cracks and numerous spalls." The seriousness and the amount of all deficiencies must be clearly stated in an inspection report.

In addition to inspection findings about the various bridge components, other important items to be included in the report are any load, speed, or traffic restrictions on the bridge; unusual loadings; high water marks; clearance diagram; channel profile; and work or repairs done to the bridge since the last inspection.

When some improvement or maintenance work alters the dimensions of the structure, new dimensions should be obtained and reported. When the structure plans are not in the history file, it may be necessary to prepare plans using field measurements. These measurements will later be used to perform the rating analysis of the structure.

Photographs and sketches are the most effective ways of describing a defect or the condition of structural elements. It is therefore recommended to include sketches and/or photographs to describe or illustrate a defect in a structural element. At least two photographs for each bridge for the record are recommended.

Other tips on photographs are

- Place some recognizable items that will allow the reviewer to visualize the scale of the detail;
- Include plumb bob to show the vertical line; and
- Include surrounding details so one could relate other details with the specific detail.

After inspecting a bridge, the inspector should come to a reasonable conclusion. When the inspector cannot interpret the inspection findings and determine the cause of a specific finding (defect), the advice of more-experienced personnel should be sought. Based on the conclusion, the inspector may need to make a practical recommendation to correct or preclude bridge defects or deficiencies. All instructions for maintenance work, stress analysis, posting, further inspection, and repairs should be included in the recommendation. Whenever recommendations call for bridge repairs, the inspector must carefully describe the type of repairs, the scope of the work, and an estimate of the quantity of materials.

49.4 Inspection Guidelines

49.4.1 Timber Members

Common damage in timber members is caused by fungi, parasites, and chemical attack. Deterioration of timber can also be caused by fire, impact or collisions, abrasion or mechanical wear, overstress, and weathering or warping.

Timber members can be inspected by both visual and physical examination. Visual examination can detect the following: fungus decay, damage by parasites, excessive deflection, checks, splits, shakes, and loose connections. Once the damages are detected visually, the inspector should investigate the extent of each damage and properly document them in the inspection report. Deterioration of timber can also be detected using sounding methods — a nondestructive testing method. Tapping on the outside surface of the member with a hammer detects hollow areas, indicating internal decay. There are a few advanced nondestructive and destructive techniques available. Two of the commonly used destructive tests are boring or drilling and probing. And, two of the nondestructive tests are Pol-Tek and ultrasonic testing. The Pol-Tek method is used to detect low-density regions and ultrasonic testing is used to measure crack and flaw size.

49.4.2 Concrete Members

Common concrete member defects include cracking, scaling, delamination, spalling, efflorescence, popouts, wear or abrasion, collision damage, scour, and overload. Brief descriptions of common damages are given in this section.

Cracking in concrete is usually large enough to be seen with the naked eye, but it is recommended to use a crack gauge to measure and classify the cracks. Cracks are classified as hairline, medium, or wide cracks. Hairline cracks cannot be measured by simple means such as pocket ruler, but simple means can be used for the medium and wide cracks. Hairline cracks are usually insignificant to the capacity of the structure, but it is advisable to document them. Medium and wide cracks are significant to the structural capacity and should be recorded and monitored in the inspection reports. Cracks can also be grouped into two types: structural cracks and nonstructural cracks. Structural cracks are caused by the dead- and live-load stresses. Structural cracks need immediate attention, since they affect the safety of the bridge. Nonstructural cracks are usually caused by thermal expansion and shrinkage of the concrete. These cracks are insignificant to the capacity, but these cracks may lead to serious maintenance problems. For example, thermal cracks in a deck surface may allow water to enter the deck concrete and corrode the reinforcing steel.

Scaling is the gradual and continuing loss of surface mortar and aggregate over an area. Scaling is classified into four categories: light, medium, heavy, and severe.

Delamination occurs when layers of concrete separate at or near the level of the top or outermost layer of reinforcing steel. The major cause of delamination is the expansion or the corrosion of reinforcing steel due to the intrusion of chlorides or salts. Delaminated areas give off a hollow sound when tapped with a hammer. When a delaminated area completely separates from the member, a roughly circular or oval depression, which is termed as spall, will be formed in the concrete.

The inspection of concrete should include both visual and physical examination. Two of the primary deteriorations noted by visual inspections are cracks and rust stains. An inspector should recognize the fact that not all cracks are of equal importance. For example, a crack in a prestressed concrete girder beam, which allows water to enter the beam, is much more serious than a vertical crack in the backwall. A rust stain on the concrete members is one of the signs of corroding reinforcing steel in the concrete member. Corroded reinforcing steel produces loss of strength within concrete due to reduced reinforced steel section, and loss of bond between concrete and reinforcing steel. The length, direction, location, and extent of the cracks and rust stains should be measured and reported in the inspection notes.

Some common types of physical examination are hammer sounding and chain drag. Hammer sounding is used to detect areas of unsound concrete and usually used to detect delaminations. Tapping the surfaces of a concrete member with a hammer produces a resonant sound that can be used to indicate concrete integrity. Areas of delamination can be determined by listening for hollow sounds. The hammer sounding method is impractical for the evaluation of larger surface areas. For larger surface areas, chain drag can be used to evaluate the integrity of the concrete with reasonable accuracy. Chain drag surveys of decks are not totally accurate, but they are quick and inexpensive.

There are other advanced techniques — destructive and nondestructive — available for concrete inspection. Core sampling is one of the destructive techniques of concrete inspection. Some of the nondestructive inspection techniques are

- Delamination detection machinery to identify the delaminated deck surface;
- Copper sulfate electrode, nuclear methods to determine corrosion activity;
- Ground-penetrating radar, infrared thermography to detect deck deterioration;
- Pachometer to determine the position of reinforcement; and
- Rebound and penetration method to predict concrete strength.

49.4.3　Steel and Iron Members

Common steel and iron member defects include corrosion, cracks, collision damage, and overstress. Cracks usually initiate at the connection detail, at the termination end of a weld, or at a corroded location of a member and then propagate across the section until the member fractures. Since all of the cracks may lead to failure, bridge inspectors need to look at each and every one of these potential crack locations carefully. Dirt and debris usually form on the steel surface and shield the defects on the steel surface from the naked eye. Thus, the inspector should remove all dirt and debris from the metal surface, especially from the surface of fracture-critical details, during the inspection of defects.

The most recognizable type of steel deterioration is corrosion. The cause, location, and extent of the corrosion need to be recorded. This information can be used for rating analysis of the member and for taking preventive measures to minimize further deterioration. Section loss due to corrosion can be reported as a percentage of the original cross section of a component. The corrosion section loss is calculated by multiplying the width of the member and the depth of the defect. The depth of the defect can be measured using a straightedge ruler or caliper.

One of the important types of damage in steel members is fatigue cracking. Fatigue cracks develop in bridge structures due to repeated loadings. Since this type of cracking can lead to sudden and catastrophic failure, the bridge inspector should identify fatigue-prone details and should perform a thorough inspection of these details. For painted structures, breaks in the paint accompanied by rust staining indicate the possible existence of a fatigue crack. If a crack is suspected, the area should be cleaned and given a close-up visual inspection. Additionally, further testing such as dye penetrant can be done to identify the crack and to determine its extent. If fatigue cracks are discovered, inspection of all similar fatigue details is recommended.

Other types of damage may occur due to overstress, vehicular collision, and fire. Symptoms of damage due to overstress are inelastic elongation (yielding) or decrease in cross section (necking) in tension members, and buckling in compression members. The causes of the overstress should be investigated. The overstress of a member could be the result of several factors such as loss of composite action, loss of bracing, loss of proper load-carrying path, and failure or settlement of bearing details.

Damage due to vehicular collision includes section loss, cracking, and shape distortion. These types of damage should be carefully documented and repair work process should be initiated. Until the repair work is completed, restriction of vehicular traffic based on the rating analysis results is recommended.

Similar to timber and concrete members, there are advanced destructive and nondestructive techniques available for steel inspection. Some of the nondestructive techniques used in steel bridges are

- Acoustic emissions testing to identify growing cracks;
- Computer tomography to render the interior defects;
- Dye penetrant to define the size of the surface flaws; and
- Ultrasonic testing to detect cracks in flat and smooth members.

49.4.4　Fracture-Critical Members

Fracture-critical members (FCM) or member components are defined as tension components of members whose failure would be expected to result in collapse of a portion of a bridge or an entire bridge [7,8]. A redundant steel bridge that has multiple load-carrying mechanisms is seldom categorized as a fracture-critical bridge.

Since the failure to locate defects on FCMs in a timely manner may lead to catastrophic failure of a bridge, it is important to ensure that FCMs are inspected thoroughly. Hands-on involvement of the team leader is necessary to maintain the proper level of inspection and to make independent checks of condition appraisals. In addition, adequate time to conduct a thorough inspection should be allocated by the team leader. Serious problems in FCMs must be addressed immediately by restricting traffic on the bridge and repairing the defects under an emergency contract. Less serious problems requiring repairs or retrofit should be placed on the programmed repair work so that they will be incorporated into the maintenance schedule.

Bridge inspectors need to identify the FCMs using the guidelines provided in the Inspection of Fracture Critical Bridge Members [7,8]. There are several vulnerable fracture-critical locations in a bridge. Some of the obvious locations are field welds, nonuniform welds, welds with unusual profile, and intermittent welds along the girder. Other possible locations are insert plate termination points, floor beam to girder connections, diaphragm connection plates, web stiffeners, areas that are vulnerable to corrosion, intersecting weld location, sudden change in cross section, and coped sections. Detailed descriptions of each of these fracture-critical details are listed in the Inspection of Fracture Critical Bridge Members [7,8]. Once the FCM is identified in a bridge structure, information such as location, member components, likelihood to have fatigue- or corrosion related damage, needs to be gathered. The information gathered on the member should become a permanent record and the condition of the member should be updated on every subsequent inspection.

FCMs can be inspected by both visual and physical examination. During the visual inspection, the inspector performs a close-up, hands-on inspection using standard, readily available tools. During the physical examination, the inspector uses the most-sophisticated nondestructive testing methods. Some of the FCMs may have details that are susceptible to fatigue cracking and others may be in poor condition due to corrosion. The inspection procedures of corrosion- and fatigue-prone members are described in Section 49.4.3.

49.4.5 Scour-Critical Bridges

Bridges spanning over waterways, especially rivers and streams, sometimes provide major maintenance challenges. These bridges are susceptible to scour of the riverbed. When the scoured riverbed elevation falls below the top of the footing, the bridge is referred to as scour critical.

The rivers, whether small or large, could significantly change their size over the period of the lifetime of a bridge. A riverbed could be altered in several ways and thereby jeopardize the stability of the bridges. A few of the possible types of riverbed alterations are scour, hydraulic opening, channel misalignment, and bank erosion. Scour around the bridge substructures poses potential structural stability concerns. Scour at bridges depends on the hydraulic features upstream and downstream, riverbed sediments, substructure section profile, shoreline vegetation, flow velocities, and potential debris. The estimation of the overall scour depth will be used to identify scour-prone and scour-critical bridges. Guidance for the scour evaluation process is provided in Evaluating Scour at Bridges [11].

A typical scour evaluation process falls into two phases: inventory phase and evaluation phase. The main goal of the inventory phase is to identify those bridges that are vulnerable to scour (scour-prone bridges). Evaluation during this phase is made using the available bridge records, inspection records, history of the bridge, original stream location, evidence of scour, deposition of debris, geology, and general stability of the streambed. Once the scour-prone bridges are identified, the evaluation phase needs to be performed. The scour evaluation phase requires in-depth field review to generate data for estimation of the hydraulics and scour depth. The procedure of scour estimation is outlined in Evaluating Scour at Bridges [11]. The scour depths are then compared with the existing foundation condition. When the scour depth is above the top of the footing, the bridge would require no action. However, when the scour depth is within the limits of the footing or piles, a structural stability analysis is needed. If the scour depth is below the pile tips or spread footing base, monitoring of the bridge is required. These results obtained from the scour evaluation process are entered into the bridge inventory.

49.4.6 Underwater Components

Underwater components are mostly substructure members. Since the accessibility of these members is difficult, special equipment is necessary to inspect these underwater components. Also, visibility during the underwater inspection is generally poor, and therefore a thorough inspection of the members will not be possible. Underwater inspection is classified as visual (Level 1), detailed (Level 2), and comprehensive (Level 3) to specify the level of effort of inspection. Details of these various levels of inspection are discussed in the Manual for Maintenance Inspection of Bridges [2] and Evaluating Scour at Bridges [11].

Underwater steel structure components are susceptible to corrosion, especially in the low to high water zone. Some of the defects observed in underwater timber piles are splitting, decay or rot, marine borers, decay of timber at connections, and corrosion of connectors. It is important to recognize that the timber piles may appear sound on the outside shell but be severely damaged inside. Some of the most common defects in underwater concrete piles are cracking, spalls, exposed reinforcing, sulfate attack, honeycombing, and scaling. When cracking, spalls, and exposed reinforcing are detected, structural analysis may be required to ensure the safety of the bridge.

49.4.7 Decks

The materials typically used in the bridge structures are concrete, timber, and steel. Sections 49.4.1 to 49.4.3 discuss some of the defects associated with each of these materials. In this section, the damage most likely to occur in bridge decks is discussed.

Common defects in steel decks are cracked welds, broken fasteners, corrosion, and broken connections. In a corrugated steel flooring system, section loss due to corrosion may affect the load-carrying capacity of the deck and thus the actual amount of remaining materials needs to be evaluated and documented.

Common defects in timber decks are crushing of the timber deck at the supporting floor system, flexure damages such as splitting, sagging, and cracks in tension areas, and decay of the deck due to biological organisms, especially in the areas exposed to drainage.

Common defects in concrete decks are wear, scaling, delamination, spalls, longitudinal flexure cracks, transverse flexure cracks in the negative moment regions, corrosion of the deck rebars, cracks due to reactive aggregates, and damage due to chemical contamination. The importance of a crack varies with the type of concrete deck. A large to medium crack in a noncomposite deck may not affect the load-carrying capacity of the main load-carrying member. On the other hand, several cracks in a composite deck will affect the structural capacity. Thus, an inspector must be able to identify the functions of the deck while inspecting it.

Sometimes a layer of asphalt concrete (AC) overlay will be placed to provide a smooth driving and wearing surface. Extra care is needed during the inspection, because AC overlay prevents the inspector's ability to inspect the top surface of the deck visually for damage.

49.4.8 Joint Seals

Damage to the joint seals is caused by vehicle impact, extreme temperature, and accumulation of dirt and debris. Damage from debris and vehicles such as snowplows could cause the joint seals to be torn, pulled out of anchorage, or removed altogether. Damage from extreme temperature could break the bond between the joint seal and deck and consequently result in pulling out the joint seal altogether.

The primary function of deck joints is to accommodate the expansion and contraction of the bridge superstructure. These deck joints also provide a smooth transition from the approach roadway to the bridge deck. Deck joints are placed at hinges between two decks of adjacent structures, and between the deck sections and abutment backwall. The joint seals used in the bridge industry can be divided into two groups: open joints and closed joints. Open joints allow water and debris to pass through the joints. Dripping water through open joints usually damages the bearing details. Closed joints do not allow water and debris to pass through them. A few of the closed joints are compression seal, poured joint seal, sliding plate joint, plank seal, sheet seal, and strip seal.

In the case of closed joints, damage to the joint seal material will cause the water to drip on the bearing seats and consequently damage the bearing. Accumulation of dirt and debris may prevent normal thermal expansion and contraction, which may in turn cause cracking in the deck, backwall, or both. Cracking in the deck may affect the ride quality of the bridge, may produce larger impact load from vehicles, and may reduce the live-load-carrying capacity of the bridge.

49.4.9 Bearings

Bearings used in bridge structures could be categorized into two groups: metal and elastomeric. Metal bearings sometimes become inoperable (sometimes referred as "frozen") due to corrosion, mechanical bindings, buildup of debris, or other interference. Frozen bearings may result in bending, buckling, and improper alignment of members. Other types of damage are missing fasteners, cracked welds, corrosion on the sliding surface, sole plate rests only on a portion of the masonry plate, and binding of lateral shear keys.

Damage in elastomeric bearing pads is excessive bulging, splitting or tearing, shearing, and failure of bond between sole and masonry plate. Excessive bulging indicates that the bearing might be too tall. When the pad is under excessive strain for a long period, the pad will experience shearing failure.

Inspectors need to assess the exact condition of the bearing details and to recommend corrective measures that allow the bearing details to function properly. Since the damage to the bearings will affect the other structural members as time passes, repair of bearing damage needs to be considered as a preventive measure.

49.5 Fundamentals of Bridge Rating

49.5.1 Introduction

Once a bridge is constructed, it becomes the property of the owner or agency. The evaluation or rating of existing bridges is a continuous activity of the agency to ensure the safety of the public. The evaluation provides necessary information to repair, rehabilitate, post, close, or replace the existing bridge.

In the United States, since highway bridges are designed for the AASHTO design vehicles, most U.S. engineers tend to believe that the bridge will have adequate capacity to handle the actual present traffic. This belief is generally true if the bridge was constructed and maintained as shown in the design plan. However, changes in a few details during the construction phase, failure to attain the recommended concrete strength, unexpected settlements of the foundation after construction, and unforeseen damage to a member could influence the capacity of the bridge. In addition, old bridges might have been designed for a lighter vehicle than is used at present, or a different design code. Also, the live-load-carrying capacity of the bridge structure may have altered as a result of deterioration, damage to its members, aging, added dead loads, settlement of bents, or modification to the structural member.

Sometimes, an industry would like to transport their heavy machinery from one location to another location. These vehicles would weigh much more than the design vehicles and thus the bridge owner may need to determine the current live-load-carrying capacity of the bridge. In the following sections, establishing the live load-carrying capacity and the bridge rating will be discussed.

49.5.2 Rating Principles

In general, the resistance of a structural member (R) should be greater than the demand (Q) as follows:

$$R \geq Q_d + Q_l + \sum_i Q_i \qquad (49.1)$$

where Q_d is the effect of dead load, Q_l is the effect of live load, and Q_i is the effect of load i.

Eq. (49.1) applies to design as well as evaluation. In the bridge evaluation process, maximum allowable live load needs to be determined. After rearranging the above equation, the maximum allowable live load will become

$$Q_l \leq R - \left(Q_d + \sum_i Q_i \right) \qquad (49.2)$$

Maintenance engineers always question whether a fully loaded vehicle (rating vehicle) can be allowed on the bridge and, if not, what portion of the rating vehicle could be allowed on a bridge. The portion of the rating vehicle will be given by the ratio between the available capacity for live-load effect and the effect of the rating vehicle. This ratio is called the rating factor (RF).

$$\text{RF} = \frac{\text{Available capacity for the live-load effect}}{\text{Rating vehicle load demand}} = \frac{R - \left(Q_d + \sum_i Q_i \right)}{Q_l} \qquad (49.3)$$

When the rating factor equals or exceeds unity, the bridge is capable of carrying the rating vehicle. On the other hand, when the rating factor is less than unity the bridge may be overstressed while carrying the rating vehicle.

The capacity of a member is usually independent of the live-load demand. Thus, Eq. (49.3) is generally a linear expression. However, there are cases where the capacity of a member dependent on the live-load forces. For example, available moment capacity depends on the total axial load in biaxial bending members. In a biaxially loaded member, the Eq. (49.3) will be a second-order expression.

Thermal, wind, and hydraulic loads may be neglected in the evaluation process because the likelihood of occurrence of extreme values during the relatively short live-load loading is small. Thus, the effects of the dead and live loads are the only two loads considered in the evaluation process.

49.5.3 Rating Philosophies

During the structural evaluation process, the location and type of critical failure modes are first identified; Eq. (49.3) is then solved for each of these potential failures. Although the concept of evaluation is the same, the mathematical relationship of this basic equation for allowable stress design (ASD), load factor design (LFD), and Load and resistance factor design (LRFD) differs. Since the resistance and load effect can never be established with certainty, engineers use safety factors to give adequate assurance against failure. ASD includes safety factors in the form of allowable stresses of the material. LFD considers the safety factors in the form of load factors to account for the uncertainty of the loadings and resistance factors to account for the uncertainty of structural response. LRFD treats safety factors in the form of load and resistance factors that are based on the probability of the loadings and resistances.

For ASD, the rating factor expression Eq. (49.3) can be written as

$$RF = \frac{R - \left(\sum D + \sum_i L_i(1+I) \right)}{L(1+I)} \tag{49.4}$$

For LFD, the rating factor expression Eq. (49.3) can be written as

$$RF = \frac{\phi R_n - \sum \gamma_D D - \sum_{i=1}^n \gamma_{Li} L_i(1+I)}{\gamma_L L(1+I)} \tag{49.5}$$

For LFRD, the rating factor expression Eq. (49.3) can be written as

$$RF = \frac{\phi R_n - \sum \gamma_D D - \sum_{i=1}^n \gamma_{Li} L_i(1+I)}{\gamma_L L(1+I)} \tag{49.6}$$

where R is the allowable stress of the member; ϕR_n is nominal resistance; D is the effect of dead loads; L_i is the live-load effect for load i other than the rating vehicle; L the nominal live-load effect of the rating vehicle; I is the impact factor for the live-load effect; γ_D, γ_{Li}, and γ_L are dead- and live-load factors, respectively.

Researchers are now addressing the LRFD method, and thus the LRFD approach may be revised in the near future. Since the LRFD method is being developed at this time, the LRFD method is not discussed further in this chapter.

In order to use the above equations (Eqs. 49.4 to 49.6) in determining the rating factors, one needs to estimate the effects of individual live-load vehicles. The effect of individual live-load vehicles on structural member could only be obtained by analyzing the bridge using a three-dimensional analysis. Thus, obtaining the rating factor using the above expressions is very difficult and time-consuming.

To simplify the above equations, it is assumed that similar rating vehicles will occupy all the possible lanes to produce the maximum effect on the structure. This assumption allows us to use the AASHTO live-load distribution factor approach to estimate the live-load demand and eliminate the need for the three-dimensional analysis.

And the simplified rating factor equations become as follows:

$$\text{For ASD:} \quad RF = \frac{R - D}{L(1+I)} \tag{49.7}$$

$$\text{For LFD:} \quad RF = \frac{\phi R_n - \gamma_D D}{\gamma_L L(1+I)} \tag{49.8}$$

$$\text{For LRFD:} \quad RF = \frac{\phi R_n - \gamma_D D}{\gamma_L L(1+I)} \tag{49.9}$$

In the derivation of the above equations (Eqs. 49.7 to 49.9), it is assumed that the resistance of the member is independent of the loads. A few exceptions to this assumption are beam–column members and beams with high moment and shear. In a beam–column member, axial capacity or moment capacity depends on the applied moment or applied axial load on the member. Thus, as the live-load forces in the member increase, the capacity of the member would decrease. In other words, the numerator of the above equations (available live-load capacity) will drop as the live load increases. Thus, the rating factor will no longer be a constant value, and will be a function of live load.

49.5.4 Level of Ratings

There are two levels of rating for bridges: inventory and operating. The rating that reflects the absolute maximum permissible load that can be safely carried by the bridge is called an operating rating. The load that can be safely carried by a bridge for indefinite period is called an inventory rating.

The life of a bridge depends on the fatigue life or serviceability limits of bridge materials. Higher frequent loading and unloading may affect the fatigue life or serviceability of a bridge component and thereby the life of the bridge. Thus, in order to maintain a bridge for an indefinite period, live-load-carrying capacity available for frequently passing vehicles needs to be estimated at service. This process is referred to as inventory rating.

Less frequent vehicles may not affect the fatigue life or serviceability of a bridge, and thus live-load-carrying capacity available for less frequent vehicles need not be estimated using serviceability criteria. In addition, since less frequent vehicles do not damage the bridge structure, bridge structures could be allowed to carry higher loads. This process is referred to as operating rating.

49.5.5 Structural Failure Modes

In the ASD approach, when a portion of a structural member is stressed beyond the allowable stress, the structure is considered failed. In addition, since any portion of the structural member material never reaches its yield, the deflections or vibrations will always be satisfied. Thus, the serviceability of a bridge is assured when the allowable stress method is used to check a bridge member. In other words, in the ASD approach, serviceability and strength criteria are satisfied automatically. The inventory and operating allowable stresses for various types of failure modes are given in the AASHTO Manual for Condition Evaluation of Bridges 1994 [12] (Rating Manual).

In the LFD approach, failure could occur at two different limit states: serviceability and strength. When the load on a member reaches the ultimate capacity of the member, the structure is considered failed at its ultimate strength limit state. When the structure reaches its maximum allowable serviceability limits, the structure is considered failed at its serviceability limit state. In LFD approach, satisfying one of the limit states will not automatically guarantee the satisfaction of the other limit state. Thus, both serviceability and strength criteria need to be checked in the LFD method. However, when the operating rating is estimated, the serviceability limits need not be checked.

49.6 Superstructure Rating Examples

In this section, several problems are illustrated to show the bridge rating procedures. In the following examples, AASHTO *Standard Specification for Highway Bridges,* 16th ed.[13] is referred to as Design Specifications and AASHTO *Manual for Condition Evaluation of Bridges* 1994 [12] is referred to as Rating Manual. All the notations used in these examples are defined in either the Design Specifications or the Rating Manual.

49.6.1 Simply Supported Timber Bridge

Given
Typical cross section of a 16-ft (4.88-m) long simple-span timber bridge is shown in Figure 49.1. 13.4 × 16 in. (101.6 × 406.4 mm) timber stringers are placed at 18 in. (457 mm) spacing. 4 × 12 in.

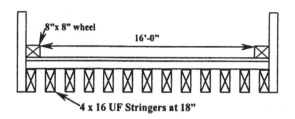

FIGURE 49.1 Typical cross section detail of simply supported timber bridge example.

(101.6 × 305 mm) timber planks are used as decking. 8 × 8 in. (203 × 203 mm) timber is used as wheel guard. Barrier rails (10 lb/ft or 0.1 N/mm) are placed at either side of the bridge. The traffic lane width of the bridge is 16 ft (4.88 m). Assume that the allowable stresses at operating level are as follows: F_b for stringer as 1600 psi (11 MPa) and F_v of stringer level as 115 psi (0.79 MPa).

Requirement
Determine the critical rating factors for interior stinger for HS20 vehicle using the ASD approach.

Solution
For this simply supported bridge, the critical locations for ratings will be the locations where shear and moments are higher.

According to Design Specifications Section 13.6.5.2, shear needs to be checked at a distance (s) $3d$ or $0.25L$ from the bearing location for vehicle live loads; thus,

$$s = 3d = 3 \times 16 \text{ in.}/12 = 4.0 \text{ ft or}$$

$$= 0.25L = 0.25 \times 16 \text{ ft} = 4.0 \text{ ft. Thus, } s \text{ is taken as } 4.0 \text{ ft } (1.22 \text{ m}).$$

Maximum dead- and live-load shear will occur at this point and thus in the following calculations, shear is estimated at this critical location.

1. Dead Load Calculations

Self-weight of the stringer $= 0.05 \times 4 \times 16 \times \dfrac{1}{144} = 0.022 \text{ kips/ft}$

Weight of deck (using tributary area) $= 1.5 \times 4 \times 12 \times \dfrac{1}{144} \times 0.05 = 0.025 \text{ kips/ft}$

Weight of 1.5 in. AC on the deck $= 1.5 \times \dfrac{1.5}{12} \times 0.144 = 0.027 \text{ kips/ft}$

Barrier rail and curb $= \left(10 + 50 \times \dfrac{8 \times 8}{144}\right) \times \dfrac{2}{13 \times 1000} = 0.004 \text{ kips/ft}$

Total uniform dead load on the stringer $= 0.022 + 0.025 + 0.027 + 0.004$

$= \underline{0.078 \text{ kips/ft}}$

Maximum dead load moment at midspan $= \dfrac{wl^2}{8} = \dfrac{0.078 \times 16^2}{8} = 2.5 \text{ kip-ft } (3390 \text{ N-m})$

Maximum dead load shear at this critical point $\quad = w \times (0.5L - s)$

$$= 0.078 \times (0.5 \times 16 - 4)$$

$$= 0.31 \text{ kips } (1.38 \text{ kN})$$

2. Live-Load Calculations

The travel width is less than 18 ft. Thus, according to Section 6.7.2.2 of the Rating Manual, this bridge needs to be rated for one traffic lane. From Designs Specifications Table 3.23.1,

$$\text{Number of wheels on the stringer} = \frac{S}{4} = \frac{1.5}{4} = 0.38$$

Maximum moment due to HS20 loading (Appendix A3, Rating Manual)

$= (64)(0.38)$
$= 24.32$ kip-ft (33,000 N-m)

In order to estimate the live-load shear, we need to estimate the shear due to undistributed and distributed HS20 loadings. (See Design Specifications 13.6.5.2 for definition of V_{LU} and V_{LD})

Shear due to undistributed HS20 loadings $\quad = V_{LU} \qquad\qquad = 16 \times 12/16 = 12 \text{ kips } (53.4 \text{ kN})$

Shear due to distributed HS20 loading $\;= V_{LD} \qquad\qquad = 16 \times 12/16 \times (0.38)$

$$= 4.56 \text{ kips } (20.3 \text{ kN})$$

Thus, shear due to HS20 live load $\qquad = 0.5(0.6V_{LU} + V_{LD}) \;\; = 5.88 \text{ kips } (26.1 \text{ kN})$

3. Capacity Calculations

a. *Moment capacity at midspan:*
   Moment capacity of the timber stringer at *Operating level* =

$$F_b S_x = 1600 \times \frac{1}{6} \times 4 \times 16^2 \times \frac{1}{12,000} = 22.8 \text{ kip-ft } (30,900 \text{ N-m})$$

According to Section 6.6.2.7 of Rating Manual, the operating level stress of a timber stringer can be taken as 1.33 times the inventory level stress.
   Thus, moment capacity of the timber stringer at *Inventory level*

$= 22.8/1.33$
$= 17.1$ kip-ft (23,200 N-m)

b. *Shear capacity at support:*
   Shear capacity of the timber section (controlled by horizontal shear) $= (\tfrac{2}{3})bdf_v$:

$$V_c \text{ at } operating \text{ level} = (2/3) \times 4 \times 16 \times 115 \text{ psi} \times 1/1000 = 4.91 \text{ kips } (21.8 \text{ kN})$$

$$V_c \text{ at } inventory \text{ level} = 14.91/1.33 = 3.69 \text{ kips } (16.4 \text{ kN})$$

4. Rating Calculations

$$\text{Rating factor based on ASD method} = RF = \frac{R - D}{L(1 + I)}$$

By substituting appropriate values, the rating factor can be determined.

a. *Based on moment at midspan:*

$$\text{Inventory rating factor } RF_{INV\text{-}MOM} = \frac{17.1 - 2.5}{24.32} = 0.600$$

$$\text{Operating rating factor } RF_{OPR\text{-}MOM} = \frac{22.8 - 2.5}{24.32} = 0.835$$

b. *Based on shear at the support:*

$$\text{Inventory rating factor } RF_{INV\text{-}SHE} = \frac{3.69 - 0.31}{5.88} = 0.575$$

$$\text{Operating rating factor } RF_{OPR\text{-}SHE} = \frac{4.91 - 0.31}{5.88} = 0.782$$

5. Summary
It is found that the critical rating factor is controlled by shear in the stringers. The critical inventory and operating rating of the bridge will be 0.575 and 0.782, respectively.

49.6.2 Simply Supported T-Beam Concrete Bridge

Given
A bridge, which was built in 1929, consists of three simple-span reinforced concrete T-beams on concrete bents and abutments. The span lengths are 16 ft (4.88 m), 50 ft (15.24 m), and 10 ft (3.05 m). Typical cross section and girder details are shown in Figure 49.2. General notes given in the plan indicate that $f_c = 1000$ psi (6.9 MPa) and $f_s = 18,000$ psi (124.1 MPa). Assume the weight of each barrier rail as 250 lb/ft (3.6 N/mm).

Requirement
Determine the critical rating factor of the interior girder of the second span (50 ft. or 15.24 m) for HS20 vehicles assuming no deterioration of materials occurred.

Solution
1. Dead-Load Calculations

Self-weight of the girder = (3.5) (1.333) (0.15)	= 0.700 kips/ft
(4 × 4 in.) Fillets between girder and slab = 2(1/2) (4/12) (4/12) (0.15)	= 0.017 kips/ft
Slab weight (based on tributary area) = (6.667)(8/12) (0.15)	= 0.667 kips/ft
Contribution from barrier rail (equally distributed among girders) = 2 (0.25/3)	= 0.167 kips/ft

Thus, total uniform load on the interior girder = 1.551 kips/ft (22.6 N/mm)

Dead-load moment at midspan = 484.6 kips/ft (0.657 MN/m)

Dead load shear at a distance d from support = 32.31 kips (143.7 kN)

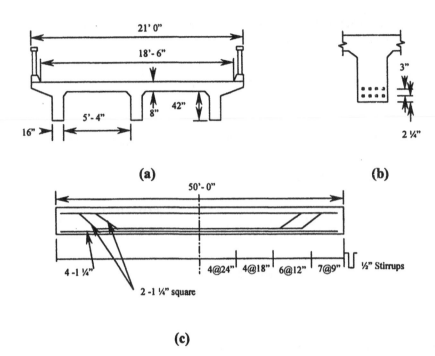

FIGURE 49.2 Details of simply supported T-beam concrete bridge example. (a) Typical cross section; (b) reinforcement locations; (c) T-beam girder details.

2. Live-Load Calculations

The traffic lane width of this bridge is 18.5 ft. According to Design Specifications, any bridge with a minimum traffic lane width of 18 ft needs to carry two lanes. Hence, the number of live-load wheels will be based on two traffic lanes. From Table 3.23.1A of Design Specifications for two traffic lanes for T-beams is given by $S/6.0$

Number of live-load wheel line	= 6.667/6.0	= <u>1.111</u>
AASHTO standard impact factor for moment	= 50/(125 + 50) =	<u>0.286</u>
AASHTO standard impact factor for shear at support	= 50/(125 + 50) =	<u>0.286</u>

The live-load moments and shear tables listed in the Rating Manual are used to determine the live-load demand.

Maximum HS20 moment for 50 ft span without impact/wheel line	= 298.0 kips-ft
Thus, HS20 moment with impact at midspan	= (1.286) (1.111) (298.0)
	= <u>425.7 kips-ft (0.58 MN-m)</u>
Maximum HS20 shear at a distance d from the support/wheel line	= 28.32 kips
Thus, maximum HS20 shear = (1.286) (1.111) (28.32)	= <u>40.46 kips (180.0 kN)</u>

3. Capacity Calculations

Strengths of concrete and rebars are first determined (see Rating Manual Section 6.6.2.3):

$$f'_c = \frac{f_c}{0.4} \quad \text{thus} \quad f'_c = 2500 \text{ psi}$$

and $f_s = 18,000$ psi and thus $f_y = 33,000$ psi.

a. *Moment capacity at midspan:*

Total area of the steel (note these bars are 1¼ square bars) $= (8)(1.25)(1.25) = 12.5$ in.^{2}

Centroid of the rebars from top deck $= 42 + 8 - 3.75 = 46.250$ in.

Effective width of the deck b_{eff} = minimum of $12t_s + b_w$ $= 112$ in.

Span/4 $= 150$ in.

Spacing $= \underline{80 \text{ in.}}$ (Controls)

Uniform stress block depth $= a = \dfrac{A_s f_y}{0.85 f'_c b_{eff}} = 2.426$ in. $< t_s = 8$ in.

$$M_u = \phi A_s f_y \left(d - \frac{a}{2} \right) = 0.9 \times 12.5 \times 33 \left(46.25 - \frac{2.426}{2} \right) \times \left(\frac{1}{12} \right) = 1393.3 \text{ kips-ft (1.88 MN-m)}$$

b. *Shear capacity at support:*

According to AASHTO specification, shear at a distance d (50 in.) from the support needs to be designed. Thus, the girder is rated at a distance d from the support. From the girder details, it is estimated that ½ in. ϕ stirrups were placed at a spacing of 12 in. and two 1¼ square bars were bent up. The effects of these bent-up bars are ignored in the shear capacity calculations.

Shear capacity due to concrete section:

$$V_c = 2\sqrt{f'_c}\, b_w d = 2\sqrt{2500} \times 16 \times 46.25 \times \left(\frac{1}{1000} \right) = 74 \text{ kips (329 kN)}$$

Shear capacity due to shear reinforcement:

$$V_s = 2 A_v \frac{F_y d_s}{S} = 2 \times 0.20 \times \frac{33 \times 46.25}{12} = 50.88 \text{ kips (226 kN)}$$

Total shear capacity:

$$V_u = \phi (V_s + V_c) = 0.85 (74.0 + 50.88) = 106.2 \text{ kips (472 kN)}$$

4. Rating Calculations

$$\text{Rating factor} = \frac{\phi R_n - \gamma_D D}{\gamma_L \beta_L L(1+I)}$$

TABLE 49.1 Rating Calculations of Simply Supported T-Beam Concrete Bridge Example

Location	Description	Inventory Rating	Operating Rating
Midspan	Moment	$\dfrac{1393.3 - 1.3 \times 484.6}{1.3 \times 1.67 \times 425.7} = 0.825$	$\dfrac{1393.3 - 1.3 \times 484.6}{1.3 \times 425.7} = 1.38$
At support	Shear	$\dfrac{106.2 - 1.3 \times 32.31}{1.3 \times 1.67 \times 40.46} = 0.731$	$\dfrac{106.2 - 1.3 \times 32.31}{1.3 \times 40.46} = 1.22$

According to Rating Manual, γ_D is 1.3, γ_L is 1.3, and β_L is 1.67 and 1.0 for inventory and operating factors, respectively. By substituting these values and appropriate load effect values, the moment and shear rating could be estimated. The calculations and results are given in Table 49.1.

5. **Summary**

Critical rating of the interior girder will then be 0.731 at inventory level and 1.22 at operating rating level for HS20 vehicle.

49.6.3 Two-Span Continuous Steel Girder Bridge

Given

Typical section of a two-span continuous steel girder bridge, which was built in 1967, is shown in Figure 49.3a. Steel girder profile is given in Figure 49.3b. The general plan states that $f_s = 20,000$ psi (137.9 MPa) and $f_c = 1200$ psi (8.28 MPa). Assume that (a) each barrier rail weighs 250 lb/ft (3.6 N/mm); (b) girders were not temporarily supported during the concrete pour; (c) girder is composite for live loads; (d) girder is braced every 15 ft and the weight of bracing per girder is 330 lb.

Requirement

Determine the rating factors of interior girders using ASD method.

Solution

1. **Dead Load Calculations**

Deck weight (tributary area approach) = (6.625/12) (6.625) (0.15)	= 0.549 kips/ft
Average uniform self-weight for the analysis = 1431 kips/90 ft	= 0.159 kips/ft
Average diaphragm load (uniformly distributed) = (0.33) (4/90)	= 0.015 kips/ft
Thus, total uniform dead load on the girder	= 0.723 kips/ft (10.5 N/mm)
Barrier rail load (equally distributed among all girders) = (2)(250)/14 =	0.0358 kips/ft
Thus, total additional dead load on the girder	= 0.0358 kips/ft (0.56 N/mm)

2. **Live Load Calculations**

Number of wheels per girder (for two or more lanes) = $S/5.5$ = 6.625/5.5 = 1.206

Analysis Results: Analysis is done using two-dimensional program and the moments and shears at critical locations are listed in the Table 49.2.

Section properties at 0.4th and 1.0th points are estimated and the results are given in Table 49.3.

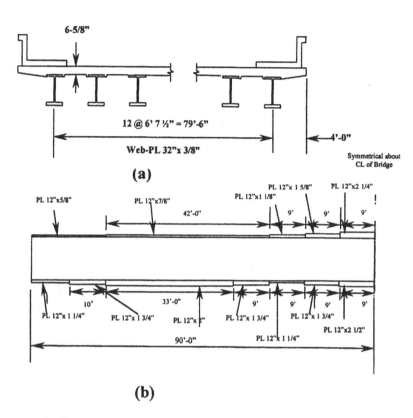

FIGURE 49.3 Details of two-span continuous steel girder bridge example. (a) Typical section; (b) girder elevation.

TABLE 49.2 Load Demands at 0.4th and 1.0th Point for Steel Girder Bridge Example

Description	0.4th Point (36.1 ft)	At Support (90 ft)
Dead load moments in kip-ft	410.0	−732.0
Dead load shear in kips	−1.7	−40.7
Additional dead load moment in kip-ft	22.0	−39.0
Additional dead load shear in kips	−0.1	−2.2
HS20 maximum positive moment in kip-ft	807.0	0.0
HS20 maximum negative moment in kip-ft	−177.0	−714.0
HS20 max. positive shear force in kips	21.8	0.0
HS20 max. negative shear force in kips	−19.3	−49.9

TABLE 49.3 Section Properties of Girder Sections for Steel Girder Bridge Example

	I_{gg} (in.^{4})	Y_b (in.)	Y_t (in.)	S_{xb} (in.^{3})	S_{xt} (in.^{3})
Section at 0.4th Point					
For dead loads	9,613.9	12.94	21.94	743.11	438.24
For additonal dead loads	17,406.5	19.69	15.19	884.18	1,146.10
For live loads	24,782.7	26.02	8.86	952.59	2,797.50
Section at 1.0th Point					
For dead loads	17,852.1	17.70	19.05	1,008.50	937.20
For additional dead loads	17,852.1	17.70	19.05	1,008.50	937.20
For live loads	17,852.1	17.70	19.05	1,008.50	937.20

3. Allowable Stress Calculations

Strengths of concrete and rebars are first determined (see Rating Manual Section 6.6.2.3):

$$f_c' = \frac{f_c}{0.4} \quad \text{and, thus} \quad f_c' = 3000 \text{ psi}$$

and $f_s = 20,000$ psi and thus $F_y = 36,000$ psi.

a. *Compression and tensile stresses at 0.4th point:*

Note that the section is fully braced at this location.

  i. Allowable compressive stress at inventory level = 0.55 F_y= 20 ksi (137.9 MPa)
  ii. Allowable compressive stress at operating level = 0.75 F_y= 27 ksi (186.2 MPa)
  iii. Allowable tensile stress at inventory level = 0.55 F_y= 20 ksi (137.9 MPa)

b. *Compression and tensile stresses at 1.0th point:*

  i. allowable tensile stress at inventory level = 0.55 F_y= 20 ksi (137.9 MPa)
  ii. allowable compressive stress at inventory level: Girder is braced 15 ft away from the support and thus $L_b = 15 \times 12 = 180$ in. It can be shown that $S_{xc} = 1008.3$ in³; $d = 36.75$ in.; $J = 108.63$ in.⁴; $I_{yc} = 360$ in.⁴

Then allowable stress at inventory level (Table 6.6.2.1-1 of Rating Manual):

$$F_b = \frac{91 \times 10^6 \, C_b}{1.82 \times S_{xc}} \left(\frac{I_{yc}}{L_b}\right) \sqrt{0.772\left(\frac{J}{I_{yc}}\right) + 9.87\left(\frac{d}{L_b}\right)^2}$$

$$= \frac{91 \times 10^6 \,(1.00)}{1.82 \times 1008.3}\left(\frac{360}{180}\right)\sqrt{0.772\left(\frac{108.63}{360}\right)+9.87\left(\frac{36.75}{180}\right)^2}\left(\frac{1}{1000}\right)$$

$$= 79.5 > 0.55F_y = 20 \text{ ksi (Note that } C_b \text{ is conservatively assumed as 1.0.)}$$

Thus, $F_b = 20$ ksi (137.9 MPa)

  iii. Allowable compressive stress at operating level: The allowable stress at operating level is given:

$$F_b = \frac{91 \times 10^6 \, C_b}{1.34 \times S_{xc}} \left(\frac{I_{yc}}{L_b}\right) \sqrt{0.772\left(\frac{J}{I_{yc}}\right) + 9.87\left(\frac{d}{L_b}\right)^2}$$

(Table 6.6.2.1-2, Rating Manual)

$$= 108.0 > 0.75 \, F_y = 27 \text{ ksi}$$

Thus, $F_b = 27$ ksi (186.2 MPa)

c. *Allowable inventory shear stresses at 0.4th and 1.0th point:*

$$D/t_w = 32/0.375 = 85.33$$

Girder is unstiffened and thus $k = 5$;

$$\frac{6000\sqrt{k}}{\sqrt{F_y}} = 70.71 < D/t_w < \frac{7500\sqrt{k}}{\sqrt{F_y}} = 88.3$$

TABLE 49.4 Estimated Stress Demands for Steel Girder Bridge Example

Load Description	At 0.4th Point	At 1.0th Point	Fiber Location
DL moment	−6.62	8.71	At bottom fiber
ADL moment	−0.30	0.463	At bottom fiber
LL + *I* moment	−10.16	8.49	At bottom fiber
DL moment	11.23	−9.37	At top fiber
ADL moment	0.23	−0.49	At top fiber
LL + *I* moment	3.46	−9.14	At top fiber
DL shear	0.129	2.95	Shear stress
ADL shear	0.007	0.15	Shear stress
LL + *I* shear	1.667	3.62	Shear stress

Thus,

$$C = \frac{6000\sqrt{k}}{\left(\dfrac{D}{t_w}\right)\sqrt{F_y}} = 0.828$$

$$F_v = \frac{F_y}{3}\left(C + \frac{0.87(1-C)}{\sqrt{1+\left(\dfrac{d_o}{D}\right)^2}}\right) = 11.76 \text{ ksi (81.1 MPa)}$$

d. *Allowable operating shear stresses at 0.4th and 1.0th point:*

$$F_v = 0.45F_y\left(C + \frac{0.87(1-C)}{\sqrt{1+\left(\dfrac{d_o}{D}\right)^2}}\right) = 15.88 \text{ ksi (109.5 MPa)}$$

4. **Load Stress Calculations**

   Bending stress calculations are made using appropriate section modulus and moments. Results are reported in Table 49.4. The sign convention used in Table 49.4 is as follows: compressive stress is positive and tensile stress is negative. Also, estimated shear stresses are given in Table 49.4.

5. **Rating Calculations**

   The rating factor in ASD approach is given by

$$\frac{R - D}{L(1 + I)}$$

   and the rating calculations are made and given in Table 49.5.

6. **Summary**

   The critical rating factor of the girder is controlled by tensile stress on the top fiber at the 1.0th point. The critical inventory and operating rating factors are 1.11 and 1.87, respectively.

TABLE 49.5 Rating Calculations Using ASD Method for Steel Girder Bridge Example

Location	Description	Inventory Rating		Operating Rating	
0.4th point	Shear	$\dfrac{11.76-(0.129+0.007)}{1.667}$	$=6.97$	$\dfrac{15.88-(0.129+0.007)}{1.667}$	$=9.44$
	Stress at top fiber	$\dfrac{20-(11.23+0.23)}{3.46}$	$=2.46$	$\dfrac{27-(11.23+0.23)}{3.46}$	$=4.49$
	Stress at bottom fiber	$\dfrac{20-(6.62+0.30)}{10.16}$	$=1.28$	$\dfrac{27-(6.62+0.30)}{10.16}$	$=1.97$
1.0th point	Shear	$\dfrac{11.76-(2.95+0.15)}{3.62}$	$=2.39$	$\dfrac{15.88-(2.95+0.15)}{3.62}$	$=3.53$
	Stress at top fiber	$\dfrac{20-(9.37+0.50)}{9.14}$	$=1.11$	$\dfrac{27-(9.37+0.50)}{9.14}$	$=1.87$
	Stress at bottom fiber	$\dfrac{20-(8.71+0.463)}{8.49}$	$=1.28$	$\dfrac{27-(8.71+0.463)}{8.49}$	$=2.10$

49.6.4 Two-Span Continuous Prestressed, Precast Concrete Box Beam Bridge

Given

Typical section and elevation of three continuous-span precast, prestressed box-girder bridge is shown in Figure 49.4. The span length of each span is 120 ft (36.5 m), 133 ft (40.6 m), and 121 ft (36.9 m). Total width of the bridge is 82 ft (25 m) and a number of precast, prestressed box girders are placed at a spacing of 10 ft (3.1 m). The cross section of the box beam and the tendon profile of the girder are shown in Figure 49.4. Each barrier rail weighs 1268 lb/ft (18.5 N/mm). Information gathered from the plans is (a) f_c' of the girder and slab is 5500 and 3500 psi, respectively; (b) working force (total force remaining after losses including creep) = 2020 kips; (c) x at midspan = 9 in. Assume that (1) the bridge was made continuous for live loading; (2) no temporary supports were used during the erection of the precast box beams; (3) properties of the precast box are area = 1375 in²; moment of inertia = 30.84 ft⁴; Y_t = 28.58 in.; Y_b = 34.4 in.; (4) F_y of reinforcing steel is 40 ksi.

Requirement

Rate the interior girder of Span 2 for HS20 vehicle.

Solution

1. Dead Load Calculations

Self-weight of the box beam = (1375/144) (0.15)   = 1.43 kips/ft

Weight of Slab (tributary area approach) = (6.75/12) (10) (0.15)   = 0.85 kips/ft

Total dead weight on the box beam   = 2.28 kips/ft (33.2 N/mm)

Contribution of barrier rail on box beam   = 2(1.268/8) = 0.318 kips/ft

Thus, total additional dead load on the box beam   = 0.318 kips/ft (4.6 N/mm)

Girder is simply supported for dead loads; thus maximum dead load moment = (2.28) (133²/8)

     = 4926 kips/ft (6.68 MN/m)

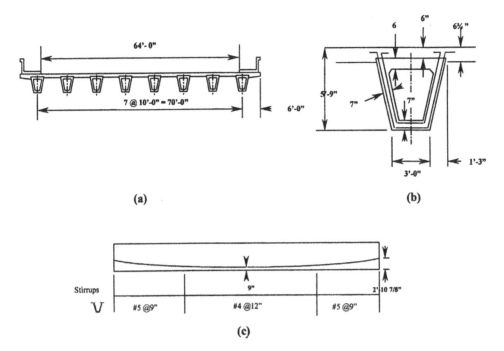

FIGURE 49.4 Details of two-span continuous prestressed box beam bridge example. (a) Typical section; (b) beam section details; (c) prestressing tendon profile.

2. Live Load Calculations

According to Article 3.28 of Design Specifications, distribution factor (DF) for interior spread box beam is given by

$$DF = \left(\frac{2\,N_L}{N_B}\right) + k\left(\frac{S}{L}\right)$$

where N_L = number of traffic lanes = 64/12 = 5 (no fractions); N_B = Number of beams = 8; S = girder spacing = 8 ft; L = span length = 133 ft; W = roadway width = 64 ft

$$k = 0.07\,W - N_L\,(0.10\,N_L - 0.26) - 0.2\,N_B - 0.12 = 1.56$$

$$\text{Thus, DF} = \left(\frac{2\times5}{8}\right) + 1.56\left(\frac{10}{133}\right) = 1.37 \text{ wheels}$$

3. Demands on the Girder

Load demands are estimated using a two-dimensional analysis, and a summary is given in Table 49.6.

4. Section Property Calculations

In order to estimate the stresses on the prestress box beam, the section properties for composite girder need to be estimated. Calculations of the composite girder properties are done separately and the final results are listed here in Table 49.7.

TABLE 49.6   Load Demands for Prestressed Precast Box Beam Bridge Example

Description	0.5L	At Bent 2	At Bent 3
Dead load moment (kip/ft)	4224	0	0
Additional dead load moment (kip/ft)	194	−506	−513
HS20 moment with impact (kip/ft)	1142	−1313	−1322
Dead load shear (kips)	0.0	153.6	−153.6
Additional dead load shear (kips)	0.0	21.1	−21.2
HS20 positive shear (moment)[a] (kips)	24.8 [1104]	61.1 [−974]	7.1 [127]
HS20 negative shear (moment)[a] (kips)	−24.8 [1104]	−7.1 [131]	−61.2 [−980]

[a] Values within brackets indicate the moment corresponds to the reported shear.

TABLE 49.7   Section Properties for Prestressed, Precast Box Beam Bridge Example

Description	Area (in.²)	Moment of Inertia (ft⁴)	Y Bottom of Girder (in.)	Y Top of Girder (in.)	Y Top of Slab (in.)
For dead loads	1375	30.84	34.42	28.58	NA
For additional dead loads	1578	39.22	38.55	24.45	30.45
For live loads	1984	50.75	44.23	18.77	24.77

TABLE 49.8   Stresses at Midspan for Prestressed, Precast Box Beam Bridge Example

Location = Midspan	Stresses in the Box Beam (psi)			
Load Description	At Top Concrete Fiber	At Bottom Concrete Fiber	At Centroid of Composite Box Beam Concrete Fiber	At Prestress Tendon
Dead load (self + slab)	2265	−2728	777	20.15
Prestress P_{eff} = 2020 kips e = 25.42 in.	−1615	3443	−108	147.1
Additional dead (barrier)	70	−110	16	0.845
Live load	244	−575	0	4.59
Live load moment for shear	236	−556	0	4.43

TABLE 49.9   Stresses at Bent 2 for Prestressed, Precast Box Beam Bridge Example

Location = Bent 2	Stresses in the Box Beam (psi)				
Load Description	At Top Concrete Fiber	At Bottom Concrete Fiber	At Centroid of Composite Box Beam Concrete Fiber	At Top of Slab Fiber	At Prestress Tendon
Dead load (self + slab)	0	0	0	0	0
Prestress P_{eff} = 2020 kips e = 12 in.	680	680	680	0	167.5
Additional dead (barrier)	−183	288	−4	−228	−0.3
Live load	−281	662	0	−371	−1.47
Live load moment for positive shear	−208	491	0	−274	−1.08

5. Stress Calculations

Stresses at different fiber locations are calculated using

$$\left(\frac{P}{A}\right) + \left(\frac{M\,c}{I}\right)$$

expression. The summary of the results at midspan and at Bent 2 locations is given in Tables 49.8 and Table 49.9, respectively.

6. Capacity Calculations

a. *Moment capacity at midspan:*
The actual area of steel could only be obtained from the shop plans. Since the shop plans are not readily available, the following approach is used. Assume the total loss including the creep loss = 35 ksi (241.3 MPa).

$$\text{Thus, the area of prestressing steel} = \frac{\text{Working force}}{0.75 \times 270 - 35} = \frac{2020}{167.5} = 12.06 \text{ in.}^2 \ (7781 \text{ mm}^2)$$

$b_{eff} = 120$ in.; $t_s = 6.75$ in.; $d_p = (5.75)(12) - 9$ in. $= 60$ in.; $b_w = 14$ in.

$$\rho^* = \frac{A_s^*}{bd} = \frac{12.06}{120 \times 60} = 0.001675$$

$$f_{su}^* = f_s \left(1 - \frac{0.5\rho^* f_s'}{f_c} \right) = 270 \left(1 - \frac{0.5 \times 0.001675 \times 270}{5.5} \right) = 258.9 \text{ ksi} \ (1785 \text{ MPa})$$

$$\text{Neutral axis location} = 1.4 \, dp^* \frac{f_{su}^*}{f_c'} = 1.4 \times 60 \times 0.001675 \times \frac{258.9}{5.5} = 6.62 \text{ in.} < t_s = 6.75 \text{ in.}$$

Since the neutral axis falls within the slab, this girder can be treated as a rectangular section for moment capacity calculations.

$$R = \phi \, M_n = \phi \, A_s^* \, f_{su}^* \, d \left(1 - 0.6 \rho \frac{f_{su}^*}{f_c'} \right) \quad \text{and} \quad \phi = 1.00$$

$$= 14873.1 \text{ kips/ft} \ (20.17 \text{ MN/m})$$

b. *Moment capacity at the face of the support:*
15 #11 bars are used at top of the bent; thus, the total area of steel = (15)(1.56) = 23.4 in.^{2} Depth of the reinforcing steel from the top of compression fiber = 69 − 1.5 − 1.41/2 = 66.795 in. (1696.6 mm). $F_y = 60$ ksi. Resistance reduction factor $\phi = 0.90$. Then, the moment capacity

$$\phi \, M_n = 6547.2 \text{ kip/ft} \ (8.88 \text{ MN/m}) \text{ (based on T section)}$$

c. *Shear capacity at midspan:*
Design Specification's Section 9.20 addresses the shear capacity of a section. Shear capacity depends on the cracking moment of the section. When the live load causes tension at bottom fiber, cracking moment is to be calculated based on the bottom fiber stress. On the other hand, when the live load causes tension at the top fiber of the beam, cracking moment is to be calculated based on the top fiber stress.

At midspan location, the moment reported with the maximum live-load shear is positive. Positive moments will induce tension at the bottom fiber and thus cracking moment is to be based on the stress at bottom fiber.

TABLE 49.10 Rating Calculations Prestressed, Precast Box Beam Bridge Example

Location	Description	Inventory Rating		Operating Rating	
Midspan	Maximum moment	$\dfrac{14873.1-1.3\times(4224+194)}{1.3\times1.67\times1142}$	$=3.69$	$\dfrac{14873.1-1.3\times(4224+194)}{1.3\times1142}$	$=6.16$
	Maximum shear	$\dfrac{179-1.3\times(0+0)}{1.3\times1.67\times24.8}$	$=3.33$	$\dfrac{179-1.3\times(0+0)}{1.3\times24.8}$	$=5.56$
Bent 2	Maximum moment	$\dfrac{6544.2-1.3\times(0+506)}{1.3\times1.67\times1313}$	$=2.06$	$\dfrac{6544.2-1.3\times(0+506)}{1.3\times1313}$	$=3.45$
	Maximum shear	$\dfrac{766-1.3\times(153.6+21.1)}{1.3\times1.67\times61.1}$	$=4.07$	$\dfrac{766-1.3\times(153.6+21.1)}{1.3\times61.1}$	$=6.80$

f_c' = 5500 psi and from Table 49.10; f_{pe} at midspan bottom fiber = 3443 psi

f_d at bottom fiber = –2728 – 110 = –2838 psi; f_{pc} at centroid = 777 – 108 + 16 = 685 psi

$$M_{cr} = \frac{I}{Y_t}\left(6\sqrt{f_c'}+f_{pe}-f_d\right) = \frac{50.75\times12^4}{44.23}\left(6\sqrt{5500}+3443-2838\right)\left(\frac{1}{12,000}\right) = 2081 \text{ kips/ft}$$

Factored total moment:

$$M_{\max} = 1.3\,M_D + (1.3)(1.67)\,M_{LL+I}$$

$$= 1.3\,(4224+194) + 2.167\,(1104) = 8136 \text{ kips-ft}$$

Factored total shear:

V_i = 1.3 (0 + 0) + 2.167 (24.8) = 53.7 kips

V_d = 0 kips; b_w = 14 in.; d = 60 in.; f_{pc} = 685 psi

$$V_{ci} = 0.6\sqrt{f_c'}\,b_w d + V_d + \frac{V_i M_{cr}}{M_{\max}} = 0.6\sqrt{5500}\times14\times60\times\left(\frac{1}{1000}\right)+0+\frac{53.7\times2081}{8136}$$

$$= 51.2 \text{ kips (227.7 kN) (Controls — since smaller than } V_{cw})$$

$$V_{cw} = \left(3.5\sqrt{f_c'}+0.3f_{pc}\right)b_w d + V_p = \left(3.5\sqrt{5500}+0.3\times685\right)14\times60\times\left(\frac{1}{1000}\right)+0$$

$$= 390 \text{ kips (1734 kN)}$$

V_c = 51.2 kips (227.7 kN) (smaller of V_{ci} and V_{cw})

$$V_s = 2A_v\frac{F_y d_s}{S} = 4\times0.20\times\frac{40\times60}{12} = 160 \text{ kips (711.7 kN)}$$

Shear capacity at midspan:

V_u = ϕ (V_c + V_s) = 0.85 (51.2 + 160) = 179 kips (796.1 kN)

d *Shear capacity at the face of support at Bent 2:*
Negative shear reported at this location is so small and thus rating will not be controlled by the negative shear at Bent 2. Moment reported with the positive shear is negative, and thus, the following calculations are based on the stress at top fiber. From the Table 49.11, f_d at top of slab fiber = −228 psi, and f_{pe} at support top of slab fiber (slab poured after prestressing) = 0 psi.

$$M_{cr} = \frac{I}{Y_t}\left(6\sqrt{f_c'} + f_{pe} - f_d\right) = \frac{50.75 \times 12^4}{44.23}\left(6\sqrt{3500} + 0 - 228\right) = 252 \text{ kips/ft}$$

$V_d = 153.6 + 21.1 = 174.7$ kips; $b_w = 14$ in.; $d = 69 - 1.5 - 1.41/2 = 66.795$ in.; $f_{pc} = 676$ psi

Factored total moment:

$$M_{max} = 1.3\, M_D + (1.3)(1.67)\, M_{LL+I}$$

$$= 1.3(0 + -506) + 2.167(-974) = -2769 \text{ kips-ft}$$

Factored total shear:

$$V_i = 1.3 \times (153.6 + 21.1) + 2.167 \times (61.1) = 360 \text{ kips}$$

$$V_{ci} = 0.6\sqrt{5500} \times 14 \times 66.795\left(\frac{1}{1000}\right) + 0 + \frac{360 \times 251.7}{2769} = 74.3 \text{ kips}$$

$$V_{cw} = \left(3.5\sqrt{f_c'} + 0.3 f_{pc}\right) b_w d + V_p = \left(3.5\sqrt{5500} + 0.3 \times 676\right)14 \times 66.795\left(\frac{1}{1000}\right) + 0 = 432 \text{ kips}$$

$V_c = 74.3$ kips (330.4 kN)   (smaller of the V_{cw} and V_{ci})

$$V_s = 2A_v\frac{F_y d_s}{S} = 4 \times 0.31 \times \frac{60 \times 66.695}{6} = 827 \text{ kips} \ (3678.5 \text{ kN})$$

Shear capacity at Bent 2:

$$V_u = \phi\,(V_c + V_s) = 0.85\,(74.3 + 827) = 766 \text{ kips} \ (3408 \text{ kN})$$

7. Rating Calculations

As discussed in Section 49.5.4, the rating calculations for load factor method need to be done using strength and serviceability limit states. Serviceability level rating needs not be done at the operating level.

a. *Rating calculations based on serviceability limit state:*
Serviceability conditions are listed in AASHTO Design Specification Sections 9.15.1 and 9.15.2.2. These conditions are duplicated in the Rating Manual.
i. Using the compressive stress under all load combination:

The general expression will be $RF_{INV\text{-}COMALL} = \dfrac{0.6 f_c' - f_d - f_p + f_s}{f_l}$

At midspan $RF_{INV\text{-}COMALL} = \dfrac{0.6 \times 5500 - (2265 + 70) - (-1615) + 0}{244} = 10.57$

At Bent 2 support $RF_{INV\text{-}COMALL} = \dfrac{0.6 \times 5500 - (0+288) - 680 + 0}{662} = 3.52$

ii. Using the compressive stress of live load, half the prestressing and permanent dead load:

The general expression will be $RF_{INV\text{-}COMLIVE} = \dfrac{0.4 f_c' - f_d - 0.5 f_p + 0.5 f_s}{f_l}$

At midspan $RF_{INV\text{-}COMLIVE} = \dfrac{0.4 \times 5500 - (2265 + 70) - 0.5(-1615) + 0.5(0)}{244} = 2.76$

At Bent 2 support $RF_{INV\text{-}COMLIVE} = \dfrac{0.4 \times 5500 - (0 + 288) - 0.5(680) + 0.5(0)}{662} = 2.37$

iii. Using the allowable tension in concrete:

The general expression will be $RF_{INV\text{-}CONTEN} = \dfrac{6\sqrt{f_c'} - f_d - f_p - f_s}{f_l}$

At midspan $RF_{INV\text{-}CONTEN} = \dfrac{6\sqrt{5500} - (2728 + 110) - (-3443) - 0}{575} = 1.826$

At Bent 2 support $RF_{INV\text{-}CONTEN} = \dfrac{6\sqrt{5500} - (0 + 183) - (-680) - 0}{281} = 3.352$

iv. Using the allowable prestressing steel tension at service level:

The general expression will be $RF_{INV\text{-}PRETEN} = \dfrac{0.8 f_y^* - f_d - f_p - f_s}{f_l}$

At midspan $RF_{INV\text{-}PRETEN} = \dfrac{0.8 \times 270 - 20.99 - (147.1) - 0}{4.59} = 10.43$

At Bent 2 support $RF_{INV\text{-}PRETEN} = \dfrac{0.8 \times 270 - (-3.08) - 167.5 - 0}{1.468} = 30.94$

b. Rating calculations based on strength limit state:

The general expression for Rating factor $= \dfrac{\phi R_n - \gamma_D D}{\gamma_L \beta_L L(1+I)}$

According to AASHTO Rating Manual, γ_D is 1.3, γ_L is 1.3, and β_L is 1.67 and 1.0 for inventory and operating factor, respectively. Rating calculations are made and given in Table 49.10.

8. Summary

The critical inventory rating of the interior girder is controlled by the tensile stress on concrete at midspan location. The critical operating rating of the girder is controlled by moment at Bent 2 location.

49.6.5 Bridges without Plans

There are some old bridges in service without plans. Establishing safe live-load-carrying capacity is essential to have a complete bridge document. When an inspector comes across a bridge without plans, sufficient field physical dimensions of each member and overall bridge geometry should be taken and recorded. In addition, information such as design year, design vehicle, designer, live-load history, and field condition of the bridge needs to be collected and recorded. This information will be very helpful to determine the safe live-load-carrying capacity. Also, bridge inspectors need to establish the material strength either using the design year or coupon testing.

Design vehicle information could be established based on the designer (state or local agency) and the design year. For example, all state bridges have been designed using the HS20 vehicle since 1944 and all local agency bridges have been designed using the H15 vehicle since 1950.

In steel girder bridges, section properties of the members could be determined based on the field dimensions. During the estimation of the moment capacity, it is recommended to assume that the steel girders are noncomposite with the slab unless substantial evidence is gathered to prove otherwise.

In concrete girder bridges, field dimensions help to estimate the dead loads on the girders. Since the area of reinforcing steels is not known or is difficult to establish, determining the safe live load poses challenges to bridge owners. The live-load history and field condition of a bridge could be used to establish the safe load capacity of the bridge. For example, if a particular bridge has been carrying several heavy vehicles for years without damaging the bridge, this bridge could be left open for all legal vehicles.

49.7 Posting of Bridges

Bridge inspection and the strength evaluation process are two integral parts of bridge posting. The purpose of bridge inspection is to obtain the information that is necessary to evaluate the bridge capacity and the adequacy of the bridge properly. When a bridge is found to have inadequate capacity for legal vehicles, engineers need to look at several alternatives prior to closing the bridge to the public. Some of the possible alternatives are imposing speed limits, reducing vehicular traffic, limiting or posting for vehicle weight, restricting the vehicles to certain lanes, recommending possible small repairs to alleviate the problem. In addition, when the evaluations show that the structure is marginally inadequate, frequent inspections to monitor the physical condition of the bridge and traffic flow may be recommended.

Standard evaluation methods described in the Section 49.5 may be overly conservative. When a more accurate answer is required, a more-detailed analysis, such as three-dimensional analysis or physical load testing can be performed.

The weight and axle configuration of vehicles allowed to use highways without special permits is governed by the statutory law. Thus, the traffic live loads used for posting purposes should be representative of the actual vehicles using the bridge. The representative vehicles vary with each state in the United States. Several states use the three hypothetical legal vehicle configurations given in the Rating Manual [12]. Whereas a few states use their own specially developed legal truck configurations, AASHTO H or HS design trucks, or some combination of truck types. NBIS requires that posting of a bridge must be done when the operating rating for three hypothetical legal vehicles listed in the Rating Manual [12] is less than unity. Furthermore, the NBIS requirement allows the bridge owner to post a bridge for weight limits between inventory and operating level. Because of this flexible NBIS requirement, there is a considerable variation in posting practices among various state and local jurisdictions.

Although engineers may recommend one or a combination of the alternatives described above, it is the owner, not the engineer, who ultimately makes the decision. Many times, bridges are posted for reasons other than structural evaluation, such as posting at a lower weight level to limit vehicular or truck traffic, posting at a higher weight level when the owner believes a lower posting would not be prudent and is willing to accept a higher level of risk. Weight limit posting may cause inconvenience and hardship to the public. In order to reduce inconvenience to the public, the owner needs to look at the weight limit posting as a last resort. In addition, it is sometimes in the public interest to allow certain overweight vehicles such as firefighting equipment and snow removal equipment on a posted bridge. This is usually done through the use of special permits.

References

1. FHWA, Bridge Inspector's Training Manual 70, U.S. Department of Transportation, Washington, D.C., 1970.
2. AASHO, *Manual for Maintenance Inspection of Bridges*, American Association of State Highway Officials, Washington, D.C., 1970.
3. FHWA, Recording and Coding Guide for the Structure Inventory and Appraisal of the Nation's Bridges, U.S. Department of Transportation, Washington, D.C., 1972.
4. FHWA, Bridge Inspector's Manual for Movable Bridges, (Supplement to Manual 70), U.S. Department of Transportation, Washington, D.C., 1970.
5. FHWA, Culvert Inspection Manual, (Supplement to Manual 70), U.S. Department of Transportation, Washington, D.C., 1970.
6. FHWA, Inspection of Fracture Critical Bridge Members, U.S. Department of Transportation, Washington, D.C., 1970.
7. FHWA, Inspection of Fracture Critical Bridge Members, U.S. Department of Transportation, Washington, D.C., 1986.
8. FHWA, Recording and Coding Guide for the Structure Inventory and Appraisal of the Nation's Bridges, U.S. Department of Transportation, Washington, D.C., 1979.
9. FHWA, Recording and Coding Guide for the Structure Inventory and Appraisal of the Nation's Bridges, U.S. Department of Transportation, Washington, D.C., 1988.
10. FHWA, Bridge Inspector's Training Manual 90, U.S. Department of Transportation, Washington, D.C., 1991.
11. FHWA, Hydraulic Engineering Circular (HEC) No.18, Evaluating Scour at Bridges, U.S. Department of Transportation, Washington, D.C., 1990.
12. AASHTO, *Manual for Condition Evaluation of Bridges 1994*, American Association of State Highway and Transportation Officials, Washington, D.C., 1994.
13. AASHTO, *Standard Specification for Highway Bridges*, 16th ed., American Association of State Highway and Transportation Officials, Washington, D.C., 1996.

50

Strengthening and Rehabilitation

F. Wayne Klaiber
Iowa State University

Terry. J. Wipf
Iowa State University

50.1 Introduction

About one half of the approximately 600,000 highway bridges in the United States were built before 1940, and many have not been adequately maintained. Most of these bridges were designed for lower traffic volumes, smaller vehicles, slower speeds, and lighter loads than are common today. In addition, deterioration caused by environmental factors is a growing problem. According to the Federal Highway Administration (FHWA), almost 40% of the nation's bridges are classified as deficient and in need of rehabilitation or replacement. Many of these bridges are deficient because their load-carrying capacity is inadequate for today's traffic. Strengthening can often be used as a cost-effective alternative to replacement or posting.

The live-load capacity of various types of bridges can be increased by using different methods, such as (1) adding members, (2) adding supports, (3) reducing dead load, (4) providing continuity, (5) providing composite action, (6) applying external post-tensioning, (7) increasing member cross section, (8) modifying load paths, and (9) adding lateral supports or stiffeners. Some methods have been widely used, but others are new and have not been fully developed.

All strengthening procedures presented in this chapter apply to the superstructure of bridges. Although bridge span length is not a limiting factor in the various strengthening procedures presented, the majority of the techniques apply to short-span and medium-span bridges. Several of the strengthening techniques, however, are equally effective for long-span bridges. No information is included on the strengthening of existing foundations because such information is dependent on soil type and conditions, type of foundation, and forces involved.

The techniques used for strengthening, stiffening, and repairing bridges tend to be interrelated so that, for example, the stiffening of a structural member of a bridge will normally result in its being strengthened also. To minimize misinterpretation of the meaning of strengthening, stiffening, and repairing, the authors' definitions of these terms are given below. In addition to these terms, definitions of maintenance and rehabilitation, which are sometimes misused, are also given.

Maintenance: The technical aspect of the upkeep of the bridges; it is preventative in nature. Maintenance is the work required to keep a bridge in its present condition and to control potential future deterioration.

Rehabilitation: The process of restoring the bridge to its original service level.

Repair: The technical aspect of rehabilitation; action taken to correct damage or deterioration on a structure or element to restore it to its original condition.

Stiffening: Any technique that improves the in-service performance of an existing structure and thereby eliminates inadequacies in serviceability (such as excessive deflections, excessive cracking, or unacceptable vibrations).

Strengthening: The increase of the load-carrying capacity of an existing structure by providing the structure with a service level higher than the structure originally had (sometimes referred to as upgrading).

In recent years the FHWA and National Cooperative Highway Research Program (NCHRP) have sponsored several studies on bridge repair, rehabilitation, and retrofitting. Inasmuch as some of these procedures also increase the strength of a given bridge, the final reports on these investigations are excellent references. These references, plus the strengthening guidelines presented in this chapter, will provide information an engineer can use to resolve the majority of bridge strengthening problems. The FHWA and NCHRP final reports related to this investigation are References [1–13].

Four of these references, [1,2,11,12] are of specific interest in strengthening work. Although not discussed in this chapter, the live-load capacity of a given bridge can often be evaluated more accurately by using more-refined analysis procedures. If normal analytical methods indicate strengthening is required, frequently more-sophisticated analytical methods (such as finite-element analysis) may result in increased live-load capacities and thus eliminate the need to strengthen or significantly decrease the amount of strengthening required.

By load testing bridges, one frequently determines live-load capacities considerably larger than what one would determine using analytical procedures. Load testing of bridges makes it possible to take into account several contributions (such as end restraint in simple spans, structural contributions of guardrails, etc.) that cannot be included analytically. In the past few years, several states have started using load testing to establish live-load capacities of their bridges. An excellent reference on this procedure is the final report for NCHRP Project 12-28(13)A [14]. Most U.S. states have some type of bridge management system (BMS). To the authors' knowledge, very few states are using their BMS to make bridge strengthening decisions. At the present time, there are not sufficient base line data (first cost, life cycle costs, cost of various strengthening procedures, etc.) to make strengthening/replacement decisions.

Examination of National Bridge Inventory (NBI) bridge records indicates that the bridge types with greatest potential for strengthening are steel stringer, timber stringer, and steel through-truss. If rehabilitation and strengthening cannot be used to extend their useful lives, many of these bridges will require replacement in the near future. Other bridge types for which there also is potential for strengthening are concrete slab, concrete T, concrete stringer, steel girder floor beam, and concrete deck arch. In this chapter, information is provided on the more commonly used strengthening procedures as well as a few of the new procedures that are currently being researched.

50.2 Lightweight Decks

50.2.1 Introduction

One of the more fundamental approaches to increase the live-load capacity of a bridge is to reduce its dead load. Significant reductions in dead load can be obtained by removing an existing heavier concrete deck and replacing it with a lighter-weight deck. In some cases, further reduction in dead load can be obtained by replacing the existing guardrail system with a lighter-weight guardrail. The concept of strengthening by dead-load reduction has been used primarily on steel structures, including the following types of bridges: steel stringer and multibeam, steel girder and floor beam, steel truss, steel arch, and steel suspension bridges; however, this technique could also be used on bridges constructed of other materials.

Lightweight deck replacement is a feasible strengthening technique for bridges with structurally inadequate, but sound, steel stringers or floor beams. If, however, the existing deck is not in need of replacement or extensive repair, lightweight deck replacement would not be economically feasible.

Lightweight deck replacement can be used conveniently in conjunction with other strengthening techniques. After an existing deck has been removed, structural members can readily be strengthened, added, or replaced. Composite action, which is possible with some lightweight deck types, can further increase the live-load carrying capacity of a deficient bridge.

50.2.2 Types

Steel grid deck is a lightweight flooring system manufactured by several firms. It consists of fabricated, steel grid panels that are field-welded or bolted to the bridge superstructure. The steel grids may be filled with concrete, partially filled with concrete, or left open (Figure 50.1).

Open-Grid Steel Decks

Open-grid steel decks are lightweight, typically weighing 15 to 25 psf (720 to 1200 Pa) for spans up to 5 ft (1.52 m). Heavier decks, capable of spanning up to 9 ft (2.74 m), are also available; the percent increase in live-load capacity is maximized with the use of an open-grid steel deck. Rapid installation is possible with the prefabricated panels of steel grid deck. Open-grid steel decks also have the advantage of allowing snow, water, and dirt to wash through the bridge deck, thus eliminating the need for special drainage systems.

A disadvantage of the open grids is that they leave the superstructure exposed to weather and corrosive chemicals. The deck must be designed so water and debris do not become trapped in the grids that rest on the stringers. Other problems associated with open-steel grid decks include weld failure and poor skid resistance. Weld failures between the primary bearing bars of the deck and the supporting structure have caused maintenance problems with some open-grid decks. The number of weld failures can be minimized if the deck is properly erected.

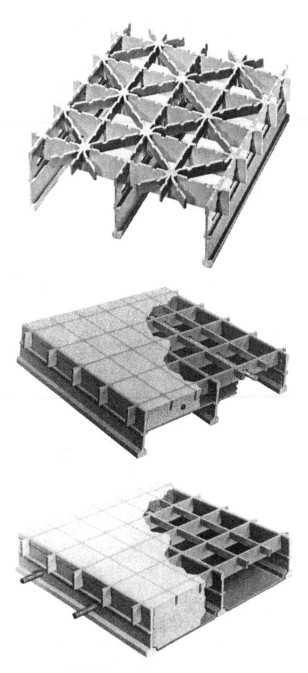

FIGURE 50.1 Steel-grid bridge deck. Top photo shows open steel grid deck; center photo shows half-filled steel grid deck; bottom photo shows filled steel grid deck. (*Source:* Klaiber, F.W. et al., NCHRP 293, Transportation Research Board, 1987. With permission.)

In an effort to improve skid resistance, most open-grid decks currently on the market have serrated or notched bars at the traffic surface. Small studs welded to the surface of the steel grids have also been used to improve skid resistance. While these features have improved skid resistance, they have not eliminated the problem entirely [12]. Open-grid decks are often not perceived favorably by the general public because of the poor riding quality and increased tire noise.

Concrete-Filled Steel Grid Decks

Concrete-filled steel grid decks weigh substantially more, but have several advantages over the open-grid steel decks, including increased strength, improved skid resistance, and better riding quality. The steel grids can be either half or completely filled with concrete. A 5-in. (130-mm)-thick, half-filled steel grid weighs 46 to 51 psf (2.20 to 2.44 kPa), less than half the weight of a reinforced concrete deck of comparable strength. Typical weights for 5-in. (130-mm) thick steel grid decks, filled to full depth with concrete, range from 76 to 81 psf (3.64 to 3.88 kPa). Reduction in the deadweight resulting from concrete-filled steel grid deck replacement alone only slightly improves the live-load capacity; however, the capacity can be further improved by providing composite action between the deck and stringers.

Steel grid panels that are filled or half-filled with concrete may either be precast prior to erection or filled with concrete after placement. With the precast system, only the grids that have been left open to allow field welding of the panels must be filled with concrete after installation. The precast system is generally used when erection time must be minimized.

A problem that has been associated with concrete-filled steel grid decks, addressed in a study by Timmer [15], is the phenomenon referred to as deck growth — the increase in length of the filled grid deck caused by the rusting of the steel I-bar webs. The increase in thickness of the webs due to rusting results in comprehensive stresses in the concrete fill. Timmer noted that in the early stages of deck growth, a point is reached when the compression of the concrete fill closes voids and capillaries in the concrete. Because of this action, the amount of moisture that reaches the resting surfaces is reduced and deck growth is often slowed down or even halted. If, however, the deck growth continues beyond this stage, it can lead to breakup of the concrete fill, damage to the steel grid deck, and possibly even damage to the bridge superstructure and substructure. Timmer's findings indicate that the condition of decks that had been covered with some type of wearing surface was superior to those that had been left unsurfaced. A wearing surface is also recommended to prevent wearing and eventual cupping of the concrete between the grids.

Exodermic Deck

Exodermic deck is a recently developed, prefabricated modular deck system that has been marketed by major steel grid deck manufacturers. The first application of Exodermic deck was in 1984 on the Driscoll Bridge located in New Jersey [16]. As shown in Figure 50.2, the bridge deck system consists of a thin upper layer, 3 in. (76 mm) minimum, of prefabricated concrete joined to a lower layer of steel grating. The deck weighs from 40 to 60 psf (1.92 to 2.87 kPa) and is capable of spanning up to 16 ft (4.88 m).

Exodermic decks have not exhibited the fatigue problems associated with open-grid decks or the growth problems associated with concrete-filled grid decks. As can be seen in Figure 50.2, there is no concrete fill and thus no grid corrosion forces. This fact, coupled with the location of the neutral axis, minimizes the stress at the top surface of the grid.

Exodermic deck and half-filled steel grid deck have the highest percent increase in live-load capacity among the lightweight deck types with a concrete surface. As a prefabricated modular deck system, Exodermic deck can be quickly installed. Because the panels are fabricated in a controlled environment, quality control is easier to maintain and panel fabrication is independent of the weather or season.

Laminated Timber Deck

Laminated timber decks consist of vertically laminated 2-in. (51-mm) (nominal) dimension lumber. The laminates are bonded together with a structural adhesive to form panels that are approximately 48 in. (1.22 m) wide. The panels are typically oriented transverse to the supporting structure of the bridge (Figure 50.3). In the field, adjacent panels are secured to each other with steel dowels or stiffener beams to allow for load transfer and to provide continuity between the panels.

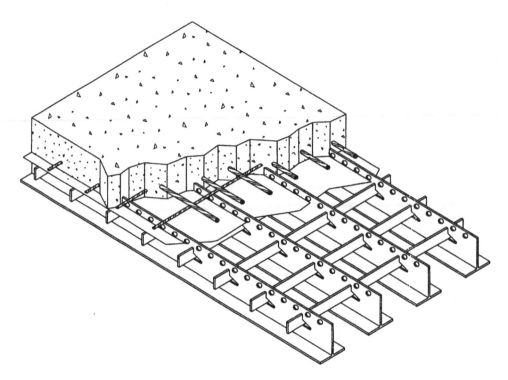

FIGURE 50.2   Exodermic deck system. (*Source:* Exodermic Bridge Deck Inc., Lakeville, CT, 1999. With permission.)

A steel–wood composite deck for longitudinally oriented laminates has been developed by Bakht and Tharmabala [17]. Individual laminates are transversely post-tensioned in the manner developed by Csagoly and Taylor [18]. The use of shear connectors provides partial composite action between the deck and stringers. Because the deck is placed longitudinally, diaphragms mounted flush with the stringers may be required for support. Design of this type of timber deck is presented in References [19–21].

The laminated timber decks used for lightweight deck replacement typically range in depth from 3⅛ to 6¾ in. (79 to 171 mm) and from 10.4 to 22.5 psf (500 to 1075 Pa) in weight. A bituminous wearing surface is recommended.

Wood is a replenishable resource that offers several advantages: ease of fabrication and erection, high strength-to-weight ratio, and immunity to deicing chemicals. With the proper treatment, heavy timber members also have excellent thermal insulation and fire resistance [22]. The most common problem associated with wood as a structural material is its susceptibility to decay caused by living fungi, wood-boring insects, and marine organisms. With the use of modern preservative pressure treatments, however, the expected service life of timber decks can be extended to 50 years or more.

Lightweight Concrete Deck

Structural lightweight concrete, concrete with a unit weight of 115 pcf (1840 kg/m³) or less, can be used to strengthen steel bridges that have normal-weight, noncomposite concrete decks. Special design considerations are necessary for lightweight concrete. Its modulus of elasticity and shear strength are less than that of normal-weight concrete, whereas its creep effects are greater [23]. The durability of lightweight concrete has been a problem in some applications.

Lightweight concrete for deck replacement can be either cast in place or installed in the form of precast panels. A cast-in-place lightweight concrete deck can easily be made to act compositely with the stringers. The main disadvantage of a cast-in-place concrete deck is the length of time required for concrete placement and curing.

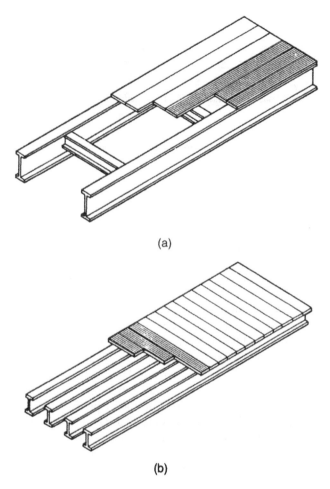

(a)

(b)

FIGURE 50.3 Laminated timber deck. (a) Longitudenal orietation; (b) transverse orientation. (*Source:* Klaiber, F.W. et al., NCHRP 293, Transportation Research Board, 1987. With permission.)

Lightweight precast panels, fabricated with either mild steel reinforcement or transverse prestressing, have been used in deck replacement projects to help minimize erection time and resulting interruptions to traffic. Precast panels require careful installation to prevent water leakage and cracking at the panel joints. Composite action can be attained between the deck and the superstructure; however, some designers have chosen not to rely on composite action when designing a precast deck system.

Aluminum Orthotropic Plate Deck

Aluminum orthotropic deck is a structurally strong, lightweight deck weighing from 20 to 25 psf (958 to 1197 Pa). A proprietary aluminum orthotropic deck system that is currently being marketed is shown in Figure 50.4. The deck is fabricated from highly corrosion-resistant aluminum alloy plates and extrusions that are shop-coated with a durable, skid-resistant, polymer wearing surface. Panel attachments between the deck and stringer must not only resist the upward forces on the panels, but also allow for the differing thermal movements of the aluminum and steel superstructure. For design purposes, the manufacturer's recommended connection should not be considered to provide composite action.

The aluminum orthotropic plate is comparable in weight to the open-grid steel deck. The aluminum system, however, eliminates some of the disadvantages associated with open grids: poor ridability and acoustics, weld failures, and corrosion caused by through drainage. A wheel-load

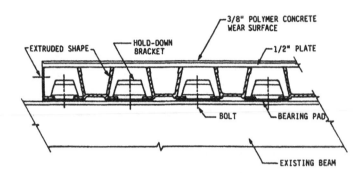

FIGURE 50.4 Aluminum orthotropic deck. (*Source:* Klaiber, F.W. et al., NCHRP 293, Transportation Research Board, 1987. With permission.)

distribution factor has not been developed for the aluminum orthotropic plate deck at this time. Finite-element analysis has been used by the manufacturer to design the deck on a project-by-project basis.

Steel Orthotropic Plate Deck

Steel orthotropic plate decks are an alternative for lightweight deck replacements, that generally have been designed on a case-by-case basis, without a high degree of standardization. The decks often serve several functions in addition to carrying and distributing vertical live loads and, therefore, a simple reinforced concrete vs. steel orthotropic deck weight comparison could be misleading.

Originally, steel orthotropic plate decks were developed to minimize steel use in 200- to 300-ft (61- to 91-m) span girder bridges. Then the decks were used in longer-span suspension and cable-stayed bridges where the deck weight is a significant part of the total superstructure design load. Although the steel orthotropic deck is applicable for spans as short as 80 to 120 ft (24.4 to 36.6 m), it is unlikely that there would be sufficient weight savings at those spans to make it economical to replace a reinforced concrete deck with a steel orthotropic plate deck. Orthotropic steel decks are heavier than aluminum orthotropic decks and usually have weights in the 45 to 130 psf (2.15 to 6.22 kPa) range.

50.2.3 Case Studies

Steel Grid Deck

The West Virginia Department of Highways was one of the first to develop a statewide bridge rehabilitation plan using open-grid steel deck [24]. By 1974, 25 bridges had been renovated to meet or exceed AASHTO requirements. Deteriorated concrete decks were replaced with lightweight, honeycombed steel grid decks fabricated from ASTM A588 steel. The new bridge floors are expected to have a 50-year life and to require minimal maintenance.

In 1981, the West Virginia Department of Highways increased the live-load limit on a 1794-ft (546.8-m)-long bridge over the Ohio River from 3 tons (26.69 kN) to 13 tons (115.65 kN) by replacing the existing reinforced concrete deck with an open steel grid deck [25, 26]. The existing deck was removed and the new deck installed in sections allowing half of the bridge to be left open for use by workers, construction vehicles, and equipment, and, if needed, emergency vehicles.

The strengthening of the 250-ft (76.2-m)-long Old York Road Bridge in New Jersey in the early 1980s combined deck replacement with the replacement of all of the main framing members and the modernization of the piers and abutments [27]. The existing deck was replaced with an ASTM A588 open-grid steel deck. The posted 10-ton (89-kN) load limit was increased to 36 tons (320 kN) and the bridge was widened from 18 ft (5.49 m) to 26 ft (7.92 m).

Exodermic Deck

The first installation of Exodermic deck was in 1984 on the 4400-ft (1340-m)-long Driscoll Bridge located in New Jersey [16]. The deck, weighing 53 psf (2.54 kPa), consisted of a 3-in. (76-mm) upper layer of prefabricated reinforced concrete joined to a lower layer of steel grating. Approximately 30,000 ft² (2790 m²) of deck was replaced at this site.

Exodermic deck was also specified for the deck replacement on a four-span bridge which overpasses the New York State Thruway [28]. The bridge was closed to traffic during deck removal and replacement. Once the existing deck has been removed, it is estimated that approximately 7500 ft² (697 m²) of Exodermic deck will be installed in 3 working days.

Lightweight Concrete Deck

Lightweight concrete was used as early as 1922 for new bridge construction in the United States. Over the years, concrete made with good lightweight aggregate has generally performed satisfactorily; however, some problems related to the durability of the concrete have been experienced. The Louisiana Department of Transportation has experienced several deck failures on bridges built with lightweight concrete in the late 1950s and early 1960s. The deck failures have typically occurred on bridges with high traffic counts and have been characterized by sudden and unexpected collapse of sections of the deck.

Lightweight concrete decks can either be cast in place or factory precast. Examples of the use of lightweight concrete for deck replacement follow.

Cast-in-Place Concrete

New York state authorities used lightweight concrete to replace the deck on the north span of the Newburgh–Beacon Bridge [8, 29]. The existing deck was replaced with 6½ in. (165 mm) of cast-in-place lightweight concrete that was surfaced with a 1½ in. (38 mm) layer of latex modified concrete. Use of the lightweight concrete allowed the bridge to be widened from two to three lanes with minimal modifications to the substructure. A significant reduction in the cost of widening the northbound bridge was attributed to the reduction in dead load.

Precast Concrete Panels

Precast modular-deck construction has been used successfully since 1967 when a joint study, conducted by Purdue University and Indiana State Highway Commission, found precast, prestressed deck elements to be economically and structurally feasible for bridge deck replacement [30, 31].

Precast panels, made of lightweight concrete, 115 pcf (1840 kg/m³), were used to replace and widen the existing concrete deck on the Woodrow Wilson Bridge, located on Interstate 95 south of Washington, D.C. [32,33]. The precast panels were transversely prestressed and longitudinally post-tensioned. Special sliding steel-bearing plates were used between the panels and the structural steel to prevent the introduction of unwanted stresses in the superstructure. The Maryland State Highway Commission required that all six lanes of traffic be maintained during the peak traffic hours of the morning and evening. Two-way traffic was maintained at night when the removal and replacement of the deck was accomplished.

Aluminum Orthotropic Plate Deck

The 104-year-old Smithfield Street Bridge in Pittsburgh, Pennsylvania has undergone two lightweight deck replacements, both involving aluminum deck [34]. The first deck replacement occurred in 1933 when the original heavyweight deck was replaced with an aluminum deck and floor framing system. The aluminum deck was coated with a 1½-in. (38-mm) asphaltic cement wearing surface. The new deck, weighing 30 psf (1.44 kPa), eliminated 751 tons (6680 kN) of deadweight and increased the live-load capacity from 5 tons (44.5 kN) to 20 tons (178 kN).

Excessive corrosion of some of the deck panels and framing members necessitated the replacement of the aluminum deck on the Smithfield Street Bridge in 1967. At that time, a new aluminum orthotropic plate deck with a ⅜-in. (9.5-mm)-thick polymer concrete wearing surface was installed. This new deck weighed 15 psf (718 Pa) and resulted in an additional 108-ton (960-kN) reduction in deadweight. The panels were originally attached to the structure with anodized aluminum bolts, but the bolts were later replaced with galvanized steel bolts after loosening and fracturing of the aluminum bolts became a problem. The aluminum components of the deck have shown no significant corrosion; however, because of excessive wear, the wearing surface had to be replaced in the mid-1970s. The new wearing surface consisted of aluminum-expanded mesh filled with epoxy resin concrete. This wearing surface has also experienced excessive wear, and thus early replacement is anticipated.

Steel Orthotropic Plate Deck

Steel orthotropic plate decks were first conceived in the 1930s for movable bridges and were termed battledecks. Steel orthotropic decks were rapidly developed in the late 1940s in West Germany for replacement of bridges destroyed in World War II during a time when steel was in short supply, and replacement of bridge decks with steel orthotropic plate decks became a means for increasing the live-load capacity of medium- to long-span bridges in West Germany in the 1950s.

In 1956 Woelting and Bock [35] reported the rebuilding of a wrought iron, 536-ft (163-m) span bridge near Kiel. The two-hinged, deck arch bridge, which carried both rail and highway traffic, was widened and strengthened through rebuilding essentially all of the bridge except the arches and abutments. The replacement steel orthotropic deck removed approximately 190 tons of dead load from the bridge, improved the deck live-load capacity, and was constructed in such a way as to replace the original lateral wind bracing truss.

The live-load class of a bridge near Darmstadt was raised by means of a replacement steel orthotropic deck also in the mid-1970s [36]. The three-span, steel-through-truss bridge had been repaired and altered twice since World War II, but the deck had finally deteriorated to the point where it required replacement. The existing reinforced concrete deck was then replaced with a steel orthotropic plate deck, and the reduction in weight permitted the bridge to be reclassified for heavier truck loads.

50.3 Composite Action

50.3.1 Introduction

Modification of an existing stringer and deck system to a composite system is a common method of increasing the flexural strength of a bridge. The composite action of the stringer and deck not only reduces the live-load stresses but also reduces undesirable deflections and vibrations as a result of the increase in the flexural stiffness from the stringer and deck acting together. This procedure can also be used on bridges that only have partial composite action, because the shear connectors originally provided are inadequate to support today's live loads.

The composite action is provided through suitable shear connection between the stringers and the roadway deck. Although numerous devices have been used to provide the required horizontal shear resistance, the most common connection used today is the welded stud.

50.3.2 Applicability and Advantages

Inasmuch as the modifications required for providing composite action for continuous spans and simple spans are essentially the same, this section is written for simple spans. Composite action can effectively be developed between steel stringers and various deck materials, such as normal-weight reinforced concrete (precast or cast in place), lightweight reinforced concrete (precast or cast in place), laminated timber, and concrete-filled steel grids. These are the most common materials used in composite decks; however, there are some instances in which steel deck plates have been made

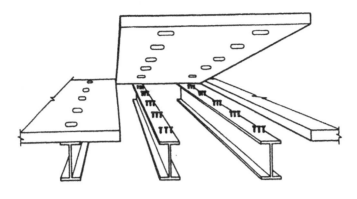

NOTE: SHEAR STUDS SHOWN ARE
ACTUALLY ADDED AFTER
PRECAST DECK IS POSITIONED.

FIGURE 50.5 Precast deck with holes. (*Source:* Klaiber, F.W. et al., NCHRP 293, Transportation Research Board, 1987. With permission.)

composite with steel stringers. In the following paragraphs these four common deck materials will be discussed individually.

Because steel stringers are normally used for support of all the mentioned decks, they are the only type of superstructure reviewed. The condition of the deck determines how one can obtain composite action between the stringers and an existing concrete deck. If the deck is badly deteriorated, composite action is obtained by removing the existing deck, adding appropriate shear connectors to the stringers, and recasting the deck. This was done in Blue Island, Illinois, on the 1500-ft (457-m)-long steel plate girder Burr Oak Avenue Viaduct [37].

If it is desired to reduce interruption of traffic, precast concrete panels are one of the better solutions. The panels are made composite by positioning holes formed in the precast concrete directly over the structural steel. Welded studs are then attached through the preformed holes. This procedure was used on an I-80 freeway overpass near Oakland, California [38]. As shown in Figure 50.5, panels 30 ft (9.1 m) to 40 ft (12.2 m) long, with oblong holes 12 in. (305 mm) × 4 in. (100 mm) were used to replace the existing deck. Four studs were welded to the girders through each hole. Composite action was obtained by filling the holes, as well as the gaps between the panels and steel stringers, with fast-curing concrete.

If the concrete deck does not need replacing, composite action can be obtained by coring through the existing concrete deck to the steel superstructure. Appropriate shear connectors are placed in the holes; the desired composite action is then obtained by filling the holes with nonshrink grout. This procedure was used in the reconstruction of the Pulaski Skyway near the Holland Tunnel linking New Jersey and New York [38]. After removing an asphalt overlay and some of the old concrete, the previously described procedure with welded studs placed in the holes was used. The holes were then grouted and the bridge resurfaced with latex-modified concrete.

Structural lightweight concrete has been used in both precast panels and in cast-in-place bridge decks. Comments made on normal-weight concrete in the preceding paragraphs essentially apply to lightweight concrete also. However, since the shear strength, fatigue strength, and modulus of elasticity of lightweight concrete are less than that of normal-weight concrete, these lesser values must be taken into account in design.

The advantages of composite action can be seen in Figure 50.6. Shown in this graph is the decrease in the top flange stress as a result of providing composite action on a simply supported single-span bridge with steel stringers and an 8-in. concrete deck. As may be seen in this figure, two stringer spacings, 6 ft (1.8 m) and 8 ft (2.4 m) are held constant, while the span length was varied from 20 ft (6.1 m) to 70 ft (21.3 m). These stresses are based on the maximum moment that results from either

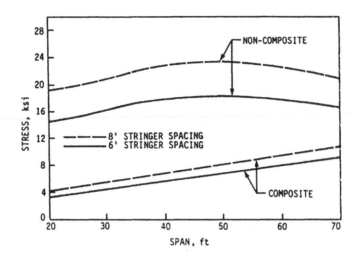

FIGURE 50.6 Stress in top flange of stringer, composite action vs. noncomposite action. (*Source:* Klaiber, F.W. et al., NCHRP 293, Transportation Research Board, 1987. With permission.)

the standard truck loading (HS20-44) or the standard lane loading, whichever governs. Concrete stresses were considerably below the allowable stress limit; composite action reduced the stress in the bottom flange 15 to 30% for long and short spans, respectively. As may be seen in Figure 50.6 for a 40-ft (12.2-m) span with 8-ft (2.4-m) stringer spacing, composite action will reduce the stress in the top flange 68%, 22 ksi (152 Mpa) to 7 ksi (48 MPa). Composite action is slightly more beneficial in short spans than in long spans, and the larger the stringer spacings, the more stress reduction when composite action is added. Results for other types of deck are similar but will depend on the type and size of deck, amount of composite action obtained, type of support system, and the like.

50.3.3 Types of Shear Connectors

As previously mentioned, in order to create composite action between the steel stringers and the bridge deck some type of shear connector is required. In the past, several different types of shear connectors were used in the field; these connectors can be seen in Figure 50.7. Of these, because of the advancements and ease in application, welded studs have become the most commonly used shear connector today. In the strengthening of an existing bridge, frequently one of the older types of shear connectors will be encountered. A strength evaluation must be undertaken to ensure that the shear connectors present are adequate. The following references can be used to obtain the ultimate strength of various types of shear connectors. A method for calculating the strength of a flat bar can be found in Cook [39]; also, work done by Klaiber et al. [40] can be used in evaluating the strength of stiffened angles. Older AASHTO standard specifications can be used to obtain ultimate strength of shear connectors; for example, values for spirals can be found in the AASHTO standard specifications from 1957 to 1968. The current AASHTO specifications only give ultimate-strength equations for welded studs and channels; thus, if shear connectors other than these two are encountered, the previously mentioned references should be consulted.

The procedure employed for using high-strength bolts as shear connectors (Figure 50.8) is very similar to that used for utilizing welded studs in existing concrete, except for the required holes in the steel stringer. To minimize slip, the hole in the steel stringer is made the same size as the diameter of the bolt. Dedic and Klaiber [41] and Dallam [42,43] have shown that the strength and stiffness of high-strength bolts are essentially the same as those of welded shear studs. Thus, existing AASHTO ultimate-strength formulas for welded stud connectors can be used to estimate the ultimate capacity of high-strength bolts.

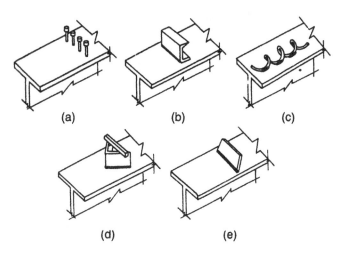

FIGURE 50.7   Common shear connectors. (a) Welded studs; (b) channel; (c) spiral; (d) stiffened angle; (e) inclined flat bar. (*Source:* Klaiber, F.W. et al., NCHRP 293, Transportation Research Board, 1987. With permission.)

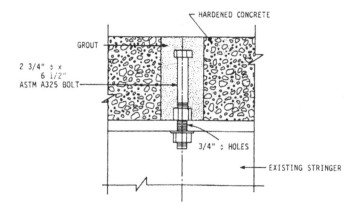

FIGURE 50.8   Details of double-nutted high-strength bolt shear connector. (*Source:* Klaiber, F.W. et al., NCHRP 293, Transportation Research Board, 1987. With permission.)

50.3.4   Design Considerations

The means of obtaining composite action will depend on the individual bridge deck. If the deck is in poor condition and needs to be replaced, the following variables should be considered: (1) weldability of steel stringers, (2) type of shear connector, and (3) precast vs. cast in place.

To determine the weldability of the shear connector, the type of steel in the stringers must be known. If the type of steel is unknown, coupons may be taken from the stringers to determine their weldability. If it is found that welding is not possible, essentially the only alternative for shear connection is high-strength bolts. Although the procedure is rarely done, bolts could be used to attach channels to the stringers for shear connection. When welding is feasible, either welded studs or channels can be used. Because of the ease of application of the welded studs, channels are rarely used today. In older constructions where steel cover plates were riveted to the beam flanges, an option that may be available is to remove the rivets connecting the top cover plate to the top flange of the beam and replace the rivets with high-strength bolts in a manner similar to that which is shown in Figure 50.8.

According to the current AASHTO manual, *Standard Specifications for Highway Bridges*, in new bridges shear connectors should be designed for fatigue and checked for ultimate strength. However, in older bridges, the remaining fatigue life of the bridge will be considerably less than that of the new shear connectors; thus, one only needs to design the new shear connectors for ultimate strength. If an existing bridge with composite action requires additional shear connectors, the ultimate strength capacity of the original shear connector (connector #1) and new shear connectors (connectors #2) can be simply added even though they are different types of connectors. Variation in the stiffness of the new shear connectors and original shear connectors will have essentially no effect on the elastic behavior of the bridge and nominal effect on the ultimate strength [44].

The most common method of creating composite action when one works with precast concrete decks is to preform slots in the individual panels. These slots are then aligned with the stringers for later placement of shear connectors (see Figure 50.5). Once shear connectors are in place, the holes are filled with nonshrink concrete. A similar procedure can be used with laminated timber except the holes for the shear connectors are drilled after the panels are placed.

When it is necessary to strengthen a continuous span, composite action can still be employed. One common approach is for the positive moment region to be designed using the same procedure as that for simple span bridges. When designing the negative moment region, the engineer has two alternatives. The engineer can continue the shear connectors over the negative moment region, in which case the longitudinal steel can be used in computing section properties in the negative moment region. The other alternative is to discontinue the shear connectors over the negative moment region. As long as the additional anchorage connectors in the region of the point of dead-load contraflexure are provided, as required by the code, continuous shear connectors are not needed. When this second alternative is used, the engineer cannot use the longitudinal steel in computing the section properties in the negative moment region. If shear connectors are continued over the negative moment region, one should check to be sure that the longitudinal steel is not overstressed. Designers should consult the pertinent AASHTO standards to meet current design guidelines.

50.4 Improving the Strength of Various Bridge Members

50.4.1 Addition of Steel Cover Plates

Steel Stringer Bridges

Description
One of the most common procedures used to strengthen existing bridges is the addition of steel cover plates to existing members. Steel cover plates, angles, or other sections may be attached to the beams by means of bolts or welds. The additional steel is normally attached to the flanges of existing sections as a means of increasing the section modulus, thereby increasing the flexural capacity of the member. In most cases the member is jacked up during the strengthening process, relieving dead-load stresses on the existing member. The new cover plate section is then able to accept both live-load and dead-load stresses when the jacks are removed, which ensures that less steel will be required in the cover plates. If the bridge is not jacked up, the cover plate will carry only live-load stresses, and more steel will be required.

Applicability, Advantages, and Disadvantages
The techniques described in this section are widely applicable to steel members whose flexural capacity is inadequate. Members in this category include steel stringers (both composite and non-composite), floor beams, and girders on simply supported or continuous bridges. Note, however, that cover plating is most effective on composite members.

There are a number of advantages to using steel cover plates as a method of strengthening existing bridges. This method can be quickly installed and requires little special equipment and minimal labor and materials. If bottom flange stresses control the design, cover plating is effective even if the deck is not replaced. In this case, it is more effective when applied to noncomposite construction. In addition, design procedures are straightforward and thus require minimal time to complete.

In certain instances these advantages may be offset by the costly problems of traffic control and jacking of the bridge. As a minimum, the bridge may have to be closed or separate traffic lanes established to relieve any stresses on the bridge during strengthening. In addition, significant problems may develop if part of the slab must be removed in order to add cover plates to the top of the beams. When cover plates are attached to the bottom flange, the plates should be checked for underclearance if the situation requires it. Still another potential problem if welding is used is that the existing members may not be compatible with current welding materials.

The most commonly reported problem encountered with the addition of steel cover plates is fatigue cracking at the top of the welds at the ends of the cover plates. In a study by Wattar et al. [45], it was suggested that bolting be used at the cover plate ends. Tests showed that bolting the ends raises the fatigue category of the member from stress Category E to B and also results in material savings by allowing the plates to be cut off at the theoretical cutoff points.

Another method for strengthening this detail is to grind the transverse weld to a 1:3 taper [46]. This is a practice of the Maryland State Highway Department. Using an air hammer to peen the toe of the weld and introduce compressive residual stresses is also effective in strengthening the connection [46]. The fatigue strength can be improved from stress category E to D by using this technique. Either solution has been shown to reduce significantly the problem of fatigue cracking at the cover plate ends.

Materials other than flange cover plates may be added to stringer flanges for strengthening. For example, the Iowa Department of Transportation prefers to attach angles to the webs of steel I-beam bridges (either simply supported or continuous spans) with high-strength bolts as a means to reduce flexural live-load stresses in the beams. Figure 50.9 shows a project completed by the Iowa Department of Transportation involving the addition of angles to steel I-beams using high-strength bolts. In some instances the angles are attached only near the bottom flange. Normally, the bridge is not jacked up during strengthening, and only the live loads are removed from the particular I-beam being strengthened. Because the angles are bolted on, problems of fatigue cracking that could occur with welding are eliminated. This method does have one potential problem, however: the possibility of having to remove part of a web stiffener should one be crossed by an angle.

Another method of adding material to existing members for strengthening is shown in Figure 50.10 where structural Ts were bolted to the bottom flanges of the existing stringers using structural angles. This idea represented a design alternative recommended by Howard, Needles, Tammen and Bergendoff as one method of strengthening a bridge comprising three 50-ft (15.2-m) simple spans. Each of the four stringers per span was strengthened in a similar manner.

Design Procedure

The basic design steps required in the design of steel cover plates follow:

1. Determine moment and shear envelopes for desired live-load capacity of each beam.
2. Determine the section modulus required for each beam.
3. Determine the optimal amount of steel to achieve desired section modulus–strength requirement, fatigue requirement.
4. Design connection of cover plates to beam strength requirement, fatigue requirement.
5. Determine safe cutoff point for cover plates.

In addition to the foregoing design steps, the following construction considerations may prove helpful:

1. Grinding the transverse weld to a 1:3 taper or bolting the ends of plates rather than welding reduces fatigue cracking at the cover plate ends [46,47].
2. In most cases a substantial savings in steel can be made if the bridge is jacked to relieve dead-load stresses prior to adding cover plates.

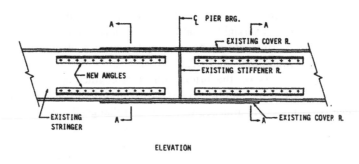

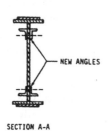

FIGURE 50.9 Iowa DOT method of adding angles to steel I-beams. (*Source:* Klaiber, F.W. et al., NCHRP 293, Transportation Research Board, 1987. With permission.)

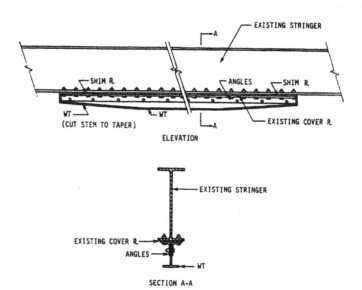

FIGURE 50.10 Strengthening of existing steel stringer by addition of structural T section. (*Source:* Klaiber, F.W. et al., NCHRP 293, Transportation Research Board, 1987. With permission.)

3. The welding of a cover plate should be completed within a working day. This minimizes the possibility of placing a continuous weld at different temperatures and inducing stress concentrations.
4. Shot blasting of existing beams to clean welding surface may be necessary.

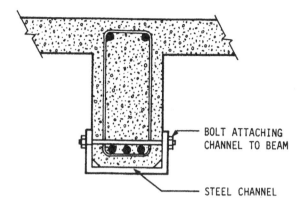

FIGURE 50.11 Addition of a steel channel to an existing reinforced concrete beam. (*Source:* Klaiber, F.W. et al., NCHRP 293, Transportation Research Board, 1987. With permission.)

Reinforced Concrete Bridges

Description

One method of increasing flexural capacity of a reinforced concrete beam is to attach steel cover plates or other steel shapes to the tension face of the beam. The plates or shapes are normally attached by bolting, keying, or doweling to develop continuity between the old beam and the new material. If the beam is also inadequate in shear, combinations of straps and cover plates may be added to improve both shear and flexural capacity. Because a large percentage of the load in most concrete structures is dead load, for cover plating to be most effective, the structure should be jacked prior to cover plating to reduce the dead-load stresses of the member. The addition of steel cover plates may also require the addition of concrete to the compression face of the member.

Applicability

A successful method of strengthening reinforced concrete beams has involved the attachment of a steel channel to the stem of a beam. This technique is shown in Figure 50.11. Taylor [48] performed tests on a section using steel channels and found it to be an effective method of strengthening. An advantage to this method is that rolled channels are available in a variety of sizes, require little additional preparation prior to attachment, and provide a ready formwork for the addition of grouting. The channels can also be easily reinforced with welded cover plates if additional strength is required. Prefabricated channels are an effective substitute when rolled sections of the required size are not available. It should be noted that the bolts are placed above the longitudinal steel so that the stirrups can carry shear forces transmitted by the channels. If additional sheer capacity is required, external stirrups should also be installed. It is also recommended that an epoxy resin grout be used between the bolts and concrete. The epoxy resin grout provides greater penetration in the bolt holes, thereby reducing slippage and improving the strength of the composite action.

Bolting steel plates to the bottom and sides of beam sections has also been performed successfully, as documented by Warner [49]. Bolting may be an expensive and time-consuming method, because holes usually have to be drilled through the old concrete. Bolting is effective, however, in providing composite action between the old and new material.

The placement of longitudinal reinforcement in combination with a concrete sleeve or concrete cover is another method for increasing the flexural capacity of the member. This method is shown in Figures 50.12a and b as outlined in an article on strengthening by Westerberg [50]. Warner [49] presents a similar method that is shown in Figure 50.12c.

Developing a bond between the old and new material is critical to developing full continuity. Careful cleaning and preparation of the old concrete and the application of a suitable epoxy-resin

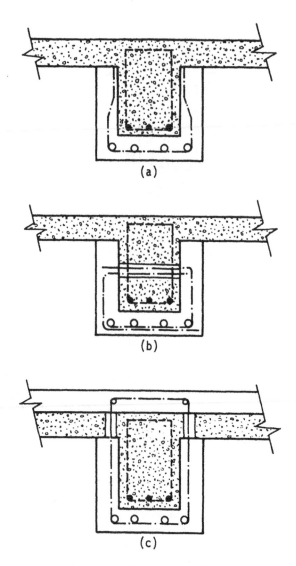

FIGURE 50.12 Techniques for increasing the flexural capacity of reinforced concrete beams with reinforced concrete sleeves. (*Source:* Klaiber, F.W. et al., NCHRP 293, Transportation Research Board, 1987. With permission.)

primer prior to adding new concrete should provide adequate bonding. Stirrups should also be added to provide additional shear reinforcement and to support the added longitudinal bars.

Design and Analysis Procedure

The design of steel cover plates for concrete members is dependent on the amount of continuity assumed to exist between the old and new material. If one assumes that full continuity can be achieved and that strains vary linearly throughout the depth of the beam, calculations are basically straightforward. As stated earlier, much of the load in concrete structures is dead load, and jacking of the deck during cover plating will greatly reduce the amount of new steel required. It should also be pointed out that additional steel could lead to an overreinforced section. This could be compensated for by additional concrete or reinforcing steel in the compression zone.

Case Studies

Steel cover plates can be used in a variety of situations. They can be used to increase the section modulus of steel, reinforced concrete, and timber beams. Steel cover plates are also an effective method of strengthening compression members in trusses by providing additional cross-sectional area and by reducing the slenderness ratio of the member.

Mancarti [51] reported the use of steel cover plates to strengthen floor beams on the Pit River Bridge and Overhead in California. The truss structure required strengthening of various other components to accommodate increased dead load. Stringers in this bridge were strengthened by applying prestressing tendons near the top flange to reduce tensile stress in the negative moment region. This prestressing caused increased compressive stresses in the bottom flanges, which in turn required the addition of steel bars to the tops of the stringer bottom flange.

In a report by Rodriguez et al. [52], a number of cases of cover-plating existing members of old railway trusses were cited. These case studies included the inspection of 109 bridges and a determination of their safety. Some strengthening techniques included steel-cover-plating beam members as well as truss members. Cover plates used to reinforce existing floor beams on a deficient through-truss were designed to carry all live-load bending moment. Deficient truss members were strengthened with box sections made up of welded plates. The box was placed around the existing member and connected to it by welding.

50.4.2 Shear Reinforcement

External Shear Reinforcement for Concrete, Steel, and Timber Beams

The shear strength of reinforced concrete beams or prestressed concrete beams can be improved with the addition of external steel straps, plates, or stirrups. Steel straps are normally wrapped around the member and can be post-tensioned. Post-tensioning allows the new material to share both dead and live loads equally with the old material, resulting in more efficient use of the material added. A disadvantage of adding steel straps is that cutting the deck to apply the straps leaves them exposed on the deck surface and thus difficult to protect. By contrast, adding steel plates does not require cutting through the deck. The steel plates are normally attached to the beam with bolts or dowels.

External stirrups may also be applied with different configurations. Figure 50.13a shows a method of attaching vertical stirrups using channels at the top and bottom of the beam. The deck (not shown in either figure) provides protection for the upper steel channel [53]. Adding steel sections at the top of the beam web and attaching stirrups is shown in Figure 50.13b. In this manner, cutting holes through the deck is eliminated. External stirrups can also be post-tensioned in most situations if desired.

Another method of increasing shear strength is shown in Figure 50.14. This method is a combination of post-tensioning and the addition of steel in the form of prestressing tendons. As recommended in a strengthening manual by the OECD [54], tendons may be added in a vertical or inclined orientation and may be placed either within the beam web or inside the box as shown in the figure. Care should be taken to avoid overstressing parts of the structure when prestressing. If any cracks exist in the member, it is a good practice to inject them with an epoxy before applying the prestressing forces. Documentation of this type of reinforcement technique is made also by Audrey and Suter [55] and Dilger and Ghali [56]. Figure 50.15 illustrates the technique used by Dilger and Ghali [56] where web thickening was added to the inside of the box web before adding external reinforcement consisting of stressed steel bars. The thickening was required to reduce calculated tensile stresses at the outside of the web due to prestressing the reinforcement.

West [57] makes reference to a number of methods of attaching steel plates to deficient steel I-beam girder webs as a means of increasing their shear strength. The steel plates are normally of panel size and are attached between stiffeners by bolting or welding. Where shear stresses are high, the plates should fit tightly between the stiffeners and girder flanges. West indicates that one advantage of this method is that it can be applied under traffic conditions.

Timber stringers with inadequate shear capacity can be strengthened by adding steel cover plates. NCHRP Report 222 [11] demonstrates a method of repairing damaged timber stringers with

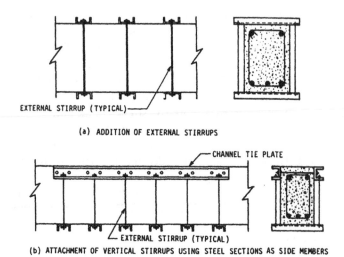

FIGURE 50.13   Methods of adding external shear reinforcement to reinforced concrete beams. (*Source:* Klaiber, F.W. et al., NCHRP 293, Transportation Research Board, 1987. With permission.)

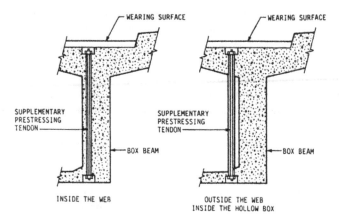

FIGURE 50.14   External shear reinforcement of box beam girders. (*Source:* Klaiber, F.W. et al., NCHRP 293, Transportation Research Board, 1987. With permission.)

inadequate shear capacity. The procedure involves attaching steel plates to the bottom of the beam in the deficient region and attaching it with draw-up bolts placed on both sides of the beam. Holes are drilled through the top of the deck, and a steel strap is placed at the deck surface and at the connection to the bolts.

Epoxy Injection and Rebar Insertion

The Kansas Department of Transportation has developed and successfully used a method for repairing reinforced concrete girder bridges. The bridges had developed shear cracks in the main longitudinal girders [58]. The procedure used by the Kansas Department of Transportation not only prevented further shear cracking but also significantly increased the shear strength of the repaired girders.

The method involves locating and sealing all of the girder cracks with silicone rubber, marking the girder centerline on the deck, locating the transverse deck reinforcement, vacuum drilling 45° holes that avoid the deck reinforcement, pumping the holes and cracks full of epoxy, and inserting reinforcing bars into the epoxy-filled holes. A typical detail is shown in Figure 50.16.

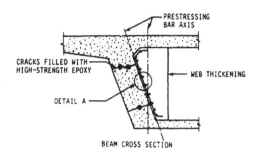

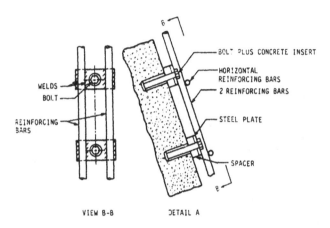

FIGURE 50.15 Details of web reinforcement to strengthen box beam in shear. (*Source:* Klaiber, F.W. et al., NCHRP 293, Transportation Research Board, 1987. With permission.)

An advantage of using the epoxy repair and rebar insertion method is its wide application to a variety of bridges. Although the Kansas Department of Transportation reported using this strengthening method on two-girder, continuous, reinforced concrete bridges, this method can be a practical solution on most types of prestressed concrete beam and reinforced concrete girder bridges that require additional shear strength. The essential equipment requirements needed for this strengthening method may limit its usefulness, however. Prior to drilling, the transverse deck steel must be located. The drilling unit and vacuum pump required must be able to drill quickly straight holes to a controlled depth and keep the holes clean and free of dust.

Addition of External Shear Reinforcement

Strengthening a concrete bridge member that has a deficient shear capacity can be performed by adding external shear reinforcement. The shear reinforcement may consist of steel side plates or steel stirrup reinforcement. This method has been applied on numerous concrete bridge systems.

A method proposed by Warner [49] involves adding external stirrups. The stirrups consist of steel rods placed on both sides of the beam section and attached to plates at the top and bottom of the section. In some applications, channels are mounted on both sides at the top of the section to attach the stirrups. This eliminates drilling through the deck to make the connection to a plate.

In a study by Dilger and Ghali [56], external shear reinforcement was used to repair webs of prestressed concrete bridges. Although the measures used were intended to bring the deficient members to their original flexural capacity, the techniques applied could be used for increasing the shear strength of existing members. Continuous box girders in the 827-ft (252-m)-long bridges had become severely cracked when prestressed. The interior box beam webs were strengthened by the addition of 1-in. (25-mm)-diameter steel rods placed on both sides of the web. Holes were drilled

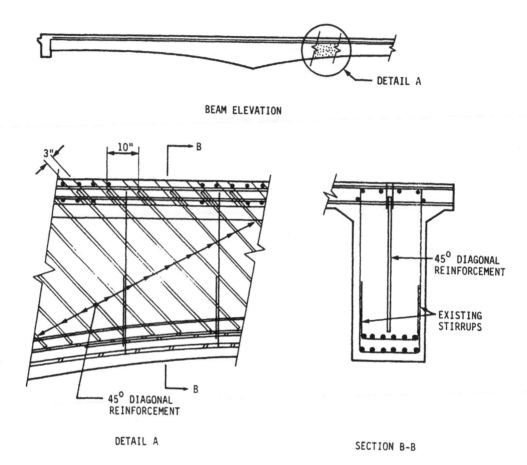

BEAM ELEVATION

DETAIL A

SECTION B-B

FIGURE 50.16 Kansas DOT shear strengthening procedure. (*Source:* Klaiber, F.W. et al., NCHRP 293, Transportation Research Board, 1987. With permission.)

in the upper and lower slabs as close as possible to the web to minimize local bending stresses in the slabs. Post-tensioning tendons were placed through the holes, stressed, and then anchored.

The slanted outside webs were strengthened with reinforcing steel. Before the bars were added, the inside of the web was "thickened" and the reinforcement was attached with anchor bolts placed through steel plates that were welded to the reinforcement. The web thickening was necessary because the prestressing would have produced substantial tensile stresses at the outside face of the web.

50.4.3 Jacketing of Timber or Concrete Piles and Pier Columns

Improving the strength of timber or concrete piles and pier columns can be achieved by encasing the column in concrete or steel jackets. The jacketing may be applied to the full length of the column or only to severely deteriorated sections. The jacketing increases the cross-sectional area of the column and reduces the slenderness ratio of the column. Partial encasement of a column can also be particularly effective when an unbalanced moment acts on the column. Figure 50.17 illustrates two such concepts for member addition that were noted from work on strengthening reinforced concrete structures in Europe [50].

Completely encasing the existing column in a concrete jacket has been a frequently used method of strengthening concrete pier columns. Normally, the reinforcement is placed around the existing column perimeter inside the jacket and "ramset" to the existing member [50]. The difficulty most often observed with this technique is developing continuity between the old and new material. This

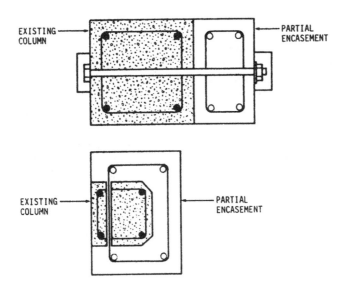

FIGURE 50.17 Partial jacketing of an existing column. (*Source:* Klaiber, F.W. et al., NCHRP 293, Transportation Research Board, 1987. With permission.)

is critical if part of the load is to be transferred to the new material. Work by Soliman [59] on repair of reinforced concrete columns by jacketing has included an experimental investigation of the bond stresses between the column and jacket. The first step is normally surface preparation of the existing concrete column. Consideration should also be given at this time to jacking of the superstructure and placing temporary supports on either side of the column. Solimann [59] concludes that this is an important step, since the shrinkage phenomenon causes compressive stresses on the column that will be reduced if the existing column is unloaded. In addition, supports will be necessary if the column shows significant signs of deterioration. This procedure will also allow the new material to share equally both dead and live loads after the supports are removed. Additional longitudinal reinforcing bars and stirrups are then placed around the column. Spiral stirrup reinforcement should be used because it will provide greater strength and ductility than normal stirrups [59]. An epoxy resin is then applied to the old concrete to increase the bonding action between the old concrete and the concrete to be added. Formwork is then erected to form the jacket, and concrete is placed and compacted.

Jacketing techniques have been used extensively for seismic retrofitting of existing pier columns and this topic is discussed in Chapter 43. A recent report by Wipf et al. [60] provides an extensive list and discussion of various retrofit methods for reinforced concrete bridge columns, including the use of steel jackets and fiber-reinforced polymer wraps.

Modification Jacketing

Increasing the load-carrying capacity of bridge pier columns or timber piles supporting bent caps is normally achieved through the addition of material to the existing cross section. Jacketing or adding a sleeve around the column perimeter can be performed a number of ways.

In a paper by Karamchandani [61], various concepts for jacketing existing members are illustrated. These include addition of reinforcement and concrete around three sides of rectangular beams as well as placement only at the bottom of the beam web. Additional schemes are also illustrated for column members. The effectiveness of this method depends on the degree of adhesion between new and existing concrete, which can vary between 30 and 80% of the total strength of the *in situ* concrete. The author suggests welding new reinforcing to the existing reinforcement and using concrete with a slump of 3 to 4 in. (75 to 100 mm). The use of rapid hardening cements is not recommended, since it results in a lower strength of concrete on the contact surface because of high contraction stresses.

The addition of concrete collars on reinforced concrete columns is performed most efficiently by using circular reinforcement rather than dowels or shear keys according to Klein and Gouwens [62]. While the other methods may require costly and time-consuming drilling and/or cutting, circular reinforcement does not. When this method is used, shear-friction is the primary load-transfer mechanism between the collar and the existing column. Klein and Gouwens have outlined a design procedure for this strengthening method.

In a paper by Syrmakezis and Voyatzis [63], an analytical method for calculating the stiffness coefficients of columns strengthened by jacketing is presented. The procedure uses compatibility conditions for the deformations of the strengthened system and the analysis can consider rigid connections between the jacket and column on a condition where relative slip is allowed.

50.5 Post-Tensioning Various Bridge Components

50.5.1 Introduction

Since the 19th century, timber structures have been strengthened by means of king post and queen post-tendon arrangements; these forms of strengthening by post-tensioning are still used today. Since the 1950s, post-tensioning has been applied as a strengthening method in many more configurations to almost all common bridge types. The impetus for the recent surge in post-tensioning strengthening is undoubtedly a result of its successful history of more than 40 years and the current need for strengthening of bridges in many countries.

Post-tensioning can be applied to an existing bridge to meet a variety of objectives. It can be used to relieve tension overstresses with respect to service load and fatigue-allowable stresses. These overstresses may be axial tension in truss members or tension associated with flexure, shear, or torsion in bridge stringers, beams, or girders.

Post-tensioning also can reduce or reverse undesirable displacements. These displacements may be local, as in the case of cracking, or global, as in the case of excessive bridge deflections. Although post-tensioning is generally not as effective with respect to ultimate strength as with respect to service-load-allowable stresses, it can be used to add ultimate strength to an existing bridge. It is possible to use post-tensioning to change the basic behavior of a bridge from a series of simple spans to continuous spans. All of these objectives have been fulfilled by post-tensioning existing bridges, as documented in the engineering literature.

Most often, post-tensioning has been applied with the objective of controlling longitudinal tension stresses in bridge members under service-loading conditions. Figure 50.18 illustrates the axial forces, shear forces, and bending moments that can be achieved with several simple tendon configurations. The concentric tendon in Figure 50.18a will induce an axial compression force that, depending on magnitude, can eliminate part or all of an existing tension force in a member or even place a residual compression force sufficient to counteract a tension force under other loading conditions. The amount of post-tensioning force that can safely be applied, of course, is limited by the residual-tension dead-load force in the member.

The tendon configuration in Figure 50.18a is generally used only for tension members in trusses, whereas the remaining tendon configurations in Figure 50.18 would be used for stringers, beams, and girders. The eccentric tendon in Figure 50.18b induces both axial compression and negative bending. The eccentricity of the tendon may be varied to control the proportions of axial compression vs. bending applied to the member. Length of the tendon also may be varied to apply post-tensioning only to the most highly stressed portion of the member. The polygonal tendon profile in Figure 50.18c also induces axial compression and negative bending, but the negative bending is nonuniform within the post-tensioned region. Locations of bends on the tendon and eccentricities of the attachments at the bends can be set to control the moments caused by the post-tensioning.

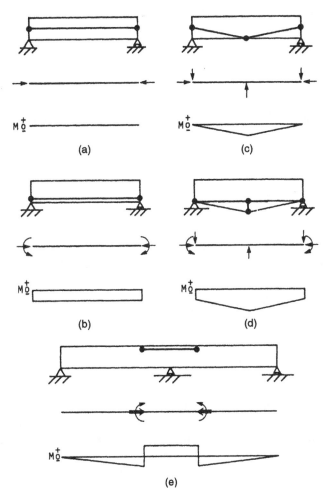

FIGURE 50.18 Forces and moment induced by longitudinal post-tensioning. (a) Concentric tendon; (b) eccentric tendon; (c) polygonal tendon; (d) king post; (e) eccentric tendon, two-span member. (*Source:* Klaiber, F.W. et al., NCHRP 293, Transportation Research Board, 1987. With permission.)

The polygonal tendon also induces shear forces that are opposite to those applied by live and dead loads.

The king post tendon configuration in Figure 50.18d is a combination of the eccentric and polygonal tendon configurations. Because the post is beyond the profile of the original member, the proportion of moment to axial force induced in the member to be strengthened will be large.

The tendon configuration in Figure 50.18e is an eccentric tendon attached over the central support of a two-span member. In this configuration, the amount of positive moment applied in the central support region depends not only on the force in the tendon and its eccentricity, but also on the locations of the anchorages on the two spans. If the anchorages are moved toward the central support, the amount of positive moment applied will be greater than if the anchorages are moved away from the central support. This fact and the fact that there is some distribution of moment and force among parallel post-tensioned members have not always been correctly recognized, and there are published errors in the literature.

The axial force, shear force, and bending moment effects of post-tensioning described above have enough versatility in application so as to meet a wide variety of strengthening requirements. Probably this is the only strengthening method that can actually reverse undesirable behavior in an existing bridge rather than provide a simple patching effect. For both these reasons, post-tensioning has become a very commonly used repair and strengthening method.

50.5.2 Applicability and Advantages

Post-tensioning has many capabilities: to relieve tension, shear, bending, and torsion-overstress conditions; to reverse undesirable displacements; to add ultimate strength; to change simple span to continuous span behavior. In addition, post-tensioning has some very practical advantages. Traffic interruption is minimal; in some cases, post-tensioning can be applied to a bridge with no traffic interruption. Few site preparations, such as scaffolding, are required. Tendons and anchorages can be prefabricated. Post-tensioning is an efficient use of high strength steel. If tendons are removed at some future date, the bridge will generally be in no worse condition than before strengthening.

To date, post-tensioning has been used to repair or strengthen most common bridge types. Most often, post-tensioning has been applied to steel stringers, floor beams, girders, and trusses, and case histories for strengthening of steel bridges date back to the 1950s. Since the 1960s, external post-tensioning has been applied to reinforced concrete stringer and T bridges. In the past 20 years, external post-tensioning has been added to a variety of prestressed, concrete stringer and box beam bridges. Many West German prestressed concrete bridges have required strengthening by post-tensioning due to construction joint distress. Post-tensioning even has been applied to a reinforced concrete slab bridge by coring the full length of the span for placement of tendons [63].

Known applications of post-tensioning will be idealized and summarized as Schemes A through L in Figures 50.19 through 50.22. Typical schemes for stringers, beams, and girders are contained in Figure 50.19. The simplest and, with the exception of the king post, the oldest scheme is Scheme A: a straight, eccentric tendon shown in Figure 50.19a. Lee reported use of the eccentric tendon for strengthening of British cast iron and steel highway and railway bridges in the early 1950s [64]. Since then, Scheme A has been applied to many bridges in Europe, North America, and other parts of the world. Scheme A is most efficient if the tendon has a length less than that of the member, so that the full post-tensioning negative moment is not applied to regions with small dead-load moments. The variation on Scheme A for continuous spans, Scheme AA in Figure 50.19e, has been reported in use for deflection control or strengthening in Germany [65] and the United States [66] since the late 1970s.

The polygonal tendon, Scheme B in Figure 50.19b and its extension to continuous spans, Scheme BB in Figure 50.19f, has been in use since at least the late 1960s. Vernigora et al. [67] reported the use of Scheme BB for a five-span, reinforced-concrete T-beam bridge in 1969 [67]. The bridge over the Welland Canal in Ontario, Canada, was converted from simple-span to continuous-span behavior by means of external post-tensioning cables.

Scheme C in Figure 50.19c provided the necessary strengthening for a steel plate, girder railway bridge in Czechoslovakia in 1964 [68]. The tendons and compression struts for the bridge were fabricated from steel T sections, and the tendons were stressed by deflection at bends rather than by elongation as is the usual case. The tendons for the plate girder bridge were given a three-segment profile to apply upward forces at approximately the third points of the span, so that the existing dead-load moments could be counteracted efficiently. In the late 1970s in the United States, Kandall [69] recommended use of Scheme C for strengthening because it does not place additional axial compression in the existing structure. For other schemes, the additional axial compression induced by post-tensioning will add compressive stress to regions that may be already overstressed in compression.

Scheme D in Figure 50.19d, was used in Minnesota in 1975 to strengthen temporarily a steel stringer bridge [70]. It was possible to strengthen that bridge economically with scrap timber and cable for the last few years of its life before it was replaced.

The tendon schemes in Figure 50.19, in general, appear to be very similar to reinforcing bar patterns for concrete beams. Thus, it is not surprising that post-tensioning also has been used for

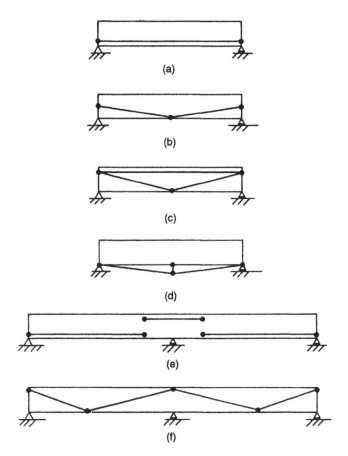

FIGURE 50.19 Tendon configurations for flexural post-tensioning of beams. (a) Scheme A, eccentric tendon; (b) Scheme B, polygonal tenton; (c) Scheme C, polygonal tendon with compression strut; (d) Scheme D, king post; (e) Scheme AA, eccentric tendons; (f) Scheme BB, polygonal tendons. (*Source:* Klaiber, F.W. et al., NCHRP 293, Transportation Research Board, 1987. With permission.)

shear strengthening, in patterns very much like those for stirrups in reinforced concrete beams. Scheme E in Figure 50.20a illustrates a pattern of external stirrups for a beam in need of shear strengthening. Types of post-tensioned external stirrups have been used or proposed for timber beams [11], reinforced concrete beams and, as illustrated in Figure 50.20b, for prestressed concrete box-girder bridges [71].

Post-tensioning was first applied to steel trusses for purposes of strengthening in the early 1950s [64], at about the same time that it was first applied to steel stringer and steel girder, floor beam bridges. Typical strengthening schemes for trusses are presented in Figure 50.21. Scheme F, concentric tendons on individual members, shown in Figure 50.21a, was first reported for the proposed strengthening of a cambered truss bridge in Czechoslovakia in 1964 [68]. For that bridge it was proposed to strengthen the most highly stressed tension diagonals by post-tensioning. Scheme F tends to be uneconomical because it requires a large number of anchorages, and very few truss members benefit from the post-tensioning.

Scheme G in Figure 50.21b, a concentric tendon on a series of members, has been the most widely used form of post-tensioning for trusses. Lee [64] describes the use of this scheme for British railway bridges in the early 1950s, and there have been a considerable number of bridges strengthened with this scheme in Europe.

The polygonal tendon in Scheme H, Figure 50.21c, has not been reported for strengthening purposes, but it has been used in the continuous-span version of Scheme I in Figure 50.21d for a

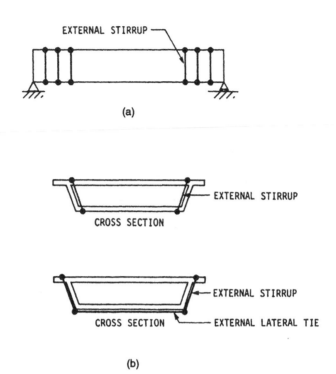

FIGURE 50.20 Tendon configurations for shear post-tensioning. (*Source:* Klaiber, F.W. et al., NCHRP 293, Transportation Research Board, 1987. With permission.)

two-span truss bridge in Switzerland [72]. In the late 1960s, a truss highway bridge in Aarwangen, Switzerland, was strengthened by means of four-segment tendons on each of the two spans. The upper chord of each truss was unable to carry the additional compression force induced by the post-tensioning, and, therefore, a free-sliding compression strut was added to each top chord to take the axial post-tensioning force.

Scheme J, the king post in Figure 50.21e, has been suggested for new as well as existing trusses [7]; however, cases of its actual use for strengthening have not been reported in the literature. Because most trusses are placed on spans greater than 100 ft (30.5 m), the posts below the bridge could extend down quite far and severely reduce clearance under the bridge. The king post or queen post would thus be in a very vulnerable position and would not be appropriate in many situations.

Most uses of post-tensioning for strengthening have been on the longitudinal members in bridges; however, post-tensioning has also been used for strengthening in the transverse direction. After the deterioration of the lateral load distribution characteristics of laminated timber decks was noted in Canada in the mid-1970s [73], Scheme K in Figure 50.22a was used to strengthen the deck. A continuous-steel channel waler at each edge of the deck spreads the post-tensioning forces from threadbar tendons above and below the deck, thereby preventing local overstress in the timber. A similar tendon arrangement, shown in Figure 50.22b, was used in an Illinois bridge [74] to tie together spreading, prestressed concrete box beams.

The overview of uses of post-tensioning for bridge strengthening given above identifies the most important concepts that have been used in the past and indicates the versatility of post-tensioning as a strengthening method.

50.5.3 Limitations and Disadvantages

When post-tensioning is used as a strengthening method, it increases the allowable stress range by the magnitude of the applied post-tensioning stress. If maximum advantage is taken of the increased

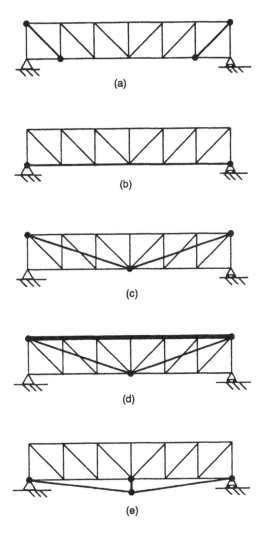

FIGURE 50.21 Tendon configurations for post-tensioning trusses. (a) Scheme F, concentric tendons on individual members; (b) Scheme G, concentric tendon on a series of members; (c) Scheme H, polygonal tendon; (d) Scheme I, polygonal tendon with compression strut; (e) Scheme J, king post. (*Source:* Klaiber, F.W. et al., NCHRP 293, Transportation Research Board, 1987. With permission.)

allowable-stress range, the factor of safety against ultimate load will be reduced. The ultimate-load capacity thus will not increase at the same rate as the allowable-stress capacity. For short-term strengthening applications, the reduced factor of safety should not be a limitation, especially in view of the recent trend toward smaller factors of safety in design standards. For long-term strengthening applications, however, the reduced factor of safety may be a limitation.

At anchorages and brackets where tendons are attached to the bridge structure, there are high local stresses that require consideration. Any cracks initiated by holes or expansion anchors in the structure will spread with live-load dynamic cycling.

Because post-tensioning of an existing bridge affects the entire bridge (beyond the members that are post-tensioned), consideration must be given to the distribution of the induced forces and moments within the structure. If all parallel members are not post-tensioned, if all parallel members are not post-tensioned equally, or if all parallel members do not have the same stiffness, induced forces and moments will be distributed in some manner different from what is assumed in a simple analysis.

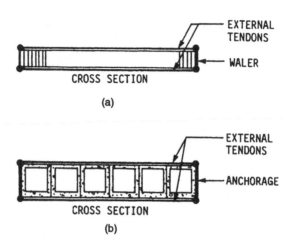

FIGURE 50.22   Tendon configuration for transverse post-tensioning of decks. (a) Scheme K, concentric tendons and walers, laminated timber deck; (b) Scheme K, concentric tendons, box beams. (*Source:* Klaiber, F.W. et al., NCHRP 293, Transportation Research Board, 1987. With permission.)

Post-tensioning does require relatively accurate fabrication and construction and relatively careful monitoring of forces locked into the tendons. Either too much or too little tendon force can cause overstress in the members of the bridge being strengthened.

Tendons, anchorages, and brackets require corrosion protection because they are generally in locations that can be subjected to saltwater runoff or salt spray. If tendons are placed beyond the bridge profile, they are vulnerable to damage from overheight vehicles passing under the bridge or vulnerable to damage from traffic accidents. Exposed tendons also are vulnerable to damage from fires associated with traffic accidents.

50.5.4 Design Procedures

In general, strengthening of bridges by post-tensioning can follow established structural analysis and design principles. The engineer must be cautious, however, in applying empirical design procedures as they are established only for the conditions of a particular strengthening problem.

Every strengthening problem requires careful examination of the existing structure. Materials in an existing bridge were produced to some previous set of standards and may have deteriorated due to exposure over many years. The existing steel in steel members may not be weldable with ordinary procedures, and steel shapes are not likely to be dimensioned to current standards. Shear connectors and other parts may have unknown capacities due to unusual configurations.

Strengthening an existing bridge involves more than strengthening individual members. Even a simple-span bridge is indeterminate, and post-tensioning and other strengthening will affect the behavior of the entire bridge. If the indeterminate nature of the bridge is not recognized during analysis, the post-tensioning applied for strengthening purposes may not have the desired stress-relieving effects and may actually cause overstress.

Post-tensioning involves application of relatively large forces to regions of a structure that were not designed for such large forces. There is more likelihood of local overstress at tendon anchorages and brackets than at conventional member connections. Brackets need to be designed to distribute the concentrated post-tensioning forces over sufficiently large portions of the existing structure.

Members and bridges subjected to longitudinal post-tensioning will shorten axially and, depending on the tendon configuration, also will shorten and elongate with flexural stresses. These shortening and elongation effects must be considered, so that the post-tensioning has its desired effect. Frozen bridge bearings require repair and lubrication, and support details should be checked for restraints.

External tendons, whether cable or threadbar, are relatively vulnerable to corrosion, damage from overheight vehicles, traffic accidents, or fires associated with accidents. Corrosion protection and placement of the tendons are thus very important with respect to the life of the post-tensioning. Safety is also a consideration because a tendon that ruptures suddenly can pose a hazard.

For the past few years, the authors and other Iowa State University colleagues have been investigating the use of external post-tensioning (Scheme A and AA in Figure 50.19) for strengthening existing single-span and continuous-span steel stringer bridges. The research, which has been recently completed, involved laboratory testing, field implementation, and the development of design procedures. The strengthening procedures that were developed are briefly described in the following sections.

50.5.5 Longitudinal Post-Tensioning of Stringers

Simple Spans

Essentially all single-span composite steel stringer bridges constructed in Iowa between 1940 and 1960 have smaller exterior stringers. These stringers are significantly overstressed for today's legal loads; interior stringers are also overstressed to a lesser degree. Thus, the post-tension system developed is only applied to the exterior stringers; through lateral load distribution a stress reduction is also obtained in the interior stringers.

By analyzing an undercapacity bridge, an engineer can determine the overstress in the interior and exterior stringers. This overstress is based on the procedure of isolating each bridge stringer from the total structure. The amount of post-tensioning required to reduce the stress in the stringers can then be determined if the amount of post-tensioning force remaining on the exterior stringers is known; this force can be quantified with force and moment fractions. A force fraction, FF, is the ratio of the axial force that remains on a post-tensioned stringer at midspan to the sum of the axial forces for all bridge stringers at midspan, while a moment fraction, MF, is the moment remaining on the post-tensioned stringer divided by the sum of midspan moments for all bridge stringers. Knowing these fractions, the required post-tensioning force may be determined by utilizing the following relationship:

$$f = FF\left[\frac{P}{A}\right] + MF\left[\frac{Pec}{I}\right] \qquad (50.1)$$

where
f = desired stress reduction in stringer lower flange
P = post-tensioning force required on each exterior stringer
A = cross-sectional area of exterior stringers
e = eccentricity of post-tensioning force measured from the neutral axis of the bridge
c = distance from neutral axis of stringer to lower flange
I = moment of inertia of exterior stringer at section being analyzed

Force fractions and moment fractions as well as other details on the procedure may be found in Reference [75].

Span length and relative beam stiffness were determined to be the most significant variables in the moment fractions. As span length increases, exterior beams retain less moment; exterior beams that are smaller than the interior beams retain less post-tensioning moment than if the beams were all the same size.

The strengthening procedure and design methodology just described have been used on several bridges in the states of Iowa, Florida, and South Dakota. In all instances, the procedure was employed by local contractors without any significant difficulties. Application of this strengthening procedure to a 72-ft (34.0-m)-long 45° skewed bridge in Iowa is shown in Figure 50.23.

FIGURE 50.23   Single-span bridge strengthened by post-tensioning.

Continuous Spans

Similar to the single-span bridges, Iowa has a large number of continuous-span composite steel stringer bridges that also have excessive flexural stresses. Through laboratory tests, it was determined that the desired stress reduction could be obtained by post-tensioning the positive moment regions of the various stringers in most situations. In the cases in which there are excessive overstresses in the negative moment regions, it may be necessary to use superimposed trusses (see Figure 50.24) on the exterior stringers in addition to post-tensioning the positive moment regions. Similar to single-span bridges, it was decided to use force fractions and moment fractions to determine the distribution of strengthening forces in a given bridge. As one would expect, the design procedure is considerably more involved for continuous-span bridges as one has to consider transverse and longitudinal distribution of forces.

The required strengthening forces and final stringer envelopes should be calculated. The various strengthening schemes that can be used are shown in Figure 50.25. A designer selects the schemes required for obtaining the desired stress reduction. For additional details on the strengthening procedure the reader is referred to Reference [76]. Shown in Figure 50.26 is a three-span continuous bridge near Mason City, IA, that has been strengthened using the schemes shown in Figure 50.25.

50.6   Developing Additional Bridge Continuity

50.6.1   Addition of Supplemental Supports

Description

Supplemental supports can be added to reduce span length and thereby reduce the maximum positive moment in a given bridge. By changing a single-span bridge to a continuous, multiple-span bridge,

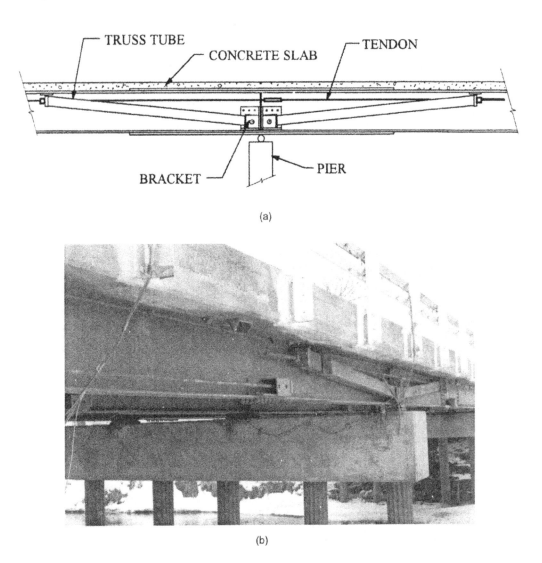

(a)

(b)

FIGURE 50.24 Superimposed truss system. (a) Superimposed truss; (b) photograph of superimposed truss.

stresses in the bridge can be altered dramatically, thereby improving the maximum live-load capacity of the bridge. Even though this method may be quite expensive because of the cost of adding an additional pier(s), it may still be desirable in certain situations.

Applicability and Advantages

This method is applicable to most types of stringer bridges, such as steel, concrete, and timber, and has also been used on truss bridges [7]. Each of these types of bridges has distinct differences.

If a supplemental center support is added to the center of an 80-ft (24.4-m)-long steel stringer bridge that has been designed for HS20-44 loading, the maximum positive live-load moment is reduced from 1164.9 ft-kips (1579.4 kN·m) to 358.2 ft-kips (485.7 kN·m), which is a reduction of over 69%. At the same time, however, a negative moment of 266.6 ft-kips (361.5 kN·m) is created which must be taken into account. In situations where the added support cannot be placed at the center, reductions in positive moments are slightly less.

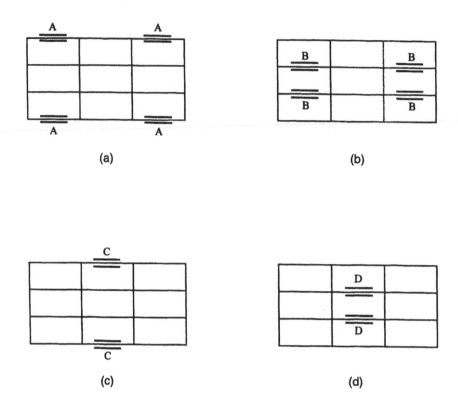

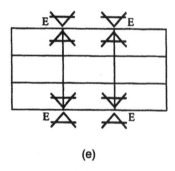

FIGURE 50.25 Strengthening schemes for continuous-span bridge. (a) Strengthening Scheme A; post-tensioning end spans of the exterior stringers; (b) strengthening Scheme B: post-tensioning end spans of the interior stringers; (c) strengthening Scheme C: post-tensioning center spans of the exterior stringers; (d) strengthening Scheme D: post-tensioning center spans of the interior strangers; (e) strengthening Scheme E: superimposed trusses at the piers of the exterior stringers.

Limitations and Disadvantages

Depending on the type of bridge, there are various limitations in this method of strengthening. First, because of conditions directly below the existing bridge, there may not be a suitable location for the pier, as, for example, when the bridge requiring strengthening passes over a roadway or railroad tracks. Other constraints, such as soil conditions, the presence of a deep gorge, or stream velocity, could greatly increase the length of the required piles, making the cost prohibitive.

FIGURE 50.26 Photograph of three-span continuous bridge strengthened with post-tensioning and superimposed trusses.

This method is most cost-effective with medium- to long-span bridges. This eliminates most timber stringer bridges because of their short lengths. In truss bridges, the trusses must be analyzed to determine the effect of adding an additional support. All members would have to be examined to determine if they could carry the change in force caused by the new support. Of particular concern would be members originally designed to carry tension, but which because of the added support must now carry compressive stresses. Because of these problems, the emphasis in this section will be on steel and concrete stringer bridges.

Design Considerations

Because the design of each intermediate pier system is highly dependent on many variables such as the load on pier, width and height of bridge, and soil conditions, it is not feasible to include a generalized design procedure for piers. The engineer should use standard pier design procedures. A brief discussion of several of the more important considerations (condition of the bridge, location of pier along bridge, soil condition, type of pier, and negative moment reinforcement) is given in the following paragraphs.

Providing supplemental support is quite expensive; therefore, the condition of the bridge is very important. If the bridge is in good to excellent condition and the only major problem is that the bridge lacks sufficient capacity for present-day loading, this method of strengthening should be considered. On the other hand, if the bridge has other deficiencies, such as a badly deteriorated deck or insufficient roadway width, a less expensive strengthening method with a shorter life should be considered.

The type of pier system employed greatly depends on the loading and also the soil conditions. The most common type of pier system used in this method is either steel H piles or timber piles with a steel or timber beam used as a pier cap. A method employed by the Florida Department of Transportation [77] can be used to install the piles under the bridge with limited modification to the existing bridge. This method consists of cutting holes through the deck above the point of application of the piles. Piles are then driven into position through the deck. The piles are then cut off so that a pier cap and rollers can be placed under the stringers. Other types of piers, such as concrete pile bents, solid piers, or hammerhead piers, can also be used; however, cost may restrict their use.

Another major concern with this method is how to provide reinforcement in the deck when the region in the vicinity of the support becomes a negative moment region. With steel stringers the bridge may either be composite or noncomposite. If noncomposite, the concrete deck is not required to carry any of the negative moment and therefore needs no alteration. On the other hand, if composite action exists, the deck in the negative moment region should be removed and replaced with a properly reinforced deck. For concrete stringer bridges the deck in the negative moment region should be removed. Reinforcement to ensure shear connection between the stringers and deck must be installed and the deck replaced with a properly reinforced deck. This method, although expensive and highly dependent on the surroundings, may be quite effective in the right situation.

50.6.2 Modification of Simple Spans

Description

In this method of strengthening, simply supported adjacent spans are connected together with a moment and shear-type connection. Once this connection is in place, the simple spans become one continuous span, which alters the stress distribution. The desired decrease in the maximum positive moment, however, is accompanied by the development of a negative moment over the interior supports.

Applicability and Advantages

This method can be used primarily with steel and timber bridges. Although it could also be used on concrete stringer bridges, the difficulties in structural connecting to adjacent reinforced concrete beams result in the method being impractical. The stringer material and the type of deck used will obviously dictate construction details. Thus, the main advantage of this procedure is that it is possible to reduce positive moments (obviously the only moments present in simple spans) by working over the piers and not near the midspan of the stringers. This method also reduces future maintenance requirements because it eliminates a roadway joint and one set of bearings at each pier where continuity is provided [12].

Limitations and Disadvantages

The main disadvantage of modifying simple spans is the negative moment developed over the piers. To provide continuity, regardless of the type of stringers or deck material, one must design for and provide reinforcement for the new negative moments and shears. Providing continuity also increases the vertical reactions at the interior piers; thus, one must check the adequacy of the piers to support the increase in axial load.

Design Considerations

The main design consideration for both types of stringers (steel and timber) concerns how to ensure full connection (shear and moment) over the piers. The following sections will give some insight into how this may be accomplished.

Steel Stringers
Berger [12] has provided information, some of which is summarized here, on how to provide continuity in a steel stringer concrete deck system. If the concrete deck is in sound condition, a portion of it must be removed over the piers. A splice, which is capable of resisting moment as well as shear, is then installed between adjacent stringers. Existing bearings are removed and a new bearing assembly is installed. In most instances, it will be necessary to add new stiffener plates and diaphragms at each interior pier. After the splice plates and bearing are in place, the reinforcement required in the deck over the piers is added and a deck replaced. Such a splice is shown in Figure 50.27.

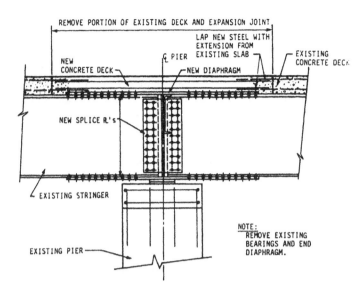

FIGURE 50.27   Conceptual details of a moment- and shear-type connection. (*Source:* Klaiber, F.W. et al., NCHRP 293, Transportation Research Board, 1987. With permission.)

Recently, the Robert Moses Parkway Bridge in Buffalo, New York [78] which originally consisted of 25 simply supported spans ranging from 63 ft (19.2 m) to 77 ft (23.5 m) in length was seismically retrofitted. Moment and shear splices were added to convert the bridge to continuous spans: one two-span element, one three-span element, and five four-span elements. This modification not only strengthened the bridge, but also provided redundancy and improved its earthquake resistance.

Timber Stringers
When providing continuity in timber stringers, steel plates can be placed on both sides and on the top and bottom of the connection and then secured in place with either bolts or lag screws. When adequate plates are used, this provides the necessary moment and shear transfer required. Additional strength can be obtained at the joint by injecting epoxy into the timber cracks as is suggested by Avent et al. [79]. Although adding steel plates requires the design and construction of a detailed connection, significant stress reduction can be obtained through its use.

50.7   Recent Developments

50.7.1   Epoxy Bonded Steel Plates

Epoxy-bonded steel plates have been used to strengthen or repair buildings and bridges in many countries around the world including Australia, South Africa, Switzerland, the United Kingdom, Japan, to mention a few.

The principle of this strengthening technique is rather simple: an epoxy adhesive is used to bond steel plates to overstressed regions of reinforced concrete members. The steel plates are typically located in the tension zone of a beam; however, plates located in the compression and shear zones have also been utilized. The adhesive provides a shear connection between the reinforced concrete beam and the steel plate, resulting in a composite structural member. The addition of plates in the tension zone not only increases the area of tension steel, but also lowers the neutral axis, resulting in a reduction of live-load stresses in the existing reinforcement. The tension plates effectively increase the flexural stiffness, thereby reducing cracking and deflection of the member.

Although this procedure has been used on dozens of bridges in other countries, to the authors' knowledge, it has not been used on any bridges in the United States due to concerns with the method. Some of these concerns are plate corrosion, long-term durability of the bond connection, plate peeling, and difficulties in handling and installing heavy plates.

In recent years, the steel plates used in this strengthening procedure have been replaced with fiber-reinforced plastic sheets; the most interest has been in carbon fiber-reinforced polymer (CFRP) strips. Although CFRP strips have been used to strengthen various types of structures in Europe and Japan for several years, in the United States there have only been laboratory investigations and some field demonstrations. Discussion in the following section is limited to the use of CFRP in plate strengthening. For information on the use of FRP for increasing the shear strength and ductility of reinforced concrete columns in seismic area, the reader is referred to Reference [91]. This reference is a comprehensive literature review of the various methods of seismic strengthening of reinforced columns.

50.7.2 CFRP Plate Strengthening

CFRP strips have essentially replaced steel plates as CFRP has none of the previously noted disadvantages of steel plates. Although CFRP strips are expensive, the procedure has many advantages: less weight, strengthening can be added to the exact location where increased strength is required, strengthening system takes minimal space, material has high tensile strength, no corrosion problems, easy to handle and install, and excellent fatigue properties. As research is still in progress in Europe, Japan, Canada, and the United States on this strengthening procedure, and since the application of CFRP strips obviously varies from structure to structure, rather than providing details on this procedure, several examples of its application will be described in the following paragraphs.

In 1994, legal truck loads in Japan were increased by 25% to 25 tons. After a review of several concrete slab bridges, it was determined that they were inadequate for this increased load. Approximately 50 of these bridges were strengthened using CFRP sheets bonded to the tension face. The additional material not only reduced the stress in the reinforcing bars, it also reduced the deflections in the slabs due to the high modulus of elasticity of the CFRP sheets.

Recently, a prestressed concrete (P/C) beam in West Palm Beach, Florida, which had been damaged by being struck by an overheight vehicle, was repaired using CFRP. This repair was accomplished in 15 hours by working three consecutive nights with minimal disruption of traffic. The alternative to this repair technique was to replace the damaged P/C with a new P/C beam. This procedure would have taken close to 1 month, and would have required some road closures.

The Oberriet–Meiningen three-span continuous bridge was completed in 1963. This bridge over the Rhine River connects Switzerland and Austria. Due to increased traffic loading, it was determined that the bridge needed strengthening. Strengthening was accomplished in 1996 by increasing the deck thickness 3.1 in. (8 cm) and adding 160 CFRP strips 13.1 ft (4 m) long on 29.5-in. (75-cm) intervals to the underside of the deck. The combination of these two remedies increased the capacity of the bridge so that it is in full compliance with today's safety and load requirements.

Three severely deteriorated 70-year-old reinforced concrete frame bridges near Dreselou, Germany, have recently been strengthened (increased flexure and shear capacity) using CFRP plates. Prior to strengthening, the bridges were restricted to 2-ton vehicles. With strengthening, 16-ton vehicles are now permitted to use the bridges. Prior to implementing the CFRP strengthening procedure, laboratory tests were completed on this strengthening technique at the Technical University in Brauwschweigs, Germany.

50.8 Summary

The purpose of this chapter is to identify and evaluate the various methods of strengthening existing highway bridges and to a lesser extent railroad bridges. Although very few references have been made to railroad bridges, the majority of the strengthening procedures presented could in most situations be applied to railroad bridges.

In this chapter, information on five strengthening procedures (lightweight deck replacement, composite action, strengthening of various bridge members, post-tensioning, and development of bridge continuity) have been presented. A brief introduction to using CFRP strips in strengthening has also been included.

In numerous situations, strengthening a given bridge, rather than replacing it or posting it, is a viable economical alternative which should be given serious consideration.

For additional information on bridge strengthening/rehabilitation, the reader is referred to References [1,2,60] which have 208, 379, and 199 references, respectively, on the subject.

Acknowledgments

This chapter was primarily based on NCHRP 12-28(4) "Methods of Strengthening Existing Highway Bridges" (NCHRP Report 293 [2]) which the authors and their colleagues, Profs. K. F. Dunker and W. W. Sanders, completed several years ago. Information from this investigation was supplemented with information from NCHRP Synthesis 249 "Methods for Increasing Live Load Capacity of Existing Highway Bridges" [1], literature reviews, and the results of research projects the authors have completed since submitting the final report to NCHRP 12-28(4).

We wish to gratefully acknowledge the Transportation Research Board, National Research Council, Washington, D.C., who gave us permission to use material from NCHRP 293 and NCHRP Synthesis 249. Information, opinions, and recommendations are from the authors. The Transportation Research Board, the American Association of State Highway and Transportation Officials, and the Federal Highway Administration do not necessarily endorse any particular products, methods, or procedures presented in this chapter.

References

1. Dorton, R. A. and Reel, R., Methods of Increasing Live Load Capacity of Existing Highway Bridges, Synthesis of Highway Practice 249, National Academy Press, Washington, D.C., 1997, 66 pp.
2. Klaiber, F. W., Dunker, K. F., Wipf, T. J., and Sanders, W. W., Jr., Methods of Strengthening Existing Highway Bridges, NCHRP 293, Transportation Research Board, 1987, 114 pp.
3. Sprinkle, M. M., Prefabricated Bridge Elements and Systems. NCHRP Synthesis of Highway Practice 119, Aug. 1985, 75 pp.
4. Shanafelt, G. O. and Horn, W. B., Guidelines for Evaluation and Repair of Prestressed Concrete Bridge Members, NCHRP Report 280, Dec. 1985, 84 pp.
5. Shanafelt, G. O. and Horn, W. B., Guidelines for Evaluation and Repair of Damaged Steel Bridge Members, NCHRP Report 271, June 1984, 64 pp.
6. Applied Technology Council (ATC), Seismic Retrofitting Guidelines for Highway Bridges. Federal Highway Administration, Final Report, Report No. FHWA-RD-83-007, Dec. 1983, 219 pp.
7. Sabnis, G. M., Innovative Methods of Upgrading Structurally and Geometrically Deficient through Truss Bridges, FHWA, Report No. FHWA-RD-82-041, Apr. 1983, 130 pp.
8. University of Virginia Civil Engineering Department, Virginia Highway and Transportation Research Council, and Virginia Department of Highways and Transportation, Bridges on Secondary Highways and Local Roads — Rehabilitation and Replacement, NCHRP Report 222, May 1980, 73 pp.
9. Mishler, H. W. and Leis, B. N., Evaluation of Repair Techniques for Damaged Steel Bridge Members: Phase I, Final Report, NCHRP Project 12-17, May 1981, 131 pp.
10. Shanafelt, G. O. and Horn, W. B., Damage Evaluation and Repair Methods for Prestressed Concrete Bridge Members, NCHRP Report 226, Nov. 1980, 66 pp.
11. University of Virginia Civil Engineering Department, Virginia Highway and Transportation Research Council, and Virginia Department of Highways and Transportation, Bridges on Secondary Highways and Local Roads — Rehabilitation and Replacement. NCHRP Report 222, May 1980, 73 pp.

12. Berger, R. H., Extending the Service Life of Existing Bridges by Increasing Their Load Carrying Capacity, FHWA Report No. FHWA-RD-78-133, June 1978, 75 pp.

13. Fisher, J. W., Hausamann, H., Sullivan, M. D., and Pense, A. W., Detection and Repair of Fatigue Damage in Welded Highway Bridges, NCHRP Report 206, June 1979, 85 pp.

14. Lichenstein, A. G., Bridge Rating through Nondestructive Load Testing, NCHRP Project 12-28(13)A, 1993, 117 pp.

15. Timmer, D. H., A study of the concrete filled steel grid bridge decks in Ohio, in *Bridge Maintenance and Rehabilitation Conference*, Morgantown, WV, Aug. 13–16, 1980, 422–475.

16. DePhillips, F. C., Bridge deck installed in record time, *Public Works*, 116(1), 76–77, 1985.

17. Bakht, B. and Tharmabala, T., Steel-wood composite bridges and their static load performance, in *Can. Soc. Civ. Eng. Annual Conference*, Saskatoon, May 27–31, 1985, 99–118.

18. Csagoly, P. F. and Taylor, R. J., A structural wood system for highway bridges, in *International Association for Bridge and Structural Engineering 11th Congress*, Final Report, Italy, Aug. 31–Sept. 5, 1980, 219–225.

19. Taylor, R. J., Batchelor, B. D., and Vandalen, K., Prestressed wood bridges, *Proc. International Conference on Short and Medium Span Bridges*, Toronto, Aug. 8–12, 1982, 203–218.

20. Ministry of Transportation and Communications, Design of Wood Bridge Using the Ontario Highway Bridge Design Code, Ontario, Canada, 1983, 72 pp.

21. Ministry of Transportation and Communications, Design of Prestressed Wood Bridges Using the Ontario Highway Bridge Design Code, Ontario, Canada, 1983, 30 pp.

22. Muchmore, F. W., Techniques to bring new life to timber bridges, *ASCE J. Struct. Eng.*, 110(8), 1832–1846, 1984.

23. Mackie, G. K., Recent uses of structural lightweight concrete, *Concrete Constr.*, 30(6), 497–502, 1985.

24. Steel grids rejuvenate old bridges, in *The Construction Advisor*, Associated General Contractors of Missouri, Jefferson City, MO, May 1974, 18–19.

25. "Lightweight decking rehabs downrated Ohio River span, *Rural Urban Roads*, 20(4), 25–26, 1982.

26. CAWV Members Join Forces to Reinforce Bridge, Reprint from *West Virginia Construction News*, 1982, 3 pp.

27. The rehabilitation of the Old York Road Bridge, *Rural Urban Roads*, 21(4), 22–23, 1983.

28. Campisi, V. N., Exodermic deck systems: a recent development in replacement bridge decks, *Modern Steel Constr.*, 26(3), 28–30, 1986.

29. Holm, T. A., Structural lightweight concrete for bridge redecking, *Concrete Constr.*, 30(8), 667–672, 1985.

30. Ford, J. H., Use of Precast, Prestressed Concrete for Bridge Decks, Progress Report, Purdue University and Indiana State Highway Commission, Joint Highway Research Project No. C-36-56N, No. 20, July 1969, 161 pp.

31. Kropp, P. K., Milinski, E. L., Gutzwiller, M. J., and Lee, R. H., Use of Precast-Prestressed Concrete for Bridge Decks, Final Report, Purdue University and Indiana State Highway Commission, Joint Highway Research Project No. C-36-56N, July 1975, 85 pp.

32. Greiner Engineering Sciences, Inc., Widening and replacement of concrete deck of Woodrow Wilson Memorial Bridge, Paper presented at Session 187 (Modular Bridge Decks), 62nd Annual Transportation Research Board Meeting, Jan. 20, 1983, 16 pp.

33. Nickerson, R. L., Bridge rehabilitation–construction view expediting bridge redecking, in *Proc. 2nd Annual International Bridge Conference*, Pittsburgh, PA, June 17–19, 1985, 5–9.

34. Stemler, J. R., Aluminum orthotropic bridge deck system, Paper presented at 3rd Annual International Bridge Conference, Pittsburgh, PA, June 2–4, 1986, 62–65.

35. Woeltinger, O. and Bock, F., The alteration of the High Bridge Levensau over the North-East Canal [Der Umbau der Hochbruecke Levensau über den Nord-Ostsee-Kanal.] *Stahlbau*, West Germany, 23(12), 295–303, 1956 [in German].

36. Freudenberg, G., Raising the load capacity of an old bridge, [Erhoehung der Tragfaehigkeit einer alten Bruecke], *Stahlau*, West Germany, 48(3), 76–78, 1979 [in German].

37. Bridge rebuilt with composite design, *Eng. News Rec.*, 165(20), 91, 1960.
38. Collabella, D., Rehabilitation of the Williamsburg and Queensboro Bridges — New York City, *Municipal Eng. J.*, 70 (Summer), 1984, 30 pp.
39. Cook, J. P., The shear connector, *Composite Construction Methods,* John Wiley & Sons, New York, 1977, 168–172.
40. Klaiber, F. W., Dedic, D. J., Dunker, K. F., and Sanders, W. W., Jr., Strengthening of Existing Single Span Steel Beam and Concrete Deck Bridges, Final Report — Part I, Engineering Research Institute Project 1536, ISU-ERI-Ames-83185, Iowa State University, Feb. 1983, 185 pp.
41. Dedic, D. J. and Klaiber, F. W., High strength bolts as shear connectors in rehabilitation work, *Concrete Int. Des. Constr.* 6(7), 41–46, 1984.
42. Dallam, L. N., Pushout Tests with High-Strength Bolt Shear Connectors, Missouri Cooperative Highway Research Program Report 68-7, Engineering Experiment Station, University of Missouri-Columbia, 1968, 66 pp.
43. Dallam, L. N., Static and Fatigue Properties of High-Strength Bolt Shear Connectors, Missouri Cooperative Highway Research Program Report 70-2, Engineering Experiment Station, University of Missouri-Columbia, 1970, 49 pp.
44. Dunker, K. F., Klaiber, F. W., Beck, B. L., and Sanders, W. W., Jr., Strengthening of Existing Single-Span Steel-Beam and Concrete Deck Bridges, Final Report — Part II, Engineering Research Institute Project 1536, ISU-ERI-Ames-85231, Iowa State University, March 1985, 146 pp.
45. Watter, F., Albrecht, P., and Sahli, A. H., End bolted cover plates, *ASCE J. Struct. Eng.*, 111(6), 1235–1249, 1985.
46. Park, S. H., *Bridge Rehabilitation and Replacement (Bridge Repair Practice)*, S. H. Park, Trenton, NJ, 1984, 818 pp.
47. Albrecht, P., Watter, F., and Sahli, A., Toward fatigue-proofing cover plate ends, in *Proc. W. H. Munse Symposium on Behavior of Metal Structures, Research to Practice,* ASCE National Convention, Philadelphia, PA, May 17, 1983, 24–44.
48. Taylor, R., Strengthening of reinforced and prestressed beams, *Concrete*, 10(12), 28–29, 1976.
49. Warner, R. F., Strengthening, stiffening and repair of concrete structures, *Int. Assoc. Bridge Struct. Eng. Surv.*, 17 May. 25–41, 1981.
50. Westerberg, B., Strengthening and repair of concrete structures, [Forstarkning och reparation av betongkonstruktioner], *Nord. Betong*, Sweden, 7–13, 1980 [in Swedish].
51. Mancarti, G. D., Resurfacing, restoring and rehabilitating bridges in California, in *Proc. International Conference on Short and Medium Span Bridges*, Toronto, Aug. 8–12, 1982, 344–355.
52. Rodriguez, M., Giron H., and Zundelevich, S., Inspection and design for the rehabilitation of bridges for Mexican railroads, *Proc. 2nd Annual International Bridge Conference*, Pittsburgh, PA, June 17–19, 1985, 12 pp.
53. Warner, R. F., Strengthening, stiffening and repair of concrete structures, in *Proc. International Symposium on Rehabilitation of Structures*, Maharastra Chapter of the American Concrete Institute, Bombay, Dec. 1981, 187–197.
54. OECD, Organization for Economic Co-Operation and Development Scientific Expert Group, *Bridge Rehabilitation and Strengthening*, Paris, 1983, 103 pp.
55. Suter, R. and Andrey, D., Rehabilitation of bridges, [Assainissement de ponts] *Inst. Statique et Structures Beton Arme et Precontraint*, Switzerland, 106 (Mar.) 105–115, 1985, [in French].
56. Dilger, W. H. and Ghali, A., Remedial measures for cracked webs of prestressed concrete bridges, *J. PCI*, 19(4), 76–85, 1984.
57. West, J. D., Some methods of extending the life of bridges by major repair or strengthening, *Proc. ICE*, 6, (Session 1956–57) 183–215, 1957.
58. Stratton, F. W., Alexander, R., and Nolting, W., Development and Implementation of Concrete Girder Repair by Post-Reinforcement, Kansas Department of Transportation, May 1982, 31 pp.
59. Soliman, M. I., Repair of distressed reinforced concrete columns, in *Canadian Society for Civil Engineering Annual Conference*, Saskatoon, Canada, May 27–31, 1985, 59–78.

60. Wipf, T. J., Klaiber, F. W., and Russo, F. M., Evaluation of Seismic Retrofit Methods for Reinforced Concrete Bridge Columns, Technical Report NCEER-97-0016 National Center for Earthquake Engineering Research, Buffalo, NY, Dec. 1997, 168 pp.

61. Karamchandani, K. C., Strengthening of Reinforced Concrete Members, in *Proc. International Symposium on Rehabilitation of Structures*, Maharastra Chapter of the American Concrete Institute, Bombay, Dec. 1981, 157–159.

62. Klein, G. J. and Gouwens, A. J., Repair of columns using collars with circular reinforcement, *Concrete Int. Des. Constr.*, 6(7) 23–31, 1984.

63. Rheinisches Strassenbauamt Moenchengladbach, Rehabilitation of Structure 41 in Autobahnkreuz Holz, [Erfahrungsbericht–Sanierung des Bauwerks Nr. 41 in Autobahnkreuz Holz] Rheinisches Strassenbauamt Moenchengladbach, Germany, 1983, 11 pp. [in German].

64. Lee, D. H., Prestressed concrete bridges and other structures, *Struct. Eng.*, 30(12), 302–313, 1952.

65. Jungwirth, D. and Kern, G., Long-term maintenance of prestressed concrete structures — prevention, detection and elimination of defects, [Langzeitverhalten von Spannbeton — Konstruktionen Verhueten, Erkennen und Beheben von Schaden], *Beton- Stahlbetonbau*, West Germany, 75(11),262–269, 1980, [in German].

66. Mancarti, G. D., Strengthening California's Steel Bridges by Prestressing, TRB Record 950, Transportation Research Board (1984) 183–187.

67. Vernigora, E., Marcil, J. R. M., Slater, W. M., and Aiken, R. V., Bridge rehabilitation and strengthening by continuous post-tensioning, *J. PCI*, 14(2) 88–104, 1969.

68. Ferjencik, P. and Tochacek, M., *Prestressing in Steel Structures*, [Die Vorspannung in Stahlbau], Wilhelm Ernst & Sohn, West Germany, 1975, 406 [in German].

69. Kandall, C., Increasing the load-carrying capacity of existing steel structures, *Civ. Eng.*, 38(10), 48–51, 1968.

70. Benthin, K., Strengthening of Bridge No. 3699, Chaska, Minnesota, Minnesota Department of Transportation, 1975, 11 pp.

71. Andrey, D. and Suter, R., *Maintenance and Repair of Construction Works*, [Maintenance et reparation de ouvrages d'art], Ecole Polytechnique Federale de Lausanne, Lausanne, Switzerland, 1986, [in French].

72. Mueller, T., Alteration of the highway bridge over the Aare River in Aarwangen, [Umbau der Strassenbruecke über die Aare in Aarwangen] *Schweiz. Bauz.*, 87(11), 199–203, 1969, [in German].

73. Taylor, R. J. and Walsh, H., Prototype Prestressed Wood Bridge, *TRB Report 950*, Transportation Research Board, 1984, 110–122.

74. Lamberson, E. A., Post-Tensioning Concepts for Strengthening and Rehabilitation of Bridges and Special Structures: Three Case Histories of Contractor Initiated Bridge Redesigns, Dywidag Systems International, Lincoln Park, NJ, 1983, 48 pp.

75. Dunker, K. F., Klaiber, F. W., and Sanders, W. W., Jr., Post-tensioning distribution in composite bridges, *J. Struct. Eng. ASCE*, 112 (ST11), 2540–2553, 1986.

76. El-Arabaty, H. A., Klaiber, F. W., Fanous, F. S., and Wipf, T. J., Design methodology for strengthening of continuous-span composite bridges, *J. Bridge Eng., ASCE*, 1(3), 104–111, 1996.

77. Roberts, J., Manual for Bridge Maintenance Planning and Repair Methods, Florida Department of Transportation, 1978, 282 pp.

78. Malik, A. H., Seismic Retrofit of the Robert Moses Parkway Bridge, in *Proc. 12th U.S.–Japan Bridge Engineering Workshop*, Buffalo, NY, Oct. 1996, 215–228.

79. Avent, R. R., Emkin, L. Z., Howard, R. H., and Chapman, C. L., Epoxy-repaired bolted timber connections, *J. Struct. Div. Proc. ASCE*, 102(ST4), 821–838, 1976.

Section VI
Special Topics

51

Applications of Composites in Highway Bridges

Joseph M. Plecnik
California State University,
Long Beach

Oscar Henriquez
California State University,
Long Beach

51.1   Introduction

Building a functional transportation infrastructure is a high priority for any nation. Equally important is maintaining and upgrading its integrity to keep pace with increasing usage, higher traffic loads, and new technologies. At present, in the United States a great number of bridges are considered structurally deficient, and many are restricted to lighter traffic loads and lower speeds. Such bridges need to be repaired or replaced. This task may be achieved by using the same or similar technologies and materials used originally for their initial construction many years ago. However, new materials and technologies may provide beneficial alternatives to traditional materials in upgrading existing bridges, and in the construction of new bridges. Composite materials offer unique properties that may justify their gradual introduction into bridge repair and construction.

The difference between industrial or commercial composites and advanced composites is vague but based primarily on the quality of materials. Advanced composites utilize fibers, such as graphite and Kevlar®, and matrix materials of higher strength and modulus of elasticity than industrial composites, which usually are fabricated with E-glass or S-glass fibers and with polyester or vinyl ester matrices. Advanced composites use polymer matrix materials, such as modified epoxies and polyimides, or ceramic and metal matrices. More-sophisticated manufacturing techniques are generally required to produce advanced composites. Industrial composites require little or no special curing processes, such as the use of autoclaves or vacuum techniques for advanced composites.

Composites are herein limited to materials fabricated with thin fibers or filaments and bonded together in layers or lamina with a polymer matrix. The polymers discussed are primarily polyesters and epoxies. The fibers considered include glass (E and S type), graphite, and Kevlar. The filaments may be of short fiber length (such as chopped fibers which may be less than 25 mm long) or continuous filaments. The ability to orient the fibers in any desired direction is one of the truly great advantages of composite materials as opposed to isotropic materials such as steel. The anisotropic nature of composite materials enables an engineer literally to custom design each element within a structure to achieve the optimum use of material properties.

Composite materials have been successfully utilized in many other industries in the past 50 years. The leisure industry, primarily boating, was probably the first industry to successfully and overwhelmingly adopt composite materials in the construction of pleasure craft and small ships. In the industrial application fields, pipes, tanks, pressure vessels, and a variety of other components manufactured primarily with fiberglass, composite materials have been used for over 50 years. In the defense and aerospace industry, more-advanced composites have been increasingly used since the early 1960s. In all of these industries, one or several unique properties of composites were successfully exploited to replace conventional materials.

The initial study [1] on the use of composites in bridges was performed for the U.S. Federal Highway Administration in the early 1980s and had as its main objective the determination of the feasibility of adapting composites to highway bridges. This study considered the adaptability of composites to major bridge components, given the unique characteristics of these types of materials. The study concluded that bridge decks and cables are the most suitable bridge components for use of composite materials.

The purpose of this chapter is to introduce the current and future technologies and the most feasible applications of composites in highway bridge infrastructure. Basic composite material properties are presented and their advantages and disadvantages discussed. The applications of composites in bridges presented in this chapter include beams and girders, cables, reinforcing bars, decks and wearing surface, and techniques to repair or retrofit existing bridge structures. The methods and significance of nondestructive evaluation techniques are also discussed relative to the feasibility of incorporating composite components into bridge systems.

51.2 Material Properties

51.2.1 Reinforcing Fibers

Fibers provide the reinforcement for the matrix of composite materials. Fiber reinforcement can be found in many forms, from short fibers to very long strands, and from individual fibers to cloth and braided material. The fibers provide most of the strength of the composites since most matrix materials have relatively low strength properties. Thus, fibers in composites function as steel in reinforced concrete.

The most typical fiber materials used in civil engineering composite structures are glass, aramid (Kevlar), and graphite (carbon). A variation in mechanical properties can be achieved with different types of fiber configurations. A comparison of typical values of mechanical properties for common reinforcing fibers is provided in Table 51.1.

TABLE 51.1 Typical Fiber Design Properties

Property	E-Glass (Strand)	S-Glass (Strand)	Kevlar-49 (Yarn)	High-Modulus Graphite (Tow)	High-Strength Graphite (Tow)
Tensile strength (MPa)	3100	3800	3400	2200	3600
Tensile modulus (GPa)	72	86	124	345	235
Specific gravity	2.60	2.50	1.44	1.90	1.80
Tensile elongation (%)	4.9	5.7	2.8	0.6	1.4

TABLE 51.2 Typical Properties of Polymer Resins

Property	Polyester	Epoxy	Phenolic
Tensile strength (MPa)	55	27–90	35–50
Tensile modulus (GPa)	2.0	0.70–3.4	7.0–9.7
Specific gravity	1.25–1.45	1.1–1.4	1.4–1.9
Elongation (%)	5–300	3–50	—
Coefficient of thermal expansion (10^{-6} m/m/K)	70–145	18–35	27–40
Water absorption (% in 24 h)	0.08–0.09	0.08–0.15	0.30–0.50

Glass fiber has been the most common type of reinforcement for polymer matrix. Glass fibers, which are silica based, were the first synthetic fibers commercially available with relatively high modulus. Two common types of glass fibers are designated as E-glass and S-glass. E-glass fibers are good electrical insulators. S-glass fibers, which have a higher silica content, possess slightly better mechanical properties than E-glass. Some applications require fibers with better strength or elastic modulus than glass. Graphite (carbon) and aramid fibers can provide these desired properties. The use of these fibers is generally selective in civil engineering applications, given their higher cost compared with glass fibers.

51.2.2 Matrix Materials

Thermosetting polymer resins are the type of matrix material commonly used for civil engineering applications. Polymers are chainlike molecules built up from a series of monomers. The molecular size of the polymer helps to determine its mechanical properties. Thermosetting polymers, unlike thermoplastic polymers, do not soften or melt on heating, but they decompose. Other matrix materials, such as ceramics and metals, are used for more-specialized applications.

The most common thermosetting resins used in civil engineering applications are polyesters, epoxies, and to a lesser degree, phenolics. A summary of typical properties for resins is provided in Table 51.2. Polyester resins are relatively inexpensive, and provide adequate resistance to a variety of environmental factors and chemicals. Epoxies are more expensive but also have better properties than polyesters. Some of the advantages of epoxies over polyesters are higher strength, slightly higher modulus, low shrinkage, good resistance to chemicals, and good adhesion to most fibers. Phenolic resin is generally used for high-temperature (150 to 200°C) applications and relatively mild corrosive environments.

51.3 Advantages and Disadvantages of Composites in Bridge Applications

The rapid rise in the use of composites in many industries, such as aerospace, leisure, construction, and transportation, is due primarily to significant advantages of composites over conventional materials, such as metals, concrete, or unreinforced plastics. The following presents a brief discussion on the probable advantages and disadvantages of composites in highway bridge type applications.

The first and primary advantage of composites in bridge structures will probably be a significant reduction in weight, due to the higher specific strength (strength/density) of composites over conventional materials, such as steel and concrete. The lightweight advantage of composites in bridge decks is clearly illustrated in Table 51.3. In most short bridge applications, the lighter structural system, if adequate from the structural point of view, will probably not affect the dynamic performance of the bridge. In longer bridges, it is conceivable that a lighter-weight system may require additional design considerations to avoid dynamic behavioral problems.

The second and equally important advantage of composites is their superior corrosion resistance in all environments typically experienced by bridges throughout the world. Corrosion resistance of composites can be further enhanced by the use of premium resin systems, such as vinyl esters or epoxies in comparison with conventional resins, such as polyesters. The excellent corrosion resistance characteristics of composites, and the lower maintenance costs, may result in lower life-cycle costs than those of bridge components manufactured with steel or concrete materials. The lower life-cycle costs may be the third significant advantage of composite bridge components. However, it is anticipated that the initial cost of such composite bridge components will be considerably greater than that of conventional materials.

The fourth significant advantage of composites in bridge applications is their modular construction. It is envisioned that composite bridge deck components will be fabricated in large modules, either in the shop or at the bridge site, then assembled at the bridge site to form a desired structural system. Such modular construction will not only reduce construction costs, but also reduce the time of construction. Fifth, it is envisioned that the initial usage of composites in bridges will involve rehabilitation or retrofitting of existing bridges in large urban areas. The modular construction described above will greatly reduce the time required for retrofitting, thus reducing traffic congestion, accidents, and time delays for commuters in heavily traveled urban areas. The layered structure of composites is also an advantage that may be highly beneficial for fatigue-type loads in bridges. By placing fibers in appropriate directions, both the strength and fatigue resistance of the composite laminate is greatly enhanced. The fatigue behavior of composites when properly designed is superior to that of ductile materials, such as the conventional A36 steel.

The disadvantages of utilizing composites in infrastructure applications such as bridges are considerable, but not overwhelming. The first, but not necessarily the most significant, disadvantage of composites is their relatively high initial costs. This topic was discussed in the previous section relative to initial vs. life-cycle costs. Although graphite and other advanced fibers will probably reduce in cost with increased volume of consumption, it is very doubtful that the cost of glass fibers can be significantly reduced with increasing volume of consumption. The cost of matrices, such as polymer-based resins, will also not be reduced significantly with increased consumption.

The second disadvantage of composite structural systems is the lack of highly efficient mechanical connections. The mechanical bolted connections in composite applications are not as efficient or as easily designed as in the case of steel-type welded and bolted connections used in steel structures. To reduce mechanical-type connections, adhesive-type joints are required. However, adhesion of one part to another requires detailed knowledge of the adhesive and the bond surfaces, as well as quality control. All of these factors generally result in relatively low allowable adhesive stresses. Furthermore, many engineers tend to dislike adhesive-type connections in the presence of fatigue and vibration-type loads.

TABLE 51.3 Comparison of Dead Load (D.L.) of Deck Systems and Superstructure

Bridge Type	Bascule with 1.22 m Stringer Spacing (span = 76.2 m, width = 18.9 m)				Deck on Steel I-Girder 2.13 m Girder Spacing (span = 16.3 m, width = 8.5 m)		Deck on AASHTO Type III Prestressed Girders Spaced at 2.13 m (span = 16.3 m, width = 8.5 m)	
Deck Type	127 mm Open Steel Grid	152 mm Deep X-Shaped FRP with Sand Layer Wearing Surface	127 mm Concrete-Filled Steel Grid	152 mm Deep X-Shaped FRP with Sand Layer Wearing Surface	165 mm Thick Concrete with 5 mm Wearing Surface	229 mm Deep X-Shaped FRP with Sand Layer Wearing Surface	178 mm Thick Concrete with 5 mm Wearing Surface	229 mm Deep X-Shaped FRP with Sand Layer Wearing Surface
Deck weight only (KN)	1379	1155	5654	1155	541.4	157	583.2	157
Deck D.L. % reduction	16		80		71		73	
Girder weight (KN)	3610	3610	3610	3610	111.8	111.8	693.5	693.5
Curbs and railing (KN)	300.3	300.3	300.3	300.3	166.6	166.6	166.6	166.6
Future wearing surface (KN)	0	17.8	0	17.8	166.6	2.2	166.6	2.2
Details, stiffeners, etc. (KN)	290.9	290.9	290.9	290.9	35.7	35.7	23.8	23.8
Inspection walkway (KN)	66.7	66.7	66.7	66.7	None	None	None	None
Total D.L. (KN)	5647	5441	9922	5441	1022	473.3	1634	1043
Total D.L. % reduction	3.6		45		54		36	

The third disadvantage is the relatively low modulus of glass fiber composites. Unless all fibers are oriented in a single direction, the modulus of elasticity of glass-type composites (E- or S-glass) will be somewhat similar to that of concrete. Since design of bridges is often governed by deflection or stiffness criteria, as opposed to strength, the cross-sectional properties of the fiberglass component would have to be nearly identical to that of concrete. The use of high-modulus fibers, such as graphite, enhances the modulus or stiffness characteristics of composites. However, even if all the graphite fibers are placed in the same direction (unidirectional laminate), the modulus of elasticity of the composite may not approach that of steel. Only with the use of very high modulus fibers (above 350 GPa), will the tensile modulus of the composite approach that of steel. To alleviate this very significant stiffness disadvantage, composite structural systems must generally be designed differently when stiffness criteria govern the design.

The fourth significant disadvantage is the relatively low fire resistance of structural composites where polymer-based matrices are used, which represent the bulk of the composites utilized outside of the aerospace industry. This disadvantage has effectively disallowed the use of polymer-based composites in fire-critical applications such as buildings. In bridge applications, fire is a relatively infrequent phenomenon. Elevated temperatures, such as in the southwestern part of the United States, may, however, affect the structural properties of composites on bridge applications.

Several additional disadvantages of composites include relatively complex material properties and current lack of codes and specifications, which tend to dissuade engineers from understanding and utilizing such materials. The presence of local defects, which are difficult and perhaps impossible to detect on a large structural system, are also viewed as a significant quality control problem.

51.4 Pultruded Composite Shapes and Composite Cables

51.4.1 Pultruded Composite Shapes

Composites are commercially available in a variety of pultruded shapes [2–4]. Some of the most common shapes available for construction purposes are I-beams, W-sections, angles, channels, square and rectangular tubes, round tubes, and solid bars. However, almost any shape of constant cross section can be pultruded.

Composite pultruded beam shapes have a potential use in bridges. However, the relatively low modulus of glass and graphite composite shapes limits their use. The effect of modulus of elasticity can be seen with the following comparison between A36 steel beams with modulus of elasticity $E = 200$ GPa, fiber-reinforced polymer (FRP) beams with $E = 17.2$ GPa, graphite beams with $E = 103.5$ GPa, and glass fiber–reinforced polymer (GFRP)/graphite hybrid beams for a two-lane, 16.76-m-span bridge. Assuming full lateral support, a total of five beams spaced at 2.29 m, and a 178-mm-thick concrete slab, the following results are obtained.

For the case of noncomposite action between beams and concrete slab, a steel beam, W36 × 194, with cross-sectional area, $A = 36{,}770$ mm², satisfies all AASHTO requirements for HS20-44 loading [5,6]. Using GFRP beams, the deflection requirement of $L/800$, which controls the design, cannot possibly be satisfied using a depth of 914 mm and a flange width of 457 mm. A GFRP I48 × 24 × 3.25 with $A = 187{,}700$ mm² or I60 × 30 × 1.5 with $A = 113{,}200$ mm² beam is necessary. If an all-graphite beam is used, an I36 × 18 × 1.25 with $A = 56{,}060$ mm² will satisfy all requirements. A hybrid beam with a 1320-mm total depth, 660-mm flange width, and 45.7-mm web and flange thickness also satisfies stiffness requirements. For this example, the hybrid beam should have a 7.6-mm-thick layer of graphite in the center of both flanges, and for the total width of the flange.

If composite action is achieved between the concrete slab and the beams, a W30 × 99 with $A = 18{,}770$ mm² steel beam is adequate for this bridge. An all-FRP I48 × 24 × 1.0 beam with $A = 60{,}650$ mm² or I42 × 21 × 1.5 with $A = 78{,}390$ mm² will also meet stiffness and stress requirements when composite action is included. A comparison of sizes for all the beam cross sections used in this example is presented in Figure 51.1.

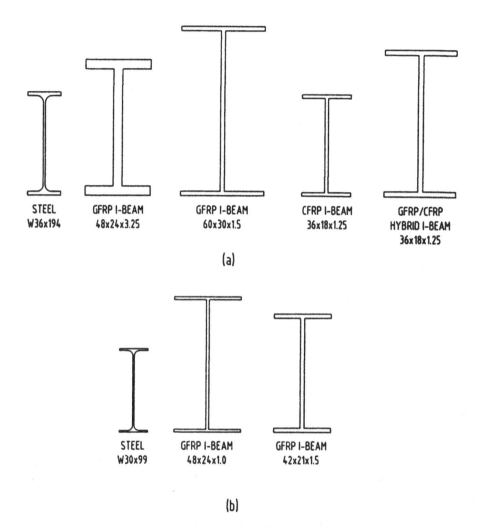

FIGURE 51.1 Comparison of different beams for a two-lane 16.80-m-span bridge, with a total of five beams spaced at 2.30 m, and a 180-mm-thick concrete slab. (a) Noncomposite action between beams and slab; (b) composite action between beams and slabs.

51.4.2 Composite Cables

Composites in the form of cables, strands, and rods have potential applications in bridges. Among these applications are suspension and stay cables and prestressing tendons. High tensile strength, corrosion resistance, and light weight are the most important characteristics that make composites strong candidates to replace steel for these types of applications. Corrosion of traditional steel cables and tendons may impose a significant maintenance cost for bridges. Composite cables, with proper selection of materials and design, may exceed the useful life of traditional bridge cables.

Carbon fiber–reinforced polymer (CFRP) composite cables have been used for cable-stay bridges [7]. Compared with steel, carbon composites can provide the equivalent tensile strength with only a fraction of the weight. GFRP tendons have been used to prestress concrete bridge girders. The computation of section strength using GFRP tendons is very similar to the methods used for steel tendons. In the case of post-tensioned structures, an adequate anchorage system must be used to minimize prestressing losses. One of the advantages of GFRP compared with steel tendons is that a lower modulus of elasticity is translated into lower prestressing losses, due to creep and shrinkage of the concrete.

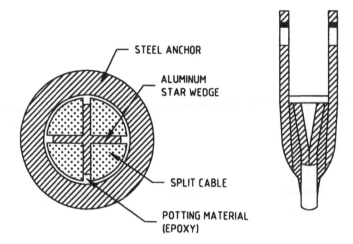

STEEL ANCHOR

ALUMINUM
STAR WEDGE

SPLIT CABLE

POTTING MATERIAL
(EPOXY)

FIGURE 51.2 Potted end anchorage assembly for composite cables; cross sectional view (left) and longitudinal view (right)

A key issue in the design of composite cables is the anchoring system. Development of the full tensile strength of the composite cable is not yet possible; however, a good design of the anchors can allow the development of a large percentage of the total available cable strength. A potted-type anchor is shown in Figure 51.2 [1]. This assembly utilizes a metal end socket into which the composite cable is fitted and subsequently potted with various polymers such as epoxies. The load is transferred from the cable to the metal anchor through the potting material by shear and radial compressive stresses. The aluminum wedge is used to split the cable into four equal sectors to create greater wedging action, but this also creates large radial compressive stresses. Since the largest stresses at the stress transfer region occur at the cable perimeter, several related parameters affect the strength property of such potted anchors and, therefore, the ultimate strength of the cable system.

Another type of anchoring system [7], specifically designed for CFRP cables, utilizes a conical cavity filled with a variable ceramic/epoxy mix (Figure 51.3). The variable formulation is designed to control creep and rupture of the cable.

51.5 FRP Reinforcing Bars for Concrete

Fibers such as glass, aramid, and carbon can be used as reinforcing bars (rebars) for concrete beams. The use of these fibers can increase the longevity of this type of structural element, given the corrosive deterioration of steel reinforcement in reinforced concrete members. Tests have shown that a higher ultimate strength can be achieved with FRP rebars than with mild steel rebars. This strength can be achieved due to the high tensile strength of most fibers. The lower stiffness of FRP fibers, such as glass, will result in larger deflections compared with steel-reinforced concrete.

An important factor in the use of FRP bars is the bond between the bar and the concrete [8]. The use of smooth FRP bars results in a significant reduction of flexural capacity. Thus, smooth FRP bars must be surface-treated to improve bonding by methods such as sand coating. Test results have also shown that smaller-diameter FRP rebars are more effective for flexural capacity than larger-diameter bars. However, in general, bond characteristics are variable due to the variations in FRP reinforcing bar products. Other factors that affect the bond characteristics are concrete strength, concrete confinement, type of loading, time-dependent effects, amount of concrete cover, and type and volume of fiber and matrix. In the State-of-the-Art Report 440R-96 on FRP Reinforcement for Concrete Structures [9], the American Concrete Institute (ACI) recognizes the need for additional testing data to develop expressions that will be valid for different conditions, and can be included in a design code. Some expressions for FRP bar development lengths have been proposed recently.

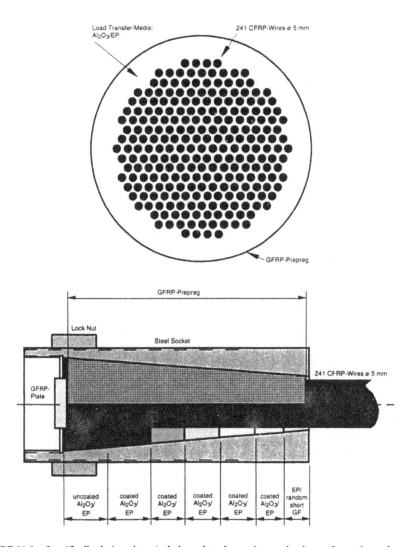

FIGURE 51.3 Specifically designed conical-shaped anchor enhances load transfer to the carbon cable.

51.6 Composite Bridge Decks

51.6.1 Advantages and Disadvantages

The bridge deck appears to be one of the most suitable bridge components for use of structural composites in highway applications [1]. The primary advantages of composite bridge decks are their relative lighter weight, corrosion resistance, and fabrication in modular units which may be rapidly installed without the need for shoring and formwork.

Reference [1] provides the results of a study of bridge dead loads with various types of conventional and composite bridge decks. Table 51.3 provides a summary of this comparison for a 76.2-m-long bascule bridge and two 16.3-m-span conventional bridges. The deck used for the comparison is the X-shaped cross section yielding truss-type deck behavior. The last row of Table 51.3 indicates that for bascule-type bridges with an open steel grid deck, the composite deck would not appreciably reduce the total dead loads of the bridge superstructure (deck, stringers, and girders). However, if the steel grid is filled with concrete, the total dead-load reduction with a composite deck is 45%. Similarly, for conventional bridges, the composite deck reduces the total dead load of the bridge superstructure by up to 54%. If a comparison of bridge decks alone is considered, composite decks are typically about 20 to 30% of the weight of conventional concrete decks as shown in row 3 of Table 51.3. The reduced weights of bridge decks could be translated into:

1. Increased allowable live loads that result from moving traffic on the bridge;
2. Increased number of lanes with the same girders, columns, or piers, resulting in the same total dead and live loads;
3. Continued use of bridge without reducing its load capacity;
4. Reduced construction costs, because a lighter bridge deck requires less construction time and effort than heavier conventional decks.

A composite deck may be made of prefabricated modular units quickly assembled at the bridge site. Due to economics and the need for minimization of joints, it would be desirable that deck sections could be fabricated as large as possible. Modular construction may also translate into relatively short erection time. The quick field assembly will greatly reduce traffic routing costs, a significant advantage in urban areas.

The disadvantages of composite decks include possible higher initial costs, greater deck and girder deflections, and lower bridge stiffness. Although the lighter composite decks will reduce dead loads on the girders, columns, and piers, other structural factors must also be considered. First, the reduced mass of the deck will result in different bridge vibrational characteristics. For long bridges, the reduced mass and stiffness may result in possible vibrational problems and excessive deflections. For short spans, such problems should not occur. On the positive side, composite materials provide higher damping, thereby reducing these vibrational tendencies.

51.6.2 Composite Deck Systems

The choice of a deck configuration should be made on structural and economic feasibility considerations. Structurally, the deck should carry dead loads and specified live loads, and also satisfy deflection requirements. Economically, a composite deck should be cost-effective if it is to replace conventional decks.

The transfer of traffic loads through a composite deck can be achieved mainly by flexure or truss action. The effectiveness of these load-transfer systems depends greatly on the mechanical properties of the materials. Studies [1] have shown that AASHTO stiffness or deflection requirements for decks are difficult to satisfy with low-modulus materials such as GFRP. It was also shown that truss-type load-transfer elements (Figure 51.4a through e) are preferable to sandwich or flexural-type structural elements (Figure 51.4f).

51.6.3 Truss-Type Deck System

A composite deck system that transfers the traffic loads to the stringers and girders mainly by truss action in the transverse direction of the bridge has been developed [1]. Several shapes were studied and evaluated based on AASHTO requirements in order to determine an economical and structurally efficient deck cross section. A deck with a total depth of 229 mm satisfied the stress and stiffness requirements for all the cross sections considered and shown in Figure 51.4. The X-shaped cross section (Figure 51.4b) is the optimum design from the viewpoint of stiffness and dead load. The X-shaped deck transfers live loads primarily by truss action, which provides less deflection than flexure-type members.

In addition to analytical studies, an extensive experimental program has proved the feasibility of the X-shaped cross section for a composite deck [1,10,11]. Specimens were fabricated and subjected to static and fatigue testing under AASHTO loads and the heavier "alternate military load." These specimens sustained over 30 million fatigue cycles without failure or degradation, and suffered only minor overall stiffness loss.

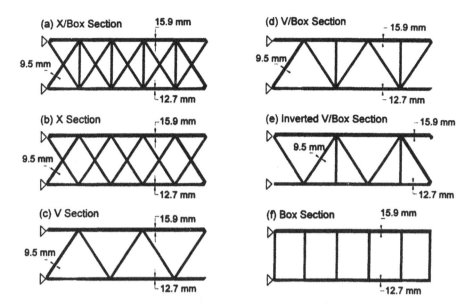

FIGURE 51.4 Shapes considered for design of a truss-type composite bridge deck. (a) X/box section; (b) X section; (c) V section; (d) V/box section; (e) inverted V/box section; (f) box section.

51.7 Wearing Surface for a Composite Deck

The wearing surface of a composite bridge deck should provide laminate protection, adequate skid resistance, and safety against hydroplaning. A thin layer of sand–epoxy mix has been developed for this purpose. The mix consists of sand retained between No. 8 and No. 30 sieves and a matrix-type epoxy. The mix is applied directly to the top surface of the composite FRP deck to a thickness of 1.5 to 3 mm after surface preparation.

The performance of this wearing surface has been evaluated with a series of tests. Freeze/thaw cycling and high-temperature tests were performed to determine the response of the system to weather conditions expected in most parts of the United States.

Simulated truck traffic was applied to evaluate the performance in terms of particle loss and abrasion. Specimens were tested using the accelerated loading facility (ALF), which consists of a frame with a set of truck tires that run along a stretch of pavement. Specimens with the sand–epoxy wearing surface were embedded in the pavement, and their integrity observed during the test. When there is a loss of sand particles, the skid resistance and texture depth of the wearing surface decreases, and the risk of skidding and hydroplaning increases. The surface deterioration was monitored using the British Pendulum method and the sand patch test [12]. Variation of the average British Pendulum Number (BPN) with number of tire passes is shown in Figure 51.5. A stabilized BPN above the minimum acceptable BPN of 60, after 1 million cycles, may indicate that the wearing surface will maintain its serviceability for an extended amount of time. The sand patch readings, shown in Figure 51.6, also indicate a reduction in the rate of mean average texture height loss at an acceptable level.

51.8 Composite Bridge Structural Systems

The use of composite materials to build an entire bridge superstructure is a possibility that is being explored by engineers. An all-composite bridge, designed to meet AASHTO HS25 loading, was built and installed near Russell, Kansas in 1996 [13]. The net span is 7.08 m and the width is 8.45 m. Three side-by-side panels connected by interlocking longitudinal joints were used to construct the

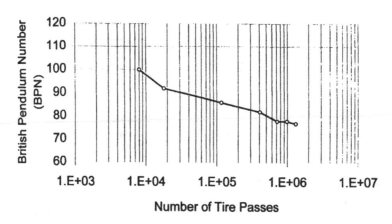

FIGURE 51.5   British pendulum readings for sand–epoxy wearing surface specimen.

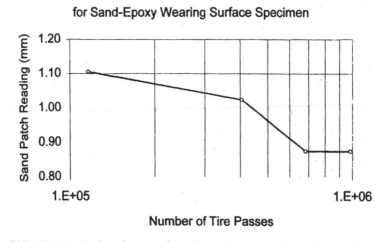

FIGURE 51.6   Sand patch test readings for sand–epoxy wearing surface specimen.

bridge. A 56-cm-deep sandwich construction was used for the panels, whose facing thicknesses were 13 and 19 mm for the top and bottom, respectively. A combination of chopped strand mat and uniaxial fibers was used with polyester resin to build the facings, which are attached by a honeycomb core. The wearing surface was a 19-mm-thick gravel–polyester resin mix.

Another all-composite 10-m-wide by 30-m-long bridge was installed in Hamilton, Ohio in 1997 [14]. This bridge, designed to meet the AASHTO HS20 specification, was fabricated with polyester resin matrix and E-glass fiber reinforcement. The bridge was delivered to the site in three sections, which consisted of two main components that formed the final box beam type of structure. The first component was a tapered U-shaped beam of approximately 0.6 m depth. The webs and the lower flange of the beam were reinforced with stitched triaxial and biaxial fabrics. The flange was also reinforced with additional unidirectional fibers. Beams, the second main component of the structure, were integral with the composite deck. A sandwich panel construction (approximately 15 cm deep) was used for the deck, with flat composite facing plates and a core of pultruded rectangular tubes oriented in the transverse direction of the bridge. The total weight of the composite bridge was approximately 100 kN, including the guardrails, but excluding the asphalt wearing surface.

51.9 Column Wrapping Using Composites

A unique application of composite materials in bridge infrastructure is bridge column wrapping or jacketing. This procedure involves the application of multiple layers of a composite around the perimeter of columns. Since the late 1980s, column wrapping with composites was seen as an alternative to the conventional steel jackets used to retrofit reinforced concrete columns of bridges in California. Column wrapping may also be used to repair columns that suffered a limited amount of damage. Column wrapping with composites may have some advantages over steel jacketing, such as reduced maintenance, improved durability, speed of installation, and reduced interference with ongoing operations, including traffic.

A reinforced concrete column can be retrofitted using the wrapping technique to increase its flexural ductility and shear strength. A proper confinement of the concrete core and longitudinal reinforcement is highly desirable for a ductile design. Confinement has been used to prevent the longitudinal bars from buckling, even after a plastic hinge has formed in the confined region, and to improve the performance of lapped longitudinal reinforcement in regions of plastic hinge formation. The shear capacity of the column can also be increased by wrapping a column using composites.

Several materials have been used to retrofit bridges with the wrapping method. The most common types of fibers used are glass and carbon. Glass fibers have been used with a polyester, vinyl ester, or epoxy matrix, and carbon fibers are used with epoxy resins. Fibers may be applied in various forms, such as individual rovings, mats, and woven fabrics. The California Department of Transportation (Caltrans) uses its Composite Specification to establish standard procedures for selection of the system to be used, material properties, and application.

The most common techniques for application of composite column jackets are wet wrap, prepreg wrap, and precured shells. In the wet wrap technique, a fiberglass fabric is wrapped around the column as many times as required to achieve the design thickness. The fabric is saturated with resin just before or during the application process, and then allowed to cure at ambient temperature. This system is manually applied and does not require special equipment.

The second technique, prepreg wrap, involves the use of continuous prepreg carbon fiber/epoxy tow that is mechanically wound onto the column. External heating equipment may be used to cure the composite.

Another column-wrapping system uses a precured shell. These shells, usually with glass fiber reinforcement, are fabricated with the same curvature as the column. A longitudinal cut is made on one side of the shell, or it may be cut in two longitudinal sections. The cut shell is then fitted and bonded onto the column.

Experimental studies [15] have been performed to determine the effectiveness of the column-wrapping systems using composite materials as compared to steel jacketing. However, even though favorable results have been published, acceptance by bridge owners and design engineers is not yet universal.

51.10 Strengthening of Bridge Girders Using CFRP Laminates

Strengthening and repair of bridge girders have been recently achieved with the use of CFRP. This technique was initially developed in the late 1980s [16] and applied to bridge structures in the early 1990s.

The use of CFRP for girder strengthening is similar to the attachment of steel plates onto concrete girders. However, CFRP presents the advantages of easier handling, higher corrosion resistance, elimination of welded connections, and excellent fatigue behavior. An important factor to be considered when using this technique is the adhesion between the beam and the CFRP strip. The contact surfaces must be adequately prepared, and an effective bond must be developed [17].

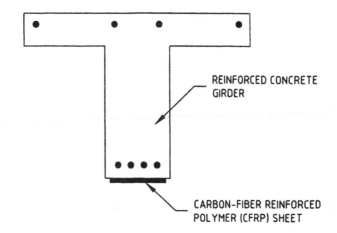

FIGURE 51.7 Strengthening of reinforced concrete girder using CFRP laminate.

In principle, reinforcing with CFRP laminates consists of a CFRP strip bonded onto the tension surface of the girder to increase or restore its original flexural capacity. The CFRP strip can be applied either nontensioned or tensioned. The high strength and stiffness of the CFRP allow the use of very thin layers to achieve the desired capacity, which can be calculated using procedures similar to those used to design traditional concrete beams. Figure 51.7 illustrates this technique.

51.11 Composite Highway Light Poles

Composite light poles were originally developed for nonhighway applications such as parking lots. Circular uniform taper along its length is the most common design. Typical pole lengths vary from 4.30 to 13.7 m. The outside pole diameter typically varies from 73 to 133 mm at the top, and from 122 to 311 mm at the base.

The most common manufacturing process for light poles is filament winding. In this process, the fiberglass filaments are wrapped around a tapered steel mandrel at specified angles to obtain the design shape with a polyester matrix. The filament winding angles vary according to the design requirements.

The primary advantages of FRP light poles are reduced weight and higher corrosion and weathering resistance compared with steel or aluminum poles. The reduced weight allows for lower shipping costs and lower installation costs. The higher corrosion resistance results in reduced maintenance costs and longer life expectancy as opposed to metal poles.

The primary disadvantages of the FRP poles are the complexity of the design and manufacturing process. For light poles, the maximum stresses are in the axial direction. Hence, the filament winding process requires more material than a process which places most or all of the fibers in the axial direction. Nevertheless, fiberglass is the most cost-effective material for use in highway light poles for heights of 7.60 to 10.7 m.

51.12 Nondestructive Evaluation of Composite Bridge Systems

The success or failure of composites in adaptation to various components in bridges greatly depends on the ability to evaluate both the short- and the long-term behavior of such composite structural elements using nondestructive evaluation (NDE) techniques. NDE and visual inspection are routinely performed on existing bridges, and new NDE technology is being developed for conventional bridge materials such as concrete and steel. Due to the greater complexity of composite materials in comparison with conventional steel and concrete, it is anticipated that existing or new NDE technology will have to be adapted to composite bridge components in order to justify their usage in critical structural elements.

NDE techniques in composites have been developed primarily in the defense and advanced technology industries. These techniques have evolved over the past 30 to 40 years, and their effectiveness in evaluating the performance of composite materials is quite impressive. However, most of these techniques require relatively sophisticated equipment, and are generally localized. Such localized NDE techniques will only describe the current and possibly predict the future behavior of the composite in a very small or localized region. In large-scale structures or structural components such as bridges, such localized techniques have limited significance in terms of the overall behavior of the structure. Therefore, NDE techniques which can evaluate the behavior and performance on a large or global scale are preferable. However, it is unfortunate that at the present time, such global techniques are not sufficiently accurate, too expensive, or not well developed technologically.

The strain gauge method of determining localized stresses and strains is well understood in civil engineering, and is widely used in the analysis of structures such as bridges, both in the laboratory and the field. The strain gauge technique is a localized type of an NDE technique and, therefore, may yield only the stress and strain levels in the lamina to which the strain gauge is attached. Strain gauge data may reveal very little about the possible delamination of inner lamina or the presence of defects within the laminate. The strain gauge method is currently used for evaluation of stresses and strains in composite tanks, pressure vessels, buildings, and various composite bridge applications which have been described here. As in other materials, the strain gauge technique is extremely beneficial in determining stress concentrations at critical locations such as the radii regions of the sections shown in Figure 51.4.

The second NDE technique is acoustic emissions (AE), which was developed more than 40 years ago, but has been successfully adapted to the evaluation of composite materials only within the last 20 years. Although this technique is described as nondestructive, the sounds or the energy emitted by the composite occur when some form of degradation of the laminate is occurring at the time of the applied loads. In simplistic terms, if no AEs are recorded, no degradation of the composite laminate is occurring. This method has been successfully utilized in the evaluation of many aerospace composite components, as well as civil engineering types of composite elements or structures, such as stacks, tanks, pressure vessels, building components, and tanker trucks. In the last application, composite tanker trucks have been evaluated on a regular basis for a period of 15 years using the AE technique. The results from these AE studies have shown the feasibility of predicting the future behavior of composite systems under fatigue loading. The bridge deck in Figure 51.4b has also been evaluated with the AE technique and the results have indicated that it is possible to predict the future behavior of such a composite bridge element utilizing AE data collected at different fatigue cycles as discussed in Section 51.6. Since AE sensors are attached at localized points on a composite structure, the data that are gathered only define the behavior of the bridge deck in a relatively localized region. However, the significant advantage of AE over strain gauges is that the behavior of the entire thickness of the laminate can be evaluated.

The continuous graphite filaments technique is a relatively simple and a global NDE method. This method essentially utilizes graphite filaments, which are electrically conductive, embedded in a glass type of composite. Glass composites are nonelectrically conductive. Since the graphite fibers can be chosen with a modulus of more than 10 times that of glass, and with a strain at failure of much less than the corresponding glass filaments, the graphite fiber will fail first within a glass fiber composite. This method is relatively inexpensive and global. The graphite filaments may be embedded anywhere within the glass fiber laminate during the fabrication of the composite structure. Additional graphite filaments may be bonded onto the surfaces of the glass fiber composite before, during, and/or after installation of the structure. This technique has been successfully utilized in determining the critical stress locations on a global scale for the extremely complex bridge deck system shown in Figure 51.4b.

When the graphite filament is broken, the electrical circuit is also broken, thus indicating high stress levels. However, the location of these high stress levels on any single graphite filament circuit is nearly impossible to predict at this time. The open electrical circuit indicates that the strain level within the fiberglass structure is excessive, but failure of the overall structure will not occur. Thus, the primary intent of such a graphite filament technique is to signal existing degradation, and possible future failure of the bridge, many truck cycles before it actually occurs. This technique may be utilized to provide a warning to the public, and cause a bridge to be shut down prior to impending failure.

The visual inspection method for composites has also been codified into an ASTM specification. Such inspections would be very similar to current periodic visual bridge inspections of steel and concrete bridges. Visual inspections are global in nature but cannot detect any possible degradation of the interior lamina within a laminate. This is a distinctive drawback to any method that involves visual inspection of external surfaces only.

Ultrasonic NDE has been widely used as an NDE technique in advanced and aerospace composites. In industrial composites, such as tanks, pipes, etc., the ultrasonic technique has been limited to determining thicknesses, detecting localized defects or voids, crack formations, and delaminations. The ultrasonic technique is a localized NDE technique and relatively time-consuming and labor-intensive. Extensive computer imaging is possible with this and other techniques discussed below which can greatly enhance the accuracy of this method.

The fiber-optics technique is analogous to the continuous graphite filament concept. The fiber-optics technology utilizes continuous fiber-optic cables which can be embedded in the laminate or on exterior surfaces. The presence of localized stress concentrations results in reduced transmission of light through the fiber-optic cable which can be related to the level of localized stresses. This technique may also predict the location of high stress levels, which continuous graphite filaments cannot do.

A variety of advanced NDE techniques are utilized in evaluation of advanced composites but are seldom used in industrial composites. Thermal NDE methods essentially use the theory of heat flow in laminates where the presence of defects, voids, or delaminations will alter the heat transfer properties. Radiographic NDE techniques utilize the transmission of electromagnetic waves through materials, and the knowledge that the presence of defects, voids, or delaminations will result in alteration of such wave transmissions. Both of these two methods may be used in local or global applications. Computer imaging may greatly enhance the effectiveness of both these techniques. However, due to the current cost, both of these techniques are economically prohibitive for periodic evaluation of large composite components as envisioned for bridges.

Other advanced NDE techniques are also available in the advanced composites industry. However, most of these techniques are currently cost-prohibitive or impractical for field evaluation of composite bridge components under less than laboratory type conditions.

51.13 Summary

The inclusion of composites into highway bridges will probably occur gradually over the next decade. Due to the strict stiffness and safety requirements, the use of composites in all structural elements of highway bridges may not be feasible in the near future. Therefore, the initial use of composites in bridges will probably be limited to those bridge elements where the unique properties of composites will result in more favorable design than with the use of conventional materials.

References

1. Plecnik, J. M. and Ahmad, S. H., Transfer of Composites Technology to Design and Construction of Bridges, Final Report Prepared for the U.S. Department of Transportation, Federal Highway Administration, Sept. 1989.
2. Creative Pultrusions, Inc., Design Guide — Standard and Custom Fiberglass-Reinforced Structural Shapes, Alum Bank, PA.
3. Morrison Molded Fiberglass Company (MMFG), Engineering Manual — EXTREN® Fiberglass Structural Shapes, Bristol, VA, 1989.
4. NUPLA Corporation, Pultruded Shapes Catalog, Sun Valley, CA, 1995.
5. American Association of State Highway and Transportation Officials, *Standard Specifications for Highway Bridges*, 16th ed., AASHTO, Washington, D.C., 1996.
6. American Institute of Steel Construction, *Manual of Steel Construction — Allowable Stress Design*, 8th ed., AISC, Chicago, IL, 1989.
7. Meier, U. and Meier, H., CFRP finds use in cable support for bridge, *Mod. Plast.*, 73(4), 87, 1996.
8. Ehsani, M. R., Saadatmanesh, H., and Tao, S., Design recommendations for bond of GFRP rebars to concrete, *J. Struct. Eng.*, 122(3), 247, 1996.
9. American Concrete Institute, State-of-the-Art Report on Fiber Reinforced Plastic (FRP) Reinforcement for Concrete Structures, Reported by ACI Committee 440, 1996.
10. Plecnik, J. and Henriquez, O., Composite bridges and NDE applications, in *Proceedings: Conference on Nondestructive Evaluation of Bridges*, Arlington, VA, August 22–27, 1992.
11. Plecnik, J. M., Henriquez, O. E., Cooper, J., and Munley, E., Development of an FRP system for bridge deck replacement, paper presented at U.S.–Canada–Europe Workshop on Bridge Engineering, Zurich, Switzerland, July 1997.
12. Lemus, J., Wearing Surface Studies on Accelerated Loading Facility (ALF), California State University, Long Beach, Report No. 97-10-16, Long Beach, CA, 1997.
13. Plunkett, J. D., Fiber-Reinforced Polymer Honeycomb Short Span Bridge for Rapid Installation, IDEA Project Final Report, Transportation Research Board, National Research Council, June 1997.
14. Dumlao, C., Lauraitis, K., Abrahamson, E., Hurlbut, B., Jacoby, M., Miller, A., and Thomas, A., Demonstration low-cost modular composite highway bridge, paper presented at 1st International Conference on Composites in Infrastructure, Tucson, AZ, 1996.
15. Seible, F., Priestley, M. J. N., Hegemier, G. A., and Innamorato, D., Seismic retrofit of RC columns with continuous carbon fiber jackets, *J. Composites Constr.*, 1(2), 52, 1997.
16. Meier, U., Deuring, M., Meier, H., and Schwegler, G., Strengthening of structures with CFRP laminates: research and applications in Switzerland, in *Advanced Composite Materials in Bridges and Structures*, Canadian Society for Civil Engineers, 1992, 243.
17. Arduini, M. and Nanni, A., Behavior of precracked RC beams strengthened with carbon FRP sheets, *J. Composites Constr.*, 1(2), 63, 1997.

52

Effective Length of Compression Members

Lian Duan
California Department of Transportation

Wai-Fah Chen
Purdue University

52.1 Introduction   *

The concept of *effective length factor* or *K factor* plays an important role in compression member design. Although great efforts have been made in the past years to eliminate the K factor in column design, K factors are still popularly used in practice for routine design [1].

Mathematically, the effective length factor or the *elastic K* factor is defined as

$$K = \sqrt{\frac{P_e}{P_{cr}}} = \sqrt{\frac{\pi^2 E I}{L^2 P_{cr}}} \tag{52.1}$$

where P_e is Euler load, elastic buckling load of a pin-ended column, P_{cr} is elastic buckling load of an end-restrained framed column, E is modulus of elasticity, I is moment of inertia in the flexural buckling plane, and L is unsupported length of column.

---

* Much of the material of this chapter was taken from Duan, L. and Chen, W. F., Chapter 17: Effective length factors of compression members, in *Handbook of Structural Engineering,* Chen, W. F., Ed., CRC Press, Boca Raton, FL, 1997.

0-8493-7434-0/00/$0.00+$.50
© 2000 by CRC Press LLC

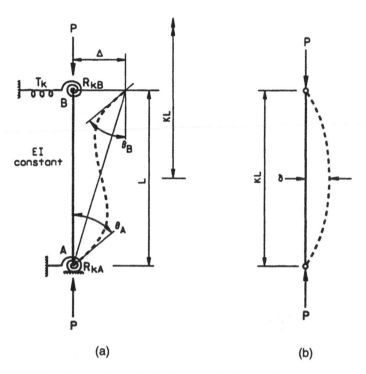

FIGURE 52.1   Isolated columns. (a) End-restrained columns; (b) pin-ended columns.

Physically, the *K* factor is a factor that, when multiplied by actual length of the end-restrained column (Figure 52.1a), gives the length of an equivalent pin-ended column (Figure 52.1b) whose buckling load is the same as that of the end-restrained column. It follows that the *effective length KL* of an end-restrained column is the length between adjacent inflection points of its pure flexural buckling shape.

Practically, design specifications provide the resistance equations for pin-ended columns, while the resistance of framed columns can be estimated through the *K* factor to the pin-ended column strength equations. Theoretical *K* factor is determined from an elastic eigenvalue analysis of the entire structural system, while practical methods for the *K* factor are based on an elastic eigenvalue analysis of selected subassemblages. This chapter presents the state-of-the-art engineering practice of the effective length factor for the design of columns in bridge structures.

52.2   Isolated Columns

From an eigenvalue analysis, the general *K* factor equation of an end-restrained column as shown in Figure 52.1 is obtained as

$$
\mathbf{det}
\begin{vmatrix}
C + \dfrac{R_{kA}L}{EI} & S & -(C+S) \\[2.5ex]
S & C + \dfrac{R_{kB}L}{EI} & -(C+S) \\[2.5ex]
-(C+S) & -(C+S) & 2(C+S) - \left(\dfrac{\pi}{K}\right)^2 + \dfrac{T_k L^3}{EI}
\end{vmatrix}
= 0
\qquad (52.2)
$$

	(a)	(b)	(c)	(d)	(e)	(f)
Buckled shape of column is shown by dashed line	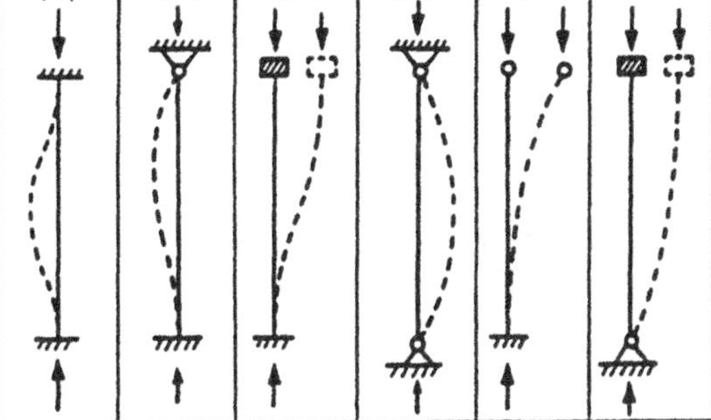					
Theoretical K value	0.5	0.7	1.0	1.0	2.0	2.0
Recommended design value when ideal conditions are approximated	0.65	0.80	1.2	1.0	2.10	2.0
End condition code	Rotation fixed and translation fixed					
	Rotation free and translation fixed					
	Rotation fixed and translation free					
	Rotation free and translation free					

FIGURE 52.2 Theoretical and recommended *K* factors for isolated columns with idealized end conditions. (*Source: American Institute of Steel Construction. Load and Resistance Factor Design Specification for Structural Steel Buildings*, 2nd ed., Chicago, IL, 1993. With permission. Also from Johnston, B. G., Ed., Structural Stability Research Council, *Guide to Stability Design Criteria for Metal Structures*, 3rd ed., John Wiley & Sons, New York, 1976. With permission.)

where the stability function *C* and *S* are defined as

$$C = \frac{(\pi/K)\sin(\pi/K) - (\pi/K)^2 \cos(\pi/K)}{2 - 2\cos(\pi/K) - (\pi/K)\sin(\pi/K)} \tag{52.3}$$

$$S = \frac{(\pi/K)^2 - (\pi/K)\sin(\pi/K)}{2 - 2\cos(\pi/K) - (\pi/K)\sin(\pi/K)} \tag{52.4}$$

The largest value of *K* satisfying Eq. (52.2) gives the elastic buckling load of an end-retrained column.

Figure 52.2 summarizes the theoretical *K* factors for columns with some idealized end conditions [2,3]. The recommended *K* factors are also shown in Figure 52.2 for practical design applications. Since actual column conditions seldom comply fully with idealized conditions used in buckling analysis, the recommended *K* factors are always equal or greater than their theoretical counterparts.

52.3 Framed Columns — Alignment Chart Method

In theory, the effective length factor *K* for any columns in a framed structure can be determined from a stability analysis of the entire structural analysis — eigenvalue analysis. Methods available for stability analysis include slope–deflection method [4], three-moment equation method [5], and energy methods [6]. In practice, however, such analysis is not practical, and simple models are often used to determine the effective length factors for farmed columns [7~10]. One such practical procedure that provides an approximate value of the elastic *K* factor is the alignment chart method [11]. This procedure has been adopted by the AASHTO [2] and AISC [3]. Specifications and the

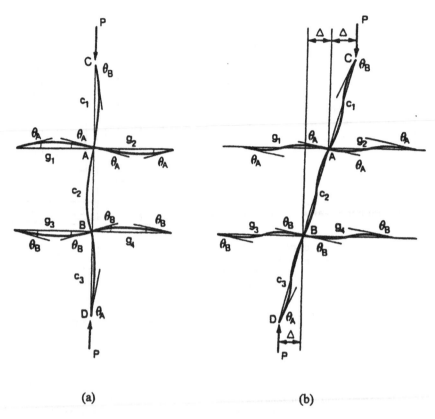

FIGURE 52.3 Subassemblage models for *K* factors of framed columns. (a) Braced frames; (b) unbraced frames.

ACI-318-95 Code [12], among others. At present, most engineers use the alignment chart method in lieu of an actual stability analysis.

52.3.1 Alignment Chart Method

The structural models employed for determination of *K* factors for framed columns in the alignment chart method are shown in Figure 52.3 The assumptions [2,4] used in these models are

1. All members have constant cross section and behave elastically.
2. Axial forces in the girders are negligible.
3. All joints are rigid.
4. For braced frames, the rotations at near and far ends of the girders are equal in magnitude and opposite in direction (i.e., girders are bent in single curvature).
5. For unbraced frames, the rotations at near and far ends of the girders are equal in magnitude and direction (i.e., girders are bent in double curvature).
6. The stiffness parameters $L\sqrt{P/EI}$, of all columns are equal.
7. All columns buckle simultaneously.

By using the slope–deflection equation method and stability functions, the effective length factor equations of framed columns are obtained as follows:

For columns in braced frames:

$$\frac{G_A G_B}{4}(\pi/K)^2 + \left(\frac{G_A + G_B}{2}\right)\left(1 - \frac{\pi/K}{\tan(\pi/K)}\right) + \frac{2\tan(\pi/2K)}{\pi/K} - 1 = 0 \qquad (52.5)$$

For columns in unbraced frames:

$$\frac{G_A G_B (\pi/K)^2 - 36}{6(G_A + G_B)} - \frac{\pi/K}{\tan(\pi/K)} = 0 \qquad (52.6)$$

where G is stiffness ratios of columns and girders, subscripts A and B refer to joints at the two ends of the column section being considered, and G is defined as

$$G = \frac{\sum(E_c I_c / L_c)}{\sum(E_g I_g / L_g)} \qquad (52.7)$$

where Σ indicates a summation of all members rigidly connected to the joint and lying in the plane in which buckling of the column is being considered; subscripts c and g represent columns and girders, respectively.

Eqs. (52.5) and (52.6) can be expressed in form of alignment charts as shown in Figure 52.4. It is noted that for columns in braced frames, the range of K is $0.5 \le K \le 1.0$; for columns in unbraced frames, the range is $1.0 \le K \le \infty$. For column ends supported by but not rigidly connected to a footing or foundations, G is theoretically infinity, but, unless actually designed as a true friction-free pin, may be taken as 10 for practical design. If the column end is rigidly attached to a properly designed footing, G may be taken as 1.0.

Example 52.1

Given
A four-span reinforced concrete bridge is shown in Figure 52.5. Using the alignment chart, determine the K factor for Column DC. $E = 25,000$ MPa.

Section Properties are

Superstructure:	$I = 3.14\ (10^{12})$ mm^4	$A = 5.86\ (10^6)$ mm^2
Columns:	$I = 3.22\ (10^{11})$ mm^4	$A = 2.01\ (10^6)$ mm^2

Solution

1. **Calculate G factor for Column DC.**

$$G_D = \frac{\displaystyle\sum_D (E_c I_c / L_c)}{\displaystyle\sum_D (E_g I_g / L_g)} = \frac{3.22(10^{12})/12,000}{2(3.14)(10^{12})/55,000} = 0.235$$

$G_D = 1.0$ (Ref. [3])

2. **From the alignment chart in Figure 52.4b, $K = 1.21$ is obtained.**

52.3.2 Requirements for Braced Frames

In stability design, one of the major decisions engineers have to make is the determination of whether a frame is braced or unbraced. The AISC-LRFD [3] states that a frame is braced when "lateral stability is provided by diagonal bracing, shear walls or equivalent means." However, there is no specific provision for the "amount of stiffness required to prevent sidesway buckling" in the AISC, AASHTO, and other specifications. In actual structures, a completely braced frame seldom exists.

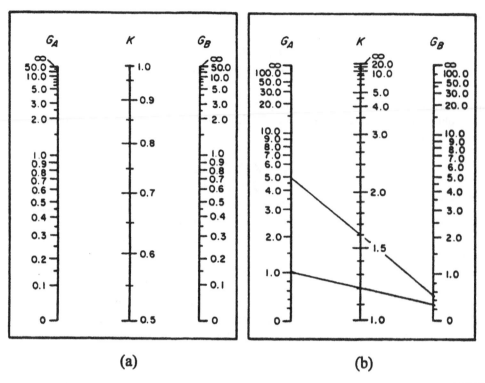

(a)   (b)

FIGURE 52.4 Alignment charts for effective length factors of framed columns. (a) Braced frames; (b) unbraced frames. (*Source:* American Institute of Steel Construction, *Load and Resistance Factor Design Specifications for Structural Steel Buildings,* 2nd ed., Chicago, IL, 1993. With permission. Also from Johnston, B. G., Ed., Structural Stability Research Council, *Guide to Stability Design Criteria for Metal Structures,* 3rd ed., John Wiley & Sons, New York, 1976. With permission.)

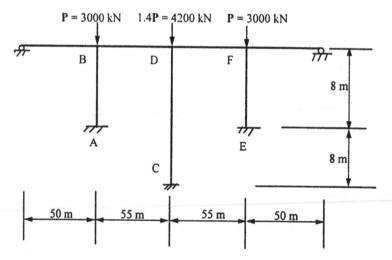

FIGURE 52.5 A four-span reinforced concrete bridge.

But in practice, some structures can be analyzed as braced frames as long as the lateral stiffness provided by bracing system is large enough. The following brief discussion may provide engineers with the tools to make engineering decisions regarding the basic requirements for a braced frame.

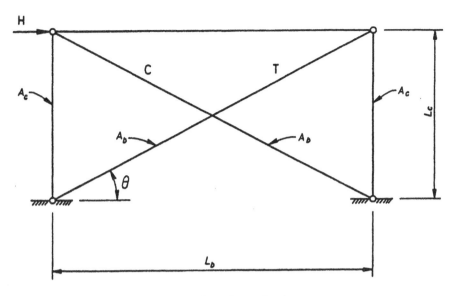

FIGURE 52.6 Diagonal cross-bracing system.

52.3.2.1 Lateral Stiffness Requirement

Galambos [13] presented a simple conservative procedure to estimate the minimum lateral stiffness provided by a bracing system so that the frame is considered braced.

$$\text{Required Lateral Stiffness} \quad T_k = \frac{\sum P_n}{L_c} \tag{52.8}$$

where $\sum$ represents summation of all columns in one story, P_n is nominal axial compression strength of column using the effective length factor $K = 1$, and L_c is unsupported length of the column.

52.3.2.2 Bracing Size Requirement

Galambos [13] employed Eq. (52.8) to a diagonal bracing (Figure 52.6) and obtained minimum requirements of diagonal bracing for a braced frame as

$$A_b = \frac{\left[1 + (L_b / L_c)^2\right]^{3/2} \sum P_n}{(L_b / L_c)^2 E} \tag{52.9}$$

where A_b is cross-sectional area of diagonal bracing and L_b is span length of beam.

A recent study by Aristizabal-Ochoa [14] indicates that the size of diagonal bracing required for a totally braced frame is about 4.9 and 5.1% of the column cross section for "rigid frame" and "simple farming," respectively, and increases with the moment inertia of the column, the beam span and with beam to column span ratio L_b/L_c.

52.3.3 Simplified Equations to Alignment Charts

52.3.3.1. Duan–King–Chen Equations

A graphical alignment chart determination of the K factor is easy to perform, while solving the chart Eqs. (52.5) and (52.6) always involves iteration. To achieve both accuracy and simplicity for design purpose, the following alternative K factor equations were proposed by Duan, King, and Chen [15].

For braced frames:

$$K = 1 - \frac{1}{5+9G_A} - \frac{1}{5+9G_B} - \frac{1}{10+G_A G_B} \qquad (52.10)$$

For unbraced frames:

For $K < 2$ $\quad K = 4 - \frac{1}{1+0.2G_A} - \frac{1}{1+0.2G_B} - \frac{1}{1+0.01G_A G_B} \qquad (52.11)$

For $K \geq 2$ $\quad K = \frac{2\pi a}{0.9 + \sqrt{0.81 + 4ab}} \qquad (52.12)$

where

$$a = \frac{G_A G_B}{G_A + G_B} + 3 \qquad (52.13)$$

$$b = \frac{36}{G_A + G_B} + 6 \qquad (52.14)$$

52.3.3.2 French Equations

For braced frames:

$$K = \frac{3G_A G_B + 1.4(G_A + G_B) + 0.64}{3G_A G_B + 2.0(G_A + G_B) + 1.28} \qquad (52.15)$$

For unbraced frames:

$$K = \sqrt{\frac{1.6G_A G_B + 4.0(G_A + G_B) + 7.5}{G_A + G_B + 7.5}} \qquad (52.16)$$

Eqs. (52.15) and (52.16) first appeared in the French Design Rules for Steel Structure [16] in 1966, and were later incorporated into the *European Recommendations for Steel Construction*[17]. They provide a good approximation to the alignment charts [18].

52.4 Modifications to Alignment Charts

In using the alignment charts in Figure 52.4 and Eqs. (52.5) and (52.6), engineers must always be aware of the assumptions used in the development of these charts. When actual structural conditions differ from these assumptions, unrealistic design may result [3,19,20]. SSRC Guide [19] provides methods enabling engineers to make simple modifications of the charts for some special conditions, such as, for example, unsymmetrical frames, column base conditions, girder far-end conditions, and flexible conditions. A procedure that can be used to account for far ends of restraining columns being hinged or fixed was proposed by Duan and Chen [21~23], and Essa [24]. Consideration of effects of material inelasticity on the *K* factor for steel members was developed originally by Yura

[25] and expanded by Disque [26]. LeMessurier [27] presented an overview of unbraced frames with or without leaning columns. An approximate procedure is also suggested by AISC-LRFD [3]. Several commonly used modifications for bridge columns are summarized in this section.

52.4.1 Different Restraining Girder End Conditions

When the end conditions of restraining girders are not rigidly jointed to columns, the girder stiffness (I_g/L_g) used in the calculation of G factor in Eq. (52.7) should be multiplied by a modification factor α_k given below:

For a braced frame:

$$\alpha_k = \begin{cases} 1.0 & \text{rigid far end} \\ 2.0 & \text{fixed far end} \\ 1.5 & \text{hinged far end} \end{cases} \tag{52.17}$$

For a unbraced frame:

$$\alpha_k = \begin{cases} 1.0 & \text{rigid far end} \\ 2/3 & \text{fixed far end} \\ 0.5 & \text{hinged far end} \end{cases} \tag{52.18}$$

52.4.2 Consideration of Partial Column Base Fixity

In computing the K factor for monolithic connections, it is important to evaluate properly the degree of fixity in foundation. The following two approaches can be used to account for foundation fixity.

52.4.2.1. Fictitious Restraining Beam Approach

Galambos [28] proposed that the effect of partial base fixity can be modeled as a fictitious beam. The approximate expression for the stiffness of the fictitious beam accounting for rotation of foundation in the soil has the form:

$$\frac{I_s}{L_B} = \frac{q\,BH^3}{72\,E_{\text{steel}}} \tag{52.19}$$

where q is modulus of subgrade reaction (varies from 50 to 400 lb/in.³, 0.014 to 0.109 N/mm³); B and H are width and length (in bending plane) of foundation, and E_{steel} is modulus of elasticity of steel.

Based on Salmon et al. [29] studies, the approximate expression for the stiffness of the fictitious beam accounting for the rotations between column ends and footing due to deformation of base plate, anchor bolts, and concrete can be written as

$$\frac{I_s}{L_B} = \frac{b\,d^2}{72\,E_{\text{steel}}\,/\,E_{\text{concrete}}} \tag{52.20}$$

where b and d are width and length of the base plate, subscripts concrete and steel represent concrete and steel, respectively. Galambos [28] suggested that the smaller of the stiffness calculated by Eqs. (52.25) and (52.26) be used in determining K factors.

52.4.2.2 AASHTO-LRFD Approach

The following values are suggested by AASHTO-LRFD [2]:

$G = 1.5$ footing anchored on rock
$G = 3.0$ footing not anchored on rock
$G = 5.0$ footing on soil
$G = 1.0$ footing on multiple rows of end bearing piles

Example 52.2

Given
Determine K factor for the Column AB as shown in Figure 52.5 by using the alignment chart with the necessary modifications. Section and material properties are given in Example 52.1 and spread footings are on soil.

Solution

1. **Calculate G factor with Modification for Column AB.**
   Since the far end of restraining girders are hinged, girder stiffness should be multiplied by 0.5. Using section properties in Example 52.1, we obtain:

$$G_B = \frac{\sum_B (E_c I_c / L_c)}{\sum_B \alpha_k (E_g I_g / L_g)}$$

$$= \frac{3.22(10^{12})/8,000}{(3.14)(10^{12})/55,000 + 0.5(3.14)(10^{12})/50,000} = 0.454$$

$$G_A = 5.0 \quad (\text{Ref. [2]})$$

2. **From the alignment chart in Figure 52.4b, $K = 1.60$ is obtained.**

52.4.3 Column Restrained by Tapered Rectangular Girders

A modification factor α_T was developed by King et al. [30] for those framed columns restrained by tapered rectangular girders with different far-end conditions. The following modified G factor is introduced in connection with the use of alignment charts:

$$G = \frac{\sum (E_c I_c / L_c)}{\sum \alpha_T (E_g I_g / L_g)} \tag{52.21}$$

where I_g is moment of inertia of the girder at the near end. Both closed-form and approximate solutions for modification factor α_T were derived. It is found that the following two-parameter power-function can describe the closed-form solutions very well:

$$\alpha_T = \alpha_k (1 - r)^\beta \tag{52.22}$$

in which the parameter α_k is a constant (Eqs. 52.17 and 52.18) depending on the far-end conditions, and β is a function of far-end conditions and tapering factor a and r as defined in Figure 52.7.

1. For a linearly tapered rectangular girder (Figure 52.7a):

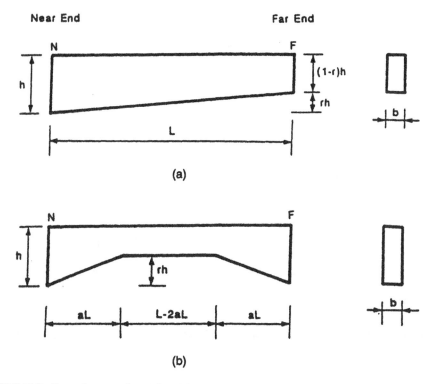

FIGURE 52.7 Tapered rectangular girders. (a) Linearly tapered girder. (b) symmetrically tapered girder.

For a braced frame:

$$\beta = \begin{cases} 0.02 + 0.4r & \text{rigid far end} \\ 0.75 - 0.1r & \text{fixed far end} \\ 0.75 - 0.1r & \text{hinged far end} \end{cases} \qquad (52.23)$$

For an unbraced frame:

$$\beta = \begin{cases} 0.95 & \text{rigid far end} \\ 0.70 & \text{fixed far end} \\ 0.70 & \text{hinged far end} \end{cases} \qquad (52.24)$$

2. For a symmetrically tapered rectangular girder (Figure 52.7b)
   For a braced frame:

$$\beta = \begin{cases} 3 - 1.7a^2 - 2a & \text{rigid far end} \\ 3 + 2.5a^2 - 5.55a & \text{fixed far end} \\ 3 - a^2 - 2.7a & \text{hinged far end} \end{cases} \qquad (52.25)$$

For an unbraced frame:

$$\beta = \begin{cases} 3 + 3.8a^2 - 6.5a & \text{rigid far end} \\ 3 + 2.3a^2 - 5.45a & \text{fixed far end} \\ 3 - 0.3a & \text{hinged far end} \end{cases} \qquad (52.26)$$

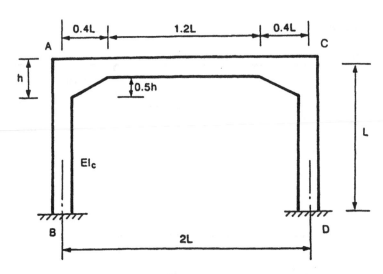

FIGURE 52.8 A simple frame with rectangular sections.

Example 52.3

Given

A one-story frame with a symmetrically tapered rectangular girder is shown in Figure 52.8. Assuming $r = 0.5$, $a = 0.2$, and $I_g = 2I_c = 2I$, determine K factor for Column AB.

Solution

1. **Use the Alignment Chart with Modification**

   For joint A, since the far end of girder is rigid, use Eqs. (52.26) and (52.22)

   $$\beta = 3 + 3.8(0.2)^2 - 6.5(0.2) = 1.852$$

   $$\alpha_T = (1 - 0.5)^{1.852} = 0.277$$

   $$G_A = \frac{\sum E_c I_c / L_c}{\sum \alpha_T E_g I_g / L_g} = \frac{EI/L}{0.277 \, E(2I)/2L} = 3.61$$

   $$G_B = 1.0 \quad (\text{Ref. [3]})$$

   From the alignment chart in Figure 52.4b, $K = 1.59$ is obtained

2. **Use the Alignment Chart without Modification**

   A direct use of Eq. (52.7) with an average section $(0.75h)$ results in

   $$I_g = 0.75^3 \, (2I) = 0.844 \, I$$

   $$G_A = \frac{EI/L}{0.844 \, EI/2L} = 2.37$$

   $$G_B = 1.0$$

   From the alignment chart in Figure 52.4b, $K = 1.50$, or $(1.50 - 1.59)/1.59 = -6\%$ in error on the less conservative side.

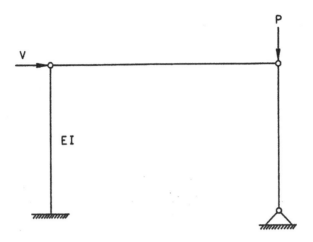

FIGURE 52.9 Subassemblage of LeMessurier method.

52.5 Framed Columns — Alternative Methods

52.5.1 LeMessurier Method

Considering that all columns in a story buckle simultaneously and strong columns will brace weak columns (Figure 52.9), a more accurate approach to calculate K factors for columns in a side-sway frame was developed by LeMessurier [27]. The K_i value for the ith column in a story can be obtained by the following expression:

$$K_i = \sqrt{\frac{\pi^2 EI_i}{L_i^2 P_i}\left(\frac{\sum P + \sum C_L P}{\sum P_L}\right)}$$
(52.27)

where P_i is axial compressive force for member i, and subscript i represents the ith column and ΣP is the sum of axial force of all columns in a story.

$$P_L = \frac{\beta EI}{L^2}$$
(52.28)

$$\beta = \frac{6(G_A + G_B) + 36}{2(G_A + G_B) + G_A G_B + 3}$$
(52.29)

$$C_L = \left(\beta \frac{K_o^2}{\pi^2} - 1\right)$$
(52.30)

in which K_o is the effective length factor obtained by the alignment chart for unbraced frames and P_L is only for those columns that provide side-sway stiffness.

Example 52.4

Given
Determine K factors for bridge columns shown in Figure 52.5 by using the LeMessurier method. Section and material properties are given in Example 52.1.

TABLE 52.1   Example 52.4 — Detailed Calculations by LeMessurier Method

Members	AB and EF	CD	Sum	Notes
I (mm$^4 \times 10^{11}$)	3.217	3.217	—	
L (mm)	8,000	12,000	—	
G_{top}	0.454	0.235	—	Eq. (52.7)
G_{bottom}	0.0	0.0	—	Eq. (52.7)
β	9.91	10.78	—	Eq. (52.29)
K_{io}	1.082	1.045	—	Alignment chart
C_L	0.176	0.193	—	Eq. (52.30)
P_L	50,813E	24,083E	123,709E	Eq. (52.28)
P	P	1.4P	3.4P	$P = 3,000$ kN
$C_L P$	0.176P	0.270P	0.622P	$P = 3,000$ kN

Solutions

The detailed calculations are listed in Table 52.1 By using Eq. (52.32), we obtain:

$$K_{AB} = \sqrt{\frac{\pi^2 E I_{AB}}{L_{AB}^2 P_{AB}}\left(\frac{\sum P + \sum C_L P}{\sum P_L}\right)}$$

$$= \sqrt{\frac{\pi^2 E (3.217)(10^{11})}{(8.000)^2 (P)}\left(\frac{3.4P + 0.622P}{123,709\,E}\right)} = 1.270$$

$$K_{CD} = \sqrt{\frac{\pi^2 E I_{CD}}{L_{CD}^2 P_{CD}}\left(\frac{\sum P + \sum C_L P}{\sum P_L}\right)}$$

$$= \sqrt{\frac{\pi^2 E (3.217)(10^{11})}{(12,000)^2 (1.4P)}\left(\frac{3.4P + 0.622P}{123,709\,E}\right)} = 0.715$$

52.5.2 Lui Method

A simple and straightforward approach for determining the effective length factors for framed columns without the use of alignment charts and other charts was proposed by Lui [31]. The formulas take into account both the member instability and frame instability effects explicitly. The K factor for the ith column in a story was obtained in a simple form:

$$K_i = \sqrt{\left(\frac{\pi^2 E I_i}{P_i L_i^2}\right)\left[\left(\sum \frac{P}{L}\right)\left(\frac{1}{5\sum \eta} + \frac{\Delta_1}{\sum H}\right)\right]} \tag{52.31}$$

where $\sum(P/L)$ represents the sum of axial-force-to-length ratio of all members in a story; $\sum H$ is the story lateral load producing Δ_1, Δ_1 is the first-order interstory deflection; η is member stiffness index and can be calculated by

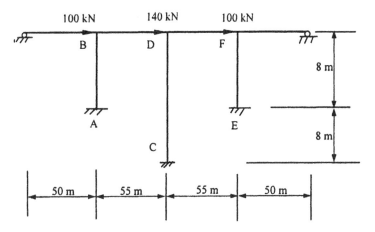

FIGURE 52.10 A bridge structure subjected to fictitious lateral loads.

TABLE 52.2 Example 52.5 — Detailed Calculations by Lui Method

Members	AB and EF	CD	Sum	Notes
I (mm⁴ × 10¹¹)	3.217	3.217	—	
L (mm)	8,000	12,000	—	
H (kN)	150	210	510	
Δ_1 (mm)	0.00144	0.00146	—	
$\Delta_1/\Sigma H$ (mm/kN)	—	—	2.843 (10^{-6})	Average
M_{top} (kN-m)	−476.9	−785.5	—	
M_{bottom} (kN-m)	−483.3	−934.4	—	
m	0.986	0.841	—	
η (kN/mm)	185,606	46,577	417,789	Eq. (52.32)
P/L (kN/mm)	$P/8,000$	$1.4\,P/12,000$	$1.1P/3,000$	$P = 3,000$ kN

$$\eta = \frac{(3 + 4.8m + 4.2m^2)\,EI}{L^3} \tag{52.32}$$

in which m is the ratio of the smaller to larger end moments of the member; it is taken as positive if the member bends in reverse curvature, and negative for single curvature.

It is important to note that the term ΣH used in Eq. (52.36) is not the actual applied lateral load. Rather, it is a small disturbing or fictitious force (taken as a fraction of the story gravity loads) to be applied to each story of the frame. This fictitious force is applied in a direction such that the deformed configuration of the frame will resemble its buckled shape.

Example 52.5

Given
Determine the K factors for bridge columns shown in Figure 52.5 by using the Lui method. Section and material properties are given in Example 52.1.

Solutions
Apply fictitious lateral forces at B, D, and F (Figure 52.10) and perform a first-order analysis. Detailed calculation is shown in Table 52.2.

By using Eq. (52.31), we obtain

$$
K_{AB} = \sqrt{\left(\frac{\pi^2 E I_{AB}}{P_{AB} L_{AB}^2}\right)\left[\left(\sum\frac{P}{L}\right)\left(\frac{1}{5\sum\eta} + \frac{\Delta_1}{\sum H}\right)\right]}
$$

$$
= \sqrt{\left(\frac{\pi^2(25,000)(3.217)(10^{11})}{P(8,000)^2}\right)\left[\left(\frac{1.1P}{3,000}\right)\left(\frac{1}{5(417,789)} + 2.843(10^{-6})\right)\right]}
$$

$$
= 1.229
$$

$$
K_{CD} = \sqrt{\left(\frac{\pi^2 E I_{CD}}{P_{CD} L_{CD}^2}\right)\left[\left(\sum\frac{P}{L}\right)\left(\frac{1}{5\sum\eta} + \frac{\Delta_1}{\sum H}\right)\right]}
$$

$$
= \sqrt{\left(\frac{\pi^2(25,000)(3.217)(10^{11})}{1.4 P(12,000)^2}\right)\left[\left(\frac{1.1P}{3,000}\right)\left(\frac{1}{5(417,789)} + 2.843(10^{-6})\right)\right]}
$$

$$
= 0.693
$$

52.5.3 Remarks

For a comparison, Table 52.3 summarizes the K factors for the bridge columns shown in Figure 52.5 obtained from the alignment chart, LeMessurier and Lui methods, as well as an eigenvalue analysis. It is seen that errors of alignment chart results are rather significant in this case. Although the K factors predicted by Lui's formulas and LeMessurier's formulas are almost the same in most cases, the simplicity and independence of any chart in the case of Lui's formula make it more desirable for design office use [32].

TABLE 52.3   Comparison of K Factors for Frame in Figure 52.5

Columns	Theoretical	Alignment Chart	Lui Eq. (52.31)	LeMessurier Eq. (52.27)
AB	1.232	1.082	1.229	1.270
CD	0.694	1.045	0.693	0.715

52.6 Crossing Bracing Systems

Picard and Beaulieu [33,34] reported theoretical and experimental studies on double diagonal cross-bracings (Figure 52.6) and found that

1. A general effective length factor equation is given as

$$
K = \sqrt{0.523 - \frac{0.428}{C/T}} \geq 0.50 \tag{52.33}
$$

where C and T represent compression and tension forces obtained from an elastic analysis, respectively.

2. When the double diagonals are continuous and attached at an intersection point, the *effective length* of the compression diagonal is 0.5 times the diagonal length, i.e., $K = 0.5$, because the C/T ratio is usually smaller than 1.6.

El-Tayem and Goel [35] reported a theoretical and experimental study about the X-bracing system made from single equal-leg angles. They concluded that

1. Design of X-bracing system should be based on an exclusive consideration of one half diagonal only.
2. For X-bracing systems made from single equal-leg angles, an effective length of 0.85 times the half-diagonal length is reasonable, i.e., $K = 0.425$.

52.7 Latticed and Built-Up Members

It is a common practice that when a buckling model involves relative deformation produced by shear forces in the connectors, such as lacing bars and batten plates, between individual components, a modified effective length factor K_m or effective slenderness ratio $(KL/r)_m$ is used in determining the compressive strength. K_m is defined as

$$K_m = \alpha_v K \tag{52.34}$$

in which K is the usual effective length factor of a latticed member acting as a unit obtained from a structural analysis; and α_v is the shear factor to account for the effect of shear deformation on the buckling strength. Details of the development of the shear factor α_v can be found in textbooks by Bleich [5] and Timoshenko and Gere [36]. The following section briefly summarizes α_v formulas for various latticed members.

52.7.1 Latticed Members

By considering the effect of shear deformation in the latticed panel on buckling load, shear factor α_v of the following form has been introduced:

Laced Compression Members (Figures 52.11a and b)

$$\alpha_v = \sqrt{1 + \frac{\pi^2 EI}{(KL)^2} \frac{d^3}{A_d E_d ab^2}} \tag{52.35}$$

Compression Members with Battens (Figure 52.11c)

$$\alpha_v = \sqrt{1 + \frac{\pi^2 EI}{(KL)^2} \left(\frac{ab}{12 E_b I_b} + \frac{a^2}{24 EI_f} \right)} \tag{52.36}$$

Laced-Battened Compression Members (Figure 52.11d)

$$\alpha_v = \sqrt{1 + \frac{\pi^2 E I}{(KL)^2} \left(\frac{d^3}{A_d E_d ab^2} + \frac{b}{aA_b E_b} \right)} \tag{52.37}$$

Compression Members with Perforated Cover Plates (Figure 52.11e)

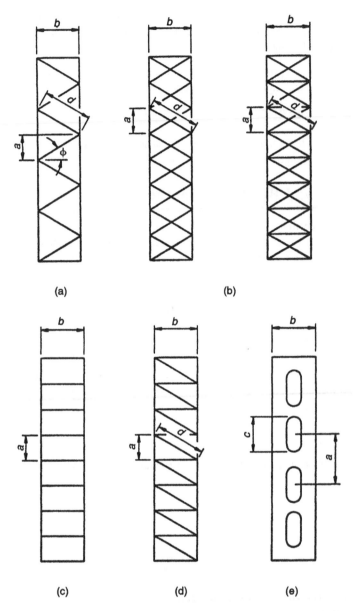

FIGURE 52.11 Typical configurations of latticed members. (a) Single lacing; (b) double lacing; (c) battens; (d) lacing-battens; (e) perforated cover plates.

$$\alpha_v = \sqrt{1 + \frac{\pi^2 EI}{(KL)^2}\left(\frac{9c^3}{64aEI_f}\right)} \tag{52.38}$$

where E_d is modulus of elasticity of materials for lacing bars; E_b is modulus of elasticity of materials for batten plates; A_d is cross-sectional area of all diagonals in one panel; I_b is moment inertia of all battens in one panel in the buckling plane, and I_f is moment inertia of one side of main components taken about the centroid axis of the flange in the buckling plane; a, b, d are height of panel, depth of member, and length of diagonal, respectively; and c is the length of a perforation.

The Structural Stability Research Council [37] suggested that a conservative estimating of the influence of 60° or 45° lacing, as generally specified in bridge design practice, can be made by modifying the overall effective length factor K by multiplying a factor α_v, originally developed by Bleich [5] as follows:

$$\text{For } \frac{KL}{r} > 40, \quad \alpha_v = \sqrt{1 + 300/(KL/r)^2} \tag{52.39}$$

$$\text{For } \frac{KL}{r} \leq 40, \quad \alpha_v = 1.1 \tag{52.40}$$

It should be pointed out that the usual K factor based on a solid member analysis is included in Eqs. (52.35) through (52.38). However, since the latticed members studied previously have pin-ended conditions, the K factor of the member in the frame was not included in the second terms of the square root of the above equations in their original derivations [5,36].

52.7.5 Built-Up Members

AISC-LRFD [3] specifies that if the buckling of a built-up member produces shear forces in the connectors between individual component members, the usual slenderness ratio KL/r for compression members must be replaced by the modified slenderness ratio $(KL/r)_m$ in determining the compressive strength.

1. *For snug-tight bolted connectors:*

$$\left(\frac{KL}{r}\right)_m = \sqrt{\left(\frac{KL}{r}\right)_o^2 + \left(\frac{a}{r_i}\right)^2} \tag{52.41}$$

2. *For welded connectors and for fully tightened bolted connectors:*

$$\left(\frac{KL}{r}\right)_m = \sqrt{\left(\frac{KL}{r}\right)_o^2 + 0.82\frac{\alpha^2}{(1+\alpha^2)}\left(\frac{a}{r_{ib}}\right)^2} \tag{52.42}$$

where $(KL/r)_o$ is the slenderness ratio of built-up member acting as a unit, $(KL/r)_m$ is modified slenderness ratio of built-up member, a/r_i is the largest slenderness ratio of the individual components, a/r_{ib} is the slenderness ratio of the individual components relative to its centroidal axis parallel to axis of buckling, a is the distance between connectors, r_i is the minimum radius of gyration of individual components, r_{ib} is the radius of gyration of individual components relative to its centroidal axis parallel to member axis of buckling, α is the separation ratio $= h/2r_{ib}$, and h is the distance between centroids of individual components perpendicular to the member axis of buckling.

Eq. (52.41) is the same as that used in the current Italian code, as well as in other European specifications, based on test results [38]. In this equation, the bending effect is considered in the first term in square root, and shear force effect is taken into account in the second term. Eq. (52.42) was derived from elastic stability theory and verified by test data [39]. In both cases, the end connectors must be welded or slip-critical-bolted.

52.8 Tapered Columns

The state-of-the-art design for tapered structural members was provided in the SSRC guide [37]. The charts as shown in Figure 52.12 can be used to evaluate the effective length factors for tapered

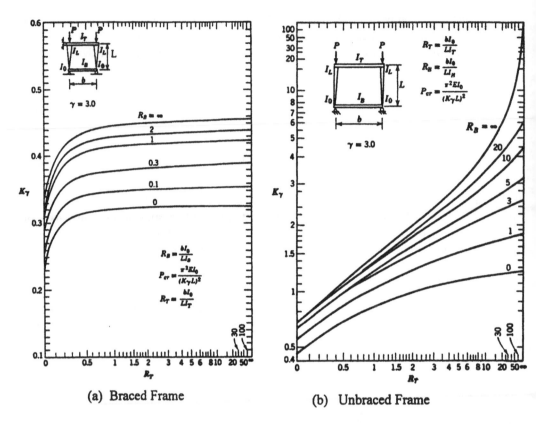

(a) Braced Frame (b) Unbraced Frame

FIGURE 52.12 Effective length factor for tapered columns. (a) Braced frame; (b) unbraced frame. (*Source:* Galambos, T. V., Ed., Structural Stability Research Council Guide to Stability Design Criteria for Metal Structures, 4th ed., John Wiley & Sons, New York, 1988. With permission.)

column restrained by prismatic beams [37]. In these figures, I_T and I_B are the moment of inertia of top and bottom beam, respectively; b and L are length of beam and column, respectively; and γ is tapering factor as defined by

$$\gamma = \frac{d_1 - d_o}{d_o} \tag{52.43}$$

where d_o and d_1 are the section depth of column at the smaller and larger end, respectively.

52.9 Summary

This chapter summarizes the state-of-the-art practice of the effective length factors for isolated columns, framed columns, diagonal bracing systems, latticed and built-up members, and tapered columns. Design implementation with formulas, charts, tables, and various modification factors adopted in current codes and specifications, as well as those used in bridge structures, are described. Several examples are given to illustrate the steps of practical applications of these methods.

References

1. McGuire, W., Computers and steel design, *Modern Steel Constr.*, 32(7), 39, 1992.
2. AASHTO, *LRFD Bridge Design Specifications*, American Association of State Highway and Transportation Officials, Washington, D.C., 1994.
3. AISC, *Load and Resistance Factor Design Specification for Structural Steel Buildings*, 2nd ed., American Institute of Steel Construction, Chicago, IL, 1993.
4. Chen, W. F. and Lui, E. M., *Stability Design of Steel Frames*, CRC Press, Boca Raton, FL, 1991.
5. Bleich, F., *Buckling Strength of Metal Structures*, McGraw-Hill, New York, 1952.
6. Johnson, D. E., Lateral stability of frames by energy method, *J. Eng. Mech. ASCE*, 95(4), 23, 1960.
7. Lu, L. W., A survey of literature on the stability of frames, *Weld. Res. Counc. Bull.*, New York, 1962.
8. Kavanagh, T. C., Effective length of framed column, *Trans. ASCE*, 127(II) 81, 1962.
9. Gurfinkel, G. and Robinson, A. R., Buckling of elasticity restrained column, *J. Struct. Div. ASCE*, 91(ST6), 139, 1965.
10. Wood, R. H., Effective lengths of columns in multi-storey buildings, *Struct. Eng.*, 50(7–9), 234, 295, 341, 1974.
11. Julian, O. G. and Lawrence, L. S., Notes on J and L Nomograms for Determination of Effective Lengths, unpublished report, 1959.
12. ACI, *Building Code Requirements for Structural Concrete* (ACI 318-95) and Commentary (ACI 318R-95), American Concrete Institute, Farmington Hills, MI, 1995.
13. Galambos, T. V., Lateral support for tier building frames, *AISC Eng. J.*, 1(1), 16, 1964.
14. Aristizabal-Ochoa, J. D., K-factors for columns in any type of construction: nonparadoxical approach, *J. Struct. Eng. ASCE*, 120(4), 1272, 1994.
15. Duan, L., King, W. S., and Chen, W. F., *K* factor equation to alignment charts for column design, *ACI Struct. J.*, 90(3), 242, 1993.
16. *Regles de Cacul des Constructions en acier*, CM66, Eyrolles, Paris, 1975.
17. ECCS, *European Recommendations for Steel Construction*, European Convention for Construction Steelworks, 1978.
18. Dumonteil, P., Simple equations for effective length factors, *AISC Eng. J.*, 29(3), 111, 1992.
19. Johnston, B. G., Ed., Structural Stability Research Council, *Guide to Stability Design Criteria for Metal Structures*, 3rd ed., John Wiley & Sons, New York, 1976.
20. Liew, J. Y. R., White, D. W., and Chen, W. F., Beam-column design in steel frameworks — insight on current methods and trends, *J. Constr. Steel. Res.*, 18, 269, 1991.
21. Duan, L. and Chen, W. F., Effective length factor for columns in braced frames, *J. Struct. Eng. ASCE*, 114(10), 2357, 1988.
22. Duan, L. and Chen, W. F., Effective length factor for columns in unbraced frames, *J. Struct. Eng. ASCE*, 115(1), 150, 1989.
23. Duan, L. and Chen, W. F., 1996. Errata of paper: effective length factor for columns in unbraced frames, *J. Struct. Eng. ASCE*, 122(1), 224, 1996.
24. Essa, H. S., Stability of columns in unbraced frames, *J. Struct. Eng., ASCE*, 123(7), 952, 1997.
25. Yura, J. A., The effective length of columns in unbraced frames, *AISC Eng. J.*, 8(2), 37, 1971.
26. Disque, R. O., Inelastic *K* factor in design, *AISC Eng. J.*, 10(2), 33, 1973.
27. LeMessurier, W. J., A practical method of second order analysis, part 2 — rigid frames, *AISC Eng. J.*, 14(2), 50, 1977.
28. Galambos, T. V., Influence of partial base fixity on frame instability, *J. Struct. Div. ASCE*, 86(ST5), 85, 1960.
29. Salmon, C. G., Schenker, L., and Johnston, B. G., Moment-rotation characteristics of column anchorage, *Trans. ASCE*, 122, 132, 1957.
30. King, W. S., Duan, L., Zhou, R. G., Hu, Y. X., and Chen, W. F., *K* factors of framed columns restrained by tapered girders in U.S. codes, *Eng. Struct.*, 15(5), 369, 1993.

31. Lui, E. M., A novel approach for K-factor determination. *AISC Eng. J.*, 29(4), 150, 1992.
32. Shanmugam, N. E. and Chen, W. F., An assessment of K factor formulas, *AISC Eng. J.*, 32(3), 3, 1995.
33. Picard, A. and Beaulieu, D., Design of diagonal cross bracings, part 1: theoretical study, *AISC Eng. J.*, 24(3), 122, 1987.
34. Picard, A. and Beaulieu, D., Design of diagonal cross bracings, part 2: experimental study, *AISC Eng. J.*, 25(4), 156, 1988.
35. El-Tayem, A. A. and Goel, S. C., Effective length factor for the design of X-bracing systems, *AISC Eng. J.*, 23(4), 41, 1986.
36. Timoshenko, S. P. and Gere, J. M., *Theory of Elastic Stability*, 2nd ed., McGraw-Hill, New York, 1961.
37. Galambos, T. V., Ed., *Structural Stability Research Council, Guide to Stability Design Criteria for Metal Structures*, 4th ed., John Wiley & Sons, New York, 1988.
38. Zandonini, R., Stability of compact built-up struts: experimental investigation and numerical simulation, *Constr. Met.*, 4, 1985 [in Italian].
39. Aslani, F. and Goel, S. C., An analytical criteria for buckling strength of built-up compression members, *AISC Eng. J.*, 28(4), 159, 1991.

53

Fatigue and Fracture

Robert J. Dexter
University of Minnesota

John W. Fisher
Lehigh University

53.1 Introduction

Bridges do not usually fail due to inadequate load capacity, except when an overweight truck is illegally driven onto an old bridge with very low load rating. When bridge superstructures "fail," it is usually because of excessive deterioration by corrosion and/or fatigue cracking rather than inadequate load capacity. Although most deterioration can be attributed to lack of proper maintenance, there are choices made in design that also can have an impact on service life. Yet the design process for bridges is focused primarily on load capacity rather than durability.

This chapter of the handbook will inform the reader about a particular aspect of durability, i.e., the fatigue and fracture failure mode, and about detailing for improved resistance to fatigue and fracture. Only aspects of fatigue and fracture that are relevant to design or assessment of bridge deck and superstructure components are discussed. Concrete and aluminum structural components are discussed briefly, but the emphasis of this section is on steel structural components.

The fatigue and fracture design and assessment procedures outlined in this chapter are included in the American Association of State Highway and Transportation Officials (AASHTO) specifications for bridges [1]. Some of the bridges built before the mid-1970s (when the present fatigue-design specifications were adopted) may be susceptible to fatigue cracking. There are valuable lessons that can be learned from the problems that these bridges experienced, and several examples will be used in this chapter to illustrate various points. These lessons have been incorporated into the present AASHTO specifications [2,3]. As a result, steel bridges that have been built in the last few decades have not and will not have any significant problems with fatigue and fracture [2].

These case histories of fatigue cracking should not create the false impression that there is an inherent fatigue problem with steel bridges. The problems that occur are confined to older bridges. These problems are, for the most part, relatively minor and can be corrected with inexpensive retrofits. The problems are even easier to avoid in new designs. Therefore, because there are some

0-8493-7434-0/00/$0.00+$.50
© 2000 by CRC Press LLC

FIGURE 53.1 View of cracked girder of Lafayette Street Bridge in St. Paul, MN showing fatigue crack originating from backing bars and lack of fusion on the weld attaching the lateral bracing attachment plate to the web and to the transverse stiffener.

fatigue problems with older bridges, one should not get the impression that there are ongoing fatigue problems with modern bridges designed by the present fatigue-design specifications.

Detailing rules are perhaps the most important part of the fatigue and fracture design and assessment procedures. The detailing rules are intended to avoid notches and other stress concentrations. These detailing rules are useful for the avoidance of brittle fracture as well as fatigue. Because of the detailing rules, modern steel bridges are detailed in a way that appears much cleaner than those built before the 1970s. There are fewer connections and attachments in modern bridges, and the connections use more fatigue-resistant details such as high-strength bolted joints.

For example, AWS D1.5 (Bridge Welding Code) does not permit backing bars to be left in place on welds. This rule is a result of experience such as that shown in Figure 53.1. Figure 53.1 shows lateral gusset plates on the Lafayette St. Bridge in St. Paul, MN that cracked and led to a fracture of a primary girder in 1976 [3,4]. In this detail, backing bars were left in place under the groove welds joining the lateral gusset plate to the transverse stiffener and to the girder web. The backing bars create a cracklike notch, often accompanied by a lack-of-fusion defect. Fatigue cracks initiate from this cracklike notch and the lack of fusion in the weld to the transverse stiffener because, in this case, the plane of the notch is perpendicular to the primary fluctuating stress.

The Bridge Welding Code AWS D1.5 at present requires that backing bars be removed from all bridge welds to avoid these notches. Prior to 1994, this detailing rule was not considered applicable to seismic moment-resisting building frames. Consequently, many of these frames fractured when the 1994 Northridge earthquake loaded them. Backing bars left on the beam flange-to-column welds of these frames created a built-in cracklike notch. This notch contributed to the Northridge fractures, along with lack-of-fusion defects and low-toughness welds [5–7].

Figure 53.2 shows a detail where a primary girder flange penetrates and is continuous through the web of a cross girder of the Dan Ryan Elevated structures in Chicago [8]. In this case, the short vertical welds at the sides of the flange were defective. Fatigue cracks initiated at these welds,

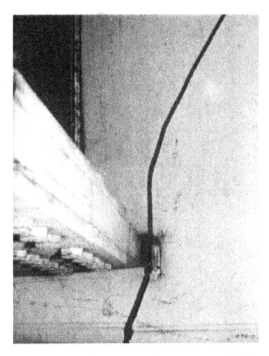

FIGURE 53.2   View of cracked cross girder of Dan Ryan elevated structure in Chicago showing cracking originating from short vertical welds which are impossible to make without lack of fusion defects.

which led to fracture of the cross girder. It is unlikely that good welds could have been made for
this detail. A better alternative would have been to have cope holes at the ends of the flange. Note
that in Figures 53.1 and 53.2, the fractures did not lead to structural collapse. The reason for this
reserve tolerance to large cracks will be discussed in Section 53.2.

In bridges, there are usually a large number of cycles of significant live load, and fatigue will
almost always precede fracture. Therefore, controlling fatigue is practically more important than
controlling fracture. The civil engineering approach for fatigue is explained in Section 53.3. The
fatigue life (N) of particular details is determined by the nominal stress range (S) from S–N curves.
The nominal stress S–N curves are the lower-bound curves to a large number of full-scale fatigue
test data. The full-scale tests empirically take into account a number of variables with great
uncertainty, e.g., residual stress, weld profile, environment, and discontinuities in the material from
manufacturing. Consequently, the variability of fatigue life data at a particular stress range is
typically about a factor of 10.

Usually, the only measures taken in design that are primarily intended to assure fracture resistance
are to specify materials with minimum specified toughness values, such as a Charpy V-Notch (CVN)
test requirement. As explained in Section 53.4, toughness is specified so that the structure is resistant
to brittle fracture despite manufacturing defects, fatigue cracks, and/or unanticipated loading. These
material specifications are less important for bridges than the S–N curves and detailing rules, however.

Steel structures have exhibited unmatched ductility and integrity when subjected to seismic
loading. Modern steel bridges in the United States which are designed to resist fatigue and fracture
from truck loading have not exhibited fractures in earthquakes. It would appear that the modern
bridge design procedures which consider fatigue and fracture from truck loading are also adequate
to assure resistance to brittle fracture under seismic loading. Although rare, fractures of bridge
structural elements have occurred during earthquakes outside the United States. For example, brittle
fractures occurred on several types of steel bridge piers during the 1995 Hyogo-ken Nanbu earthquake in Japan [9].

FIGURE 53.3 View of cracked girder of I-79 Bridge at Neville Island in Pittsburgh as an example of a bridge that is sufficiently redundant to avoid collapse despite a fracture of the tension flange and the web.

These fatigue and fracture design and assessment procedures for bridges are also applicable to many other types of cyclically loaded structures which use similar welded and bolted details, e.g., cranes, buildings, chimneys, transmission towers, sign, signal, and luminaire support structures, etc. In fact, these procedures are similar to those in the American Welding Society AWS D1.1, "Structural Welding Code — Steel" [10], which is applicable to a broad range of welded structures.

This "civil engineering" approach to fatigue and fracture could also be applied to large welded and bolted details in structures outside the traditional domain of civil engineering, including ships, offshore structures, mobile cranes, and heavy vehicle frames. However, the civil engineering approach to fatigue presented here is different from traditional mechanical engineering approaches. The mechanical engineering approaches are well suited to smooth machined parts and other applications where a major portion of the fatigue life of a part is consumed in forming an initial crack. In the mechanical engineering approaches, the fatigue strength is proportional to the ultimate tensile strength of the steel. The experimental data show this is not true for welded details, as discussed below.

53.2   Redundancy, Ductility, and Structural Collapse

Fatigue is considered a serviceability limit state for bridges because the fatigue cracks and fractures that have occurred have mostly not been significant from the standpoint of structural integrity. Redundancy and ductility of steel bridges have prevented catastrophic collapse. Only in certain truly nonredundant structural systems can fatigue cracking lead to structural collapse.

The I-79 Bridge at Neville Island in Pittsburgh is an example of the robustness of even so-called fracture critical or nonredundant two-girder bridges. In 1977, one of the girders developed a fatigue crack in the tension flange at the location of a fabrication repair of an electroslag weld splice [3]. As shown in Figure 53.3, the crack completely fractured the bottom flange and propagated up the web of this critical girder. A tugboat captain happened to look up and notice the crack extending as he passed under the bridge.

Although two-girder bridges are considered nonredundant, other elements of the bridge, particularly the deck, are usually able to carry the loads and prevent collapse as in the case of the I-79 bridge. Today, because of the penalties in design and fabrication for nonredundant or fracture-critical members, simple and low-cost two-girder bridges are seldom built. Note that the large cracks shown in the bridges in Figures 53.1 and 53.2 also did not lead to structural collapse. Unfortunately, this built-in redundancy shown by these structures is difficult to predict and is not explicitly recognized in design.

The beneficial effects of redundancy on fatigue and fracture are best explained in terms of the boundary conditions on the structural members. The truck loads and wind loads on bridges are essentially "fixed-load" or "load-control" boundary conditions. On a local scale, however, most individual members and connections in redundant structures are essentially under "displacement-control" boundary conditions. In other words, because of the stiffness of the surrounding structure, the ends of the member have to deform in a way that is compatible with nearby members. A cracked member in parallel with other similar but uncracked members will experience a decreasing load range and nominal stress range as the stiffness of the cracked member decreases. This behavior under displacement control is referred to as load shedding and it can slow down the rate of fatigue crack propagation.

If a fatigue crack forms in one element of a bolted or riveted built-up structural member, the crack cannot propagate directly into neighboring elements. Usually, a riveted member will not fail until additional cracks form in one or more additional elements. Therefore, riveted built-up structural members are inherently redundant. Once a fatigue crack forms, it can propagate directly into all elements of a continuous welded member and cause failure at service loads. Welded structures are not inferior to bolted or riveted structures; they require more attention to design, detailing, and quality.

Ductility is required in order for redundancy to be completely effective. As the net section of a cracking bridge member decreases, the plastic moment capacity of the member decreases. If a member is sufficiently ductile, it can tolerate a crack so large that the applied moment exceeds the plastic moment for the net section and a mechanism will form in the member [11–13]. If the member can then deform to several times the yield rotation, the load will be shed to the deck and other members.

Minimum levels of fracture toughness are necessary to achieve ductility, but are not sufficient. The fracture toughness assures that brittle fracture does not occur before general yielding of the net cross section. However, net section yielding is not very ductile unless the yielding can spread to the gross section, which requires strain hardening in the stress–strain relationship of the steel, or a reasonably low yield-to-tensile ratio [12–14].

53.3 Fatigue Resistance

Low-cycle fatigue is a possible failure mode for structural members or connections which are cycled into the inelastic region for a small number of cycles (less that 1000) [15,16]. For example, bridge pier structures may be subjected to low-cycle fatigue in an earthquake [9]. Brittle fractures occurred in Japan in steel piers that underwent large plastic strain cycles during the 1995 Hyogo-ken Nanbu earthquake in Japan [9]. However, in order to focus on the more common phenomenon of high-cycle fatigue, low-cycle fatigue is not discussed further in this section.

Truck traffic is the primary cause of high-cycle fatigue of bridges. Wind loads may also be a fatigue design consideration in bridges. Wind-induced vibration has caused numerous fatigue problems in sign, signal, and luminaire support structures [17].

Although cracks can form in structures cycled in compression, they arrest and are not structurally significant. Therefore, only members or connections for which the stress cycle is at least partially in tension need to be assessed.

In most bridges, the ratio of the fatigue design truck load to the strength design load is large enough that fatigue may control the design of much of the structure. In long-span bridges, the load on much of the superstructure is dominated by the dead load, with the fluctuating live-load part relatively small. These members will not be sensitive to fatigue. However, the deck, stringers, and floor beams of bridges are subjected to primarily live load and therefore will be controlled by

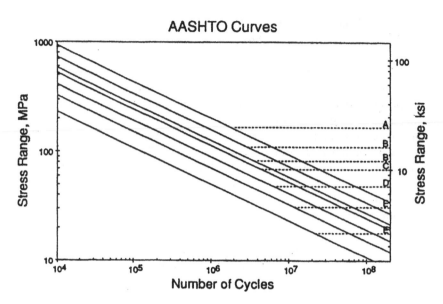

FIGURE 53.4 The lower-bound *S–N* curves for the seven primary fatigue categories from the AASHTO, AREA, AWS, and AISC specifications. The dotted lines are the CAFL and indicate the detail category.

fatigue. Fortunately, the deck, stringers, and floor beams are secondary members which, if they failed, would not lead to structural collapse.

When information about a specific crack is available, a fracture mechanics crack growth rate analysis should be used to calculate remaining life [23–25]. However, in the design stage, without specific initial crack size data, the fracture mechanics approach is not any more accurate than the *S–N* curve approach [25]. Therefore, the fracture mechanics crack growth analysis will not be discussed further.

Welded and bolted details for bridges and buildings are designed based on the nominal stress range rather than the local "concentrated" stress at the weld detail. The nominal stress is usually obtained from standard design equations for bending and axial stress and does not include the effect of stress concentrations of welds and attachments. Since fatigue is typically only a serviceability problem, fatigue design is carried out using service loads as discussed in Section 53.3.5. Usually, the nominal stress in the members can be easily calculated without excessive error. However, the proper definition of the nominal stresses may become a problem in regions of high stress gradients [26,27].

It is standard practice in fatigue design of welded structures to separate the weld details into categories having similar fatigue resistance in terms of the nominal stress. Each category of weld details has an associated *S–N* curve. The *S–N* curves for steel in the AASHTO [1], AISC [28], ANSI/AWS [10], and American Railway Engineers Association (AREA) provisions are shown in Figure 53.4. *S–N* curves are presented for seven categories of weld details — A through E', in order of decreasing fatigue strength. These *S–N* curves are based on a lower bound to a large number of full-scale fatigue test data with a 97.5% survival limit.

The slope of the regression line fit to the test data for welded details is typically in the range 2.9 to 3.1 [20,21]. Therefore, in the AISC and AASHTO codes as well as in Eurocode 3 [29], the slopes have been standardized at 3.0. The effect of the welds and other stress concentrations are reflected in the ordinate of the *S–N* curves for the various detail categories.

Figure 53.4 shows the fatigue threshold or constant amplitude fatigue limits (CAFL) for each category as horizontal dashed lines. When constant-amplitude tests are performed at stress ranges below the CAFL, noticeable cracking does not occur. The number of cycles associated with the CAFL is whatever number of cycles corresponds to that stress range on the *S–N* curve for that

TABLE 53.1 Constant-Amplitude Fatigue Limits for
AASHTO and Aluminum Association *S–N* Curves

Detail Category	CAFL for Steel (MPa)	CAFL for Aluminum (MPa)
A	165	70
B	110	41
B′	83	32
C	69	28
D	48	17
E	31	13
E′	18	7

category or class of detail. The CAFL occurs at an increasing number of cycles for lower fatigue categories or classes. Sometimes, different details, which share a common *S–N* curve (or category) in the finite-life regime, have different CAFL.

Typically, small-scale specimen tests will result in longer apparent fatigue lives. Therefore, the *S–N* curve must be based on tests of full-size structural components such as girders. Testing on full-scale welded members has indicated that the primary effect of constant-amplitude loading can be accounted for in the live-load stress range; i.e., the mean stress is not significant [18–21]. The reason that the dead load has little effect on the lower bound of the results is that, locally, there are very high residual stresses from welding. Mean stress may be important for some details that are not welded, such as anchor bolts [17,22]. In order to be conservative for nonwelded details, in which there may be a significant effect of the mean stress, the fatigue test data should be generated under loading with a high tensile mean stress.

The strength and type of steel have only a negligible effect on the fatigue resistance expected for a particular detail [18–21]. The welding process also does not typically have an effect on the fatigue resistance [18–21]. The independence of the fatigue resistance from the type of steel greatly simplifies the development of design rules for fatigue since it eliminates the need to generate data for every type of steel.

The full-scale fatigue experiments have been carried out in moist air and therefore reflect some degree of environmental effect or corrosion fatigue. Full-scale fatigue experiments in seawater do not show significantly lower fatigue lives [30], provided that corrosion is not so severe that it causes pitting. The fatigue lives seem to be more significantly influenced by the stress concentration at the toe of welds and the initial discontinuities. Therefore, these lower-bound *S–N* curves can be used for design of bridges in any natural environmental exposure, even near salt spray. However, pitting from severe corrosion may become a fatigue-critical condition and should not be allowed [31,32].

Similar *S–N* curves have been proposed by the Aluminum Association [56] for welded aluminum structures. Table 53.1 summarizes the CAFL for steel and aluminum for categories A through E′. The design procedures are based on associating weld details with specific categories. For both steel and aluminum, the separation of details into categories is approximately the same.

The categories in Figure 53.4 range from A to E′ in order of decreasing fatigue strength. There is an eighth category, F, in the specifications (not shown in Figure 53.4) which applies to fillet welds loaded in shear. However, there have been very few if any failures related to shear, and the stress ranges are typically very low such that fatigue rarely would control the design. Therefore, the shear stress Category F will not be discussed further. In fact, there have been very few if any failures which have been attributed to details which have a fatigue strength greater than Category C.

53.3.1   Classification of Details in Metal Structural Components

Details must be associated with one of the drawings in the specification [1] to determine the fatigue category. The following is a brief, simplified overview of the categorization of fatigue details. In some cases, this overview has left out some details so the specification should always be checked for the appropriate detail categorization. The AISC specification [28] has a somewhat better presentation of the sketches and explanation of the detail categorization than the AASHTO specifications. Also, several reports have been published which show a large number of illustrations of details and their categories [33,34]. In addition, the Eurocode 3 [29] and the British Standard 7608 [35] have more detailed illustrations for their categorization than does the AISC or AASHTO specifications. A book by Maddox [36] discusses categorization of many details in accordance with BS 7608, from which roughly equivalent AISC categories can be inferred.

Small holes are considered Category D details. Therefore, riveted and mechanically fastened joints (other than high-strength bolted joints) loaded in shear are evaluated as Category D in terms of the net section nominal stress. Properly tensioned high-strength bolted joints loaded in shear may be classified as Category B. Pin plates and eyebars are designed as Category E details in terms of the stress on the net section.

Welded joints are considered longitudinal if the axis of the weld is parallel to the primary stress range. Continuous longitudinal welds are Category B or B′ details. However, the terminations of longitudinal fillet welds are more severe (Category E). (The termination of full-penetration groove longitudinal welds requires a ground transition radius but gives greater fatigue strength, depending on the radius.) If longitudinal welds must be terminated, it is better to terminate at a location where the stress ranges are less severe.

Attachments normal to flanges or plates that do not carry significant load are rated Category C if less than 51 mm long in the direction of the primary stress range, D if between 51 and 101 mm long, and E if greater than 101 mm long. (The 101 mm limit may be smaller for plate thinner than 9 mm). If there is not at least 10-mm edge distance, then Category E applies for an attachment of any length. The Category E′, slightly worse than Category E, applies if the attachment plates or the flanges exceed 25 mm in thickness.

Transverse stiffeners are treated as short attachments (Category C). Transverse stiffeners that are used for cross-bracing or diaphragms are also treated as Category C details with respect to the stress in the main member. In most cases, the stress range in the stiffener from the diaphragm loads is not considered, because these loads are typically unpredictable. However, the detailing of attachment plates is critical to avoid distortion-induced fatigue, as discussed in Section 53.3.2.

In most other types of load-carrying attachments, there is interaction between the stress range in the transverse load-carrying attachment and the stress range in the main member. In practice, each of these stress ranges is checked separately. The attachment is evaluated with respect to the stress range in the main member, and then it is separately evaluated with respect to the transverse stress range. The combined multiaxial effect of the two stress ranges is taken into account by a decrease in the fatigue strength; i.e., most load-carrying attachments are considered Category E details.

The fatigue strength of longitudinal attachments can be increased if the ends are given a radius and the fillet or groove weld ends are ground smooth. For example, a longitudinal attachment (load-bearing or not) with a transition radius greater than 50 mm can be considered Category D. If the transition radius of a groove-welded longitudinal attachment is increased to greater than 152 mm (with the groove-weld ends ground smooth), the detail (load bearing or not) can be considered Category C.

53.3.2   Detailing to Avoid Distortion-Induced Fatigue

It is clear from the type of cracks that occur in bridges that a significant proportion of the cracking is due to distortion that results from such secondary loading [37]. The solution to the problem of fatigue cracking due to secondary loading usually relies on the qualitative art of good detailing [38]. Often, the best solution to distortion cracking problems may be to stiffen the structure. Typically, the better connections are more rigid.

FIGURE 53.5 View of floor beam of Throgg's Neck Bridge in New York showing crack in the cope that has been repaired by drilling a stop hole. The crack is caused by incompatibility between the curvature of the superstructure and the orthotropic steel deck that is bolted onto the floor beams.

One of the most overlooked secondary loading problems occurs at the interface of structures with different flexural rigidities and curvatures [39,40]. Figure 53.5 shows a typical crack at the floor beam flange cope in the Throg's Neck Bridge in New York. One of the closed trapezoidal ribs of an orthotropic steel deck is visible in Figure 53.5. The orthotropic deck was added to the structure to replace a deteriorating deck by bolting onto the floor beams. However, the superstructure has curvature that is incompatible with the stiff deck. The difference in curvature manifests as out-of-plane rotation of the flange of the floor beam. The crack is caused by out-of-plane bending of the floor beam web at the location of the cope, which has many built-in discontinuities due to the flame cutting.

Another example of secondary loading from out-of-plane distortion may occur at attachment plates for transverse bracing or for a floor beam. These attachment plates, which may have distortion-induced out-of-plane loads, should be welded directly to both flanges as well as the web. In older bridges, it was common practice not to weld transverse stiffeners and attachment plates to the tension flange of welded I-girders and box girders. The practice of not allowing transverse fillet welds on the tension flange is not necessary and is due to unwarranted concern about brittle fracture of the tension flange [37,38]. Unfortunately, this practice is not harmless, because numerous fatigue cracks have occurred due to distortion in the "web gap," i.e., the narrow gap between the termination of the attachment plate fillet welds and the flange [37,38]. Figure 53.6 shows an example of a crack that formed along the fillet weld that attaches a diaphragm connection plate to the web of a box girder.

In most cases, these web-gap-cracking problems can be solved by rigidly attaching the attachment plate to the tension flange. To retrofit existing bridges, a very thick T or angle may be high-strength-bolted in to join the attachment plate to the tension flange [38]. The cracked detail shown in Figure 53.6 was retrofit this way. In other cases a better solution is to make the detail more flexible. This flexibility can be accomplished by increasing the size of the gap, allowing the distortion to take place over a greater length so that lower stresses are created.

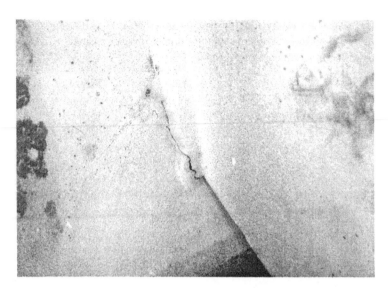

FIGURE 53.6   View of crack in the fillet weld joining the diaphragm connection plate to the web of a box section in the Washington Metro elevated structures. The crack is caused by distortion in the small gap between the bottom of the attachment plate and the box girder flange.

FIGURE 53.7   Connection angle from stringer to floor beam connection of a bridge over the St. Croix River on I-94 that cracked because it was too stiff.

    The flexibility approach is used to prevent cracking at the terminations of transverse stiffeners that are not welded to the bottom flange. If there is a narrow web gap between the end of a transverse stiffener and the bottom flange, cracking can occur due to distortion of the web gap from inertial loading during handling and shipping. To prevent this type of cracking, the gap between the flange and the end of the stiffener should be between four and six times the thickness of the web [38].

    Another example where the best details are more flexible is connection angles for "simply supported" beams. Despite our assumptions, such simple connections transmit up to 40% of the theoretical fixed-end moment, even though they are designed to transmit only shear forces. This unintentional end moment may crack the connection angles. The cracked connection angle shown in Figure 53.7 was from the stringer to floor beam connection on a bridge that formerly was on I-94 over the St. Croix River (it was recently replaced).

FIGURE 53.8 Close-up view of a crack originating at the termination of a flange on a floor beam with a "simple" connection to an attachment plate in the Dresbach Bridge in Minnesota. The crack is an indication of an end moment in the simple connection that is not predicted in design.

For a given load, the moment in the connection decreases significantly as the rotational stiffness of the connection decreases. The increased flexibility of connection angles allows the limited amount of end rotation to take place with reduced bending stresses. A criterion has been developed for the design of these angles to provide sufficient flexibility [41]. The criterion states that the angle thickness (t) must be

$$t < 12 \, (g^{2}/L) \tag{53.1}$$

where (g) is the gauge and L is the span length. For example, using S.I. units, for connection angles with a gauge of 76 mm and a beam span of 7000 mm, the angle thickness should be just less than 10 mm. To solve a connection-angle-cracking problem in service, the topmost rivet or bolt may be removed and replaced with a loose bolt to ensure the shear capacity. For loose bolts, steps are required to ensure that the nuts do not back off.

Another result of unintended end moment on a floor beam is shown in Figure 53.8 from the Dresbach Bridge 9320 in Minnesota. In this bridge, the web of the welded built-up floor beam extends beyond the flanges in order to bolt to the floor beam connection plate on the girder. The unintended moment at the end of the floor beam is enough to cause cracking in the web plate at the termination of the flange fillet welds. There is a large stress concentration at this location caused by the abrupt change in section at the end of the flange; i.e., this is a Category E' detail.

Significant stresses from secondary loading are often in a different direction than the primary stresses. Fortunately, experience with multiaxial loading experiments on large-scale welded structural details indicates the loading perpendicular to the local notch or the weld toe dominates the fatigue life. The cyclic stress in the other direction has no effect if the stress range is below 83 MPa and only a small influence above 83 MPa [26,37]. The recommended approach for multiaxial loads [26] is

1. To decide which loading (primary or secondary) dominates the fatigue cracking problem (typically the loading perpendicular to the weld axis or perpendicular to where cracks have previously occurred in similar details); and,
2. To perform the fatigue analysis using the stress range in this direction (i.e., ignore the stresses in the orthogonal directions).

53.3.3 Classification of Details in Concrete Structural Components

Concrete structures are typically less sensitive to fatigue than welded steel and aluminum structures. However, fatigue may govern the design when impact loading is involved, such as pavement, bridge decks, and rail ties. Also, as the age of concrete girders in service increases, and as the applied stress ranges increase with increasing strength of concrete, the concern for fatigue in concrete structural members has also increased.

According to ACI Committee Report 215R-74 in the Manual of Standard Practice [42], the fatigue strength of plain concrete at 10 million cycles is approximately 55% of the ultimate strength. However, even if failure does not occur, repeated loading may contribute to premature cracking of the concrete, such as inclined cracking in prestressed beams. This cracking can then lead to localized corrosion and fatigue of the reinforcement [43].

The fatigue strength of straight, unwelded reinforcing bars and prestressing strand can be described (in terms of the categories for steel details described above) with the Category B *S–N* curve. ACI Committee 215 suggests that members be designed to limit the stress range in the reinforcing bar to 138 MPa for high levels of minimum stress (possibly increasing to 161 MPa for less minimum stress). Fatigue tests show that previously bent bars had only about half the fatigue strength of straight bars, and failures have occurred down to 113 MPa [44]. Committee 215 recommends that half of the stress range for straight bars be used, i.e., 69 MPa for the worst-case minimum stress. Equating this recommendation to the *S–N* curves for steel details, bent reinforcement may be treated as a Category D detail.

Provided the quality is good, butt welds in straight reinforcing bars do not significantly lower the fatigue strength. However, tack welds reduce the fatigue strength of straight bars about 33%, with failures occurring as low as 138 MPa. Fatigue failures have been reported in welded wire fabric and bar mats [45].

If prestressed members are designed with sufficient precompression that the section remains uncracked, there is not likely to be any problem with fatigue. This is because the entire section is resisting the load ranges and the stress range in the prestressing strand is minimal. Similarly, for unbonded prestressed members, the stress ranges will be very small. However, there is reason to be concerned for bonded prestressing at cracked sections because the stress range increases locally. The concern for cracked sections is even greater if corrosion is involved. The pitting from corrosive attack can dramatically lower the fatigue strength of reinforcement [43].

Although the fatigue strength of prestressing strand in air is about equal to Category B, when the anchorages are tested as well, the fatigue strength of the system is as low as half the fatigue strength of the wire alone (i.e., about Category E). When actual beams are tested, the situation is very complex, but it is clear that much lower fatigue strength can be obtained [46,47]. Committee 215 has recommended the following for prestressed beams:

1. The stress range in prestressed reinforcement, determined from an analysis considering the section to be cracked, is not to exceed 6% of the tensile strength of the reinforcement (this is approximately equivalent to Category C).

2. Without specific experimental data, the fatigue strength of unbonded reinforcement and their anchorages is to be taken as half of the fatigue strength of the prestressing steel. (This is approximately equivalent to Category E.) Lesser values are to be used at anchorages with multiple elements.

53.3.4 Classification of Stay Cables

The Post-Tensioning Institute has issued "Recommendations for Stay Cable Design and Testing" [48]. The PTI recommends that uncoupled bar stay cables are Category B details, while coupled (glued) bar stay cables are Category D. The fatigue strengths of stay cables are verified through fatigue testing. Two types of tests are performed: (1) fatigue testing of the strand, and (2) testing of relatively short lengths of the assembled cable with anchorages. The recommended test of the system is 2 million cycles at a stress range (158 MPa) which is 35 MPa greater than the fatigue allowable for Category B at 2 million cycles. This test should pass with less than 2% wire breaks. A subsequent proof test must achieve 95% of the actual ultimate tensile strength of the tendons.

53.3.5 Characterization of Truck Loading for Fatigue

An actual service load history is likely to consist of cycles with a variety of different load ranges, i.e., variable-amplitude loading. However, the *S–N* curves that are the basis of the fatigue design provisions are based on constant-amplitude loading. A procedure is shown below to convert variable stress ranges to an equivalent constant-amplitude stress range with the same number of cycles. This procedure is based on the damage summation rule jointly credited to Palmgren and Miner (referred to as Miner's rule) [49]. If the slope of the *S–N* curve is equal to 3, then the relative damage of stress ranges is proportional to the cube of the stress range. Therefore, the effective stress range (S_{Re}) is equal to the cube root of the mean cube (rmc) of the stress ranges, i.e.,

$$S_{Re} = [(n_i / N_{total}) \, S_i^3]^{1/3} \qquad (53.2)$$

where n_i is the number of stress ranges of magnitude S_i and N_{total} is the total number of stress ranges.

The fatigue design truck in the LRFD version of the AASHTO bridge design specification and in the present 16th edition of the Standard Specifications is a three-axle HS20 truck. The front axle has a weight of 36 kN and the rear two axles each have a weight of 142 kN. It is very important to note that the single rear axles of the HS20 fatigue truck are actually intended to represent pairs of tandem axles [50,51]. This simplification eases design of main members by decreasing the number of axles (loads) which must be considered. Representation as a single axle is reasonable for design of bridge main members, since the close spacing of tandem axles (about 1.2 m) effectively generates only one stress cycle.

This simplified three-axle truck is not appropriate for the design of deck elements and even some floor beams. Each axle of the tandem axle groups creates a unique stress cycle in a deck element. When the entire distribution of trucks is considered, this results in approximately 4.5 axles or cycles of loading on average for every truck. However, the LRFD code does not clearly indicate that this should be taken into account in the fatigue design of these elements. If the HS20 tandem axle load is split, the effective axle load is 71 kN for each axle.

The HS20 truck is used in strength calculations with a load factor greater than 1.0. However, there is a load factor of 0.75 for fatigue in the LRFD specification, which implies that the actual fatigue design truck is an HS15, i.e., an axle load of 107 kN (really a pair of tandem 53 kN axles). Thus, the intent of the AASHTO LRFD Specifications is to use an HS15 loading for fatigue. The load factor was invented so that an additional and possibly confusing design truck would not have to be defined.

The use of an HS15 truck as representative of the rmc of the variable series of trucks is based on extensive weigh-in-motion (WIM) data and was recommended in NCHRP Report 299 "Fatigue Evaluation Procedures for Steel Bridges" [50]. A constant axle spacing of 30 ft was found to best approximate the axle spacing of typical four and five axle trucks responsible for most fatigue damage to bridges. This HS15 truck is supposed to represent the effective or rmc gross vehicle weight for a distribution of future trucks.

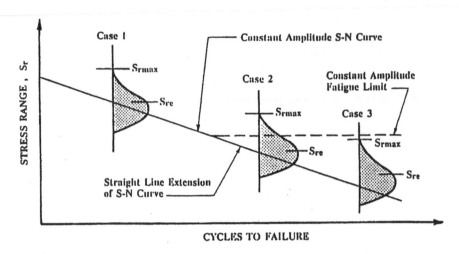

FIGURE 53.9 Schematic of three different cases of the relationship between the spectrum of applied stress ranges and the *S–N* curve. S_{re} is the effective or rmc stress range and S_{rmax} is the fatigue limit-state stress range with an exceedance of 1:10,000. Case 1 and 2 are in the finite-life regime, whereas case 3 illustrates the infinite-life regime.

In the AASHTO LRFD specification, the effective stress range is computed from this effective load, and the effective stress range is compared with an allowable stress range. The allowable stress range is obtained from the constant-amplitude *S–N* curve, using the same number of cycles as the number of variable cycles used to compute the rmc stress range.

This concept is illustrated schematically in Figure 53.9. Figure 53.9 shows the lower part of an *S–N* curve with three different variable stress-range distributions superposed. The effective stress range, shown as S_{re} in Figure 53.9, is the rmc of the stress ranges for a particular distribution, as defined in Eq. 53.2. The effective variable-amplitude stress range is used with *S–N* curves the same way as a constant-amplitude stress range is used, as shown for Case 1 in Figure 53.9.

Full-scale variable-amplitude fatigue tests show that if more than 0.01% of the stress ranges in a distribution are above the CAFL (Case 2 in Figure 53.9), fatigue cracking will still occur [52]. The few cycles that are above the CAFL seem to keep the process going, such that even the stress ranges that are below the CAFL apparently contribute to the fatigue crack growth. The effective stress range and number of cycles to failure data from these tests fell along a straight-line extrapolation of the constant-amplitude curve. Therefore, the approach in the AASHTO LRFD specification is to use the straight-line extrapolation of the *S–N* curve below the CAFL to compute the allowable stress range. There is a lower cutoff of this straight-line extrapolation, which will be discussed later.

The British fatigue design standard BS 7608 [35] uses this same approach for variable amplitude loading; i.e., an rmc stress range is used with an extrapolation of the *S–N* curve below the CAFL. Eurocode 3 [29] also uses the effective stress range concept; however, it uses a different slope for the extrapolation of the *S–N* curve below the CAFL.

Case 1 and Case 2 in Figure 53.9 are in what is called the finite-life regime because the calculations involve a specific number of cycles. Case 3 in Figure 53.9 represents what is referred to as the infinite-life regime. In the infinite-life regime, essentially all of the stress ranges are below the CAFL. The full-scale variable-amplitude fatigue tests referred to earlier [52] show that as many as 0.01% of the stress ranges can exceed the CAFL without resulting in cracking. Consequently, the stress range associated with an exceedance probability of 0.01% is often referred to as the "fatigue-limit-state" stress range.

The lower cutoff of the straight-line extrapolation of the *S–N* curve in the AASHTO LRFD specification is related to this infinite-life phenomenon. Specifically, the AASHTO LRFD specification cuts off the extrapolation of the *S–N* curve at a stress range equal to half the CAFL. The intent of this limit is actually to assure that the fatigue-limit-state stress range is just below the CAFL. The AASHTO LRFD procedure assumes that there is a fixed relationship between the rmc or effective stress range and the limit-state stress range, specifically that the effective stress range is equal to half the limit-state stress range.

Depending on the type of details, the infinite-life regime begins at about 35 million cycles. When designing for a number of cycles larger than this limit, it is no longer important to quantify the precise number of cycles. Any design where the rmc stress range is less than half the CAFL, or where the fatigue-limit-state stress range is less than the CAFL, should theoretically result in essentially infinite life.

The ratio of the fatigue-limit-state stress range to the effective stress range is assumed to be the same as the ratio of the fatigue-limit-state load range to the effective load range, although this is only an approximation at best. Therefore, the fatigue-limit-state load range is defined as having a probability of exceedance over the lifetime of the structure of 0.01%. A structure with millions of cycles is likely to see load ranges exceeding this magnitude hundreds of times, therefore, the fatigue-limit-state load range is not as large as the extreme loads used to check ultimate strength. The fatigue-limit-state load range is assumed to be about twice the effective rmc load range in the LRFD specification.

Recall that the HS20 loading with a load factor of 0.75, i.e., an HS15 loading, was defined as the effective load in the AASHTO LRFD Specification. Therefore, the LRFD Specifications implies that the fatigue-limit-state truck would be HS30. The designer computes a stress range using the HS20 truck with a load factor of 0.75. By assuring that this computed stress range is less than half the CAFL, the AASHTO LRFD specification is supposed to assure that the stress ranges from random variable traffic will not exceed the CAFL more than 0.01% of the time.

The ratio of the effective gross vehicle weight (GVW) to the GVW of the fatigue-limit-state truck in the measured spectrum is referred to in the literature as the "alpha factor" [25]. (This unfortunate choice of nomenclature should not be confused with another alpha factor related to the ratio of the actual stress ranges in the bridge to computed stress ranges.) The LRFD Specifications imply that alpha equals 0.5.

The alpha factor of 0.5 implied by the LRFD Specifications is not consistent with the findings of NCHRP Report 299 [50] and the Guide Specification for Fatigue Design of Steel Bridges [53], which imply the alpha factor should be closer to 0.33. With alpha of 0.33, the fatigue-limit-state truck is about three times heavier than the effective fatigue truck, or about HS45. This finding was based on a reliability analysis, comparison with the original AASHTO fatigue limit check, review of nationwide WIM data, and a study of alpha factors from measured stress histograms.

According to the statistics of the GVW histograms [50], this HS45 fatigue-limit-state truck has only a 0.023% probability (about 1 in 5000) of exceedance, which is almost consistent with the recommendation from NCHRP 354 that the fatigue-limit-state stress range has an exceedance of less than 1:10,000. If the alpha of 0.33 is correct, the stress range produced by the fatigue truck should be compared to ⅓ the CAFL rather than ½. The HS30 fatigue-limit-state truck implied by the AASHTO LRFD provisions clearly has a much higher probability of exceedance.

The apparent inconsistency in the exceedance level of the fatigue-limit-state truck weight was intentional and was a result of "calibrating" the LRFD bridge specifications to give fatigue design requirements which are similar to those of preceding specifications. The theoretically low exceedance level of the fatigue-limit-state truck weight implied by the LRFD code was used because it was felt that other aspects of the design process are overconservative, such as the assumptions in the structural analysis [54]. The Guide Specification for Fatigue Design of Steel Bridges [53] apparently resulted in overly conservative estimates of fatigue life when compared with observed field behavior. As a result, the LRFD Specification was "calibrated" to match existing field experience.

Measured axle load data [51,55] show axle loads which substantially exceed the 107 kN fatigue-limit-state axle load (half of the tandem axle load from the HS30 truck) implied by the AASHTO LRFD Specifications. The shape of the axle load spectra is essentially the same as the GVW spectra; i.e., the alpha factor is about 0.33. This alpha factor would suggest the fatigue-limit-state axle load with an exceedance level of about 1:10,000 would be approximately 160 kN. In fact, measured axle load data [55] show axle loads exceeding 160 kN in some cases. The appropriate axle load to use in general design specifications is very uncertain, however, because the measured axle load spectra are very site specific in this extreme tail of the distributions and the data are very sparse.

Aside from the fatigue-limit-state axle load or GVW, the rest of the loading spectrum does not matter when using the infinite-life approach. Also, the precise number of cycles does not have to be forecast. Rather, it is only necessary to establish that the total number of cycles exceeds the number of cycles associated with the CAFL. Therefore, it is not necessary to know precisely the expected life of the deck and future traffic volumes. Thus, despite the uncertainty in the appropriate value for the fatigue-limit-state axle load, the infinite-life approach is considerably simpler than trying to account for the cumulative damage of the whole distribution of future axle loads, which is even more uncertain.

The infinite-life approach relies upon the CAFL as the parameter determining the fatigue resistance. The emphasis in fatigue testing of details should therefore be on defining the CAFL. Unfortunately, there is a need for additional testing to define these CAFL better. Many of the CAFL values in Table 53.1 were based on judgment rather than specific test data at stress ranges down near the CAFL. Additional research should be performed to investigate the validity of many of the CAFL.

Clearly, most structures carry enough truck traffic to justify an infinite-life fatigue design approach, especially for the deck elements. For example, assuming 25-year life and a Category C detail it can be shown that the maximum permissible ADTT is about 850 trucks/day if the number of cycles associated with half the CAFL is just slightly exceeded in a deck element. In most cases, designing for infinite fatigue life rather than designing for a finite number of cycles adds little cost. This infinite-life approach has also recently been applied in developing AASHTO fatigue design specifications for wind-loaded sign, signal, and luminaire support structures [17] and modular bridge expansion joints [57–59].

In the fatigue design of bridge deck elements according to the AASHTO LRFD specifications, the fatigue design stress range is obtained from a static analysis where the wheel loads (half the axle loads) are applied in patches. The load patches in the LRFD Specifications are calculated in a manner which differs from the calculation in the Standard Specifications. However, in both methods, the patches increase in size as the load increases, which results in an applied pressure on continuous deck surfaces which is approximately the same as typical truck tire pressures (700 kPa). In the AASHTO LRFD Specifications, the patch has a fixed width of 508 mm and a tire pressure of 860 kPa. The patch increases in length as the load increases. At present, the LRFD specifications are not clear as to how load factors are to be used in calculating the patch size. It is not conservative to use the strength load factors in calculating the patch size, since this tends to spread the load over a greater area. Ideally, the patches for the fatigue loading should be calculated on the basis of the limit-state HS30 loading, although it would be conservative to use either the HS20 or factored HS15 loading.

53.4 Fracture Resistance

As explained previously, it is considered more important to focus on the prevention of fatigue cracks in bridges than to focus on the resistance to fracture. However, for structural components which are not subjected to significant cyclic loading, fracture could still possibly occur without prior fatigue crack growth. The primary tension chords of long-span truss would be one example. Usually, this would occur as the loads are applied for the first time during construction.

Unlike fatigue, fracture behavior depends strongly on the type and strength level of the steel or filler metal. In general, fracture toughness has been found to decrease with increasing yield strength of a material, suggesting an inverse relationship between the two properties. In practice, however, fracture toughness is more complex than implied by this simple relationship since steels with similar strength levels can have widely varying levels of fracture toughness.

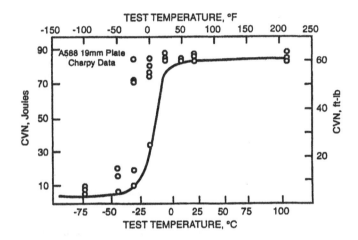

FIGURE 53.10   Charpy Energy Transition Curve for A588 Grade 50 (350 MPa yield strength) Structural Steel.

Steel exhibits a transition from brittle to ductile fracture behavior as the temperature increases. For example, Figure 53.10 shows a plot of the energy required to fracture CVN impact test specimens of A588 structural steel at various temperatures. These results are typical for ordinary hot-rolled structural steel. The transition phenomenon shown in Figure 53.10 is a result of changes in the underlying microstructural fracture mode.

There are three types of fracture with different behavior:

1. Brittle fracture is associated with cleavage of individual grains on select crystallographic planes. This type of fracture occurs at the lower end of the temperature range, although the brittle behavior can persist up to the boiling point of water in some low-toughness materials. This part of the temperature range is called the lower shelf because the minimum toughness is fairly constant up to the transition temperature. Brittle fracture may be analyzed with linear elastic fracture mechanics [2,23,24] because the plasticity that occurs is negligible.

2. Transition-range fracture occurs at temperatures between the lower shelf and the upper shelf and is associated with a mixture of cleavage and fibrous fracture on a microstructural scale. Because of the mixture of micromechanisms, transition-range fracture is characterized by extremely large variability.

3. Ductile fracture is associated with a process of void initiation, growth, and coalescence on a microstructural scale, a process requiring substantial energy. This higher end of the temperature range is referred to as the upper shelf because the toughness levels off and is essentially constant for higher temperatures. Ductile fracture is also called fibrous fracture

due to the fibrous appearance of the fracture surface, or shear fracture due the usually large slanted shear lips on the fracture surface.

Ordinary structural steel such as A36 or A572 is typically only hot-rolled, while to achieve very high toughness steels must be controlled-rolled, i.e., rolled at lower temperatures, or must receive some auxiliary heat treatment such as normalization. In contrast to the weld metal, the cost of the steel is a major part of total costs. The expense of the high-toughness steels has not been found to be warranted for most bridges, whereas the cost of high-toughness filler metal is easily justifiable. Hot-rolled steels, which fracture in the transition region at the lowest service temperatures, have sufficient toughness for the required performance of most welded buildings and bridges.

ASTM specifications for bridge steel (A709) provide for minimum CVN impact test energy levels. Structural steel specified by A36, A572, or A588, without supplemental specifications, does not require the Charpy test to be performed. The results of the CVN test, impact energies, are often referred to as "notch-toughness" values.

TABLE 53.2 Minimum Charpy Impact Test Requirements for Bridges and Buildings

	Minimum Service Temperature		
Material	−18°C, J@°C	−34°C, J@°C	−51°C, J@°C
Steel: nonfracture-critical members[a,b]	20@21	20@4	20@−12
Steel: fracture-critical members[a,b]	34@21	34@4	34@−12
Weld metal for nonfracture critical[a]	27@−18	27@−18	27@−29
Weld metal for fracture critical[a,b]	34@−29 for all service temperatures		

[a] These requirements are for welded steel with minimum specified yield strength up to 350 MPa up to 38 mm thick. Fracture-critical members are defined as those which if fractured would result in collapse of the bridge.

[b] The requirements pertain only to members subjected to tension or tension due to bending.

The CVN specification works by assuring that the transition from brittle to ductile fracture occurs at some temperature less than service temperature. This requirement ensures that brittle fracture will not occur as long as large cracks do not develop. Because the Charpy test is relatively easy to perform, it will likely continue to be the measure of toughness used in steel specifications. Often 34 J (25 ft-lbs), 27 J (20 ft-lbs), or 20 J (15 ft-lbs) are specified at a particular temperature. The intent of specifying any of these numbers is the same, i.e., to make sure that the transition starts below this temperature.

Some Charpy toughness requirements for steel and weld metal for bridges are compared in Table 53.2. This table is simplified and does not include all the requirements.

Note that the bridge steel specifications require a CVN at a temperature that is 38°C greater than the minimum service temperature. This "temperature shift" accounts for the effect of strain rates, which are lower in the service loading of bridges (on the order of 10^{-3}) than in the Charpy test (greater than 10^1) [23]. It is possible to measure the toughness using a Charpy specimen loaded at a strain rate characteristic of bridges, called an intermediate strain rate, although the test is more difficult and the results are more variable. When the CVN energies from an intermediate strain rate are plotted as a function of temperature, the transition occurs at a temperature about 38°C lower for materials with yield strength up to 450 MPa.

As shown in Table 53.2, the AWS D1.5 Bridge Welding Code specifications for weld metal toughness are more demanding than the specifications for base metal. The extra margin of fracture toughness in the weld metal is reasonable because the weld metal is always the location of discontinuities and high tensile residual stresses. Because the cost of filler metal is relatively small in comparison with the overall cost of materials, it is usually worth the cost to get high-toughness filler metal.

The minimum CVN requirements are usually sufficient to assure damage tolerance, i.e., to allow cracks to grow quite long before fracture occurs. Fatigue cracks grow at an exponentially increasing rate; therefore, most of the fatigue life transpires while the crack is very small. Additional fracture toughness, greater than the minimum specified values, will allow the crack to grow to a larger size before sudden fracture occurs. However, the crack is growing so rapidly at the end of life that the additional toughness may increase the life only insignificantly. Therefore, specification of toughness levels exceeding these minimum levels is usually not worth the increased cost of the materials.

The fractures of steel connections that occurred in the Northridge earthquake of 1994 is an example of what can happen if there are no filler metal toughness specifications. At that time, there were no requirements for weld metal toughness in AWS D1.1 for buildings, even for seismic welded steel moment frames (WSMF). This lack of requirements was rationalized because typically the weld deposits are higher toughness than the base metal. However, this is not always the case; e.g., the self-shielded flux-cored arc welds (FCAW-S) used in many of the WSMF that fractured in the Northridge earthquake were reported to be very low toughness [5–7].

ASTM A673 has specifications for the frequency of Charpy testing. The *H* frequency requires a set of three CVN specimens to be tested from one location for each heat or about 50 tons. These tests can be taken from a plate with thickness up to 9 mm different from the product thickness if it is rolled from the same heat. The *P* frequency requires a set of three specimens to be tested from one end of every plate, or from one shape in every 15 tons of that shape. For bridge steel, the AASHTO specifications require CVN tests at the *H* frequency as a minimum. For fracture-critical members, the guide specifications require CVN testing at the *P* frequency. In the AISC specifications, CVN tests are required at the *P* frequency for thick plates and jumbo sections. A special test location in the core of the jumbo section is specified, as well as the requirement that the section tested be produced from the top of the ingot.

Even the *P* testing frequency may be insufficient for as-rolled structural steel. In a recent report for NCHRP [60], CVN data were obtained from various locations on bridge steel plates. The data show that because of extreme variability in CVN across as-rolled plates, it would be possible to miss potentially brittle areas of plates if only one location per plate is sampled. For plates that were given a normalizing heat treatment, the excessive variability was eliminated.

Quantitative means for predicting brittle fracture are available, i.e., fracture mechanics [2,6,11–13,23,24,61]. These quantitative fracture calculations are typically not performed in design, but are often used in service to assess a particular defect. There is at best only about plus or minus 30% accuracy in these fracture predictions, however. Several factors contributing to this lack of accuracy include (1) variability of material properties; (2) changes in apparent toughness values with changes in test specimen size and geometry; (3) differences in toughness and strength of the weld zone; (4) complex residual stresses; (5) high gradients of stress in the vicinity of the crack due to stress concentrations; and (6) the behavior of cracks in complex structures of welded intersecting plates.

50.5 Summary

1. Structural elements where the live load is a large percentage of the total load are potentially susceptible to fatigue. Many factors in fabrication can increase the potential for fatigue including notches, misalignment, and other geometric discontinuities, thermal cutting, weld joint design (particularly backing bars), residual stress, nondestructive evaluation and weld defects, intersecting welds, and inadequate weld access holes. The fatigue design procedures in the AASHTO Specifications are based on control of the stress range and knowledge of

the fatigue strength of the various details. Using these specifications, it is possible to identify and avoid details which are expected to have low fatigue strength.

2. The simplified fatigue design method for infinite life is justified because of the uncertainty in predicting the future loading on a structure. The infinite-life fatigue design philosophy requires that essentially all the stress ranges are less than the CAFL. One advantage of this approach for structures with complex stress histories is that it is not necessary to predict accurately the entire future stress range distribution. The fatigue design procedure simply requires a knowledge of the stress range with an exceedance level of 1:10,000. The infinite-life approach relies upon the CAFL as the parameter determining the fatigue resistance. The emphasis in fatigue testing of details should therefore be on defining the CAFL. Additional research should be performed to investigate the validity of many of the CAFL in the present specification.

3. Welded connections and thermal-cut holes, copes, blocks, or cuts are potentially susceptible to brittle fracture. Many interrelated design variables can increase the potential for brittle fracture including lack of redundancy, large forces and moments with dynamic loading rates, thick members, geometric discontinuities, and high constraint of the connections. Low temperature can be a factor for exposed structures. The factors mentioned above that influence the potential for fatigue have a similar effect on the potential for fracture. In addition, cold work, flame straightening, weld heat input, and weld sequence can also affect the potential for fracture. The AASHTO Specifications require a minimum CVN "notch toughness" at a specified temperature for the base metal and the weld metal of members loaded in tension or tension due to bending. Almost two decades of experience with these bridge specifications have proved that they are successful in significantly reducing the number of brittle fractures.

References

1. "AASHTO, *LRFD Bridge Design Specifications,* American Association of State Highway Transportation Officials, Washington, D.C., 1994.
2. Fisher, J. W., The evolution of fatigue resistant steel bridges, in *1997 Distinguished Lectureship*, Transportation Research Board, 76th Annual Meeting, Washington, D.C., Jan. 12–16, 1997, Paper No. 971520: 1–22.
3. Fisher, J. W., *Fatigue and Fracture in Steel Bridges*, John Wiley & Sons, New York, 1984.
4. Fisher, J. W., Pense, A. W., and Roberts, R., Evaluation of fracture of Lafayette Street Bridge, *J. Struct. Div. ASCE*, 103(ST7), 1977.
5. Kaufmann, E. J., Fisher, J. W., Di Julio, R. M., Jr., and Gross, J. L., Failure Analysis of Welded Steel Moment Frames Damaged in the Northridge Earthquake, NISTIR 5944, National Institute of Standards and Technology, Gaithersburg, MD, January 1997.
6. Fisher, J. W., Dexter, R. J. and Kaufmann, E. J., Fracture mechanics of welded structural steel connections, in Background Reports: Metallurgy, Fracture Mechanics, Welding, Moment Connections, and Frame Systems Behavior, Report No. SAC 95-09, FEMA-288, March 1997.
7. Kaufmann, E. J., Xue, M., Lu, L.-W., and Fisher, J. W., Achieving ductile behavior of moment connections, *Mod. Steel Constr.*, 36(1), 30–39, January 1996; see also, Xue, M., Kaufmann, E. J., Lu, L.-W., and Fisher, J. W., Achieving ductile behavior of moment connections — Part II, *Mod. Steel Constr.*, 36(6), 38–42, June 1996.
8. Engineers Investigate Cracked El, *Eng. News Rec.*, 200(3), January 19, 1979.
9. Miki, C., Fractures in seismically loaded bridges, *Prog. Struct. Eng. Mat.*, 1(1), 115–121, September 1997.
10. ANSI/AWS D1.1, Structural Welding Code — Steel, American Welding Society, Miami, 1996.

11. Dexter, R. J., Load-deformation behavior of cracked high-toughness steel members, in *Proceedings of the 14th International Conference on Offshore Mechanics and Arctic Engineering Conference (OMAE)*, 18–22 June 1995, Copenhagen Denmark, Salama et al., Eds., Vol. III, *Materials Engineering*, AMSE, 1995, 87–91.

12. Dexter, R. J. and Gentilcore, M. L., Evaluation of Ductile Fracture Models for Ship Structural Details, Report SSC-393, Ship Structure Committee, Washington, D.C., 1997.

13. Dexter, R. J. and Gentilcore, M. L., Predicting extensive stable tearing in structural components, in *Fatigue and Fracture Mechanics: 29th Volume, ASTM STP 1321*, T. L. Panontin and S. D. Sheppard, Eds., American Society for Testing and Materials, Philadelphia, 1998.

14. Dexter, R. J., Significance of strength undermatching of welds in structural behaviour, in *Mis-Matching of Interfaces and Welds*, K.-H. Schwalbe and M. Kocak, Eds., GKSS Research Center, Geesthacht, Germany, 1997, 55–73.

15. Castiglioni, C. A., Cumulative damage assessment in structural steel details, in *IABSE Symposium, Extending the Lifespan of Structures*, San Francisco, IABSE, 1995, 1061–1066.

16. Krawinkler, H. and Zohrei, M., Cumulative damage in steel structures subjected to earthquake ground motion, *Comput. Struct.*, 16 (1–4), 531–541, 1983.

17. Kaczinski, M. R., Dexter, R. J., and Van Dien, J. P., Fatigue-Resistant Design of Cantilevered Signal, Sign, and Light Supports, NCHRP Report 412, National Cooperative Highway Research Program, Transportation Research Board, Washington, D.C., 1998

18. Fisher, J. W., Frank, K. H., Hirt, M. A., and McNamee, B. M., Effect of Weldments on the Fatigue Strength of Steel Beams, National Cooperative Highway Research Program (NCHRP) Report 102, Highway Research Board, Washington, D.C., 1970.

19. Fisher, J. W., Albrecht, P. A., Yen, B. T., Klingerman, D. J., and McNamee, B. M., Fatigue Strength of Steel Beams with Welded Stiffeners and Attachments, National Cooperative Highway Research Program (NCHRP) Report 147, Transportation Research Board, Washington, D.C., 1974.

20. Dexter, R. J., Fisher, J. W. and Beach, J. E., Fatigue behavior of welded HSLA-80 members, in *Proceedings, 12th International Conference on Offshore Mechanics and Arctic Engineering*, Vol. III, Part A, Materials Engineering, ASME, New York, 1993, 493–502.

21. Keating, P. B. and Fisher, J. W., Evaluation of Fatigue Tests and Design Criteria on Welded Details, NCHRP Report 286, National Cooperative Highway Research Program, September 1986.

22. VanDien, J. P., Kaczinski, M. R., and Dexter, R. J., Fatigue testing of anchor bolts, in *Building an International Community of Structural Engineers*, Vol. 1, *Proc. of Structures Congress XIV*, Chicago, 1996, 337–344.

23. Barsom, J. M. and Rolfe, S. T., *Fracture and Fatigue Control in Structures*, 2nd ed., Prentice-Hall, Englewood Cliffs, NJ, 1987.

24. Broek, D. *Elementary Fracture Mechanics*, 4th ed., Martinis Nijhoff Publishers, Dordrecht, Netherlands, 1987.

25. Kober, Dexter, R. J., Kaufmann, E. J., Yen, B. T., and Fisher, J. W., The effect of welding discontinuities on the variability of fatigue life, in *Fracture Mechanics, Twenty-Fifth Volume, ASTM STP 1220*, F. Erdogan and Ronald J. Hartranft, Eds., American Society for Testing and Materials, Philadelphia, 1994.

26. Dexter, R. J., Tarquinio, J. E., and Fisher, J. W., An application of hot-spot stress fatigue analysis to attachments on flexible plate, in *Proceedings of the 13th International Conference on Offshore Mechanics and Arctic Engineering*, American Society of Mechanical Engineers (ASME), New York, 1994.

27. Yagi, J., Machida, S., Tomita, Y., Matoba, M., and Kawasaki, T., Definition of Hot-Spot Stress in Welded Plate Type Structure for Fatigue Assessment, International Institute of Welding, IIW-XIII-1414-91, 1991.

28. AISC, *Load and Resistance Factor Design Specification for Structural Steel Buildings*, 2nd ed., American Institute of Steel Construction (AISC), Chicago, 1993.

29. ENV 1993-1-1, Eurocode 3: Design of steel structures — Part 1.1: General rules and rules for buildings, European Committee for Standardization (CEN), Brussels, April 1992.

30. Roberts, R. et al., Corrosion Fatigue of Bridge Steels, Vol. 1–3, Reports FHWA/RD-86/165, 166, and 167, Federal Highway Administration, Washington, D.C., May 1986.
31. Outt, J. M. M., Fisher, J. W., and Yen, B. T., Fatigue Strength of Weathered and Deteriorated Riveted Members, Report DOT/OST/P-34/85/016, Department of Transportation, Federal Highway Administration, Washington, D.C., October 1984.
32. Albrecht, P. and Shabshab, C., Fatigue strength of weathered rolled beam made of A588 steel, *J. Mater. Civil Eng. (ASCE)*, 6(3), 407–428, 1994.
33. Demers, C. and Fisher, J. W., Fatigue Cracking of Steel Bridge Structures, Vol. I: A Survey of Localized Cracking in Steel Bridges — 1981 to 1988, Report No. FHWA-RD-89-166; also, Volume II, A Commentary and Guide for Design, Evaluation, and Investigating Cracking, Report No. FHWA-RD-89-167, FHWA, McLean, VA, March 1990.
34. Yen, B. T., Huang, T., Lai, L.-Y., and Fisher, J. W., Manual for Inspecting Bridges for Fatigue Damage Conditions, Report No. FHWA-PA-89-022 + 85-02, Fritz Engineering Laboratory Report No. 511.1, Pennsylvania Department of Transportation, Harrisburg, PA., January 1990.
35. BS 7608, Code of Practice for Fatigue Design and Assessment of Steel Structures, British Standards Institute, London, 1994.
36. Maddox, S. J., *Fatigue Strength of Welded Structures*, 2nd ed., Abington Publishing, Cambridge, U.K., 1991.
37. Fisher, J. W., Jian, J., Wagner, D. C., and Yen, B. T., Distortion-Induced Fatigue Cracking in Steel Bridges, National Cooperative Highway Research Program (NCHRP) Report 336, Transportation Research Board, Washington, D.C., 1990.
38. Fisher, J. W. and Keating, P. B., Distortion-induced fatigue cracking of bridge details with web gaps, *J. Constr. Steel Res.*, 12, 215–228, 1989.
39. Fisher, J. W., Kaufmann, E. J., Koob, M. J., and White, G., Cracking, fracture assessment, and repairs of Green River Bridge, I-26, in *Proc. of Fourth International Bridge Engineering Conference*, San Francisco, Aug. 28–30, 1995, Vol 2, National Academy Press, Washington, D.C., 1995, 3–14.
40. Fisher, J. W., Yen, B. T., Kaufmann, E. J., Ma, Z. Z. and Fisher, T. A., Cracking evaluation and repair of cantilever bracket tie plates of Edison Bridge, in *Proc. of Fourth International Bridge Engineering Conference*, San Francisco, Aug. 28–30, 1995, Vol 2, National Academy Press, Washington, D.C. 1995, 15–25.
41. Yen, B. T. et al., Fatigue behavior of stringer-floorbeam connections, in *Proc. of the Eighth International Bridge Conference*, Paper IBC-91-19, Engineers' Society of Western Pennsylvania, 1991, 149–155.
42. ACI Committee 215, Considerations for Design of Concrete Structures Subjected to Fatigue Loading, ACI 215R-74 (Revised 1992), ACI Manual of Standard Practice, Vol. 1, 1996.
43. Hahin, C., Effects of Corrosion and Fatigue on the Load-Carrying Capacity of Structural Steel and Reinforcing Steel, Illinois Physical Research Report No. 108, Illinois Department of Transportation, Springfield, March 1994.
44. Pfister, J. F. and Hognestad, E., High strength bars as concrete reinforcement. Part 6, fatigue tests, *J. PCA Res. Dev. Lab.*, 6(1), 65–84, 1964.
45. Sternberg, F., Performance of Continuously Reinforced Concrete Pavement, I-84 Southington, Connecticut State Highway Department, June 1969.
46. Rabbat, B. G. et al., Fatigue tests of pretensioned girders with blanketed and draped strands, *J. Prestressed Concrete Inst.*, 24(4), 88–115, 1979.
47. Overnman, T. R., Breen, J. E., and Frank, K. H., Fatigue Behavior of Pretensioned Concrete Girders, Research Report 300-2F, Center for Transportation Research, The University of Texas at Austin, November 1984.
48. Ad hoc Committee on Cable-Stayed Bridges, Recommendations for Stay Cable Design and Testing, Post-Tensioning Institute, Phoenix, January 1986.
49. Miner, M. A., Cumulative damage in fatigue, *J. Appl. Mech.*, 12, A-159, 1945.

50. Moses, F., Schilling, C. G., and Raju, K. S., Fatigue Evaluation Procedures for Steel Bridges, NCHRP Report 299, National Cooperative Highway Research Program, 1987.

51. Schilling, C. G., Variable Amplitude Load Fatigue, Task A — *Literature Review*, Vol. I: — Traffic Loading and Bridge Response, Publ. No. FHWA-RD-87-059, Federal Highway Administration, July 1990.

52. Fisher, J. W., et al., Resistance of Welded Details under Variable Amplitude Long-Life Fatigue Loading, National Cooperative Highway Research Program Report 354, Transportation Research Board, Washington, D.C., 1993.

53. American Association of State Highway and Transportation Officials, Guide Specifications for Fracture Critical Non-Redundant Steel Bridge Members, American Association of State Highway and Transportation Officials, Washington, D.C., 1989 (with interims).

54. Dexter, R. J. and Fisher, J. W., The effect of unanticipated structural behavior on the fatigue reliability of existing bridge structures, in *Structural Reliability in Bridge Engineering*, D. M. Frangopol and G. Hearn, Eds., Proceedings of a Workshop, University of Colorado at Boulder, 2–4 Oct. 1996, McGraw-Hill, New York, 1996, 90–100.

55. Nowak, A. S. and Laman, J. A., Monitoring Bridge Load Spectra, in *IABSE Symposium, Extending the Lifespan of Structures*, San Francisco, 1995.

56. Menzemer, C. C. and Fisher, J. W., Revisions to the Aluminum Association fatigue design specifications, 6th Int. Conf. on Aluminum Weldments, April 3–5, 1995, Cleveland. Available from AWS, Miami, FL, 1995, 11–23.

57. Dexter, R. J., Connor, R. J., and Kaczinski, M. R., Fatigue Design of Modular Bridge Expansion Joints, NCHRP Report 402, National Cooperative Highway Research Program, Transportation Research Board, Washington, D.C., 1997.

58. Dexter, R. J., Kaczinski, M. R., and Fisher, J. W., Fatigue testing of modular expansion joints for bridges, in *IABSE Symposium, 1995, Extending the Lifespan of Structures*, Vol. 73/2, San Francisco, 1995, 1091–1096.

59. Kaczinski, M. R., Dexter, R. J., and Connor, R. J., Fatigue design and testing of modular bridge expansion joints, in *Proceedings of the Fourth World Congress on Joint Sealing and Bearing Systems for Concrete Structures*, Sacramento, September 1996.

60. Frank, K. H., et al., Notch Toughness Variability in Bridge Steel Plates, NCHRP Report 355, National Cooperative Highway Research Program, 1993.

61. PD 6493, Guidance on the Methods for Assessing the Acceptability of Flaws in Fusion Welded Structures, PD 6493: 1991, British Standards Institution (BSI), London, 1991.

54

Statistics of Steel Weight of Highway Bridges

Shouji Toma
Hokkai-Gakuen University, Japan

54.1 Introduction

In this chapter, a database of steel highway bridges is formed to assess designs by analyzing them statistically. No two bridges are exact replicas of each other because of the infinite variety of site conditions. Each bridge meets specific soil, traffic, economic, and aesthetics conditions. The structural form, the support conditions, the length, width, and girder spacing, pedestrian lanes, and the materials, all depend on a unique combination of design criteria. Even if the stipulated criteria are identical, the final bridges are not, as they naturally reflect the individual intentions of different designers. Therefore, steel weight is a major interest to engineers.

Steel weight of highway bridges is one of the most important of the many factors that influence bridge construction projects. The weight gives a good indication of structural, economic, and safety

0-8493-7434-0/00/$0.00+$.50
© 2000 by CRC Press LLC

features of the bridge. Generally, the weight is expressed by as a force per square unit of road surface area (tonf/m² or kN/m²). Stochastic distribution of the weight includes many influential factors to designs that cause scatter. The analysis of this scatter may suggest the characteristics of the bridges. As a general rule, simple bridges are lighter than more complex ones, bridges with high safety margins are heavier, and composite construction results in a lighter bridge overall. A designer thereby gets insight into the characteristics of a bridge. As bridge design also requires the estimate of steel weight in advance, the data collected here are useful.

In Japan, many steel bridges have been constructed in the past few decades. The weight of steel used in these bridges has been collected into a single database. The bridges are all Japanese, but engineers from other countries use similar structural and economic considerations and can usefully employ these in their designs. In this chapter, Japanese design criteria are presented first. The live loads and material properties are described in special detail to clarify differences that other countries may note. Then, the computer database is explained and used to make comparisons between plate and box girders, truss and frame bridges, simply supported and continuously supported bridges, reinforced concrete slab deck and steel deck, and more.

54.2 Design Criteria

54.2.1 Live Loads

The strength required for a bridge to sustain largely depends on the live load, and the live load generally differs from country to country. Since the weight information used here follows Japanese specifications, those will be the ones explained. The last version of the bridge design specification was published in 1996 [1], and is based on a truck weight of 25 tonf (245 kN). However, the bridges studied here were designed using an old version of the code [2], and thus used a truck load of 20 tonf (196 kN).

The 20 t live load (TL-20) takes the two forms shown in Figure 54.1a. The T-load is used to design local components such as the slab or the floor system and the L-load is used for global ones such as the main girders. The T-load is the concentrated wheel loads and the L-load is further subdivided. A partially distributed load (caused by the truck) and a load distributed along the length of the bridge (corresponding to the average traffic load) comprises the L-load. Most of the bridges were designed for TL-20, but on routes, such as those near harbor ports, heavy truck loads are expected and these were designed for TT-43 (Figure 54.1b). In this database the difference is not considered.

When a bridge has side lanes for pedestrian traffic, and the live load (the crowd load) is small compared to vehicular traffic loads, usually less steel is required. However, the difference of the weight for pedestrian and vehicular lanes is not considered in this database. The surface area of the sidewalk is considered equally as heavy as the area in the vehicle lanes.

54.2.2 Materials

The strength of steel varies widely. A mild steel may have a yield strength of about 235 N/mm² and is commonly used in bridge design but higher strengths of 340 or 450 N/mm² are also used, often in large bridges. Various strength of steel are considered in this study. Clearly, when higher-strength steels are used, the weight of steel required goes down. However, the difference in strength level of steel is not distinguished in the database. Aa a selection of strength level is made considering rationality of design, it will generally result in similar decisions for many bridges. In other words, similar bridge designs specify similar material strengths. The effect of strength is thus included implicitly in the database.

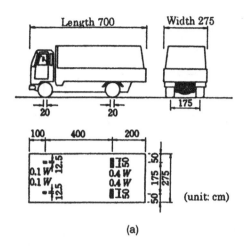

(a)

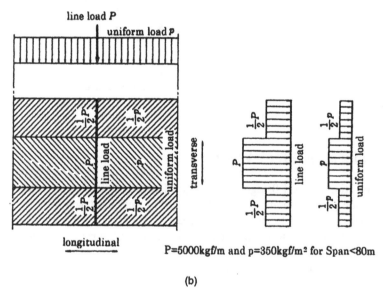

P=5000kgf/m and p=350kgf/m² for Span<80m

(b)

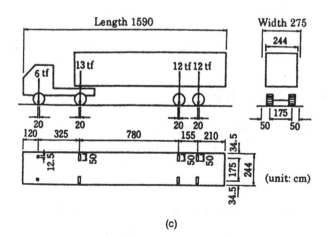

(c)

FIGURE 54.1 Live load (TL-20). (a) T-Load (W = 20 tf); (b) L-Load; (c) TT-43 (W = 43 tf).

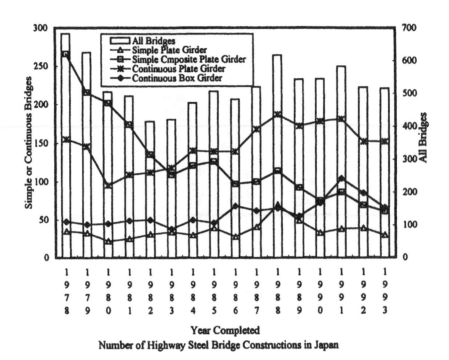

FIGURE 54.2   Number of highway steel bridge constructions in Japan.

54.3   Database of Steel Weights

The Japan Association of Steel Bridge Construction (JASBC) publishes an annual report on steel bridge construction [3]. Information about the weight of steel was taken from these reports over a period of 15 years (from 1978 to 1993). The database was collected using a personal computer [4]. The weight was expressed in terms of intensity per unit road surface area (tonf/m²). Table 54.1 shows the quantity of data available for each year relating to various types of bridges. When enough data exist to perform a reliable statistical analysis, new data are used. When the year's sample is small, all the data are included.

The data in Table 54.1 are plotted in Figure 54.2, which also shows the number of steel bridges constructed in Japan. From Figure 54.2, it can be seen that about 500 steel bridges are constructed each year. The tendency of the structural types can also be seen: simply supported composite plate girders are gradually replaced by continuous girders. This can be explained as expansion joints damage the pavement and cause vehicles to make noise as they pass over the joints.

54.4   Statistics of Steel Weights

Weight distributions for various types of bridges are shown in Figures 54.3 through 54.13. The weights are plotted against the span length which shows applicable length for the type of bridge. In the figures the mean values are shown by a line and a parabola curve; the equations are given in Table 54.2.

54.4.1   Simply Supported Noncomposite Plate Girder Bridges

In Figure 54.3 the distributions for simply supported plate girder bridges with reinforced concrete (RC) slab and steel decks are shown. The steel weight varies considerably, from which one can investigate the peculiarity of the bridge.

TABLE 54.1 Number of Input Data

Type of Bridge	Year Completed																Total
	1978	1979	1980	1981	1982	1983	1984	1985	1986	1987	1988	1989	1990	1991	1992	1993	
Simple plate girder	35	33	22	25	31	34	30	39	28	41	70	49	33	38	39	30	577
Simple plate girder (steel deck)	6	2	5	4	6	9	6	8	9	11	12	9	14	5	4	8	118
Simple composite plate girder	266	216	202	174	135	109	121	126	97	100	114	92	75	86	69	61	2043
Simple box girder	30	29	34	24	24	12	29	33	24	36	41	36	35	40	32	44	503
Simple box girder (steel deck)	15	12	6	6	4	7	6	10	16	5	16	14	14	21	20	28	200
Simple composite box girder	42	36	18	23	9	13	10	13	12	17	21	18	11	8	6	10	267
Continuous plate girder	155	146	95	109	112	118	140	139	139	168	187	172	178	180	147	150	2335
Continuous plate girder (steel deck)	0	4	4	0	0	5	6	6	6	1	4	5	6	5	0	2	54
Continuous box girder	48	44	45	49	50	38	50	46	68	62	65	55	72	104	85	66	947
Continuous box girder (steel deck)	9	18	19	16	11	16	19	24	23	17	25	20	23	27	28	42	337
Simple truss	16	26	15	7	11	16	11	14	9	15	15	10	17	8	10	11	211
Continuous truss	10	13	9	10	0	6	12	8	6	12	7	6	12	5	6	2	124
Langer	19	12	8	12	7	10	7	12	4	5	3	7	5	4	8	11	134
Trussed Langer	2	9	4	5	2	2	0	3	2	1	5	2	4	1	4	1	47
Lohse	11	12	12	10	9	11	11	8	19	7	8	11	13	17	17	7	183
Nielsen Lohse	2	0	0	0	1	4	4	2	4	3	4	5	5	7	5	7	53
Rigid frame (Rahmen)	16	12	5	15	3	9	8	10	10	12	17	15	10	8	14	18	182
Rigid frame (π type)	3	6	4	6	4	4	4	5	2	5	6	6	7	6	8	7	83
Arch bridge	—	—	—	—	—	—	—	—	—	—	—	2	4	3	4	2	15
Cable-stayed bridge (steel deck)	0	0	2	2	0	2	1	5	4	5	2	4	5	5	5	6	48
Total	685	630	509	497	419	425	475	511	482	523	622	538	543	578	511	513	8461

TABLE 54.2 Coefficients of Regression Equations

Type of Bridge	a (×10⁻²)	b	Standard Deviation (1)	α (×10⁻⁴)	β (×10⁻²)	γ	Standard Deviation (2)	Year	No. of Data	Correlation Coefficient	Fig. No.
Simple plate girder	0.5866	0.0124	0.0325	0.4621	0.2075	0.0881	0.0324	1989–1993	189	0.758	54.3a
Simple plate girder (steel deck)	0.3504	0.2499	0.0420	0.1228	−0.5853	0.4252	0.0419	1978–1993	118	0.353	54.3b
Simple composite plate girder	0.6084	−0.0306	0.0249	0.3824	0.2985	0.0307	0.0249	1989–1993	383	0.830	54.4
Simple box girder	0.5917	0.0778	0.0410	0.4350	0.1488	0.1866	0.0409	1989–1993	187	0.803	54.5a
Simple box girder (steel deck)	0.3019	0.2738	0.0709	0.0616	0.2303	0.2930	0.0709	1978–1993	200	0.556	54.5b
Simple composite box girder	0.4765	0.1007	0.0412	0.3329	0.1290	0.1887	0.0411	1981–1993	171	0.714	54.6
Continuous plate girder	0.3729	0.0533	0.0331	−0.3092	0.6425	−0.0035	0.0330	1991–1993	477	0.653	54.7a
Continuous plate girder (steel deck)	0.2329	0.2464	0.0484	−0.2413	0.4482	0.2022	0.0481	1978–1993	54	0.508	54.7b
Continuous box girder	0.3029	0.1510	0.0499	0.0099	0.2906	0.1546	0.0499	1989–1993	382	0.665	54.8a
Continuous box girder (steel deck)	0.1516	0.3110	0.0634	0.0213	0.1080	0.3307	0.0633	1978–1993	337	0.593	54.8b
Simple truss	0.2993	0.1421	0.0504	0.3711	−0.2355	0.3284	0.0493	1978–1993	211	0.592	54.9a
Continuous truss	0.2221	0.1633	0.0602	0.0959	0.4830	0.0257	0.0567	1978–1993	124	0.799	54.9b
Langer	0.2907	0.1433	0.0632	−0.0135	0.3140	0.1338	0.0632	1978–1993	134	0.675	54.10a
Trussed Langer	0.2696	0.1700	0.0609	0.1693	−0.1173	0.3794	0.0592	1978–1993	47	0.741	54.10b
Lohse	0.2372	0.1956	0.0942	0.0110	0.2128	0.2076	0.0941	1978–1993	183	0.676	54.11a
Nielsen Lohse	0.2372	0.1956	0.1019	0.0110	0.2128	0.2076	0.1018	1978–1993	53	0.735	54.11b
Rigid frame (Rahmen)	0.4326	0.0542	0.0737	0.4399	−0.1004	0.2024	0.0711	1978–1993	182	0.659	54.14a
Rigid frame (π type)	0.4982	0.0050	0.0555	0.2477	0.1528	0.1160	0.0544	1978–1993	83	0.813	54.14b
Cable-stayed bridge (steel deck)	0.2102	0.2944	0.2056	0.0407	−0.0014	0.4736	0.1937	1978–1993	48	0.784	54.15
Equations (tf/m²)	$aL + b \dots$ (1)			$\alpha L^2 + \beta L + \gamma \dots$ (2)						L = span (m)	

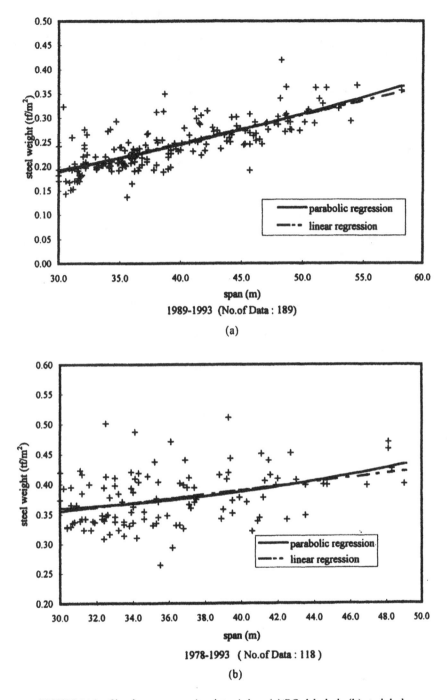

FIGURE 54.3 Simple noncomposite plate girders. (a) RC slab deck; (b) steel deck.

54.4.2 Simply Supported Composite Plate Girder Bridges

The distribution for a simply supported composite plate girder bridge is shown in Figure 54.4. Since many bridges of this type were constructed every year, only 4 years of data are used (1989 to 1993).

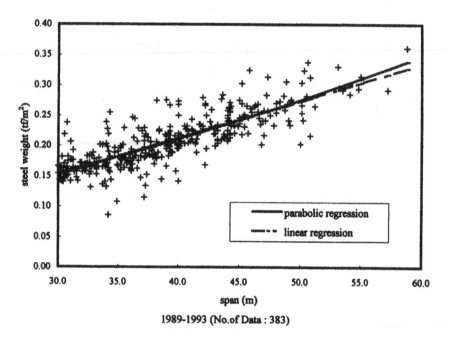

1989-1993 (No.of Data : 383)

FIGURE 54.4 Simple composite plate girders.

54.4.3 Simply Supported Box-Girder Bridges

The distribution for a simply supported box-girder bridge (noncomposite) for RC slab and steel decks is plotted in Figure 54.5. Steel deck bridges show more variation than RC deck bridges. A simply supported composite box-girder bridge is plotted in Figure 54.6.

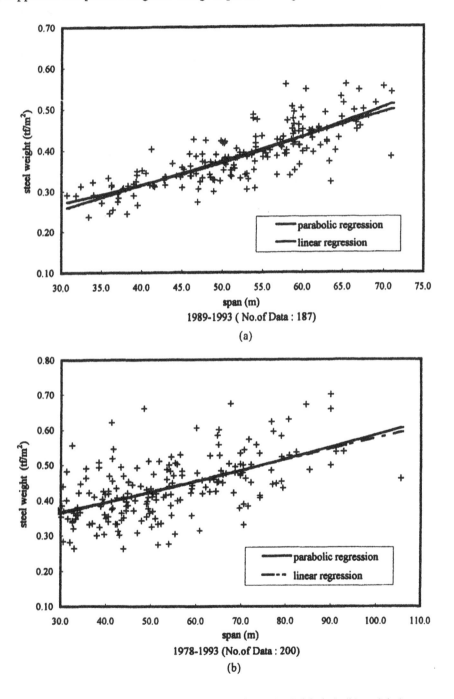

FIGURE 54.5 Simple noncomposite box girders. (a) RC slab deck; (b) steel deck.

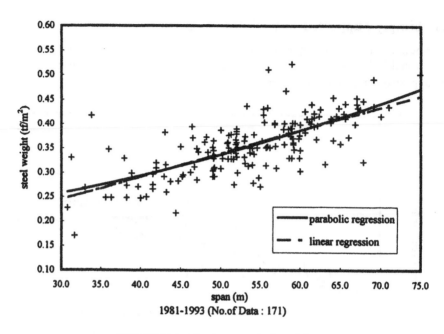

FIGURE 54.6 Simple composite box girders.

54.4.4 Continuously Supported Plate Girder Bridges

Recently, continuous bridges are gaining popularity as defects caused by expansion joints are avoided. Steel weights for continuous bridges with RC slab deck (noncomposite) constructed in the 3 years 1991 to 1993 and with steel deck constructed in the 15 years 1978 to 1993 are plotted in Figure 54.7. The steel deck has only few data and shows wide scatter.

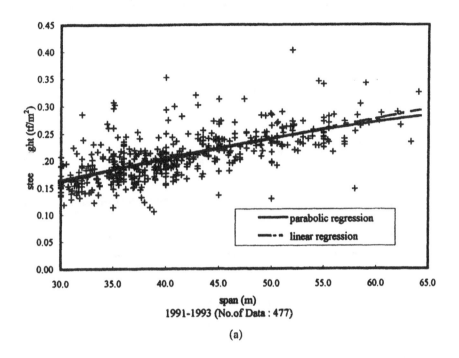

(a)

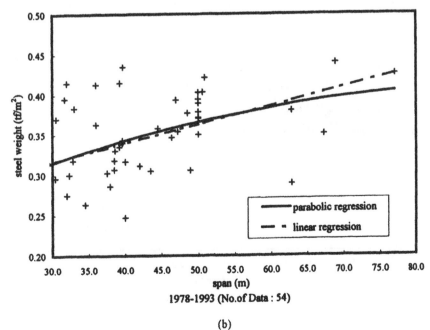

(b)

FIGURE 54.7 Continuous plate girders. (a) RC slab deck; (b) steel deck.

54.4.5 Continuously Supported Box-Girder Bridges

Figure 54.8 shows the distribution for a continuous box-girder bridge with RC slab deck and steel deck. This type has a relatively wide scatter. It can be seen that the applicable span length of steel deck bridges (Figure 54.8b) is much longer than RC slab deck bridges (Figure 54.8a).

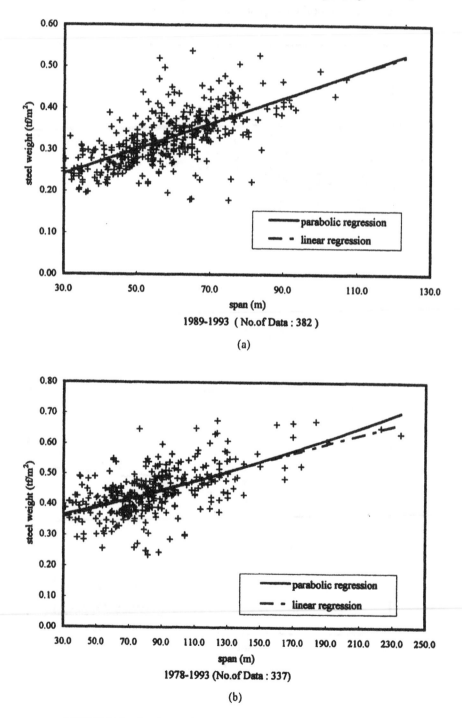

FIGURE 54.8   Continuous box girders. (a) RC slab deck; (b) steel deck.

54.4.6 Truss Bridges

Figure 54.9 is for simply and continuously supported truss bridges. The data cluster at moderate span length making prediction for the weight of truss bridges for short or long spans not accurate.

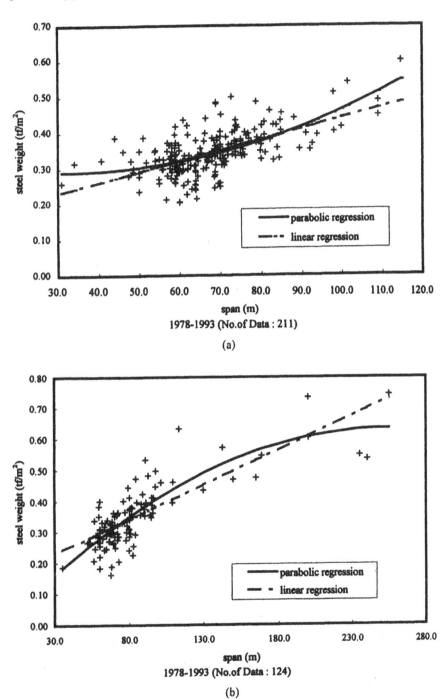

1978-1993 (No.of Data : 211)

(a)

1978-1993 (No.of Data : 124)

(b)

FIGURE 54.9 Truss bridges. (a) Simple truss; (b) continuous truss.

54.4.7 Arch Bridges

Figures 54.10 and 54.11 are the distributions for two arch types; Langer bridges and Lohse bridges. It is assumed in the structural analysis that the arch rib of Lohse bridge carries bending moment, shear force, and axial compression while Langer bridge only carries axial compression. In the Langer bridge, the main girders are stiffened by the arch rib through the vertical members. The trussed Langer uses the diagonal members for the same purpose.

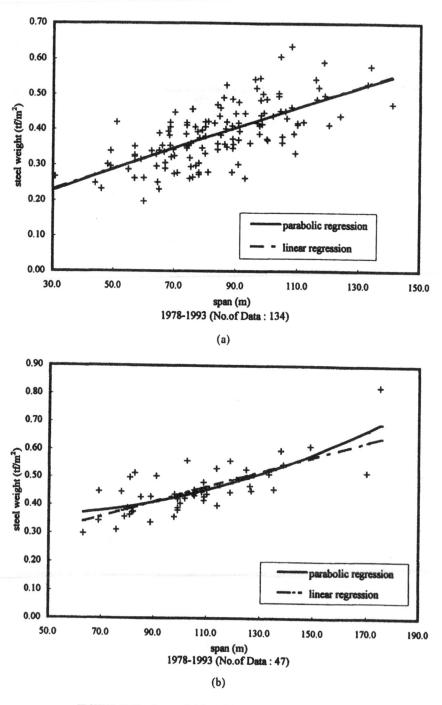

FIGURE 54.10   Langer bridges. (a) Langer; (b) trussed langer.

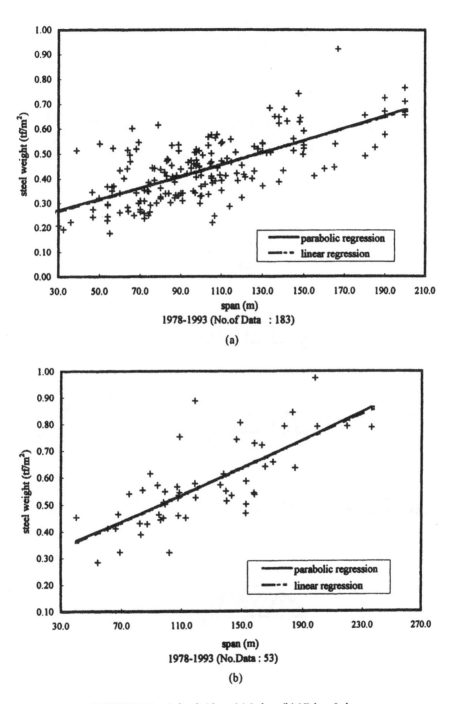

FIGURE 54.11 Lohse bridges. (a) Lohse; (b) Nielsen Lohse.

The Lohse also has vertical members between the arch and main girders, but the Nielsen Lohse has only thin rods which resist only tension and form a net. The types of arch bridges are illustrated in Figure 54.12.

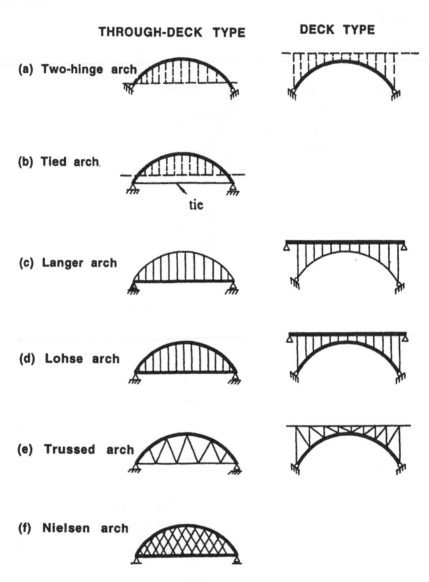

FIGURE 54.12 Types of arch bridges. (a) Two hinge; (b) tied; (c) Langer; (d) Lohse; (e) trussed; (f) Nielson.

54.4.8 Rahmen Bridges (Rigid Frames)

The Rahmen bridge is a frame structure in which all members carry bending moment and axial and shear forces. There are many variations of structural form for this type of construction as shown in Figure 54.13. Figure 54.14 shows the weight distribution for typical π-Rahmen and other types.

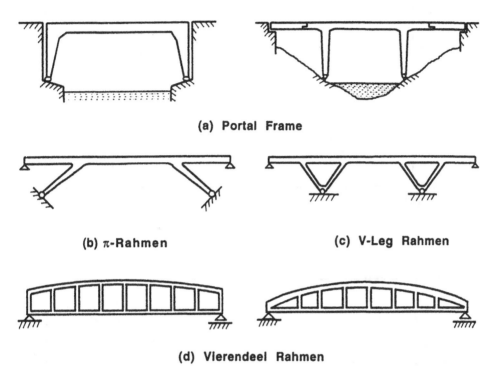

(a) Portal Frame

(b) π-Rahmen   **(c) V-Leg Rahmen**

(d) Vierendeel Rahmen

FIGURE 54.13 Types of Rahmen bridges. (a) Portal frame; (b) π-Rahmen; (c) V-leg Rahmen; (d) Vierendeel Rahmen.

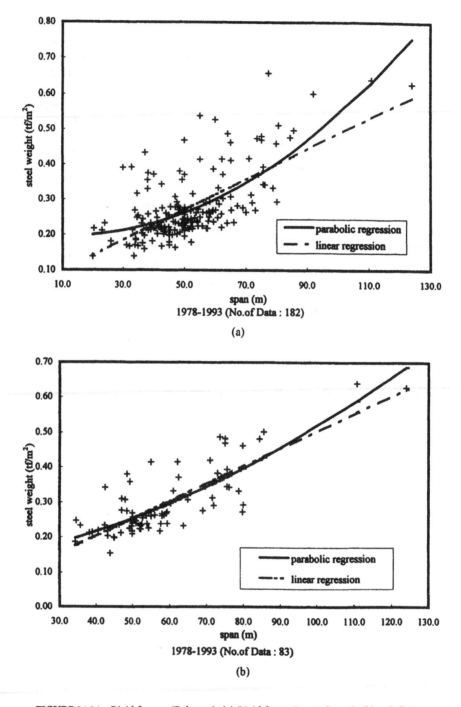

FIGURE 54.14   Rigid frames (Rahmen). (a) Rigid frame (general type); (b) π-Rahmen.

54.4.9 Cable-Stayed Bridges

Figure 54.15 shows the weight of cable-stayed bridges. The data may not be sufficient for statistical analysis. The scatter is more significant at long spans.

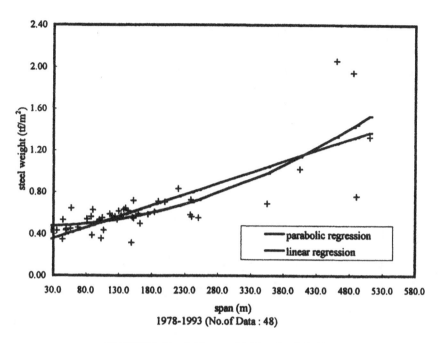

FIGURE 54.15 Cable-stayed bridges (steel deck).

54.5 Regression Equations

The two lines in the distribution figures shown previously in Figures 54.3 through 54.13 are the mean values obtained by linear regression using the least-squares method. They are linear and parabolic. It seems that the parabolic curve does not always give a better prediction. Table 54.2 gives the coefficients of the regression equations to give designers the information necessary for estimating steel weight and assessing designs.

54.6 Comparisons

The weight distributions in Figures 54.3 through 54.13 are compared from various points of view in the following.

54.6.1 Composite and Noncomposite Girders

Figure 54.16 is a comparison of the means given by the linear regression for the noncomposite plate girder bridges shown in Figure 54.3 and the composite plate girder bridges in Figure 54.4. The figure also shows a similar comparison for box-girder bridges (Figures 54.5 and 54.6). Clearly composite girders are more economical than noncomposite ones.

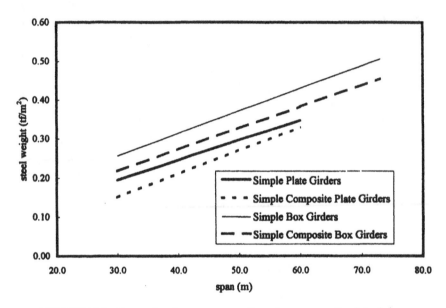

FIGURE 54.16 Comparison between composite and noncomposite plate girders.

54.6.2 Simply and Continuously Supported Girders

The difference caused by variation in support conditions is shown in Figure 54.17 for plate and box girders. The figures shown are for bridges with RC slab and steel decks. It is judged that continuous girders are more advantageous when the spans are long. There is no significant difference between simple plate and box girders for steel deck bridges. Continuous box girders can be used in long-span bridges.

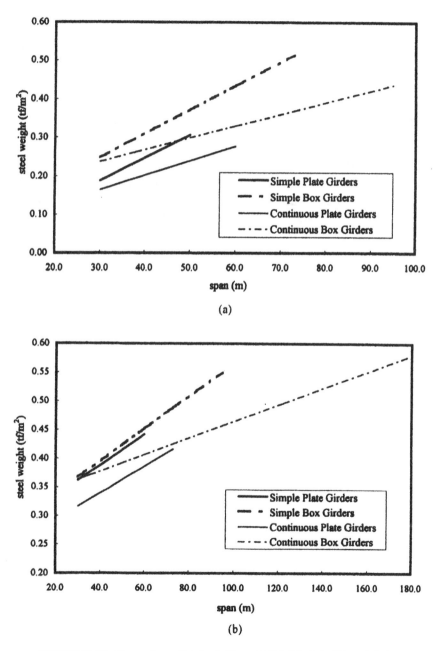

FIGURE 54.17 Comparison of girder bridges. (a) RC slab deck; (b) steel deck.

54.6.3 Framed Bridges

Six types of framed bridges are compared in Figure 54.18. The Nielsen bridge is the heaviest. The Nielsen and Lohse bridges, as well as the trussed Langer, are best suited to long spans.

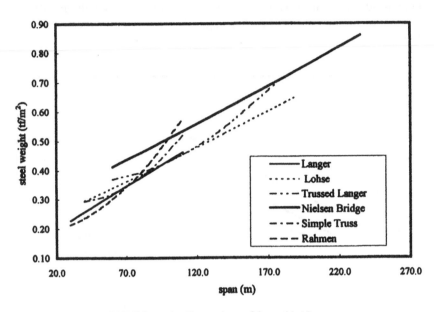

FIGURE 54.18   Comparison of framed bridges.

54.6.4 RC Slab Deck and Steel Deck

Figure 54.19a shows a comparison between the mean values of plate girder bridges with RC slab and steel decks. Bridges with steel decks are naturally much heavier than those with RC slab decks because the weight of the decks is included.

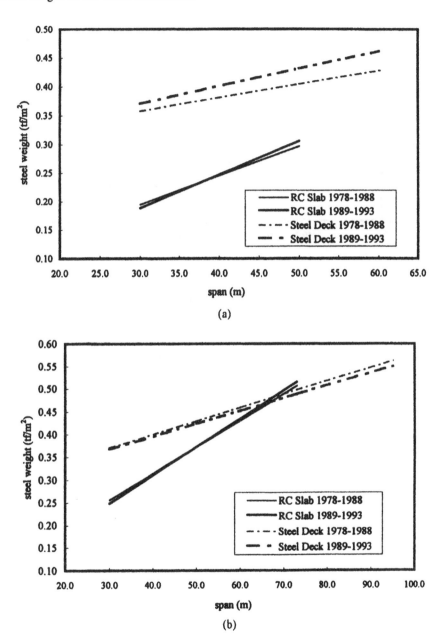

FIGURE 54.19 Comparison between RC slab and steel deck bridges. (a) Simple plate girders; (b) simple box girders; (c) continuous box girders.

A similar comparison for the box girder is shown in Figure 54.19(b). The difference gets smaller as the span length increases implying that steel deck bridges are economical when spans are long.

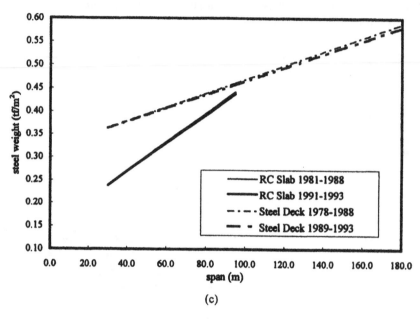

(c)

FIGURE 54.19   (continued)

54.7 Assessment of Bridge Design

54.7.1 Deviation

The distribution of the weights can be expressed by standard Gaussian techniques giving a mean value of 50 and a standard deviation of 10 as shown in Figure 54.20. The mean value $X(L)$ is calculated by the regression equations in Table 54.2 and converted to 50. The standard deviation σ can also be obtained from the regression equations table (Table 54.2), and converted to 10 using standard Gaussian procedures.

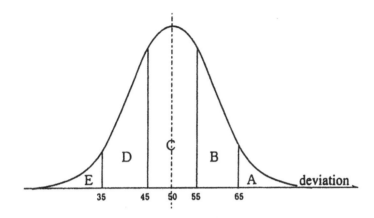

FIGURE 54.20 Classification of distribution.

The deviation (H) of the designed steel weight (X) is obtained using the equation

$$H = \frac{X - X(L)}{\sigma} \times 10 + 50 \tag{54.1}$$

H can be used as an index to compare the designs statistically and perform simple assessments of designs.

54.7.2 Assessment of Design

An example assessment of a typical design is discussed in the following. The labor and maintenance cost of bridges have become a major consideration in all countries. To solve this, a new design concept is proposed using only two girders with wide girder spacing. Figure 54.21 is one of the two-girder bridges that were constructed in Japan. It is a two-span continuous bridge with each span length 53 m. The road width is 10 m and the girder spacing 6 m. In this bridge, the section of the girder is not changed in an erection block to reduce welding length, thus reducing the labor cost.

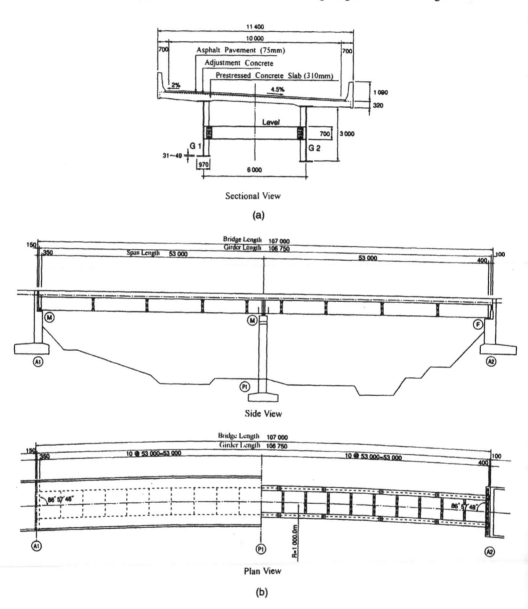

FIGURE 54.21   General plan of two-girder bridge. (a) Sectional view; (b) plan view. (*Bridges in Japan 1995-96*, JSCE)

The steel weight of this bridge is plotted in Figure 54.22. The deviation in this case is H = 62.8 (Rank B) using Eq. (54.1). In the calculation, the mean and the standard deviations are shown in Table 54.2. Note that most of the continuous bridges in Figure 54.22 are three-span continuous bridges. In addition, the design of this bridge follows the new code [1]. Those make the deviation for this case tend to be higher. From these deviation values the steel weight of a similar bridge can be estimated.

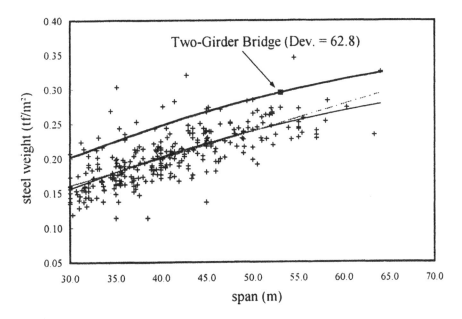

FIGURE 54.22   Two-girder bridge in continuous bridges.

54.8 Summary

The steel weight of bridges is a general indication of the design which tells an overall result. It reflects every influential design factor. A database has been put together to allow assessment of designs and prediction for the steel weight of various types of highway bridges. The distributions are plotted and shown for each type of bridge. From the figures, comparisons are made from various points of view to see the differences in each type of bridge. The regression equations for mean weight are derived, from which designers can estimate the steel weight for their own design or see economical or safety features of the bridge as compared with others.

References

1. Japan Road Association (JRA), Specifications for Highway Bridges, Vol. 1 Common Part, December 1996 [in Japanese].
2. Japan Road Association (JRA), Specifications for Highway Bridges, Vol. 1 Common Part, February 1990 [in Japanese].
3. Japan Association of Steel Bridge Construction (JASBC), Annual Report of Steel Bridge Construction, 1978 to 1993 [in Japanese].
4. Toma, S. and Honda, Y., Database of steel weight for highway bridges, *Bridge Eng.*, 29(8), 1993 [in Japanese].

55

Weigh-in-Motion Measurement of Trucks on Bridges

Andrzej S. Nowak
University of Michigan

Sangjin Kim
Kyungpook National University, Korea

55.1 Introduction

Knowledge of the past and current load spectra, together with predicted future loads, is essential in the evaluation and fatigue analysis of existing bridges. Many trucks carry loads in excess of design limits. This may lead to fatigue failure. The information concerning actual load is very important for the rating of bridges. Therefore, there is a need for accurate and inexpensive methods to determine the actual loads, the strength of the bridge, and its remaining life. There is also a need for verification of live load used for the development of a new generation of bridge design codes [1,2]. It has been confirmed that truck loads are strongly site specific [3,4,5]. Some bridges carry heavy truck traffic (volume and magnitude); others carry only lighter traffic. Furthermore, load effects such as bending moment, shear, and/or stress are component specific [6,7]. This observation is important in evaluation of the fatigue damage and prediction of remaining life. This chapter presents some of the practical procedures used for field measurement of truck weights and resulting strains.

55.2 Weigh-in-Motion Truck Weight Measurement

55.2.1 Weigh-in-Motion Equipment

The bridge live load is the load caused by truck traffic. In the past, truck data were collected by truck surveys, which had limitations. The most common survey method consisted of weighing trucks using static scales installed in weigh stations at fixed locations along major highways. The usefulness of the data obtained, however, is limited because many drivers of overloaded trucks intentionally avoid the scales, and therefore the results are biased to lighter trucks.

0-8493-7434-0/00/$0.00+$.50
© 2000 by CRC Press LLC

Therefore, research effort was focused on developing weigh-in-motion (WIM) methodology which can provide unbiased truck data, including axle weight, axle spacing, vehicle speed, multiple presence of trucks, and average daily truck traffic (ADTT). Very good results were obtained by using a WIM system with a bridge as a scale. Sensors measure strains in girders, and this is used to calculate the truck parameters at the highway speed.

The WIM system provides instrumentation invisible to truck drivers, and, therefore, the drivers do not try to avoid the scale. The system is portable and can be easily installed on a bridge to obtain site-specific truck data.

The bridge WIM system consists of three basic components: strain transducers, axle detectors (tape switches or infrared sensors), and data acquisition and processing system (Figure 55.1). The analog front end (AFE) acts as a signal conditioner and amplifier with a capacity of eight input channels. Each channel can condition and amplify signals from strain transducers. During data acquisition, the AFE maintains the strain signals at zero. The autobalancing of the strain transducers is activated when the first axle of the vehicle crosses the first axle detector. As the truck crosses the axle detectors the speed and axle spacing are determined. When the vehicle enters the bridge, the strain sampling is activated. As the last axle of the vehicle has exited the instrumented bridge span, the strain sampling is turned off. Data received from strain transducers are digitized and sent to the computer where axle weights are determined by an influence line algorithm. These data do not include dynamic loads. This process takes from 1.5 to 3.0 s, depending on the instrumented span length, vehicle length, number of axles, and speed. The data are then saved to memory.

The WIM equipment is calibrated using calibration trucks. The readings are verified and calibration constants are determined by running a truck with known axle loads over the bridge several times in each lane. The comparison of the results indicates that the accuracy of measurements is within 13% for 11-axle trucks. Gross vehicle weight (GVW) accuracy for five-axle trucks is within 5%, however, the accuracy is within 20% for axle loads [5].

55.2.2 Testing Procedure

The WIM system provides truck axle weights, gross vehicle weights, and axle spacings. Strains are measured in lower flanges of the girders and the strain time history is decomposed using influence lines to determine vehicle axle weights, as shown in Figure 55.1. The strain transducer can be clamped to the upper or lower surface of the bottom flange of the steel girder as shown in Figure 55.2. All transducers are placed on the girders at the same distance from the abutment, in the middle third of a simple span. The vehicle speed, time of arrival, and lane of travel are obtained using lane sensors on the roadway placed before the instrumented span of the bridge (Figure 55.3).

Two types of lane sensors can be used depending on the site conditions: tape switches and infrared sensors. Tape switches consist of two metallic strips that are held out of contact in the normal condition. As a vehicle wheel passes over the tape it forces the metallic strips into contact and grounds a switch. If a voltage is impressed across the switch, a signal is obtained at the instant the vehicle crosses the tape. This signal is fed to a computer whereby the speed, axle spacing, and number of axles are determined. The tape switches are placed perpendicular to the traffic flow and used to trigger the strain data collection. All cables used to connect tape switches and strain transducer to the AFE are five-pin wire cables.

The major problem with tape switches is their vulnerability to damage by moving traffic, particularly if the pavement is wet. Various alternative devices can be considered. Infrared sensors can be used to replace the tape switches. The infrared system consists of a source of infrared light beam and a reflector. The light source is installed on the side of the road. The reflector is installed in the center of the traffic lane. However, the problem of their vulnerability to damage by moving traffic has not been resolved. The infrared system is more difficult to install and trucks can easily move the reflector and interrupt the operation (the light beam must be aligned).

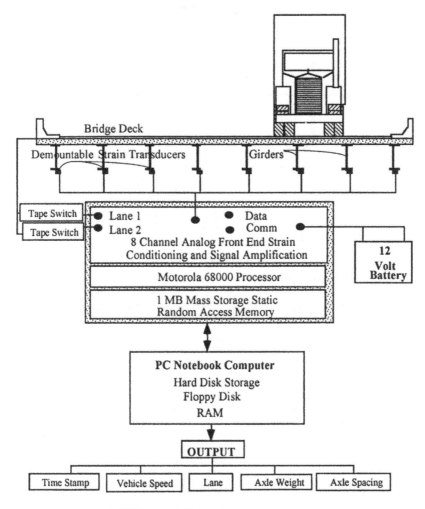

FIGURE 55.1 WIM truck measurement system.

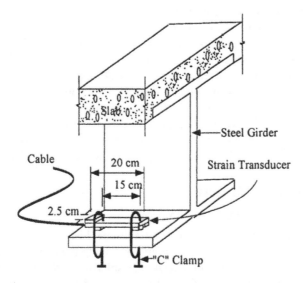

FIGURE 55.2 Demountable strain transducer mounted to the lower flange.

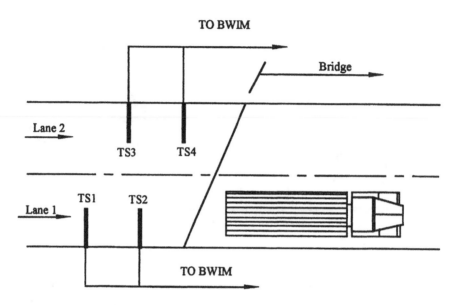

FIGURE 55.3 Plan of roadway sensor configuration.

TABLE 55.1 Parameters of Selected Bridges

Bridge Location	Span (m)	Number of Girders	Girder Spacing (m)	Number of Lanes	ADTT (One Direction)
WY/I-94	10.0	9	1.55	2	750
I-94/M-10	23.2	5	2.70	2	1,500
U.S. 12/I-94	12.0	9	1.68	2	500
DA/M-10	13.1	8	1.57	2	750
M-39/M-10	10.0	8	1.85	3	1,500
I-94/I-75	13.5	8	1.40	2	1,500
M-153/M-39	9.5	12	1.75	3	500

55.2.3 Selection of Bridges for Testing

The WIM measurements are demonstrated on seven bridges [3–5]. The selected structures are located in Michigan. Important factors considered in the selection process included accessibility from the ground, availability of space to work, low dynamic effects, and placement of tape switches or infrared sensors. The basic parameters are listed in Table 55.1. They include span length, number of girders, girder spacing, number of traffic lanes, and ADTT. ADTT was estimated on the basis of truck measurements performed for this study and it varies from 500 to 1500 in one direction. Bridge location is denoted by intersection of two roads; the first symbol stands for the road carried by the bridge, and the other one indicates the road under the bridge. Spans vary from about 10 to 25 m. The traffic volume is expressed in terms of ADTT. The selected bridges represent typical structures. The elevation and cross section of a typical bridge are shown in Figure 55.4.

55.2.4 Results from WIM Tests

55.2.4.1 Gross Vehicle Weight Distributions

The WIM results can be presented in a form of a traditional histogram (frequency or cumulative). However, this approach does not allow for an efficient analysis of the extreme values (upper or lower tails) of the considered distribution. Therefore, results of GVW WIM measurements for seven bridges are shown in Figure 55.5 in the form of cumulative distribution functions (CDFs) on the

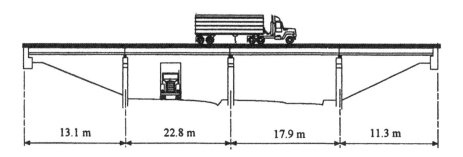

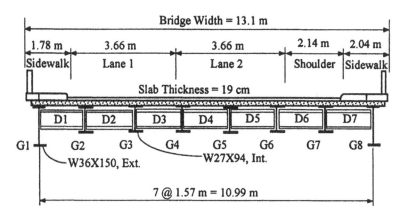

FIGURE 55.4   Bridge DA/M-10.

normal probability paper. CDFs are used to present and compare the critical extreme values of the data. They are plotted on normal probability paper [8]. The horizontal scale is in terms of the considered truck parameter (e.g., GVW, axle weight, lane moment, or shear force). The vertical scale represents the probability of being exceeded, p. Then, the probability of being exceeded (vertical scale) is replaced with the inverse standard normal distribution function, $\Phi^{-1}(p)$. For example, $\Phi^{-1}(p) = 0$, corresponds to the probability of being exceeded, $p = 0.5$; $\Phi^{-1}(p) = 1$, corresponds to $p = 0.159$; and $\Phi^{-1}(p) = -1$ corresponds to $p = 0.841$; and so on.

The distribution of truck type by number of axles will typically bear a direct relationship to the GVW distribution; the larger the population of multiple-axle vehicles (greater than five axles) the greater the GVW load spectra. Past research has indicated that 92 to 98% of trucks are four- and five-axle vehicles. The data obtained in this study indicate that between 40 and 80% of the truck population are five-axle vehicles, depending considerably on the location of the bridge. Three- and four-axle vehicles are often configured similarly to five-axle vehicles, and when included with five-axle vehicles account for between 55 and 95% of the truck population. Between 0 and 7.4% of the trucks are 11-axle vehicles in Michigan.

Most states in the United States allow a maximum GVW of 355 kN where up to five axles per vehicle are permitted. The State of Michigan legal limit allows for an 11-axle truck of up to 730 kN, depending on axle configuration. There were a number of illegally loaded trucks measured during data collection at several of the sites. Maximum WIM truck weights (1192 kN) exceeded legal limits by as much as 63%.

55.2.4.2 Axle Weight Distributions

Potentially more important for bridge fatigue and pavement design are the axle weights and spacing for trucks passing over the bridge. Figure 55.6 presents the distributions of the axle weights of the measured vehicles. All distributions include axles with weights greater than 22 kN. The maximum axle weights vary from 90 to 225 kN, and average values from 30 kN to 60 kN.

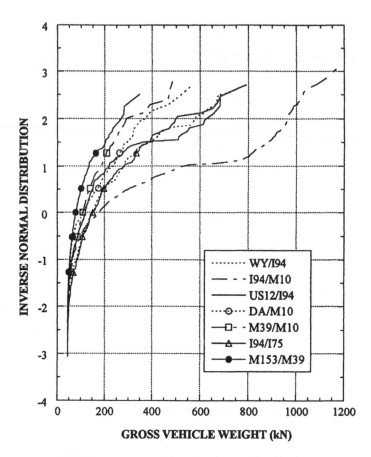

FIGURE 55.5   CDFs of GVW for the considered bridges.

55.2.4.3   Lane Moment and Shear Distributions

The structure is affected by load effects. Therefore, for the measured trucks, lane moments and shears were calculated for various spans. The resulting CDFs for 27-m span are shown in Figure 55.7 for lane moment and Figure 55.8 for lane shear. Each truck in the database is analytically driven across the bridge to determine the maximum static bending moment (shear) per lane. The CDFs of the lane moments (shears) for a span of 27 m are then determined. As a point of reference, the calculated moments (shears) are divided by design moment (shear) specified by the new AASHTO LRFD Specification [1]. The design live load according to the AASHTO LRFD is a superposition of a truck weighing 320 kN (three axles: 35, 145, and 145 kN spaced 4.25 m) and a uniformly distributed load of 9.3 kN/m. The lane moments in Figure 55.7 show a wide variation among bridges. Maximum values of the ratio of lane moment to LRFD moment vary from 0.6 at M-153/M-39 to 2.0 at I-94/M-10. All sites have a median lane moment between 0.16 and 0.34 times LRFD moment, which corresponds to an inverse normal value of 0. The variation of lane shears in Figure 55.8 is similar to that of lane moments. For I-94/M-10, the extreme value exceeds 2.0. For other bridges, the maximum shears vary from 0.65 at M-153/M-39 to 1.5 at I-94/I-75.

55.3   Fatigue Load Measurement

55.3.1   Testing Equipment

The Stress Measuring System (SMS) with the main unit manufactured by the SoMat Corporation is shown in Figure 55.9. The SMS compiles stress histograms for the girders and other components.

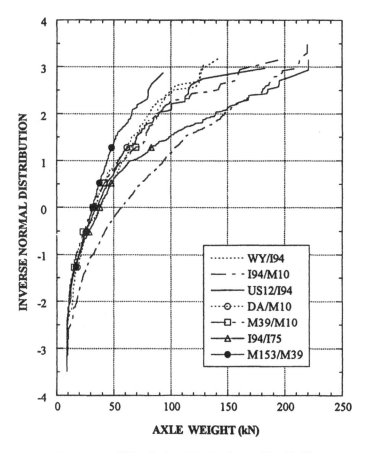

FIGURE 55.6 CDFs of axle weight for the considered bridges.

The SMS collects the strain history under normal traffic and assembles the stress cycle histogram by the rainflow method of cycle counting, and other counting methods. The data are then stored to memory and downloaded at the conclusion of the test period. The rainflow method counts the number, n, of cycles in each predetermined stress range, S_r, for a given stress history. The SMS is capable of recording up to 4 billion cycles per channel for extended periods in an unattended mode. Strain transducers were attached to all girders at the lower, midspan flanges of a bridge. Dynamic strain cycles were measured under normal traffic using the rainflow algorithm.

The SoMat Corporation system for its Strain Gauge Module is shown in Figure 55.9. It includes a power/processor/communication module, 1 MB CMOS extended memory unit, and eight strain gauge signal conditioning modules. The system is designed to collect strains through eight channels in both attended and unattended modes with a range of 2.1 to 12.5 mV. A second notebook computer is used to communicate with the SoMat system for commands regarding data acquisition mode, calibration, initialization, data display, and downloading of data. The SoMat system has been configured specifically for the purpose of collecting stress–strain histories and statistical analysis for highway bridges. This is possible due to the modular component arrangement of the system.

The data-acquisition system consists of five major components totaling 12 modules — eight strain transducer signal conditioning modules and four for Battery Pack, Power/Communications, 1-MB CMOS Extended Memory, and Model 2100 NSC 80180 Processor (see Figure 55.9). Regulated power is supplied by a rechargeable 11.3 to 13.4 V electrically isolated DC–DC converter. This unit powers all modules as well as provides excitation for strain transducers. Serial communications via RS 232C connector and battery backup for memory protection are provided by the Power/Communications

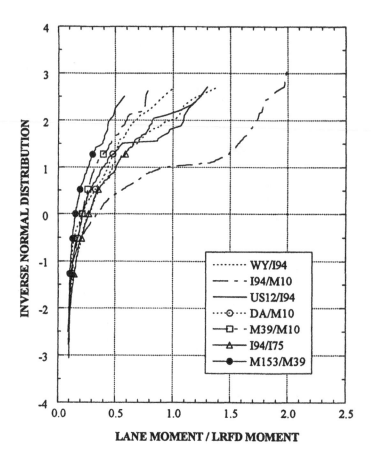

FIGURE 55.7   CDFs of lane moment for the span length of 27 m.

module. An Extended Memory Module of 1 Mbyte, high-speed, low-power CMOS RAM with backup battery for data protection is included for data storage. Eight strain gauge conditioning modules each provide 5-v strain transducer excitation, internal shunt calibration resistors, and an 8-bit, analog-to-digital converter.

Strain measurement range is ±2.1 mV minimum and ±12.5 mV maximum. The processor module consists of 32 kbytes of programmable memory and an NSC 80180 high-speed processor capable of sampling data in simultaneous mode resulting in a maximum sampling rate of 3000 Hz. Communication to the PC is via RS 232C at 57,600 baud. Data acquisition modes include time history, burst time history, sequential peak valley, time at level matrix, rainflow matrix, and peak valley matrix. Following collection, data are reviewed and downloaded to the PC hard drive for storage, processing, analysis, and plotting.

55.3.2 Rainflow Method of Cycle Counting

Development of a probabilistic fatigue load model requires collection of actual dynamic stress time histories of various members and components. Following the collection of time histories, data must be processed into a usable form. This section presents the characteristics of the dynamic stress time history commonly found in steel girder highway bridges as a random process, and rainflow method of counting fatigue damage.

Commonly occurring load histories in fatigue analysis often are categorized as either narrowband or wideband processes. Narrowband processes are characterized by an approximately constant period, such as that shown in Figure 55.10a. Wideband processes are characterized by higher

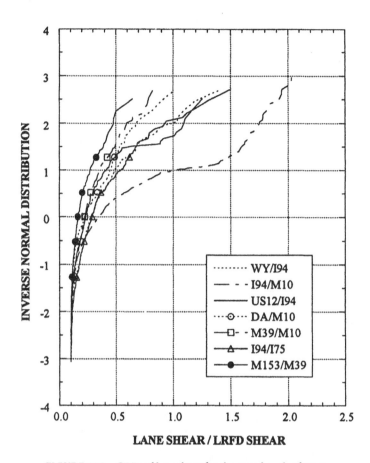

FIGURE 55.8 CDFs of lane shear for the span length of 27 m.

frequency small excursions superimposed on a lower, variable frequency process, such as that shown in Figure 55.10b. For steel girder highway bridges, where the loading is both random and dynamic, the stress histories are wideband in nature.

Stress histories that are wideband in nature do not allow for simple cycle counting. The cycles are irregular with variable frequencies and amplitudes. Several cycle-counting methods are available for the case of wideband and nonstationary processes, each successful to a degree in predicting the fatigue life of a structure. The rainflow method is preferred due to the identification of stress ranges within the variable amplitude and frequency stress histogram, which are associated with closed hysteresis loops. This is important when comparing the counted cycles with established fatigue test data obtained from constant-amplitude stress histories.

The rainflow method counts the number, n, of cycles in each predetermined stress range, S_i, for a given stress history. Rules of counting are applied to the stress history after orienting the trace vertically, positive time axis pointing downward. This convention facilitates the flow of "rain" due to gravity along the trace and is merely a device to aid in understanding of the method. Following are rules for the rainflow method (see Figure 55.11):

1. All positive peaks are evenly numbered.
2. A rainflow path is initiated at the inside of each stress peak and trough.
3. A rainflow progresses along a slope and "drips" down to the next slope.
4. A rainflow is permitted to continue unless the flow was initiated at a minimum more negative than the minimum opposite the flow and similarly for a rainflow initiated at a maximum. For example, path 1–8, 9–10, 2–3, 4–5, and 6–7.

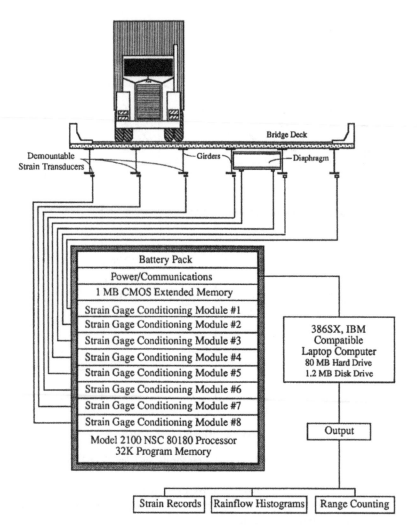

FIGURE 55.9   Somat strain data acquisition system.

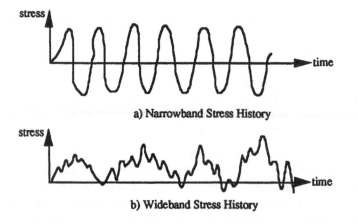

FIGURE 55.10   Example of narrowband (a) and wideband (b) stress histories.

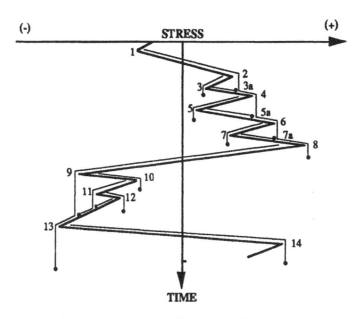

FIGURE 55.11 Rainflow counting diagram.

5. A rainflow must stop if it meets another flow that flows from above. For example, path 3–3a, 5–5a, and 7–7a.
6. A rainflow is not initiated until the preceding flow has stopped.

Following the above procedure each segment of the history is counted only once. Half cycles are counted between the most negative minimum and positive maximum, as well as the half cycles or interruptions between the maximum and minimum. As shown in Figure 55.11, all negative trough-initiated half cycles will eventually be paired with a peak initiated cycle of equal magnitude. For a more-detailed explanation and discussion of the rainflow method and others see an introductory text on fatigue analysis [9].

55.3.3 Results of Strain Spectra Testing

Strain histories were collected continuously for 1-week periods and reduced using the rainflow algorithm [5,6]. Data were collected for each girder in the bridge. The data are presented here in the form of CDFs and represent strain cycles due to 7 days of normal traffic. For an easier interpretation of results, the CDFs are plotted on normal probability paper (Figure 55.12).

For each bridge, the CDFs are shown for strains in girders numbered from 1 (exterior, on the right-hand side looking in the direction of the traffic). The number of girders for measured bridges varies from 6 to 10. The average strain is less than 50×10^{-6} for all girders and all bridges; however, the largest strains were observed in girders supporting the right traffic lane (girder numbers G3, G4, and G5) and nearest the left wheel of traffic in the right lane. As expected, the exterior girders of each bridge experience the lowest strain extremes in the spectrum.

As a means of comparison of fatigue live load, the equivalent stress, s_{eq}, is calculated for each girder using the following root mean cube (RMC) formula:

$$s_{eq} = \sqrt[3]{p_i S_i^3} \tag{55.1}$$

where S_i = midpoint of the stress interval i and p_i = the relative frequency of cycle counts for interval i. The stress, S_i, is calculated as a product of strain and modulus of elasticity of steel.

The CDFs of strain cycles and the corresponding equivalent stress values are shown in Figures 55.12 to 55.23. Stress values due to traffic load recorded in the girders are rather low. The maximum observed values are about 70 MPa. However, in most cases, the maximum values do not exceed 35 MPa for the most loaded girder, with most readings being about 7 MPa. Stress spectra considerably vary from girder to girder (component-specific). Therefore, the expected fatigue life is different depending on girder location. Exterior girders experience the lowest load spectra.

55.4 Dynamic Load Measurement

55.4.1 Introduction

The dynamic load is an important component of bridge loads. It is time variant, random in nature, and it depends on the vehicle type, vehicle weight, axle configuration, bridge span length, road roughness, and transverse position of trucks on the bridge. An example of the actual bridge response for a vehicle traveling at a highway speed is shown in Figure 55.24. For comparison, also shown is an equivalent static response, which represents the same vehicle traveling at crawling speed.

The dynamic load is usually considered as an equivalent static live load and it is expressed in terms of a dynamic load factor (DLF). There are different definitions for DLF, as summarized by Bakht and Pinjarkar [10] in their state-of-the-art report on dynamic testing of bridges. In this study, DLF is taken as the ratio of dynamic and static responses [7,11]:

$$\text{DLF} = D_{\text{dyn}}/D_{\text{stat}} \tag{55.2}$$

where D_{dyn} = the maximum dynamic response (e.g., stress, strain, or deflection) measured from the test data, $D_{\text{dyn}} = D_{\text{total}} - D_{\text{stat}}$; D_{total} = total response; and D_{stat} = the maximum static response obtained from the filtered dynamic response.

55.4.2 Measured Dynamic Load

Field measurements are performed to determine the actual truck load effects and to verify the available analytical models [7,11,12]. The tests are carried out on steel girder bridges. Measurements are taken using a WIM system with strain transducers. For each truck passage, the dynamic response is monitored by recording strain data. The truck weight, speed, axle configuration, and lane occupancy are also determined and recorded. A numerical procedure is developed to filter and process collected data. The DLF is determined under normal truck traffic for various load ranges and axle configurations.

An example of the actual static and dynamic stresses is shown in Figure 55.25. CDF of the static stress is plotted on the normal probability paper. For each value of static stress, the corresponding dynamic stress is also shown. The stress due to dynamic load is nearly constant and is not dependent on truck weight. Figure 55.26 shows DLFs as a function of static strains. Also shown in the figure is the power curve fit, which approximately represents mean values of DLFs. In general, the DLF decreases as the static strain increases. Therefore, the DLF is reduced for heavier trucks.

55.5 Summary

On the basis of WIM measurements, site-specific load spectra are presented for several bridge sites. Component stress spectra are presented for girders in the form of CDFs and the equivalent stresses are calculated for comparison. Site-specific statistics required for fatigue analysis based on current models are presented. Strains are measured to determine component-specific load spectra.

The truck load spectra for bridges are strongly site specific. Bridges located on major routes between large industrial metropolitan areas will experience the highest extreme loads. Routes where

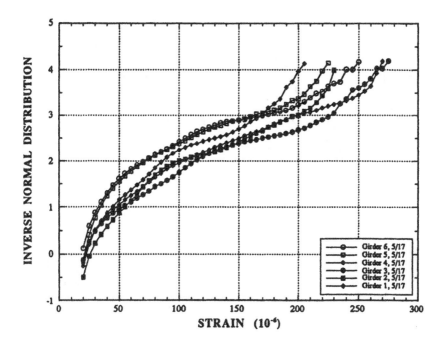

FIGURE 55.12   Strains for girders in bridge U.S. 23/HR.

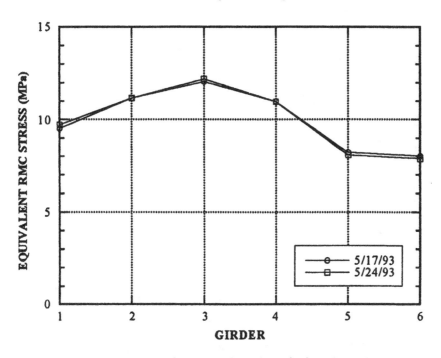

FIGURE 55.13   Equivalent stresses for girders in bridge U.S. 23/HR.

vehicles are able to circumvent stationary weigh stations will have very high extreme loads. Bridges not on a major route and those that are very near a weigh station experience lower extreme loads.

Live-load stress spectra are strongly component specific. Each component may have a different distribution of strain cycle range. The girder that is nearest the left wheel track of vehicles traveling

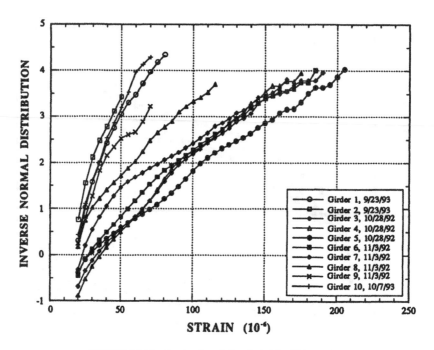

FIGURE 55.14   Strains for girders in bridge U.S. 23/SR.

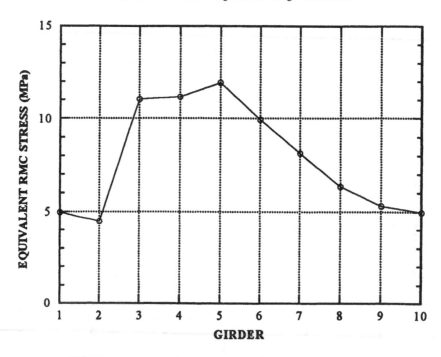

FIGURE 55.15   Equivalent stresses for girders in bridge U.S. 23/SR.

in the right lane experiences the highest stresses in the stress spectra. The stresses decrease as a function of the distance from this location. This information can be useful to target bridge inspection efforts to the critical members.

The stress due to dynamic load is nearly constant and it is not dependent on truck weight. In general, DLF decreases as the static strain increases. Therefore, the DLF is reduced for heavier trucks.

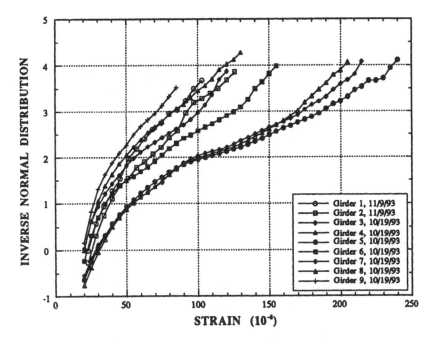

FIGURE 55.16   Strains for girders in bridge I-94/JR.

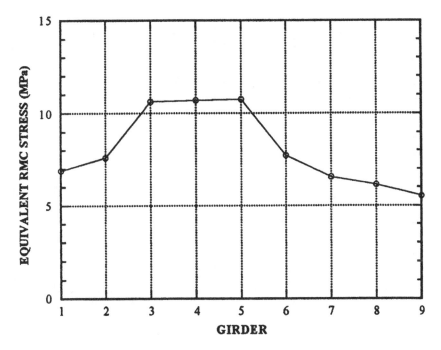

FIGURE 55.17   Equivalent stresses for girders in bridge I-94/JR.

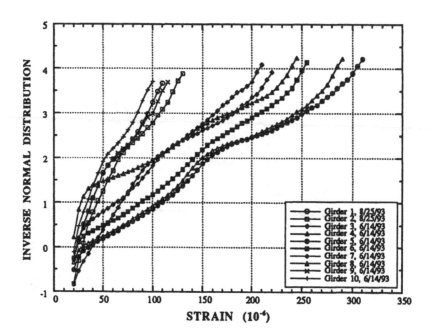

FIGURE 55.18   Strains for girders in bridge I-94/PR.

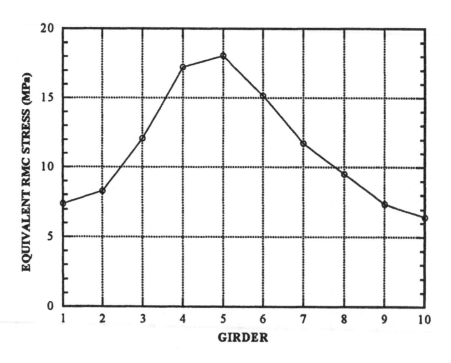

FIGURE 55.19   Equivalent stresses for girders in bridge I-94/PR.

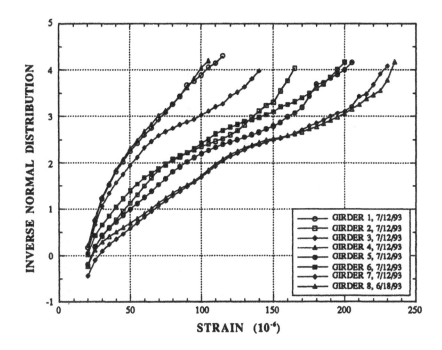

FIGURE 55.20 Strains for girders in bridge M-14/NY.

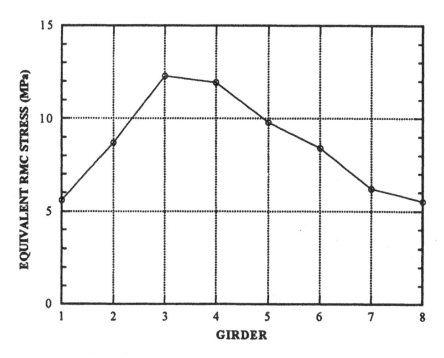

FIGURE 55.21 Equivalent stresses for girders in bridge M-14/NY.

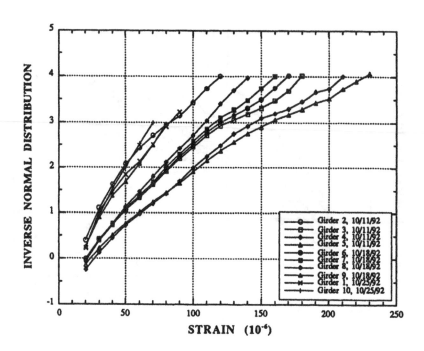

FIGURE 55.22   Strains for girders in bridge I-75/BC.

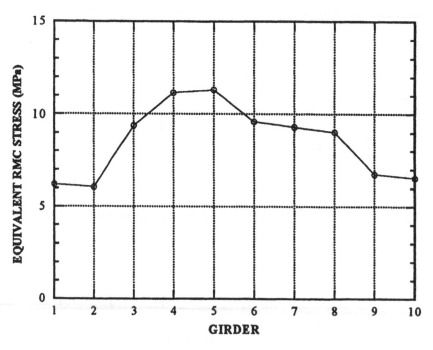

FIGURE 55.23   Equivalent stresses for girders in bridge I-75/BC.

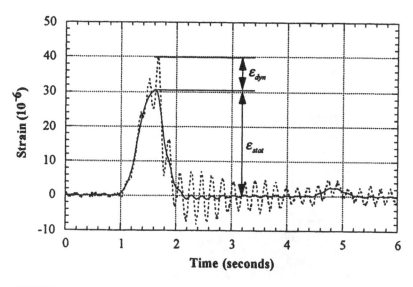

FIGURE 55.24 Dynamic and static strain under a truck traveling at highway speed.

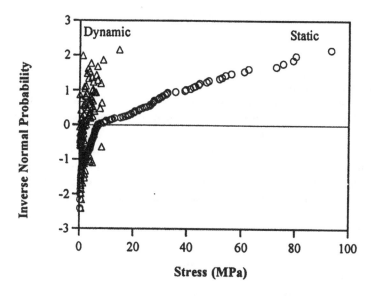

FIGURE 55.25 Typical CDF of static stress and corresponding dynamic stress.

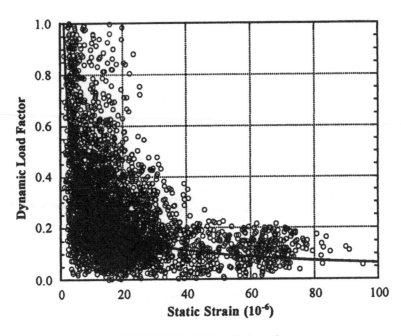

FIGURE 55.26 DLF vs. Static strain.

References

1. AASHTO, *LRFD Bridge Design Specifications*, 2nd ed., American Association of State and Transportation Officials, Washington, D.C, 1998.
2. Nowak, A. S. and Hong, Y-K., Bridge live-load models, *ASCE J. Struct. Eng.*, 117, 2757, 1991.
3. Kim, S., Sokolik, A. F., and Nowak, A. S., Measurement of truck load on bridges in the Detroit area, *Transp. Res. Rec.*, 1541, 58, 1996.
4. Nowak, A. S., Kim, S., Laman, J., Saraf, V., and Sokolik, A. F., Truck Loads on Selected Bridges in the Detroit Area, Research Report UMCE 94-34, University of Michigan, Ann Arbor, 1994.
5. Nowak, A. S., Laman, J. A., and Nassif, H., Effect of Truck Loading on Bridges, Report UMCE 94-22, Department of Civil and Environmental Engineering, University of Michigan, Ann Arbor, 1994.
6. Laman, J. A. and Nowak, A. S., Fatigue-load models for girder bridges, *ASCE J. Struct. Eng.*, 122, 726, 1996.
7. Kim, S. and Nowak, A. S., Load distribution and impact factors for I-girder bridges, *ASCE J. Bridge Eng.*, 2, 97, 1997.
8. Benjamin, J. R. and Cornell, C. A., *Probability, Statistics and Decision for Civil Engineers*, McGraw-Hill, New York, 1970.
9. Bannantine, J. A., Comer, J. J., and Handrock, J. L., *Fundamentals of Metal Fatigue Analysis*, Prentice-Hall, Englewood Cliffs, NJ, 1992.
10. Bakht, B. and Pinjarkar, S. G., Dynamic testing of highway bridges — a review, *Transp. Res. Rec.*, 1223, 93, 1989.
11. Nassif, H. and Nowak, A. S., Dynamic load spectra for girder bridges, *Transp. Res. Rec.*, 1476, 69, 1995.
12. Hwang, E. S. and Nowak, A. S., Simulation of dynamic load for bridges, *ASCE J. Struct. Eng.*, 117, 1413, 1991.

56

Impact Effect of Moving Vehicles

Mingzhu Duan
Quincy Engineering, Inc.

Philip C. Perdikaris
Case Western Reserve University

Wai-Fah Chen
Purdue University

56.1 Introduction

Vehicles such as trucks and trains passing bridges at certain speeds will cause dynamic effects, among them global vibration and local hammer effects. The dynamic loads for moving vehicles are considered "impact" in bridge engineering because of the relatively short duration. The magnitude of the dynamic response depends on the bridge span, stiffness and surface roughness, and vehicle dynamic characteristics such as moving speed and isolation system. Unlike earthquake loads which can cause vibration in bridge longitudinal, transverse, and vertical directions, moving vehicles mainly excite vertical vibration of the bridge. Impact effect has influence primarily on the superstructure and some of substructure members above the ground because the energy will be dissipated effectively in members underground by the bearing soils.

Although the interaction between moving vehicles and bridges is rather complex, the dynamic effects of moving vehicles on bridges are accounted for by a dynamic load allowance, IM, in addition to static live load (LL) in the current bridge design specifications [1–3]. According to the American Association of State Highway and Transportation Officials (AASHTO) and the American Railway Engineering Association (AREA) specifications,

$$IM = \frac{D_{dyn}}{D_{st}} - 1 \tag{56.1}$$

where D_{dyn} is the maximum dynamic response for deflection, moment, or shear of the structural members and D_{st} is the corresponding maximum static response. The total live-load effect, LL, can then be expressed as

$$LL = AF \times D_{st} \tag{56.2}$$

0-8493-7434-0/00/$0.00+$.50
© 2000 by CRC Press LLC

and

$$AF = 1 + IM \tag{56.3}$$

where *AF* is the amplification factor representing the dynamic amplification of the static load effect and *IM* is the impact factor determined by an empirical formula in design code. No dynamic analysis is thus required in the design practice.

Most early research work on the dynamic bridge behavior of bridges under moving vehicles focused on an analytical approach modeling a bridge as a simply supported beam [5] or a simply supported plate [8] under constant or pulsating moving loads (moving load model). The dynamic effects under different speeds of the moving loads and different damping ratios were studied. It was found that the speed of vehicles and the fundamental period of the bridge dominate the dynamic behavior of the bridge. Since most bridges consist of both beams and plates such as girder deck bridges, the above simplified model has limited validity. Along with analytical study, numerical methods such as finite-element analysis and the finite-difference method have been used recently in studying the dynamic response of a vehicle–bridge system [10,21]. Two sets of equations of motion were developed for the bridge and vehicle, respectively. These equations are coupled at the contacting points between bridge and vehicle and the contact points are time and space dependent due to vehicles moving along a rough surface. An iteration procedure should be used to solve the coupled equations. Field measurements are another alternative to investigate the dynamic effect [13] which disclosed the range of live-load effect for steel I-girder bridges under truck load.

Based on analytical analysis and field measurement studies, major characteristics of the bridge dynamic response under moving vehicles can be summarized as follows:

1. Measured impact factors [13], *IM*, on highway bridges vary significantly, e.g., with the mean of about 0.12 and the standard deviation of about 0.05 for steel I-girder bridges. The measured impact factors are well below those of the AASHTO specifications.
2. Impact factor increases as vehicle speed increases in most cases.
3. Impact factor decreases as bridge span increases.
4. Under the conditions of "very good" road surface roughness (amplitude of highway profile curve is less than 1 cm), the impact factor is well below that in design specifications. But the impact factor increases tremendously with increasing road surface roughness from "good" to "poor" (the amplitude of highway profile curve is more than 4 cm) and can be well beyond the impact factor in design specifications.
5. Impact factor decreases as vehicles travel in more than one lane. The chance of maximum dynamic response occurring at the same time for all vehicles is small.
6. Impact factor for exterior girders is much larger than for interior girders because the excited torsion mode shapes contribute to the dynamic response of exterior girders.
7. The first mode shape of the bridge is dominant in most cases, especially for the dynamic effect in the interior girder in single-span bridges.

The impact factor, *IM*, is a well-accepted measurement for the dynamic effect of bridges under moving vehicles and is used in design specifications worldwide. Consideration of impact effect for highway and railway bridges in design practice will be introduced through examples in Sections 56.2 and 56.3, respectively. Free vibration and forced vibration by a moving vehicle will be introduced in Sections 56.4 and 56.5 to disclose the dynamic behavior of the bridge vibration.

56.2 Consideration of Impact Effect in Highway Bridge Design

Since the impact effect on bridges by moving vehicles is influenced by factors such as bridge span, stiffness, surface roughness, and speed and suspension system of moving vehicles, the impact factor varies within a large range. While the actual modeling of this effect is complex, the calculation of impact effect is greatly simplified in the bridge design practice by avoiding any analysis of vehicle-induced vibration. In general, the dynamic effect is contributed by two sources: (1) local hammer effect by the vehicle wheel assembly riding surface discontinuities such as deck joints, cracks, delaminations, and potholes; and (2) global vibration caused by vehicles moving on long undulations in the roadway pavement, such as those caused by settlement of fill, or by resonant excitation of the bridge. The first source has a local impact effect on bridge joints and expansions. On the other hand, the second source will have influence on most of the superstructure members and some of the substructure members. A variety of considerations and design formulas are proposed worldwide for the second source, which means that the bridge community has not reached a consensus on this issue [4]. The differences among various code specifications worldwide for the dynamic amplification factor vs. the fundamental frequency are large. A large dynamic effect is considered in some countries for the bridge frequency ranging from 1.0 to 5.0 Hz, which is the frequency range of the fundamental frequencies of most truck suspension systems. It is largely an attempt to penalize the bridge designed within this frequency range. But the accurate evaluation of the first frequency of a bridge can be hardly performed in the design stage.

In the AASHTO *Standard Specifications of Highway Bridges* (1996) in the United States, the impact factor due to bridge vibration for members in Group A including superstructure, piers, and those portions of concrete and steel piles above the ground that support the superstructure, is simply expressed as a function of bridge span:

$$IM = \frac{50}{L + 125} \le 0.30 \tag{56.4}$$

where L (in ft) is the length of span loaded to create maximum stress.

In the AASHTO *LRFD Bridge Design Specifications* (1994), the static effects of the design truck or tandem shall be increased by the impact effect in the percentage specified in Table 56.1.

In Table 56.1, 75% of the impact effect is considered for deck joints for all limit states due to local hammer effect, 15% for fatigue and fracture limit states for members vulnerable to cyclic loading such as shear connectors and welding members, and 33% for all other members influenced by global vibration. Field tests indicate that in the majority of highway bridges, the dynamic component of the response does not exceed 25% of the static response to vehicles. Since the specified live-load combination of the design truck and lane load represents a group of exclusion vehicles which are at least ⅓ of those caused by the design truck alone on short- and medium-span bridges, the specified value of 33% in Table 56.1 is the product of ⅓ and the basic 25%. The impact effect is not considered for retaining walls not subjected to vertical reactions from superstructure, and for underground foundation components due to the damping effect of soil.

The dynamic effect caused by vehicles is accounted for in bridge design practice in the live-load calculation, as shown in the following example.

Example 56.1 — Live-Load Calculation in Highway Bridge Design

Given
The bridge shown in Figure 56.2 is a steel girder-concrete deck bridge with a span of 50 ft (15.24 m). The bridge carrying two lanes of traffic loads consists of four girders with girder-to-girder spacing of 7 ft (2.13 m). The steel girders are W36 × 150 and the average concrete deck thickness is 9 in. (3.54 m). The bridge is designed to carry a specified truck of HS20-44.

Solution 1
Using AASHTO *Standard Specifications of Highway Bridges* (1996) to calculate the design live-load moment for the interior steel girder under design truck HS20-44.

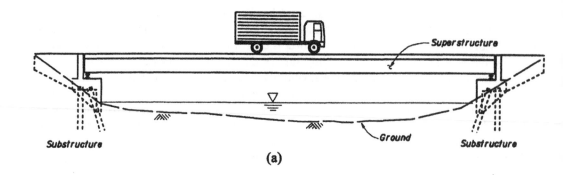

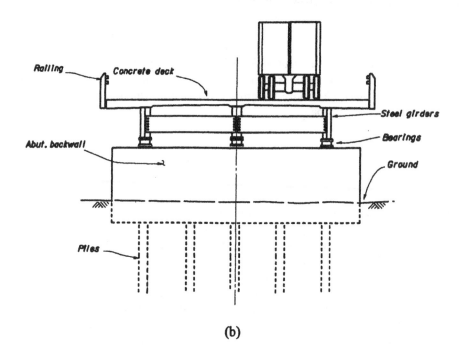

FIGURE 56.1   (a) Moving vehicle on a bridge. (b) Section view of the bridge and the vehicle.

TABLE 56.1   Dynamic Load Allowance, *IM*

Component	IM
Deck joint — All limit states	75%
All other components:	
• Fatigue and fracture limit state	15%
• All other limit states	33%

1. Calculate the Static Design Moment in the Interior Stringer under Truck HS20-44

a. *Compute the static maximum truck load moment in a lane at the critical section of the superstructure induced by truck loading:*

There are two types of vehicle live loading: truck and lane loading. For bending moments caused by HS20-44 loading for a span length of up to 140 ft, truck loading will govern.

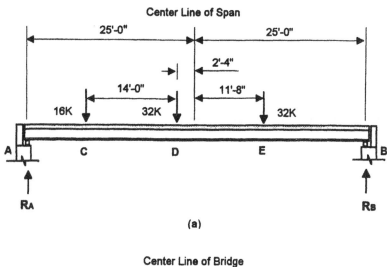

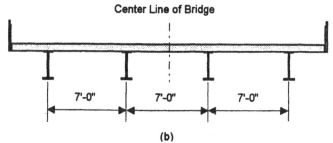

FIGURE 56.2 (a) Elevation of the steel girder bridge under truck load HS20-44. (b) Section of the steel girder bridge.

The maximum live-load moment is induced at the most adverse truck position. In fact, the maximum moment occurs at the position of the second concentrated load when the centerline of the span is midway between the center of gravity of loads and the second concentrated load as shown in Figure 56.2a. The reaction R_A at the end A is determined by using the equilibrium equation of the sum of the moments about point B:

$$\sum M_B = 0$$

Thus,

$$50\ R_A - 8 \times 41.33 - 32 \times 27.33 - 32 \times 13.33 = 0,\ \text{and}$$

$$R_A = \frac{1632}{50} = 32.64 \text{ kip (145.25 kN)}$$

The maximum live-load moment at Point D (see Figure 56.2) is

$$M_{\max} = 22.67\ R_A - 8 \times 4 = 627.95 \text{ kips-ft (851.50 kN-m)}$$

b. *Compute the axle load distribution factor, DF, for the stringers.*
   For the two lane concrete floor with steel I-beam girders spacing at 7 ft (2.13 m), the distribution factor, *DF*, according to the AASHTO-96 is

$$DF = \frac{S}{2 \times 5.5} = \frac{7}{2 \times 5.5} = 0.64$$

c. *Calculate the design static bending moment in a stringer.*

$$M_{LL} = DF(M_{max}) = 627.95 \times 0.64 = 401.89 \text{ kips-ft (544.96 kN-m)}$$

2. Determine Dynamic Amplification Factor, *AF*

$$AF = 1.0 + IM = 1.0 + \frac{50}{L + 125} = 1.0 + \frac{50}{50 + 125} = 1.29 \quad \text{(AASHTO-96)}$$

3. Calculate the Total Bending Moment under Live Load

The total live bending moment including the amplification factor is

$$M_{LL+IM} = M_{LL} AF$$

$$= 401.89 \times 1.29 = 518.44 \text{ kips-ft (703.00 kN-m)} \quad \text{(AASHTO-96)}$$

Solution 2

Using AASHTO-LRFD *Design Specifications* (1994) to calculate the design live-load moment for an interior steel girder under truck HS20-44.

1. Calculate the Static Truck Moment in the Interior Girder under Truck HS20-44

a. *The maximum moment in a lane is the same as in Method 1:*

$$M_{max} = 627.95 \text{ kips-ft} \quad (851.50 \text{ kN-m})$$

b. *Compute the moment load distribution factor, DF, for the steel girders.*
For the two-lane concrete floor with steel I-girders spacing at 7 ft (2.13 m), the distribution factor, *DF*, is

$$DF = 0.075 + \left(\frac{S}{9.5}\right)^{0.6} \left(\frac{S}{L}\right)^{0.2} \left(\frac{K_g}{12.0 \, Lt_s^3}\right)^{0.1} = 0.075 + \left(\frac{7.0}{9.5}\right)^{0.6} \left(\frac{7.0}{L}\right)^{0.2} \times 1.0 = 0.63$$

(for preliminary design use)

c. *Calculate the design static bending moment in a stringer.*

$$M_{LL} = DF(M_{max}) = 627.95 \times 0.63 = 395.64 \text{ kips-ft (538.07 kN-m)}$$

2. Determine Dynamic Amplification Factor, *AF*

$$AF = 1.0 + IM = 1.0 + 0.33 = 1.33 \quad \text{(AASHTO-94)}$$

3. Calculate the Total Bending Moment in an Interior Stringer under Truck Load

The total bending moment under truck load including the amplification factor is

$$M_{LL+IM} \text{ (truck)} = M_{LL} AF = 395.64 \times 1.33 = 526.20 \text{ kips-ft (715.64 kN-m)} \quad \text{(AASHTO-94)}$$

4. The vehicular live-load moment is the combination of the design truck load with impact allowance and design lane load without impact allowance (0.64 kip/ft over 10.0 ft per lane, 0.88 kN/m over 3.0 m):

$$M_{LL+IM} \text{ (total)} = 526.20 + 0.63 \times (0.64 \times 50/2 \times 22.67 - 1/2 \times 0.64 \times 22.67^2)$$

$$= 651.10 \text{ kips-ft (885.50 kN-m)}$$

56.3 Consideration of Impact Effect in Railway Bridge Design

For railway bridges the ratio of live load caused by moving vehicles such as locomotive and trains to the dead load is mostly higher than that in highway bridges. Similarly, as in highway bridge design, static live-load effect by vehicles should be increased by the impact factor to account for the dynamic amplification effect. The most important sources of bridge impact are

1. Initial vehicle bounce and roll,
2. Vehicle speed, and
3. Bridge dynamic properties and track-surface roughness.

In the AREA specifications [3], the impact loads specified are based on investigations and tests of railroad bridges in service under passage of locomotive and train loads. The vibration for the bridge is the most dominant dynamic effect in railway bridge design. In the vibration, the vertical vibration effect will be coupled with the rocking effect (*RE*) caused by vehicle pitch movement (transverse vehicle rotation). Thus, a couple with 10% of axle load acting down on one rail and up on the other rail should be added into the vertical impact effect. The rocking effect, *RE*, should be expressed as a percentage; either 10% of the axle load or 20% of the wheel load. The total impact effect can be calculated as

1. Percentage of live load for rolling equipment without hammer blow, such as diesels and electric locomotives, etc.,

$$IM = RE + 40 - \frac{3L^2}{1600} \qquad \text{if } L < 50 \text{ ft (15.24 m)}$$

$$IM = RE + 16 - \frac{600}{L - 30} \qquad \text{if } L \geq 80 \text{ ft (24.39 m)}$$

(56.5)

2. Percentage of live load for steam locomotives with hammer blow:
   a. For beam spans, stringers, girders, floor beams, parts of deck truss span carrying load from floor beam only:

$$IM = RE + 40 - \frac{L^2}{500} \qquad \text{if } L < 100 \text{ ft (30.48 m)}$$

$$IM = RE + 10 - \frac{1600}{L - 40} \qquad \text{if } L \geq 100 \text{ ft (30.48 m)}$$

(56.6)

   b. For truss spans:

$$IM = RE + 10 - \frac{4000}{L + 25}$$

(56.7)

where *L* is the effective span length (ft).

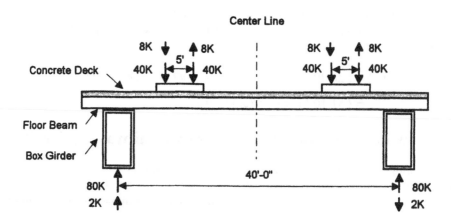

FIGURE 56.3 Rocking effect model for railway bridge.

Tests have shown that the impact load on ballasted deck bridges can be reduced to 90% of that specified for open-deck bridges because of the damping effect which results from a ballasted deck bridge. It was found from a parametric study [9] that the impact ranges from 24.9 to 26.0%, with an average of 25.6%, except for hangers. The following example shows how to account for impact effect in live-load calculation in railway bridge design.

Example 56.2 — Live Load Calculation in Railway Bridge Design

Given
A ballasted 105-ft-long (32.01-m-long) deck bridge is supported by two box girders. The box girder spacing is 40 ft (12.20 m) to carry two tracks, as shown in Figure 56.3. Live load is Cooper E-80 without hammer blow. The maximum bending moment M_{LL} at midspan under static live load is 15,585 kips-ft (21,195.60 kN-m) per track based on AREA values. The bending moment M_{LL+IM} at midspan under live load can be calculated as follows.

1. **Determine Rocking Effect (RE)**
   The maximum axle load is 80 kips (356.00 kN) (40 kips per rail, 178.00 kN per rail).
   The couple force generated by the rocking effect:

$$20\% \times 40 = 8 \text{ kip } (35.60 \text{ kN})$$

The total couple force at the box girders is:

$$\text{total couple} = 8 \times 5 \text{ ft (rail spacing)} \times 2 \text{ tracks} = 40 \text{ kips-ft } (54.24 \text{ kN-m})$$

$$\text{total couple force} = \frac{\text{total couple}}{\text{stringer spacing}} = \frac{80}{40} = 2.0 \text{ kip } (8.9 \text{ kN})$$

The rocking effect on the stringers is the ratio of rocking reaction to static reaction:

$$RE = \frac{2.0}{80} = 0.025 = 2.5\%$$

2. **Determine the Percentage of Impact Effect from AREA Specifications**
   For $L > 80$ ft (24.39 m) and no hammer effect case:

$$IM = RE + \left(16 + \frac{600}{L-30}\right)\% = 2.5 + \left(16 + \frac{600}{108-30}\right)\% = 26.21\%$$

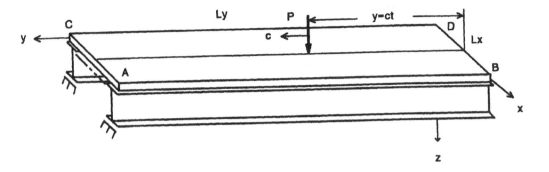

FIGURE 56.4 Girder deck bridge under moving constant load.

For ballasted deck bridges, 90% reduction should be considered, that is:

$$IM = 90\% \times 26.21\% = 0.24$$

3. Determine Live-Load Effect of the Maximum Bending Moment at Midspan

$$AF = 1.0 + IM = 1.0 + 0.24 = 1.24$$

$$M_{LL+IM} = AF(M_{LL}) = 1.24 \times 15,585 = 19,264 \text{ kips-ft (26,199.04 kN-m)}$$

56.4 Free Vibration Analysis

56.4.1 Structural Models

In the following two sections, the bridge vibration under moving load will be discussed to investigate the dynamic response of bridges. There have been basically two types of analysis methods: numerical analysis (sprung mass model) and analytical analysis (moving load model). The numerical analysis models the interaction between vehicle and bridge and expresses the dynamic behavior numerically. On the other hand, analytical analysis greatly simplifies vehicle interaction with bridge and models a bridge as a plate or beam but expresses the dynamic behavior explicitly. Good accuracy can be obtained using the analytical model when the ratio of live-load to self-weight of the superstructure is less than 0.3. The analytical analysis method will be presented for a bridge with beam-plate system.

The structure shown in Figure 56.4 represents a plate with two opposite edges AC and BD simply supported by rigid ground and the other two edges AB and CD simply supported on two beams. The assumptions made are

a. The stress–strain relationship for the beam and plate material is linear elastic;
b. There exists a neutral plane surface in the plate and the existence of the beams does not have any influence on the position of this neutral plane, i.e., the beam can only provide a vertical force reaction to the edges AB and CD for the plate. The structure is a noncomposite girder deck bridge; and
c. Thin plate and simple beam theories are applicable.

56.4.2 Free Vibration Analysis

The differential equation for free vibration of the plate shown in Figure 56.4 is

$$D \nabla^{(4)} W + m W_{tt}^{(2)} = 0 \tag{56.8}$$

where

$D = \dfrac{Eh^3}{12(1-\mu)}$ = the flexural rigidity of the plate

m = is the mass per unit area of the plate
W = is the vertical deflection of the plate

The boundary conditions for the plate are as follows:

At $y = 0$ and $y = L_y$:

$$W(x, y, t) = 0 \tag{56.9a}$$

$$M_y(x, y, t) = 0 \tag{56.9b}$$

At $x = 0$:

$$M_x(x, y, t) = 0 \tag{56.9c}$$

$$Q_x + \frac{\partial M_{xy}}{\partial y} = E_b I_b \frac{\partial^4 W}{\partial y^4} + m_b \frac{\partial^2 W}{\partial t^2} \tag{56.9d}$$

At $x = L_x$:

$$M_x(x, y, t) = 0 \tag{56.9e}$$

$$Q_x + \frac{\partial M_{xy}}{\partial y} = -E_b I_b \frac{\partial^4 W}{\partial y^4} - m_b \frac{\partial^2 W}{\partial t^2} \tag{56.9f}$$

where $E_b I_b$ and m_b is the flexural rigidity and mass per unit beam length, respectively, and M_x, M_y are the transverse and longitudinal bending moments, respectively. M_{xy} is the torque and Q_y is the shear force at the edges of the plate, respectively. Their signs are shown in Figure 56.5. The terms at the right-hand side in Eqs. (56.9d) and (56.9f) represent the interaction forces at the edges AB and CD between the plate and the beams. It is seen that the participation of the beams is taken into account in the partial differential equations through the boundary conditions for the plate.

Assuming that the structure vibrates in a mode so that the deflected shape is described by

$$W = \sin \frac{i\pi y}{L_y} X_j(x) \sin(\omega_{ij} t + q) \tag{56.10}$$

the following equations can be derived by substituting Eq. (56.10) into Eq. (56.8):

$$X_j^{(4)} - 2 \frac{i^2 \pi^2}{L_y^2} X_j^{(2)} + \frac{i^4 \pi^4}{L_y^4} X_j - \frac{m}{D} \omega_{ij} X_j = 0 \tag{56.11}$$

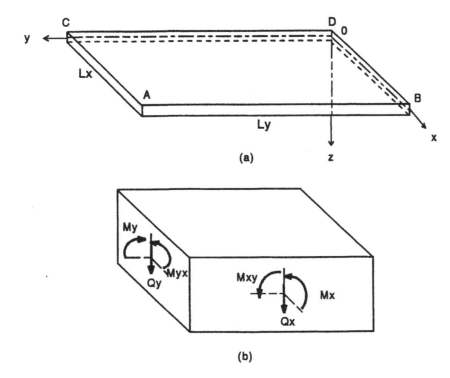

FIGURE 56.5 (a) A three-dimensional plate with length L_y and width L_x. (b) A volume element with inter forces acting on its sides.

The solution to the homogeneous differential equation above is

$$X_j = e^{\lambda \, x/L_x}$$

where λ must satisfy the characteristic equation:

$$\frac{\lambda^4}{L_x^4} - 2 \frac{i^2 \pi^2}{L_y^2} \frac{\lambda^2}{L_x^2} + \left(\frac{i^4 \pi^4}{L_y^4} - \frac{m \omega_{ij}^2}{D} \right) = 0 \qquad (56.12)$$

There are four roots for this equation. According to the signs of the roots, $X_j(x)$ will take two different forms, which will be discussed separately.

Case 1

$$i^2 > p \, r_{ij}^2$$

where

$$p \, r_{ij}^2 = \frac{\omega_{ij}^2 \, m L_y^2}{D \pi^4} \qquad (56.13)$$

We have the following solution for Eq. (56.11)

$$X_j = \sinh\lambda_1 u + A\cosh\lambda_1 u + B\sinh\lambda_2 u + C\cosh\lambda_2 u$$

$$\lambda_{1,2} = \frac{\pi}{\gamma}\left[i^2 \mp pr_{ij}^2\right] \tag{56.14}$$

where

$$u = x/L_x \tag{56.15.a}$$

$$\gamma = L_y / L_x \tag{56.15.b}$$

By substituting Eq. (56.14) and (56.10) into the boundary conditions in Eqs. (56.9a) to (56.9f), the constants A, B, and C in Eq. (56.14) can be determined by solving a group of eigenvalue equations [7].

The natural frequency ω_{ij} in Case 1 can be obtained from the following nonlinear equation from the boundary conditions as

$$2\lambda_1\lambda_2 D_1^2 D_2^2 (\cosh\lambda_1 \cosh\lambda_2 - 1) - \left(\lambda_1^2 D_1^4 + \lambda_2^2 D_2^4\right)\sinh\lambda_1 \sinh\lambda_2$$

$$-\left(D_1 + D_2\right)^2 Q^2 \sinh\lambda_1 \sinh\lambda_2 + 2Q(D_1 + D_2) \tag{56.16}$$

$$\left(\lambda_2 D_2^2 \sinh\lambda_1 \cosh\lambda_2 - \lambda_1 D_1^2 \cosh\lambda_1 \sinh\lambda_2\right) = 0$$

where

$$D_{1,2} = \frac{\pi^2}{\gamma^2}\left((1-\mu)i^2 + pr_{ij}^2\right)$$

$$Q = (E_b D_b \frac{i^4\pi^4}{L_y^4} - m_b\omega_{ij}^2)/DL_y^3$$

and Q in Eq. (56.16) represents the interaction between the plate and the beams.

Case 2

$$i^2 < pr_{ij}^2$$

In this case, the solution for Eq. (56.11) would be

$$X_j = \sin\lambda_1 u + A\cos\lambda_1 u + B\sinh\lambda_2 u + C\cosh\lambda_2 u \tag{56.17}$$

where

$$l_{1,2} = \frac{\pi}{\gamma}\left[pr_{ij}^2 \mp i^2\right]$$

By substituting Eqs. (56.17) and (56.11) into the boundary conditions in Eqs. (56.9a) to (56.9f), we can obtain the constants in Eq. (56.17) as in Case 1 by solving a group of eigenvalue equations [7].

The natural frequency ω_{ij} in Case 2 can be obtained from the following equation:

$$2\lambda_1\lambda_2 D_1^2 D_2^2 (\cos \lambda_1 \cosh \lambda_2 - 1) - (\lambda_1^2 D_1^4 + \lambda_2^2 D_2^4) \sin \lambda_1 \sinh \lambda_2$$

$$- (D_1 + D_2)^2 Q^2 \sin \lambda_1 \sinh \lambda_2 + 2Q(D_1 + D_2) \qquad (56.18)$$

$$(\lambda_2 D_2^2 \sin \lambda_1 \cosh \lambda_2 - \lambda_1 D_1^2 \cos \lambda_1 \sinh \lambda_2) = 0$$

Example 56.3 Free Vibration Analysis for a Beam-Plate Bridge

A one-lane bridge deck structure is shown in Figure 56.4. The plate is made of an isotropic material representing reinforced concrete, and the beams are made of steel with a W36 × 150 section. The bridge span length is 80 ft (24.39 m) and the bridge width is 10 ft (3.05 m). The thickness of the deck plate is 8 in. (3.15 cm). The elasticity modulus of the steel girder $E_b = 29.0 \times 10^3$ ksi (200.0×10^3 MPa) and the elasticity modulus of the concrete plate is $E = 4.38 \times 10^3$ ksi (30.18 MPa). The mass density of the plate $m = 0.8681$ lb/in.^{2} g (0.61 kN/cm^2 g) and the mass density of the beam is $m_b = 12.50$ lb/in.^{2}g (22.0 N/cm g) where g is the gravitational acceleration. The moment of inertia of the beam is $I_b = 9030$ in.^{4} (375,800 cm^4). All other properties are calculated as:

Bending rigidity of the plate in y direction:

$$L_x D = \frac{L_x E t^3}{12(1.0 - \mu)} = 4.56 \times 10^{10} \text{ lb.in.}^2 \ (1.28 \times 10^{12} \text{kN.cm}^2)$$

Bending rigidity ratio between beam and plate:

$$R_{ej} = \frac{2.0 E_b I_b}{L_x D} = 11.5$$

Mass ratio between beam and plate:

$$R_m = \frac{m_b}{L_x m} = 0.12$$

First natural frequency of the plate as a beam:

$$\omega_{11}^p = \frac{p^2}{L_y^2} \sqrt{\frac{D}{m}} = 4.40 \text{ rad / s}$$

First natural frequency of the beam:

$$\omega_1^b = \frac{p^2}{L_y^2} \sqrt{\frac{E_b I_b}{m_b}} = 30.45 \text{ rad / s}$$

Subscripts i and j represent the ith and jth mode shape in the y- and x-direction, respectively. The natural frequencies, ω_{ij}, can be determined numerically using Eqs. (56.16) and (56.18). Some of the first several natural frequencies are shown in Table 56.2.

The mode shapes of the bridge are shown in Figures 56.6 and 56.7 using normalized dimensions in all three directions. It is seen that W_{12}, W_{14}, W_{22}, and W_{24} are all asymmetric mode shapes in the

TABLE 56.2   Natural Frequencies
of One-Lane Bridge Deck (rad/s;
plate aspect ratio = 8.0)

	$j = 1$	$j = 2$	$j = 3$
$i = 1$	14.53	35.43	482.65
$i = 2$	60.09	110.62	506.27
$i = 3$	125.67	217.44	556.53
$i = 4$	206.05	354.98	647.51

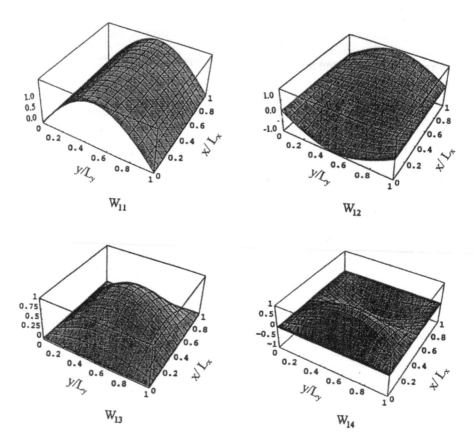

FIGURE 56.6   Normalized three-dimensional mode shapes W_{ij}, $i = 1$, $j = 1,2,3,4$, for a one-lane bridge deck. (subscripts i-j is the mode shape number in y- and x-directions, respectively).

x-direction about the center line $x = L_x/2$. When a moving load traverses the plate along this line, these mode shapes would not be excited. Thus, there are no contributions from these mode shapes.

The first mode shape of the beam–plate system is nearly a constant in the x-direction, as shown in this example of a rather high plate aspect ratio. This means that the beams have the same first mode shape as the plate. In this case, the first natural frequency of the beam–plate system can be approximately evaluated as

$$\omega_{11}^2 = \frac{2 R_m \omega_1^{b\,2} + \omega_{11}^{p\,2}}{1 + 2 R_m} \tag{56.19}$$

where ω_1^b and ω_1^p are fundamental frequencies of the beam and plate as a beam, respectively.

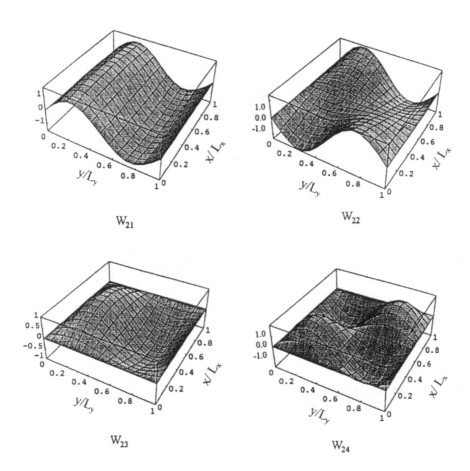

FIGURE 56.7 Normalized three-dimensional mode shapes W_{ij}, $i = 2$, $j = 1,2,3,4$, for a one-lane bridge deck (subscripts i-j is the mode shape number in y- and x-directions, respectively).

56.5 Forced Vibration Analysis under Moving Load

56.5.1 Dynamic Response Analysis

The governing equation for the plate with smooth surface supported by two beams under a moving constant load shown in Figure 56.4 is

$$\nabla^4 W + \frac{m}{D}\frac{\partial^2 W}{\partial t^2} = \frac{1}{D}p(x,y,t) \quad (0 < t < L_y/c) \tag{56.20}$$

where

$$p(x,y,t) = P\delta(y - ct)\,\delta(x - L_x/2)$$

and c is the speed of the moving load, and P the magnitude of the moving load. A Dirac delta function represents a unit concentrated force acting on the deck.

The method of modal superposition is used to get the dynamic response by assuming

$$W = \sum_{i=1}^{+\infty} \sum_{j=1}^{+\infty} W_{ij}(t) \sin\frac{i\pi y}{L_y} X_j(x) \qquad (56.21)$$

By substituting Eq. (56.21) into Eq. (56.20), the following ordinary differential equation can be derived:

$$W_{ij,t}^{(2)}(t) + 2\xi_{ij}\omega_{ij} W_{ij,t}^{(1)} + \omega_{ij}^2 W_{ij} = \frac{2P}{m L_x L_y} X_{ij}(L_x/2)\sin\frac{i\pi c}{L_y}t \quad (0 < t < L_y/c) \qquad (56.22)$$

where ξ_{ij} is the damping ratio for mode shape i–j. The initial condition is that the bridge structure is in a static state before the load enters the span and the structure is in a state of free vibration after the load traverses the bridge.

There are several parameters affecting the dynamic response of the structure. The influence of some typical parameters such as the speed of the moving load and damping ratio is presented in the following.

The vehicle speed is normalized as

$$\alpha = \frac{\pi c / L_y}{\omega_{11}} \qquad (56.23)$$

For example, if the load is moving at a speed of 60 mph, the span of the bridge is 80 ft and first natural frequency is 14 Hz, the normalized speed is $\alpha = 0.25$. A normalized speed of $\alpha = 0.5$ represents the case for which the time needed for a vehicle to traverse the span is the same as the first period of the bridge. Typical normalized dynamic deflections are shown in Figure 56.8. It is seen that for the normalized speed ranging $0 < \alpha < 1.0$, the maximum deflection occurs when the load is on the bridge while if $\alpha > 1.0$, the maximum deflection occurs after the load traverses the bridge. When the load is on the plate ($ct/L_y < 1.0$), the deflections at the center are usually positive. But the deflections are negative after the load has traversed the plate ($ct/L_y > 1.0$), especially for the case of $0 < \alpha < 0.5$. The change of sign for deflection results in a curvature change and, hence, in a stress sign change.

The normalized deflection at the center of the plate is defined as the dynamic deflection caused by moving constant load divided by the corresponding static deflection caused by the concentrated constant force. Figure 56.9 shows its spectra for an aspect ratio of 8.0. The normalized speed of the moving load affects the maximum dynamic response significantly. For small α values, e.g., less than 0.20, which refers to long-span bridges or slow-speed vehicle, there is little dynamic amplification. Very low speed values or very long bridges do not result in much dynamic response. When the normalized speed is greater than 1.3, corresponding to very short span bridges, the dynamic effect is also not significant because the duration of excitation is extremely short. The maximum dynamic response happens for normalized speeds ranging from 0.45 to 0.65, when load frequency is near the bridge fundamental frequency. It is also found that the maximum bending moment amplification factors are 1.2 for M_y and 1.1 for M_x at a normalized speed of 0.4, while the maximum deflection amplification is about 1.5 at a normalized speed of 0.55. The dynamic amplification for bending moments is less than the dynamic amplification for deflections.

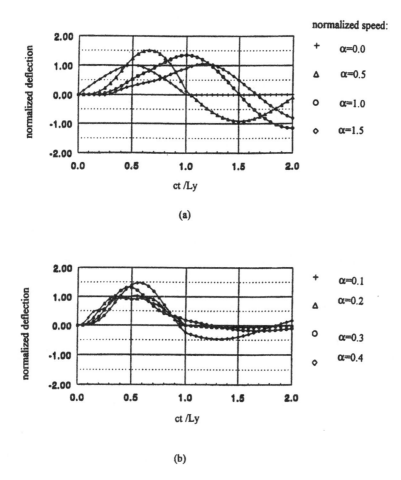

FIGURE 56.8   (a) Normalized deflections at the center of the deck with normalized speed, $\alpha = 0.0$, 0.5, 1.0, and 1.5. (b) Normalized deflections at the center of the deck with normalized speed, $\alpha = 0.1$, 0.2, 0.3, and 0.4.

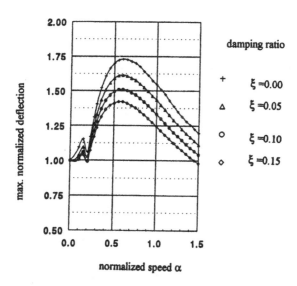

FIGURE 56.9   Normalized deflection spectra at the center of the deck for different damping ratios and normalized speeds.

56.5.2 Summary of Bridge Impact Behavior

For one-lane bridge deck structures, the conclusions of impact behavior of the bridge under moving constant loads have been drawn as:

1. The maximum deflection occurs when the moving load traverses the deck at a normalized speed less than 1.0, and the maximum deflection occurs when the load passes the deck at a normalized speed larger than 1.0.
2. The maximum impact effect is mostly expected when the duration of moving vehicle is close to the fundamental period of the bridge.
3. The aspect ratios of the deck play an important role. When they are less than 4.0, the first mode shape is dominant, when more than 8.0, other mode shapes are excited. The contributions from higher natural frequency mode shapes decrease slowly due to the fact that the natural frequency, ω_{ij}, increases slowly as subscript i increases. Thus, a sufficient number of terms of superimposed mode shapes are needed to get more accurate results.
4. Dynamic amplification for deflections is larger than for bending moments. The response curves of deflections are "smoother" than the response curves for bending moments in the time domain.
5. The dynamic response of a plate with two edges free and the other two edges simply supported is close to a beam for aspect ratios of the plate larger than 2.0.
6. The analysis of a moving constant load model usually overestimates the dynamic effect because vehicle mass is not considered in the dynamic analysis and thus overestimates the first frequency of the bridge–vehicle system which corresponds to the "shorter-span" bridge.

Acknowledgments

The authors would like to express their gratitude to Prof. Dario Gasparini for his endeavors and opinions and to the financial support from Case Western Reserve University in performing the dynamic analysis. Mr. Kang Chen, MG Engineering, Inc. provided valuable information for the railway bridge part, which is greatly appreciated.

References

1. AASHTO, *LRFD Bridge Design Specifications*, American Association of State Highway and Transportation Officials, Washington, D.C., 1994.
2. AASHTO, *Standard Specifications for Highway Bridges*, 16th ed., American Association of State Highway and Transportation Officials, Washington, D.C., 1996.
3. AREA, *Manual for Railway Engineering*, American Railway Engineering Association, Washington, D.C., 1996.
4. Barker, R. M. and Puckett, J. A., *Design of Highway Bridges — Based on AASHTO LRFD, Bridge Design Specifications*, John Wiley & Sons, New York, 1996.
5. Biggs, J. M., *Introduction to Structural Dynamics under Moving Loads*, McGraw-Hill, New York, Inc., 1964.
6. Chang, D. and Lee, H., Impact factors for simple-span highway girder bridges, *J. Struct. Eng. ASCE*, 120(3), 880–889, 1994.
7. Duan, M., Static Finite Element and Dynamic Analytical Study of Reinforced Concrete Bridge Decks, M.S. thesis, Department of Civil Engineering, Case Western Reserve University, Cleveland, OH, 1994.
8. Fryba, L., *Introduction to Structural Dynamics*, Groningen Noordhoff Futern, 1972.
9. Garg, V. K., *Dynamics of Railway Vehicle Systems*, Harcourt Brace Jovanovich, New York, 1984.
10. Huang, D., Wang, T.-L., and Shahawy, M., Impact analysis of continuous multi-girder bridges due to moving vehicles, *J. Struct. Eng. ASCE*, 118(12), 3427–3443, 1992.

11. Hino, J., Yoshimura, T., and Konishi, K., A finite element method prediction of the vibration of a bridge subjected to a moving vehicle load, *J. Sound Vibration*, 96(6), 45–53, 1984.
12. Huang, D., Wang, T.-L., and Shahawy, M., Vibration of thin-walled box-girder bridges exited by vehicles, *J. Struct. Eng. ASCE*, 121(9), 1330–1337, 1995.
13. Kim, S. and Nowak, A., Load distribution and impact factors for I-girder bridges, *J. Bridge Eng. ASCE*, 2(3), 1997.
14. Lin, Y. H. and Trethewey, M. W., Finite element analysis of elastic beams subjected to moving loads, *J. Sound Vibration*, 136(2), 323–342, 1990.
15. Petrou, M. F., Perdikaris, P. C., and Duan, M., Static behavior of noncomposite concrete bridge decks under concentrated loads, *J. Bridge Eng. ASCE*, 1(4), 143–154, 1996.
16. Scheling, D. R., Galdos, N. H., and Sahin, M. A., Evaluation of impact factors for horizontally curved steel box bridges, *J. Struct. Eng. ASCE*, 118(11), 3203–3221, 1992.
17. Timoshenko, S. P. and Woinowsky-Krieger, S., *Theory of Plates and Shells*, 2nd ed., McGraw-Hill, New York, 1959.
18. Wang, T.-L., Huang, D., and Shahawy, M., Dynamic response of multi-girder bridges, *J. Struct. Eng. ASCE*, 118(8), 2222–2238, 1992.
19. Xanthakos, P., *Theory and Design of Bridges*, John Wiley & Sons, New York, 1994.
20. Yang, Y.-B. and Lin, B.-H., Vehicle-bridge interaction analysis by dynamic condensation method, *J. Struct. Eng. ASCE*, 121(2), 1636–1643, 1995.
21. Yang, Y.-B. and Yau, J.-D., Vehicle-bridge interaction element for dynamic analysis, *J. Struct. Eng. ASCE*, 118(11), 1512–1518, 1997.

57

Wind Effects on Long-Span Bridges

Chun S. Cai
*Florida Department
  of Transportation*

Serge Montens
Jean Muller International, France

57.1 Introduction

The development of modern materials and construction techniques has resulted in a new generation of lightweight flexible structures. Such structures are susceptible to the action of winds. Suspension bridges and cable-stayed bridges shown in Figure 57.1 are typical structures susceptible to wind-induced problems.

The most renowned bridge collapse due to winds is the Tacoma Narrows suspension bridge linking the Olympic Peninsula with the rest of the state of Washington. It was completed and opened to traffic on July 1, 1940. Its 853-m main suspension span was the third longest in the world. This bridge became famous for its serious wind-induced problems that began to occur soon after it opened. "Even in winds of only 3 to 4 miles per hour, the center span would rise and fall as much as four feet..., and drivers would go out of their way either to avoid it or cross it for the roller coaster thrill of the trip. People said you saw the lights of cars ahead disappearing and reappearing as they bounced up and down. Engineers monitored the bridge closely but concluded that the motions were predictable and tolerable" [1].

On November 7, 1940, 4 months and 6 days after the bridge was opened, the deck oscillated through large displacements in the vertical vibration modes at a wind velocity of about 68 km/h. The motion changed to a torsional mode about 45 min later. Finally, some key structural members became overstressed and the main span collapsed.

0-8493-7434-0/00/$0.00+$.50
© 2000 by CRC Press LLC

Suspension Bridge

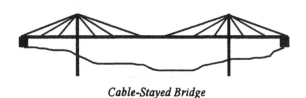

Cable-Stayed Bridge

FIGURE 57.1   Typical wind-sensitive bridges.

Some bridges were destroyed by wind action prior to the failure of the Tacoma Narrows bridge. However, it was this failure that shocked and intrigued bridge engineers to conduct scientific investigations of bridge aerodynamics. Some existing bridges, such as the Golden Gate suspension bridge in California with a main span of 1280 m, have also experienced large wind-induced oscillations, although not to the point of collapse. In 1953, the Golden Gate bridge was stiffened against aerodynamic action [2].

Wind-induced vibration is one of the main concerns in a long-span bridge design. This chapter will give a brief description of wind-induced bridge vibrations, experimental and theoretical solutions, and state-of-the-art applications.

57.2   Winds and Long-Span Bridges

57.2.1   Description of Wind at Bridge Site

The atmospheric wind is caused by temperature differentials resulting from solar radiations. When the wind blows near the ground, it is retarded by obstructions making the mean velocity at the ground surface zero. This zero-velocity layer retards the layer above and this process continues until the wind velocity becomes constant. The distance between the ground surface and the height of constant wind velocity varies between 300 m and 1 km. This 1-km layer is referred to as the boundary layer in which the wind is turbulent due to its interaction with surface friction. The variation of the mean wind velocity with height above ground usually follows a logarithmic or exponential law.

The velocity of boundary wind is defined by three components: the along-wind component consisting of the mean wind velocity, $\overline{U}$, plus the turbulent component $u(t)$, the cross-wind turbulent component $v(t)$, and the vertical turbulent component $w(t)$. The turbulence is described in terms of turbulence intensity, integral length, and spectrum [3].

The turbulence intensity I is defined as

$$I = \frac{\sigma}{\overline{U}} \tag{57.1}$$

where σ = the standard deviation of wind component $u(t)$, $v(t)$, or $w(t)$; $\overline{U}$ = the mean wind velocity.

Integral length of turbulence is a measurement of the average size of turbulent eddies in the flow. There are a total of nine integral lengths (three for each turbulent component). For example, the integral length of $u(t)$ in the x-direction is defined as

$$L_u^x = \frac{1}{\sigma_u^2} \int_0^\infty R_{u1u2}(x)dx \tag{57.2}$$

where $R_{u1u2}(x)$ = cross-covariance function of $u(t)$ for a spatial distance x.

The wind spectrum is a description of wind energy vs. wind frequencies. The von Karman spectrum is given in dimensionless form as

$$\frac{nS(n)}{\sigma^2} = \frac{4\dfrac{nL}{U}}{\left[1 + 70.8\left(\dfrac{nL}{U}\right)^2\right]^{5/6}} \tag{57.3}$$

where n = frequency (Hz); S = autospectrum; and L = integral length of turbulence. The integral length of turbulence is not easily obtained. It is usually estimated by curve fitting the spectrum model with the measured field data.

57.2.2 Long-Span Bridge Responses to Wind

Wind may induce instability and excessive vibration in long-span bridges. Instability is the onset of an infinite displacement granted by a linear solution technique. Actually, displacement is limited by structural nonlinearities. Vibration is a cyclic movement induced by dynamic effects. Since both instability and vibration failures in reality occur at finite displacement, it is often hard to judge whether a structure failed due to instability or excessive vibration-induced damage to some key elements.

Instability caused by the interaction between moving air and a structure is termed aeroelastic or aerodynamic instability. The term *aeroelastic* emphasizes the behavior of deformed bodies, and *aerodynamic* emphasizes the vibration of rigid bodies. Since many problems involve both deformation and vibration, these two terms are used interchangeably hereafter. Aerodynamic instabilities of bridges include divergence, galloping, and flutter. Typical wind-induced vibrations consist of vortex shedding and buffeting. These types of instability and vibration may occur alone or in combination. For example, a structure must experience vibration to some extent before flutter instability starts.

The interaction between the bridge vibration and wind results in two kinds of forces: motion-dependent and motion-independent. The former vanishes if the structures are rigidly fixed. The latter, being purely dependent on the wind characteristics and section geometry, exists whether or not the bridge is moving. The aerodynamic equation of motion is expressed in the following general form:

$$[M]\{\ddot{Y}\} + [C]\{\dot{Y}\} + [K]\{Y\} = \{F(Y)\}_{md} + \{F\}_{mi} \tag{57.4}$$

where $[M]$ = mass matrix; $[C]$ = damping matrix; $[K]$ = stiffness matrix; $\{Y\}$ = displacement vector; $\{F(Y)\}_{md}$ = motion-dependent aerodynamic force vector; and $\{F\}_{mi}$ = motion-independent wind force vector.

The motion-dependent force causes aerodynamic instability and the motion-independent part together with the motion-dependent part causes deformation. The difference between short-span and long-span bridge lies in the motion-dependent part. For the short-span bridges, the motion-dependent part is insignificant and there is no concern about aerodynamic instability. For flexible structures like long-span bridges, however, both instability and vibration need to be carefully investigated.

57.3 Experimental Investigation

Wind tunnel testing is commonly used for "wind-sensitive" bridges such as cable-stayed bridges, suspension bridges, and other bridges with span lengths or structure types significantly outside of the common ranges. The objective of a wind tunnel test is to determine the susceptibility of the bridges to various aerodynamic phenomena.

The bridge aerodynamic behavior is controlled by two types of parameters, i.e., structural and aerodynamic. The structural parameters are the bridge layout, boundary condition, member stiffness, natural modes, and frequencies. The aerodynamic parameters are wind climate, bridge section shape, and details. The design engineers need to provide all the information to the wind specialist to conduct the testing and analysis.

57.3.1 Scaling Principle

In a typical structural test, a prototype structure is scaled down to a scale model according to mass, stiffness, damping, and other parameters. In testing, the wind blows in different vertical angles (attack angles) or horizontal angles (skew angles) to cover the worst case at the bridge site. To obtain reliable information from a test, similarity must be maintained between the specimen and the prototype structure. The geometric scale λ_L, a basic parameter which is controlled by the size of an available wind tunnel, is denoted as the ratio of the dimensions of model (B_m) to the dimensions of prototype bridge (B_p) as [4]

$$\lambda_L = \frac{B_m}{B_p} \tag{57.5}$$

where subscripts m and p indicate model and prototype, respectively.

To maintain the same Froude number for both scale model and prototype bridge requires,

$$\left(\frac{U^2}{Bg}\right)_m = \left(\frac{U^2}{Bg}\right)_p \tag{57.6}$$

where g is the air gravity, which is the same for the model and prototype bridge. From Eqs. (57.5) and (57.6) we have the wind velocity scale λ_v as

$$\lambda_V = \frac{U_m}{U_p} = \sqrt{\lambda_L} \tag{57.7}$$

Reynolds number equivalence requires

$$\left(\frac{\rho UB}{\mu}\right)_m = \left(\frac{\rho UB}{\mu}\right)_p \tag{57.8}$$

where μ = viscosity and ρ = wind mass density. Equations (57.5) and (57.8) give the wind velocity scale as

$$\lambda_V = \frac{1}{\lambda_L} \tag{57.9}$$

which contradicts Eq. (57.7). It is therefore impossible in model scaling to satisfy both the Froude number equivalence and Reynolds number equivalence simultaneously. For bluff bodies such as bridge decks, flow separation is caused by sharp edges and, therefore, the Reynolds number is not important except it is too small. The too-small Reynolds number can be avoided by careful selection of λ_L. Therefore, the Reynolds number equivalence is usually sacrificed and Froude number equivalence is maintained.

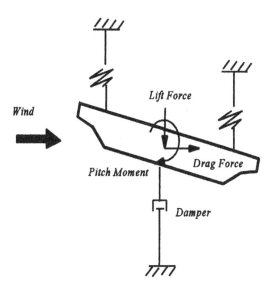

FIGURE 57.2 End view of section model.

To apply the flutter derivative information to the prototype analysis, nondimensional reduced velocity must be the same, i.e.,

$$\left(\frac{U}{NB}\right)_m = \left(\frac{U}{NB}\right)_p \tag{57.10}$$

Solving Eqs. (57.5), (57.7) and (57.10) gives the natural frequency scale as

$$\lambda_N = \frac{N_m}{N_p} = \frac{1}{\sqrt{\lambda_L}} \tag{57.11}$$

The above equivalence of reduced velocity between the section model and prototype bridge is the basis to use the section model information to prototype bridge analysis. Therefore, it should be strictly satisfied.

57.3.2 Section model

A typical section model represents a unit length of a prototype deck with a scale from 1:25 to 1:100. It is usually constructed from materials such as steel, wood, or aluminum to simulate the scaled mass and moment of inertia about the center of gravity. The section model represents only the outside shape (aerodynamic shape) of the deck. The stiffness and the vibration characteristics are represented by the spring supports.

By rigidly mounding the section in the wind tunnel, the static wind forces, such as lift, drag, and pitch moment, can be measured. To measure the aerodynamic parameters such as the flutter derivatives, the section model is supported by a spring system and connected to a damping source as shown in Figure 57.2. The spring system can be adjusted to simulate the deck stiffness in vertical and torsional directions, and therefore simulate the natural frequencies of the bridges. The damping characteristics are also adjustable to simulate different damping.

A section model is less expensive and easier to conduct than a full model. It is thus widely used in (1) the preliminary study to find the best shape of a bridge deck; (2) to identify the potential wind-induced problems such as vortex-shedding, flutter, and galloping and to guide a more-sophisticated full model study; (3) to measure wind data, such as flutter derivatives, static force coefficients for analytical prediction of actual bridge behavior; and (4) to model some less important bridges for which a full model test cannot be economically justified.

57.3.3 Full Bridge Model

A full bridge model, representing the entire bridge or a few spans, is also called an aeroelastic model since the aeroelastic deformation is reflected in the full model test. The deck, towers, and cables are built according to the scaled stiffness of the prototype bridge. The scale of a full bridge model is usually from 1:100 to 1:300 to fit the model in the wind tunnel. The full model test is used for checking many kinds of aerodynamic phenomena and determining the wind loading on bridges.

A full bridge model is more expensive and difficult to build than a section model. It is used only for large bridges at the final design stage, particularly to check the aerodynamics of the construction phase. However, a full model test has many advantages over a section model: (1) it simulates the three-dimensional and local topographical effects; (2) it reflects the interaction between vibration modes; (3) wind effects can be directly visualized at the construction and service stages; and (4) it is more educational to the design engineers to improve the design.

57.3.4 Taut Strip Model

For this model, taut strings or tubes are used to simulate the stiffness and dynamic characteristics of the bridge such as the natural frequencies and mode shapes for vertical and torsional vibrations. A rigid model of the deck is mounted on the taut strings. This model allows, for example, to represent the main span of a deck. The taut strip model falls between section model and full model with respect to cost and reliability. For less important bridges, the taut strip model is a sufficient and economical choice. The taut strip model is used to determine critical wind velocity for vortex shedding, flutter, and galloping and displacement and acceleration under smooth or turbulent winds.

57.4 Analytical Solutions

57.4.1 Vortex Shedding

Vortex shedding is a wake-induced effect occurring on bluff bodies such as bridge decks and pylons. Wind flowing against a bluff body forms a stream of alternating vortices called a von Karman vortex street shown in Figure 57.3a. Alternating shedding of vortices creates an alternative force in a direction normal to the wind flow. This alternative force induces vibration. The shedding frequency of vortices from one surface, in either torsion or lift, can be described in terms of a nondimensional Strouhal number, S, as

$$S = \frac{ND}{\overline{U}} \tag{57.12}$$

where N = shedding frequency and D = characteristic dimension such as the diameter of a circular section or depth of a deck.

The Strouhal number (ranging from 0.05 to 0.2 for bridge decks) is a constant for a given section geometry and details. Therefore, the shedding frequency (N) increases with the wind velocity to maintain a constant Strouhal value (S). The bridge vibrates strongly but self-limited when the frequency of vortex shedding is close to one of the natural frequencies of a bridge, say, N_1 as shown in Figure 57.3. This phenomenon is called lock-in and the corresponding wind velocity is called critical velocity of vortex shedding.

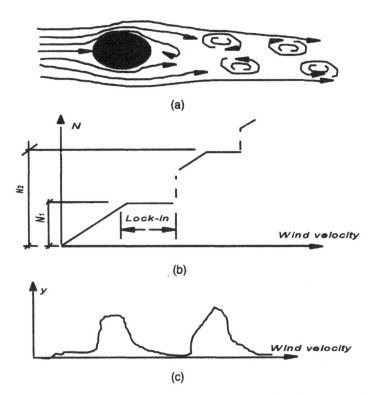

FIGURE 57.3 Explanation of vortex shedding. (a) Von Karman Street; (b) lock-in phenomenon; (c) bridge vibration

The lock-in occurs over a small range of wind velocity within which the Strouhal relation is violated since the increasing wind velocity and a fixed shedding frequency results in a decreasing Strouhal number. The bridge natural frequency, not the wind velocity, controls the shedding frequency. As wind velocity increases, the lock-in phenomenon disappears and the vibration reduces to a small amplitude. The shedding frequency may lock in another higher natural frequency (N_2) at higher wind velocity. Therefore, many wind velocities cause vortex shedding.

To describe the above experimental observation, much effort has been made to find an expression for forces resulting from vortex shedding. Since the interaction between the wind and the structure is very complex, no completely successful model has yet been developed for bridge sections. Most models deal with the interaction of wind with circular sections. A semiempirical model for the lock-in is given as [3]

$$m\ddot{y} + c\dot{y} + ky = \frac{1}{2}\rho U^2 (2D) \left[Y_1(K) \left(1 - \varepsilon \frac{y^2}{D^2} \right) \frac{\dot{y}}{D} + Y_2(K) \frac{y}{D} + \frac{1}{2} C_L(K) \sin\left(\omega t + \phi\right) \right] \quad (57.13)$$

where $k = B\omega/\overline{U}$ = reduced frequency; Y_1, Y_2, ε, and C_L = parameters to be determined from experimental observations. The first two terms of the right side account for the motion-dependent force. More particularly, the $\dot{y}$ term accounts for aerodynamic damping and y term for aerodynamic stiffness. The ε accounts for the nonlinear aerodynamic damping to ensure the self-limiting nature of vortex shedding. The last term represents the instantaneous force from vortex shedding alone which is sinusoidal with the natural frequency of bridge. Solving the above equation gives the vibration y.

Vortex shedding occurs in both laminar and turbulent flow. According to some experimental observations, turbulence helps to break up vortices and therefore helps to suppress the vortex shedding response. A more complete analytical model must consider the interaction between modes, the spanwise correlation of aerodynamic forces and the effect of turbulence.

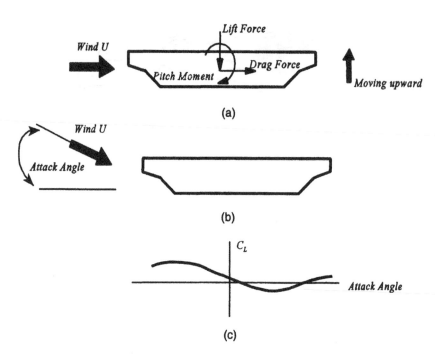

FIGURE 57.4 Explanation of galloping. (a) Section moving upward; (b) motionless section with a wind attack angle; (c) static force coefficient vs. attack angle.

For a given section shape with a known Strouhal number and natural frequencies, the lock-in wind velocities can be calculated with Eq. (57.12). The calculated lock-in wind velocities are usually lower than the maximum wind velocity at bridge sites. Therefore, vortex shedding is an inevitable aerodynamic phenomenon. However, vibration excited by vortex shedding is self-limited because of its nonlinear nature. A relatively small damping is often sufficient to eliminate, or at least reduce, the vibrations to acceptable limits.

Although there are no acceptance criteria for vortex shedding in the design specifications and codes in the United States, there is a common agreement that limiting acceleration is more appropriate than limiting deformation. It is usually suggested that the acceleration of vortex shedding is limited to 5% of gravity acceleration when wind speed is less than 50 km/h and 10% of gravity acceleration when wind speed is higher. The acceleration limitation is then transformed into the displacement limitation for a particular bridge.

57.4.2 Galloping

Consider that in Figure 57.4 (a) a bridge deck is moving upward with a velocity $\dot{y}$ under a horizontal wind U. This is equivalent to the case of Figure 57.4b that the deck is motionless and the wind blows downward with an attack angle α ($\tan(\alpha) = \dot{y}/U$). If the measured static force coefficient of this case is negative (upward), then the deck section will be pushed upward further resulting in a divergent vibration or galloping. Otherwise, the vibration is stable. Galloping is caused by a change in the effective attack angle due to vertical or torsional motion of the structure. A negative slope in the plot of either static lift or pitch moment coefficient vs. the angle of attack, shown in Figure 57.4c, usually implies a tendency for galloping. Galloping depends mainly on the quasi-steady behavior of the structure.

The equation of motion describing this phenomenon is

$$m\ddot{y} + c\dot{y} + ky = -\frac{1}{2}\rho U^2 B \left(\frac{dC_L}{d\alpha} + C_D\right)_{\alpha=0} \frac{\dot{y}}{U} \qquad (57.14)$$

The right side represents the aerodynamic damping and C_L and C_D are static force coefficients in the lift and drag directions, respectively. If the total damping is less than zero, i.e.,

$$c + \frac{1}{2}\rho UB\left(\frac{dC_L}{d\alpha} + C_D\right)_{\alpha=0} \leq 0 \tag{57.15}$$

then the system tends toward instability. Solving the above equation gives the critical wind velocity for galloping. Since the mechanical damping c is positive, the above situation is possible only if the following Den Hartog criterion [5] is satisfied

$$\left(\frac{dC_L}{d\alpha} + C_D\right)_{\alpha=0} \leq 0 \tag{57.16}$$

Therefore, a wind tunnel test is usually conducted to check against Eq. (57.16) and to make necessary improvement of the section to eliminate the negative tendency for the possible wind velocity at a bridge site.

Galloping rarely occurs in highway bridges, but noted examples are pedestrian bridges, pipe bridges, and ice-coated cables in power lines. There are two kinds of cable galloping: cross-wind galloping, which creates large-amplitude oscillations in a direction normal to the flow, and wake galloping caused by the wake shedding of the upwind structure.

57.4.3 Flutter

Flutter is one of the earliest recognized and most dangerous aeroelastic phenomena in airfoils. It is created by self-excited forces that depend on motion. If a system immersed in wind flow is given a small disturbance, its motion will either decay or diverge depending on whether the energy extracted from the flow is smaller or larger than the energy dissipated by mechanical damping. The theoretical line dividing decaying and diverging motions is called the critical condition. The corresponding wind velocity is called the critical wind velocity for flutter or simply the flutter velocity at which the motion of the bridge deck tends to grow exponentially as shown in Figure 57.5a.

When flutter occurs, the oscillatory motions of all degrees of freedom in the structure couple to create a single frequency called the flutter frequency. Flutter is an instability phenomenon; once it takes place, the displacement is infinite by linear theory. Flutter may occur in both laminar and turbulent flows.

The self-excited forces acting on a unit deck length are usually expressed as a function of the flutter derivatives. The general format of the self-excited forces written in matrix form [2,6] for finite element analysis is

$$\begin{Bmatrix} L_{se} \\ D_{se} \\ M_{se} \end{Bmatrix} = \frac{1}{2}\rho U^2(2B)\left(\begin{bmatrix} \dfrac{k^2 H_4^*}{B} & \dfrac{k^2 H_6^*}{B} & k^2 H_3^* \\ \dfrac{k^2 P_4^*}{B} & \dfrac{k^2 P_6^*}{B} & k^2 P_3^* \\ k^2 A_4^* & k^2 A_6^* & k^2 A_3^* B \end{bmatrix} \begin{Bmatrix} h \\ p \\ \alpha \end{Bmatrix} + \begin{bmatrix} \dfrac{kH_1^*}{U} & \dfrac{kH_5^*}{U} & \dfrac{kH_2^* B}{U} \\ \dfrac{kP_1^*}{U} & \dfrac{kP_5^*}{U} & \dfrac{kP_2^* B}{U} \\ \dfrac{kA_1^* B}{U} & \dfrac{kA_5^* B}{U} & \dfrac{kA_2^* B^2}{U} \end{bmatrix} \begin{Bmatrix} \dot{h} \\ \dot{p} \\ \dot{\alpha} \end{Bmatrix} \right) \tag{57.17}$$

$$= U^2\left[F_d\right]\{q\} + U^2\left[F_v\right]\{\dot{q}\}$$

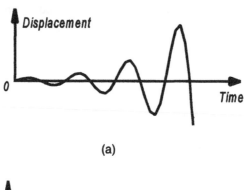

(a)

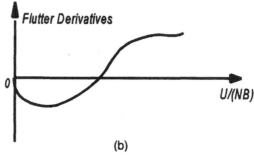

(b)

FIGURE 57.5   Explanation of flutter. (a) Bridge flutter vibration; (b) typical flutter derivations.

where L_{se}, D_{se}, and M_{se} = self-excited lift force, drag force, and pitch moment, respectively; h, p, and α = displacements at the center of a deck in the directions corresponding to L_{se}, D_{se}, and M_{se}, respectively; ρ = mass density of air; B = deck width; H_i^*, P_i^*, and A_i^* (i = 1 to 6) = generalized flutter derivatives; $k = B\omega/\overline{U}$ = reduced frequency; ω = oscillation circular frequency; $\overline{U}$ = mean wind velocity; and $[F_d]$ and $[F_v]$ = flutter derivative matrices corresponding to displacement and velocity, respectively.

While the flutter derivatives H_i^* and A_i^* have been experimentally determined for i = 1 to 4, the term P_i^* is theoretically derived in state-of-the-art applications. The other flutter derivatives (for i = 5 and 6) have been neglected in state-of-the-art analysis.

In linear analyses, the general aerodynamic motion equations of bridge systems are expressed in terms of the generalized mode shape coordinate $\{\xi\}$

$$[M^*]\{\ddot{\xi}\}+\left([D^*]-U^2[AD^*]\right)\{\dot{\xi}\}+\left([K^*]-U^2[AS^*]\right)\{\xi\}=0 \qquad (57.18)$$

where $[M^*]$, $[D^*]$, and $[K^*]$ = generalized mass, damping, and stiffness matrices, respectively; and $[AS^*]$ and $[AD^*]$ = generalized aerodynamic stiffness and aerodynamic damping matrices, respectively. Matrices $[M^*]$, $[D^*]$, and $[K^*]$ are derived the same way as in the general dynamic analysis. Matrices $[AS^*]$ and $[AD^*]$, corresponding to $[F_d]$ and $[F_v]$ in Eq. (57.17), respectively, are assembled from aerodynamic element forces. It is noted that even the structural and dynamic matrices $[K^*]$, $[M^*]$, and $[D^*]$ are uncoupled between modes, the motion equation is always coupled due to the coupling of aerodynamic matrices $[AS^*]$ and $[AD^*]$.

Flutter velocity, U, and flutter frequency, ω, are obtained from the nontrivial solution of Eq. (57.18) as

$$\left|-\omega^2[M^*]+[K^*]-\overline{U}^2[AS^*]+\omega\left([D^*]-\overline{U}^2[AD^*]\right)i\right|=0 \qquad (57.19)$$

For a simplified uncoupled single degree of freedom, the above equation reduces to

$$\omega^2 = \frac{[K*] - \overline{U}^2[AS*]}{[M*]} \tag{57.20}$$

and

$$U_{cr}^2 = \frac{[D*]}{[AD*]} \tag{57.21}$$

Since the aerodynamic force $[AS^{\cdot}]$ is relatively small, it can be seen that the flutter frequency in Eq. (57.20) is close to the natural frequency $[K^{\cdot}]/[M^{\cdot}]$. Equation (57.21) can be also derived from Eq. (57.18) as the zero-damping condition. Zero-damping cannot occur unless $[AD^{\cdot}]$ is positive. The value of $[AD^{\cdot}]$ depends on the flutter derivatives. An examination of the flutter derivatives gives a preliminary judgment of the flutter behavior of the section. Necessary section modifications should be made to eliminate the positive flutter derivatives as shown in Figure 57.5b, especially the A_2^{*} and H_1^{*}. The A_2^{*} controls the torsional flutter and the H_1^{*} controls the vertical flutter. It can be seen from Eq. (57.21) that an increase in the mechanical damping $[D^{\cdot}]$ increases the flutter velocity. It should be noted that for a coupled flutter, zero-damping is a sufficient but not a necessary condition.

A coupled flutter is also called stiffness-driven or classical flutter. An uncoupled flutter is called damping-driven flutter since it is caused by zero-damping. Since flutter of a suspension bridge is usually controlled by its first torsional mode, the terminology *flutter* was historically used for a torsional aerodynamic instability. Vertical aerodynamic instability is traditionally treated in a quasi-static approach, i.e., as is galloping. In recent literature, flutter is any kind of aerodynamic instability due to self-excited forces, whether vertical, torsional, or coupled vibrations.

Turbulence is assumed beneficial for flutter stability and is usually ignored. Some studies include turbulence effect by treating along-wind velocity U as mean velocity, $\overline{U}$, plus a turbulent component, $u(t)$. The random nature of $u(t)$ results in an equation of random damping and stiffness. Complicated mathematics, such as stochastic differentiation, need to be involved to solve the equation [7].

Time history and nonlinear analyses can be conducted on Eq. (57.18) to investigate postflutter behavior and to include the effects of both geometric and material nonlinearities. However, this is not necessary for most practical applications.

57.4.4 Buffeting

Buffeting is defined as the forced response of a structure to random wind and can only take place in turbulent flows. Turbulence resulting from topographical or structural obstructions is called oncoming turbulence. Turbulence induced by bridge itself is called signature turbulence. Since the frequencies of signature turbulence are generally several times higher than the important natural frequencies of the bridge, its effect on buffeting response is usually small.

Buffeting is a random vibration problem of limited displacement. The effects of buffeting and vortex shedding are similar, except that vibration is random in the former and periodic in the latter. Both buffeting and vortex shedding influence bridge service behavior and may result in fatigue damage that could lead to a eventual collapse of a bridge. Buffeting also influences ultimate strength behavior.

Similar to Eq. (57.17), the buffeting forces are expressed in the matrix form [2] for finite element analysis as

$$\begin{Bmatrix} L_b \\ D_b \\ M_b \end{Bmatrix} = \frac{1}{2}\rho\,\bar{U}^2 B \begin{bmatrix} 2C_L & \left(\dfrac{dC_L}{d\alpha}+C_D\right) \\ 2C_D & \dfrac{dC_D}{d\alpha} \\ 2C_M B & \dfrac{dC_M}{d\alpha}B \end{bmatrix} \begin{Bmatrix} \dfrac{u(t)}{\bar{U}} \\ \dfrac{w(t)}{\bar{U}} \end{Bmatrix} = \bar{U}^2[C_b]\{\eta\} \qquad (57.22)$$

where C_D, C_L, and C_M = static aerodynamic coefficients for drag, lift, and pitch moment, respectively; α = angle of wind attack; $[C_b]$ = static coefficient matrix; and $\{\eta\}$ = vector of turbulent wind components normalized by mean wind velocity.

The equation of motion for buffeting is similar to Eq. (57.18), but with one more random buffeting force as

$$[M^*]\{\ddot{\xi}\} + \left([D^*]-U^2[AD^*]\right)\{\dot{\xi}\} + \left([K^*]-U^2[AS^*]\right)\{\xi\} = \bar{U}^2[f^*]_b\{\eta\} \qquad (57.23)$$

Fourier transform of Eq. (57.23) yields

$$\mathcal{F}\left(\{\xi\}\right) = \bar{U}^2[G_1]\{f^*\}_b * \mathcal{F}\left(\{\eta\}\right) \qquad (57.24)$$

where

$$[G_1] = \frac{1}{\left(-\omega^2[M^*]+[K^*]-\bar{U}^2[AS^*]+\omega\left([D^*]-\bar{U}^2[AD^*]\right)\right)} \qquad (57.25)$$

Similarly, taking the conjugate transform of Eq. (57.23) yields

$$\overline{\mathcal{F}\left(\{\xi\}\right)}^T = \bar{U}^2\overline{\mathcal{F}\left(\{\eta\}\right)}^T\{f^*\}_b^T[G_2]^T \qquad (57.26)$$

where

$$[G_2] = \frac{1}{\left(-\omega^2[M^*]+[K]^*-\bar{U}^2[AS^*]-\omega\left([D^*]-\bar{U}^2[AD^*]\right)\right)} \qquad (57.27)$$

The superscript T represents for the matrix transpose, and the overbar stands for the Fourier conjugate transform for the formula above. Multiplying Eqs. (57.24) and (57.26) gives the following spectral density of generalized coordinates

$$\begin{bmatrix} S_{\xi_1\xi_1} & \cdots & S_{\xi_1\xi_m} \\ S_{\xi_i\xi_1} & \cdots & S_{\xi_i\xi_m} \\ S_{\xi_m\xi_1} & \cdots & S_{\xi_m\xi_m} \end{bmatrix} = \bar{U}^4[G_1]\{f^*\}_b \begin{bmatrix} S_{\eta_1\eta_1} & S_{\eta_1\eta_2} \\ S_{\eta_2\eta_1} & S_{\eta_2\eta_2} \end{bmatrix}\{f^*\}_b^T[G_2]^T \qquad (57.28)$$

where $S_{\eta_i\eta_j}$ = spectral density of normalized wind components. The mean square of the modal and physical displacements can be derived from their spectral densities. Once the displacement is known,

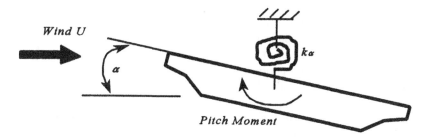

FIGURE 57.6 Explanation of torsional divergence.

the corresponding forces can be derived. The aerodynamic study should ascertain that no structural member is overstressed or overdeformed such that the strength and service limits are exceeded. For very long span bridges, a comfort criterion must be fulfilled under buffeting vibration.

57.4.5 Quasi-Static Divergence

Wind flowing against a structure exerts a pressure proportional to the square of the wind velocity. Wind pressure generally induces both forces and moments in a structure. At a critical wind velocity, the edge-loaded bridge may buckle "out-of-plane" under the action of a drag force or torsionally diverge under a wind-induced moment that increases with a geometric twist angle. In reality, divergence involves an inseparable combination of lateral buckling and torsional divergence.

Consider a small rotation angle as shown in Figure 57.6, the pitch moment resulting from wind is [3]

$$M_\alpha = \frac{1}{2}\,\rho U^2 B^2 C_M(\alpha)$$
$$= \frac{1}{2}\,\rho U^2 B^2 \left[C_{M0} + \frac{dC_m}{d\alpha}\Big|_{\alpha=0}\alpha \right] \tag{57.29}$$

When the pitch moment caused by wind exceeds the resisting torsional capacity, the displacement of the bridge diverges. Equating the aerodynamic force to the internal structural capacity gives

$$k_\alpha\alpha - \frac{1}{2}\,\rho U^2 B^2 \left[C_{M0} + \frac{dC_m}{d\alpha}\Big|_{\alpha=0}\alpha \right] = 0 \tag{57.30}$$

where k_α = spring constant of torsion. For an infinite α, we have the critical wind velocity for torsional divergence as

$$U_{cr} = \sqrt{\frac{2k_\alpha}{\rho B^2 \dfrac{dC_M}{d\alpha}\Big|_{\alpha=0}}} \tag{57.31}$$

57.5 Practical Applications

57.5.1 Wind Climate at Bridge Site

The wind climate at a particular bridge site is usually not available, but it is commonly decided according to the historical wind data of the nearest airport. The wind data are then analyzed considering the local terrain features of the bridge site to obtain the necessary information such as the maximum wind velocity, dominant direction, turbulence intensity, and wind spectrum. For large bridges, an anemometer can be installed on the site for a few months to get the characteristics of the wind on the site itself. The most important quantity is the maximum wind velocity, which is dependent on the bridge design period.

The bridge design period is decided by considering a balance between cost and safety. For the strength design of a completed bridge, a design period of 50 or 100 years is usually used. Since the construction of a bridge lasts a relatively short period, a 10-year period can be used for construction strength checking. This is equivalent to keeping the same design period but reducing the safety factor during construction.

Flutter is an instability phenomenon. Once it occurs, its probability of failure is assumed to be 100%. A failure probability of 10^{-5} per year for completed bridges represents an acceptable risk, which is equivalent to a design period of 100,000 years. Similarly, the design period of flutter during construction can be reduced to, say, 10,000 years. It should be noted that the design period does not represent the bridge service life, but a level of failure risk.

Once the design period has been decided, the maximum wind velocity is determined. Increasing the design period by one order of magnitude usually raises the wind velocity only by a few percent, depending on the wind characteristics. Wind velocity for flutter (stability) design is usually about 20% larger than that for buffeting (strength) design, although the design period for the former is several orders higher. Once wind characteristics and design velocity are available, a wind tunnel/analytical investigation is conducted.

57.5.2 Design Consideration

The aerodynamic behavior of bridges depends mainly on four parameters: structural form, stiffness, cross section shape and its details, and damping. Any significant changes that may affect these parameters need to be evaluated by a wind specialist.

1. *Structural form:* Suspension bridges, cable-stayed bridges, arch bridges, and truss bridges, due to the increase of rigidity in this order, generally have aerodynamic behaviors from worst to best. A truss-stiffened section, because it blocks less wind, is more favorable than a girder-stiffened one. But a truss-stiffened bridge is generally less stiff in torsion.
2. *Stiffness:* For long-span bridges, it is not economical to add more material to increase the stiffness. However, changing the boundary conditions, such as deck and tower connections in cable-stayed bridges, may significantly improve the stiffness. Cable-stayed bridges with A-shaped or inverted Y-shaped towers have higher torsional frequency than the bridges of H-shaped towers.
3. *Cross section shape and its details:* A streamlined section blocks less wind, thus has better aerodynamic behavior than a bluff section. Small changes in section details may significantly affect the aerodynamic behavior.
4. *Damping:* Concrete bridges have higher damping ratios than steel bridges. Consequently, steel bridges have more wind-induced problems than concrete bridges. An increase of damping can reduce aerodynamic vibration significantly.

Major design parameters are usually determined in the preliminary design stage, and then the aerodynamic behavior is evaluated by a wind specialist. Even if the bridge responds poorly under aerodynamic excitation, it is undesirable to change the major design parameters for reasons of scheduling and funding. The common way to improve its behavior is to change the section details. For example, changing the solid parapet to a half-opened parapet or making some venting slots [8] on the bridge deck may significantly improve the aerodynamic behavior. To have more choices on how to improve the aerodynamic behavior of long-span bridges, to avoid causing delays in the schedule, and to achieve an economical design, aerodynamics should be considered from the beginning.

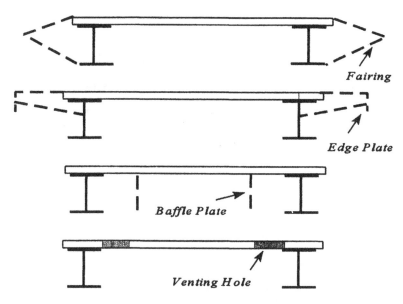

FIGURE 57.7 Typical aerodynamic modifications.

Although a streamlined section is always favorable for aerodynamic behavior, there have recently been many composite designs, due to their construction advantages, for long-span bridges. The composite section shapes, with the concrete deck on steel girders, are bluff and thus not good for aerodynamics, but can be improved by changing section details as shown in Figure 57.7 [9].

57.5.3 Construction Safety

The most common construction method for long-span bridges is segmental (staged) construction, such as balanced cantilever construction of cable-stayed bridges, tie-back construction of arch bridges, and float-in construction of suspended spans. These staged constructions result in different structural configurations during the construction time. Since some construction stages have lower stiffness and natural frequency than the completed bridges, construction stages are often more critical in terms of either strength of structural members or aerodynamic instability.

In the balanced cantilever construction of cable-stayed bridges, three stages are usually identified as critical, as shown in Figure 57.8. They are tower before completion, completed tower, and the stage with a longest cantilever arm. Reliable analytical solutions are not available yet, and wind tunnel testing is usually conducted to ensure safety. Temporary cables, tie-down, and bent are common countermeasures during the construction stages.

57.5.4 Rehabilitation

Aerodynamic design is a relatively new consideration in structural design. Some existing bridges have experienced wind problems because aerodynamic design was not considered in the original design. There are many measures to improve their aerodynamic behavior, such as structural stiffening, section streamlining, and installation of a damper. In the early days, structural stiffening was the major measure for this purpose. For example, the girders of the Golden Gate Bridge were stiffened in the 1950s, and Deer Isle Bridge in Maine has been stiffened since the 1940s by adding stays, cross-bracings, and strengthening the girders [10,11].

Although structural stiffening may have helped existing bridges survive many years of service, section streamlining has been commonly used recently. Streamlining the section is more efficient and less expensive than structural stiffening. Figure 57.9 shows the streamlined section of Deer Isle Bridge which has been proven very efficient [2,10].

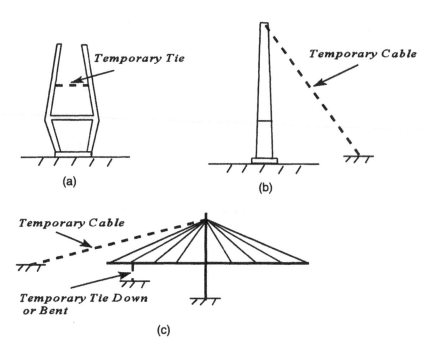

FIGURE 57.8   Typical construction stages. (a) Tower before completion; (b) completed tower; (c) stage with longest cantilever

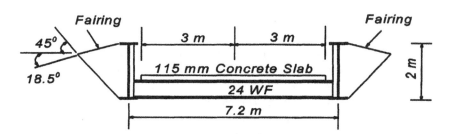

FIGURE 57.9   Deck section and fairings of Deer Isle Bridge.

57.5.5 Cable Vibration

A common wind-induced problem in long-span bridges is cable vibration. There are a number of wind-induced vibrations in cables, individually or as a group, such as vortex excitation, wake galloping, excitation of a cable by imposed movement of its extremities, rain/wind- and ice/wind-induced vibrations and buffeting of cables in strong turbulent winds.

While the causes of the cable vibrations are different from each other and the theoretical solutions complicated, some mitigating measures for these cable vibrations are shared:

1. *Raise damping.* This is an effective way for all kinds of cable vibrations. The cables are usually flexible and inherently low in damping; an addition of relatively small damping (usually at the cable ends) to the cable can dramatically reduce the vibration.
2. *Raise natural frequency.* The natural frequency depends on the cable length, the tension force, and the mass. Since the cable force and the mass are determined from the structural design, commonly the cable length is reduced by using spacers or cross cables.

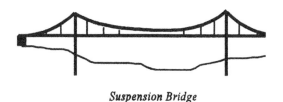

Suspension Bridge

FIGURE 57.10 Explanation of tuned mass damper.

3. *Change cable shape:* A change in the cable shape characteristics by increasing the surface roughness or adding protrusions to the cable surface reduces the rain/wind- and ice/wind-induced vibrations.
4. *Use other techniques:* Rearranging the cables or raising the cable mass density can also be used, but these are usually limited by other design constraints. Raising the mass may reduce the natural frequency, but it increases the damping and Scruton number ($m\zeta/\rho D^2$), and is overall beneficial.

57.5.6 Structural Control

Another way to improve aerodynamic behavior is to install either a passive or active control system on the bridges. A common practice in long-span bridges is the tuned mass damper (TMD). An example is the Normandy cable-stayed bridge in France. This bridge has a main span of 856 m. To reduce the horizontal vibration during construction due to buffeting, a TMD was installed. Wind tunnel testing showed that the TMD reduced the vibration by 30% [12,13].

The basic principles of a passive TMD are explained with an example shown in Figure 57.10. A TMD with spring stiffness k_2 and mass m_2 is attached to a structural mass m_1 which is excited by an external sinusoidal force $F\sin(\omega t)$. The vibration amplitude of this two-mass system is

$$X_1 = \frac{F\omega_m^2(\omega_d^2 - \omega^2)}{(\omega_d^2 - \omega^2)\left[(k_1 + k_2)\omega_m^2 - k_1\omega^2\right] - k_2\omega_d^2\omega_m^2}, \quad X_2 = \frac{F\omega_m^2\omega_d^2}{(\omega_d^2 - \omega^2)\left[(k_1 + k_2)\omega_m^2 - k_1\omega^2\right] - k_2\omega_d^2\omega_m^2} \quad (57.32)$$

where $\omega_m^2 = k_1/m_1$ and $\omega_d^2 = k_2/m_2$. It can be seen from Eq. (57.32) that by selection of the stiffness k_2 and mass m_2 such that ω_d equals ω, then the structure vibration X_1 is reduced to zero. Since the wind is not a single-frequency excitation, the TMD can reduce the vibration of bridges, but not to zero.

The performance of the passive TMD system can be enhanced by the addition of an active TMD, which can be done by replacing the passive damper device with a servo actuator system. The basic principle of active TMD is the feedback concept as used in modern control theory.

References

1. Berreby, D., The great bridge controversy, *Discover*, Feb., 26–33, 1992.
2. Cai, C. S., Prediction of Long-Span Bridge Response to Turbulent Wind, Ph.D. dissertation, University of Maryland, College Park, 1993.
3. Simiu, E. and Scanlan, R. H., *Wind Effects on Structures*, John Wiley & Sons, 2nd ed., New York, 1986.
4. Scanlan, R. H., State-of-the-Art Methods for Calculating Flutter, Vortex-Induced, and Buffeting Response of Bridge Structures, Report No. FHWA/RD-80/050, Washington, D.C., 1981.
5. Scruton, C., *An Introduction to Wind Effects on Structures*, Oxford University Press, New York, 1981.
6. Namini, A., Albrecht, P., and Bosch, H., Finite element-based flutter analysis of cable-suspended bridges, *J. Struct. Eng. ASCE*, 118(6), 1509–1526, 1992.

7. Lin, Y. K. and Ariaratnam, S. T., Stability of bridge motion in turbulent winds, *J. Struct. Mech.*, 8(1), 1–15, 1980.
8. Ehsan, F., Jones, N. J., and Scanlan, R. H., Effect of sidewalk vents on bridge response to wind, *J. Struct. Eng. ASCE*, 119(2), 484–504, 1993.
9. Irwin, P. A. and Stone, G. K., Aerodynamic improvements for plate-girder bridges, in *Proceedings, Structures Congress*, ASCE, San Francisco, CA, 1989.
10. Bosch, H. R., A Wind Tunnel Investigation of the Deer Isle-Sedgwick Bridge, Report No. FHWA-RD-87-027, Federal Highway Administration, McLean, VA, 1987.
11. Kumarasena, T., Scanlan, R. H., and Ehsan, F., Wind-induced motions of Deer Isle Bridge, *J. Struct. Eng. ASCE*, 117(11), 3356–3375, 1991.
12. Sorensen, L., The Normandy Bridge, the steel main span, in *Proc. 12th Annual International Bridge Conference*, Pittsburgh, PA, 1995.
13. Montens, S., Gusty wind action on balanced cantilever bridges, in *New Technologies in Structural Engineering*, Lisbon, Portugal, 1997.

58

Cable Force Adjustment and Construction Control

Danjian Han
*South China University
of Technology*

Quansheng Yan
*South China University
of Technology*

58.1   Introduction

Due to their aesthetic appeal and economic advantages, many cable-stayed bridges have been built over the world in the last half century. With the advent of high-strength materials for use in the cables and the development of digital computers for the structural analysis and the cantilever construction method, great progress has been made in cable-stayed bridges[1,2]. The Yangpu Bridge in China with a main span of 602 m completed in 1993, is the longest cable-stayed bridge with a composite deck. The Normandy Bridge in France, completed in 1994, with main span of 856 m is now the second-longest-span cable-stayed bridge. The Tatara Bridge in Japan, with a main span of 890 m, was opened to traffic in 1999. More cable-stayed bridges with larger spans are now in the planning.

Cable-stayed bridges are featured for their ability to have their behavior adjusted by cable stay forces [3–5]. Through the adjustment of the cable forces, the internal force distribution can be optimized to a state where the girder and the towers are compressed with little bending. Thus, the performance of material used for deck and pylons can be efficiently utilized.

0-8493-7434-0/00/$0.00+$.50
© 2000 by CRC Press LLC

During the construction of a cable-stayed bridge there are two kinds of errors encountered frequently,[6,13]: one is the tension force error in the jacking cables, and the other is the geometric error in controlling the elevation of the deck. During construction the structure must be monitored and adjusted; otherwise errors may accumulate, the structural performance may be substantially influenced, or safety concerns may arise. With the widespread use of innovative construction methods, construction control systems play a more and more important role in construction of cable-stayed bridges [18,19].

There are two ways of adjustment: adjustment of the cable forces and adjustment of the girder elevations [7]. The cable-force adjustment may change both the internal forces and the configuration of the structure, while the elevation adjustment only changes the length of the cable and does not induce any change in the internal forces of the structure.

This chapter deals with two topics: cable force adjustment and construction control. The methods for determing the cable forces are discussed in Section 58.2, then a presentation of the cable force adjustment is given in Section 58.3. A simulation method for a construction process of prestressed concrete (PC) cable-stayed bridge is illustrated in Section 58.4, and a construction control system is introduced in Section 58.5.

58.2 Determination of Designed Cable Forces

For a cable-stayed bridge the permanent state of stress in a structure subjected to dead load is determined by the tension forces in the cable stays. The cable tension can be chosen so that bending moments in the girders and pylons are eliminated or at least reduced as much as possible. Thus the deck and pylon would be mainly under compression under the dead loads [3,10].

In the construction period the segment of deck is corbeled by cable stays and each cable placed supports approximately the weight of one segment, with the length corresponding to the longitudinal distance between the two stays. In the final state the effects of other dead loads such as wearing surface, curbs, fence, etc., as well as the traffic loads, must also be taken into account. For a PC cable-stayed bridge, the long-term effects of concrete creep and shrinkage must also be considered [4].

There are different methods of determining the cable forces and these are introduced and discussed in the following.

58.2.1 Simply Supported Beam Method

Assuming that each stayed cable supports approximately the weight of one segment, corresponding to the longitudinal distance between two stays, the cable forces can be estimated conveniently [3,4]. It is necessary to take into account the application of other loads (wearing surface, curbs, fences, etc.). Also, the cable is placed in such a way that the new girder element is positioned correctly, with a view to having the required profile when construction is finished.

Due to its simplicity and easy hand calculation, the method of the simply supported beam is usually used by designers in the tender and preliminary design stage to estimate the cable forces and the area of the stays. For a cable-stayed bridge with an asymmetric arrangement of the main span and side span or for the case that there are anchorage parts at its end, the cable forces calculated by this method may not be evenly distributed. Large bending moments may occur somewhere along the deck and/or the pylons which may be unfavorable.

58.2.2 Method of Continuous Beam on Rigid Supports

By assuming that under the dead load the main girder behaves like a continuous beam and the inclined stay cables provide rigid supports for the girder, the vertical component of the forces in stay cables are equal to the support reactions calculated on this basis [4,10]. The tension in the anchorage cables make it possible to design the pylons in such a way that they are not subjected to large bending moments when the dead loads are applied.

This method is widely used in the design of cable-stayed bridges. Under the cable forces calculated by this method, the moments in the deck are small and evenly distributed. This is especially favorable for PC cable-stayed bridges because the redistribution of internal force due to the effects of concrete creep could be reduced.

58.2.3 Optimization Method

In the optimization method of determining the stresses of the stay cables under permanent loads, the criteria (objective functions) are chosen so the material used in girders and pylons is minimized [8,11]. When the internal forces, mainly the bending moments, are evenly distributed and small, the quantity of material reaches a minimum value. Also the stresses in the structure and the deflections of the deck are limited to prescribed tolerances.

In a cable-stayed bridge, the shear deformations in the girder and pylons are neglected, the strain energy can be represented by

$$U = \frac{1}{2}\int_0^L \frac{M^2}{2\,EI}dx + \frac{1}{2}\int_0^L \frac{N^2}{2\,EA}dx \tag{58.1}$$

where EI is the bending stiffness of girder and pylons and EA is the axial stiffness.

It can be given in a discrete form when the structure is simulated by a finite-element model as

$$U = \sum_{i=1}^{N} \frac{L_i}{4\,E_i}\left(\frac{M_{il}^2 + M_{ir}^2}{I_i} + \frac{N_{il}^2 + N_{ir}^2}{A_i}\right) \tag{58.2}$$

where N is the total number of the girder and pylon elements, L_i is the length of the ith element, E is the modulus of elasticity, I_i and A_i are the moment of inertia and the sections area, respectively. M_{il}, M_{ir}, N_{il}, N_{il} are the moments and the normal forces in the left and right end section of the ith element, respectively.

Under the application of dead loads and cable forces the bending moments and normal forces of the deck and pylon are given by

$$\{M\} = \{M_D\} + \{M_P\} = \{M_D\} + [S_M] * \{P_0\} \tag{58.3a}$$

$$\{N\} = \{N_D\} + \{N_P\} = \{N_D\} + [S_N] * \{P_0\} \tag{58.3b}$$

where $\{M_D\}$ and $\{M_P\}$ are the bending moment vectors induced by dead loads and the cable forces, respectively; $[S_M]$ is the moment influence matrix; $[S_N]$ is the normal force influence matrix, the component S_{ij} of influence matrix represents changes of the moment or the normal force in the ith element induced by the jth unit cable force. And $\{N_D\}, \{N_P\}$ are the normal force vectors induced by dead loads and cable forces, respectively. $\{P_0\}$ is the vector of cable forces.

The corresponding displacements in deck and pylon are given as

$$\{F\} = \{F_D\} + \{F_P\} = \{F_D\} + [S_F] * \{P_0\} \tag{58.4}$$

where $\{F\}$ is the displacement vector, $\{S_F\}$ is the displacement influence matrix, $\{F_D\}$ and $\{F_P\}$ are the displacement vectors induced by dead loads and by cable forces respectively.

Substitute Eqs. (58.3a) and (58.3b) into Eq. (58.2), and replace the variables by

$$\{\overline{M}\} = [A]\{M\}, \{\overline{N}\} = [B]\{N\} \tag{58.5}$$

in which $[A]$ and $[B]$ are diagonal matrices:

$$[A] = Diag\left[\sqrt{L_1/4E_1I_1}, \sqrt{L_2/4E_2I_2}, \ldots, \sqrt{L_n/4E_nI_n}\right]$$

$$[B] = Diag\left[\sqrt{L_1/4E_1A_1}, \sqrt{L_2/4E_2A_2}, \ldots, \sqrt{L_n/4E_nA_n}\right]$$

Then the strain energy of the cable-stayed bridge can be represented in matrix form as

$$U = \{P_0\}^T[\overline{S}]^T[\overline{S}]\{P_0\} + 2\{\overline{P}_D\}^T[\overline{S}]\{P_0\} + \{\overline{P}_D\}^T\{\overline{P}_D\} \tag{58.6}$$

in which $[\overline{S}] = (\overline{S}_M, \overline{S}_N)^T = [A, B](S_M, S_N)^T$, $\{\overline{P}_D\} = \{M_D, N_D\}^T$.

Now, we want to minimize the strain energy of structure, i.e., to let

$$\partial U/\partial P_0 = 0 \tag{58.7}$$

under the following constraint conditions:

1. The stress range in girders and pylons must satisfy

$$\{\sigma\}_L \leq \{\sigma\} \leq \{\sigma\}_U \tag{58.8}$$

   in which $\{\sigma\}$ is the maximum stress value vector. And $\{\sigma\}_L, \{\sigma\}_U$ are vectors of the lower and upper bounds.

2. The stresses in stay cables are limited so that the stays can work normally.

$$\{\sigma\}_{LC} \leq \left\{\tfrac{P_{0C}}{A_C}\right\} \leq \{\sigma\}_{UC} \tag{58.9}$$

   in which A_C is the area of a stay, P_{0C} is the cable force and $\{\sigma\}_{LC}, \{\sigma\}_{UC}$ represent the lower and upper bounds, respectively.

3. The displacements in the deck and pylon satisfy

$$\{|D_i|\} \leq \{\Delta\} \tag{58.10}$$

   in which the left hand side of Eqs. (58.10) is the absolute value of maximum displacement vector and the right-hand side is the allowable displacement vector.

Eqs. (58.6) and (58.7) in conjunction with the conditions (58.8) through (58.10) is a standard quadric programming problem with constraint conditions. It can be solved by standard mathematical methods.

Since the cable forces under dead loads determined by the optimization method are equivalent to the cable force under which the redistribution effect in the structure due to concrete creep is minimized [8], the optimization method is used more widely in the design of PC cable-stayed bridges.

58.2.4   Example

For a PC cable-stayed bridge as shown in Figures 58.1 and 58.2, the forces of cable stays under permanent loads (not taking into account the creep and shrinkage) can be determined by the above methods. The results obtained are shown in Figures 58.3 and 58.4. In these figures SB represents the "Simply Supported Beam Method," CBthe "Continuous Beam on Rigid Support Method," OPT the "Optimization Method," and M represents middle span, S side span numbering from the pylon location.

As can be seen, because the two ends of the cable-stayed bridge have anchored parts the cable forces located in these two regions obtained by the method of simply supported beam (SB) and by the method of continuous beam on rigid supports (CB) are not evenly distributed. The cable forces in the region near the pylon are very different with the three methods. In the other regions there is no prominent difference among the cable forces obtained by SB, CB, and OPT .

Generally speaking, the differences of cable forces under dead loads obtained by the above methods are not so significant. The method of simply supported beam is the most convenient and the easiest to use. The method of continuous beam on rigid supports is suitable to use in the design of PC cable-stayed bridges. The optimization method is based on a rigorous mathematical model. In practical engineering applications the choice of the above methods is very much dependent on the design stage and designer preference.

58.3   Adjustment of the Cable Forces

58.3.1   General

During the construction or service stage, many factors may induce errors in the cable forces and elevation of the girder, such as the operational errors in tensioning stays or the errors of elevation in laying forms [14,16,17]. Further, the discrepancies of parameter values between design and reality such as the modules of elasticity, the mass density of concrete, the weight of girder segments may give rise to disagreements between the real structural response and the theoretical prediction [13].If the structure is not adjusted to reduce the errors during construction, they may accumulate and the structure may deviate away from the intended design aim. Moreover, if the errors are greater than the allowable limits, they may give rise to the unfavorable effects to the structure. Through cable force adjustment, the construction errors can be eliminated or reduced to an allowable tolerance. In the service stage, because of concrete creep effects, cable force may need to be adjusted; thus, an optimal structural state can be reached or recovered.

58.3.2   Influence Matrix of the Cable Forces

Assuming that a unit amount of cable force is adjusted in one cable stay, the deformations and internal forces of the structure can be calculated by finite-element model. The vectors of change in deformations and internal forces are defined as influence vectors. In this way, the influence matrices can be formed for all the stay cables.

58.3.3   Linear Programming Method [7]

Assume that there are n cable stays whose cable forces are going to be adjusted, the adjustments are T_i $(i = 1,2,...,n)$, these values form a vector of cable force adjustment $\{T\}$ as

$$\{T\} = \{T_1, T_2, ..., T_n\}^T \tag{58.11}$$

Denote internal force influence vector $\{P_l\}$ as

$$\{P_l\} = \left(P_{l1}, P_{l2}, ..., P_{ln}\right)^T \quad (l = 1, 2, ..., m) \tag{58.12}$$

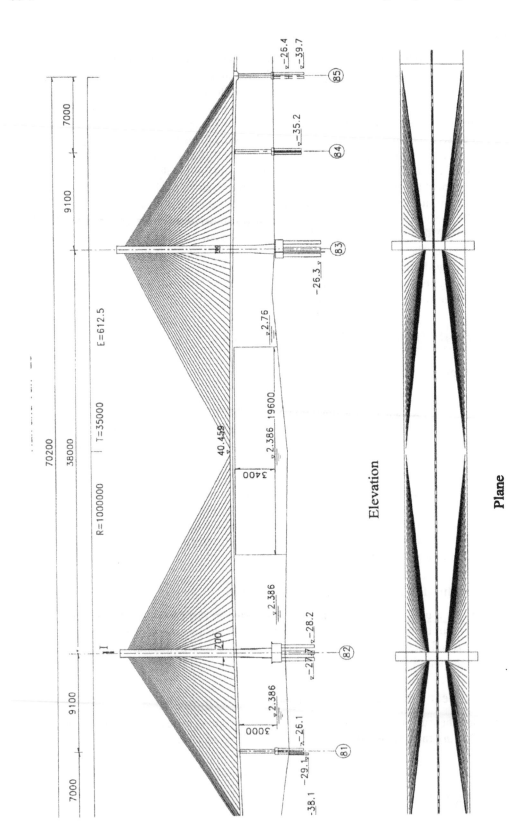

Elevation

Plane

FIGURE 58.1 General view of a PC cable-stayed bridge.

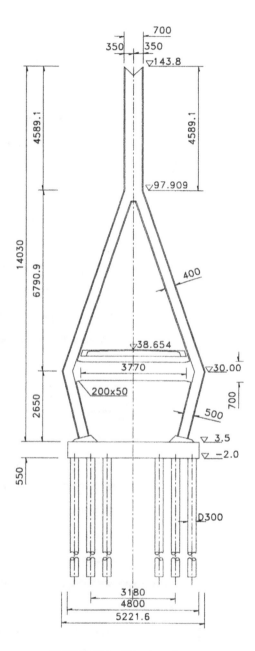

FIGURE 58.2 Side view of tower.

in which m is the number of sections of interest, P_{ij} is the internal force increment due to a unit tension of the jth cable. Denote displacement influence vector $\{D_i\}$ as

$$\{D_i\} = (D_{i1}, D_{i2}, ..., D_{in})^T \quad (i = 1, 2, ..., k) \tag{58.13}$$

in which k is the number of sections of interest, D_{ij} is the displacement increment at section i due to a unit tension of the jth cable. Thus, the influence matrices of internal forces and displacements are given by

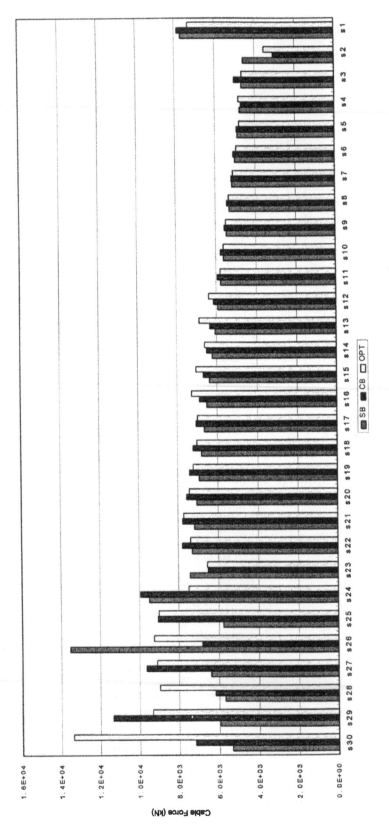

FIGURE 58.3 Comparison of the cable forces (kN) (side span).

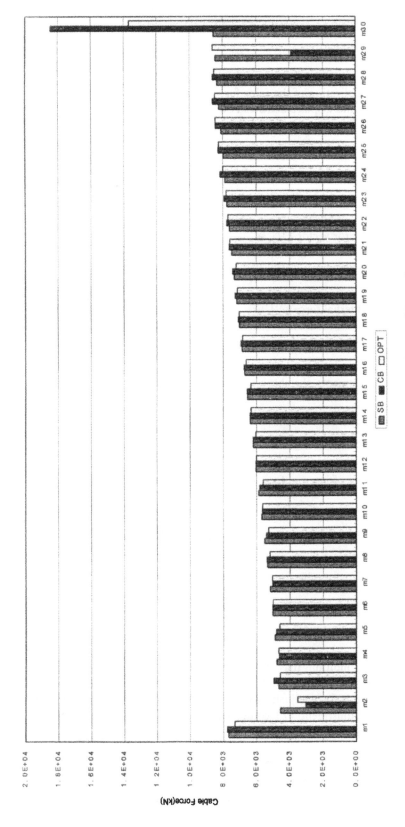

FIGURE 58.4 Comparison of the cable force (middle span).

$$\{P\} = \left(P_1, P_2, ..., P_m\right)^T \tag{58.14a}$$

$$\{D\} = \left(D_1, D_2, ..., D_k\right)^T \tag{58.14b}$$

respectively. Under application of the cable force adjustment $\{T\}$ (see Eq. (58.11)), the increment of the internal forces and displacements can be given by

$$\{\Delta P\} = \{P\}^T \{T\} \tag{58.15a}$$

$$\{\Delta D\} = \{D\}^T \{T\} \tag{58.15b}$$

respectively.

Denote deflection error vector $\{H\}$ as

$$\{H\} = \left(h_1, h_2, ..., h_m\right)^T \tag{58.16}$$

Denote vector of internal force $\{N\}$ as

$$\{N\} = \left(N_1, N_2, ..., N_m\right)^T \tag{58.17}$$

After cable force adjustments, the absolute values of the deflection errors are expressed by

$$\left|\lambda_k\right| = \left|\sum_{i=1}^{n} D_{ik} T_i - h_k\right| \tag{58.18}$$

and the absolute values of internal force errors are expressed by

$$\left|q_l\right| = \left|\sum_{i=1}^{n} P_{il} T_i - N_l\right| \tag{58.19}$$

The objective function for cable force adjustments may be defined as the errors of girder elevation, i.e.,

$$\min\left|\lambda_k\right| \tag{58.20}$$

and the constraint conditions may include limitations of the internal force errors, the upper and lower bounds of the cable forces, and the maximum stresses in girders and pylons. Then the optimum values of cable force adjustment can be determined by a linear programming model.

The value of cable adjustment $\{T\}$ could be positive for increasing or negative for decreasing of the cable forces. Introduce two auxiliary variables T_{1i}, T_{2i} as

$$T_i = T_{1i} - T_{2i} \quad T_{1i} \geq 0, \ T_{2i} \geq 0 \tag{58.21}$$

Substitute Eq. (58.18) into (58.20), then a linear program model is established by min: λ_k (58.22)

subject to: $\sum_{i=1}^{n} P_{li}(T_{1i} - T_{2i}) \geq N_l - \xi \bar{p}_l$ $(l = 1,2,...m)$ (58.23a)

$\sum_{i=1}^{n} P_{li}(T_{1i} - T_{2i}) \leq N_l + \xi \bar{p}_l$ $(l = 1,2,...m)$ (58.23b)

$\sum_{i=1}^{n} D_{ji}(T_{1i} - T_{2i}) \geq h_j - d_j$ $(j = 1,2,...k)$ (58.23c)

$\sum_{i=1}^{n} D_{ji}(T_{1i} - T_{2i}) \geq h_j - d_j$ $(j = 1,2,...k)$ (58.23d)

$T_{1i} - T_{2i} \leq \eta \bar{T}_i, \quad T_{1i} - T_{2i} \geq -\eta \bar{T}_i$ $(l = 1,2,...,n)$ (58.23e)

in which $\bar{p}_l$ is the design value of internal force at section l, ξ is the allowable tolerance in percentage of the internal force. $\bar{T}_i$ is the design value of the cable force, η is the allowable tolerance in percentage of the cable forces.

Equations (58.22) and (58.23) form a standard linear programming problem which can be solved by mathematical software.

58.3.4 Order of Cable Adjustment

The adjustment values can be determined by the above method; however, the adjustments must be applied at the same time to all cables, and a great number of jacks and workers are needed [7]. In performing the adjustment, it is preferred that the cable stays are tensioned one by one.

When adjusting the cable force individually, the influence of the other cable forces must be considered. And since any cable must be adjusted only one time, the adjustment values can be calculated through the influence matrix of cable force.

$$\{\bar{T}\} = [S]\{T\}$$ (58.24)

where $\{\bar{T}\} = \{\bar{T}_1, \bar{T}_2, ... \bar{T}_n\}$ is the vector of actual adjustment value of cable tension. $[S]$ is the influence matrix of cable tension, whose component S_{ij} represents tension change of the jth cable when the ith cable changes a unit amount of force.

58.4 Simulation of Construction Process

58.4.1 Introduction

Segmental construction techniques have been widely used in construction of cable-stayed bridges. In this technique, the pylon(s) is built first; then the girder segments are erected one by one and supported by the inclined cables hung from the pylon(s). It is evident that the profile of the main girder and the final tension forces in the cables are strongly related to the erection method and the construction scheme. It is therefore important that the designer should be aware of the construction process and the necessity to look into the structural performance at every stage of construction [9,12].

In any case, structural safety is the most important issue. Since the stresses in the girder and pylon(s) are related to the cable tensions. Thus the cable forces are of great concern. Further, during construction, the geometric profile of the girder is also very important. It is clear that if the profile of the girder were not smooth or, finally, the cantilever ends could not meet together, then the construction might experience some trouble. The profile of the girder or the elevation of the bridge segments is mainly controlled by the cable lengths. Therefore, the cable length must be appropriately set at the erection of each segment. It also should be noted that in the construction process, the internal forces of the structure and the elevation of the girder could vary because usually the bridge segments are built by a few components at a time and the erection equipment is placed at different positions during construction and because some errors such as the weight of the segment and the tension force of the cable, etc. may occur. Thus, monitoring and adjustment are absolutely needed.

To reach the design aim, an effective and efficient simulation of the construction process step by step is very necessary. The objectives of the simulation analysis are [4,12]:

1. To determine the forces required in cable stays at each construction stage;
2. To set the elevation of the girder segment;
3. To find the consequent deformation of the structure at each construction stage;
4. To check the stresses in the girder and pylon sections.

The simulation methods are introduced and discussed in detail in the following sections. In Section 58.4.2, the technique of forward analysis is presented to simulate the assemblage process. Creep effects can be considered; however, the design aim may not be successfully achieved by such simulation because it is not so easy to determine the appropriate lengths of the cable stays which make the final elevation to achieve the design profile automatically. Another technique presented in Section 58.4.3 is the backward disassemblage analysis, which starts with the final aim of the structural state and disassembles segment by segment in a reverse way. The disadvantage of this method is that the creep effects may not be able to be defined. However, values obtained from the assemblage process may be used in this analysis. These two methods may be alternatively applied until convergence is reached.

It is noted that the simulation is only limited to that of the erection of the superstructure.

58.4.2 Forward Assemblage Analysis

Following the known erection procedure, a simulation analysis can be carried out by the finite-element method. This is the so-called forward assemblage analysis. It has been used to simulate the erection process for PC bridges built by the free cantilever method.

Concerning finite-element modeling, the structure may be treated as a plane frame or a space frame [4]. A plane frame model may be good enough for construction simulation because transverse loads, such as wind, can generally be ignored. In a plane frame model, the pylon(s) and the girder are modeled by some beam elements, while the stays are modeled as two-node bar elements with Ernst modules [3,4] by which the effects of cable sag can be taken into account. The structural configuration is changed stage by stage. Typically, in one assemblage stage, a girder segment treated as one or several beam elements is connected to the existing structure, while its weight is treated as a load to apply to the element. Also, the cable force is applied. Then an analysis is performed and the structure is changed to a new configuration.

In finite-element modeling, several factors such as the construction loads (weight of equipment and traveling carriage) and effects of concrete creep and shrinkage, must be considered in detail.

Traveling carriages are specially designed for construction of a particular bridge project. Generally there are two kinds of carriages. One is cantilever type (Figure 58.5a). The traveling carriages are mounted near the ends of girders, like a cantilever to support the next girder segment. In this case, the weight of the carriage can be treated as an external load applied to the end of the girder.

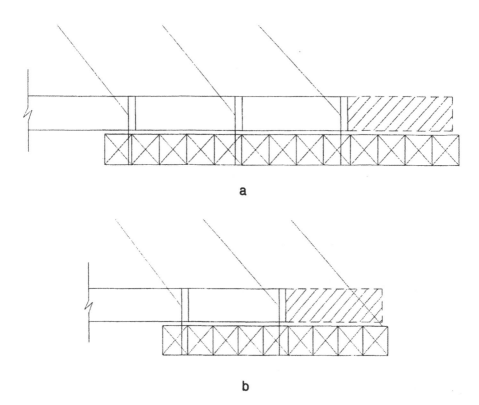

FIGURE 58.5 (a) Cantilever carriage; (b) cable-supported cantilever carriage.

With the development of multiple cable systems, the girder with lower height becomes more flexible. The girder itself is not able to carry the cantilever weights of the carriage and the segment. Then an innovative erection technique was proposed[1]. And another type of carriage is developed. This new idea is to use permanent stays to support the form traveler (Figure 58.5b) so that the concrete can be poured *in situ* [1,9]. This method enjoys considerable success at present because of the undeniable economic advantages. Its effectiveness has been demonstrated by many bridge practice. For the erecting method using the later type of carriage, the carriage works as a part of the whole structure when the segmental girder is poured *in situ*. Thus, the form traveler must be included in the finite-element model to simulate construction. A typical flowchart of forward assemblage analysis is shown in Figure 58.6.

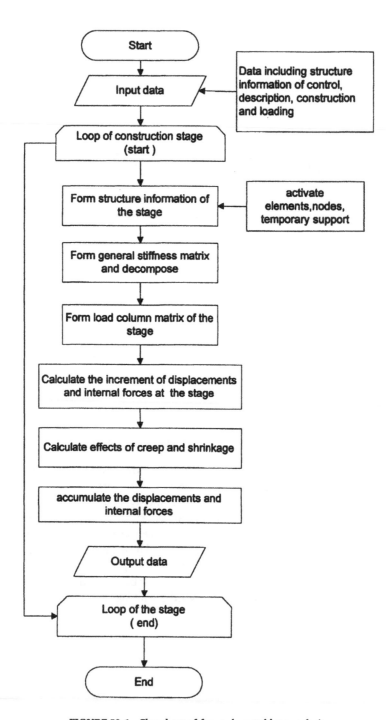

FIGURE 58.6 Flowchart of forward assemblage analysis.

With the forward assemblage analysis, the construction data can be worked out. And the actual permanent state of cable-stayed bridges can be reached. Further, if the erection scheme were modified during the construction period or in the case that significant construction error occurred, then the structural parameters or the temporary erection loads would be different from the values used in the design. It is possible to predict the cable forces and the sequential deformations at each stage by utilizing the forward assemblage analysis.

58.4.3 Backward Disassemblage Analysis

Following a reverse way to simulate the disassemblage process stage by stage, a backward analysis can be carried out also by finite-element method [11,12]. Not only the elevations of deck but also the length of cable stays and the initial tension of stays can be worked out by this method. And the completed state of structure at each stage can be evaluated.

The backward disassemblage analysis starts with a very ideal structural state in which it is assumed that all the creep and shrinkage deformation of concrete be completed, i.e., a state 5 years or 1500 days after the completion of the bridge construction. The structural deformations and internal forces at each stage are considered ideal reference states for construction of the bridge [11]. The backward analysis procedure for a PC cable-stayed bridge may be illustrated as follows:

Step 1. Compute the permanent state of the structure.
Step 2. Remove the effects of the creep and shrinkage of concrete of 1500 days or 5 years.
Step 3. Remove the second part of the dead loads, i.e., the weights of wearing surfacing, curbs and fence, etc.
Step 4. Apply the traveler and other temporary loads and supports.
Step 5. Remove the center segment, to analyze the semistructure separately.
Step 6. Move the form traveler backward.
Step 7. Remove the weight of the concrete of a pair of segments.
Step 8. Remove the cable stay.
Step 9. Remove the corresponding elements.

Repeat the Steps 6 to 9 until all the girder segments are disassembled. A flowchart for backward disassemblage analysis is shown in Figure 58.7.

As mentioned above, for the erecting method using conventional form traveler cast-in-place or precast concrete segments, the crane or the form traveler may be modeled as external loads. Thus the carriage moving is equivalent to a change of the loading position. However, for an erection method utilizing cable-supported traveling carriage the cable stays first work as supports of the carriage and later, after curing is finished, the cable stays are connected with the girder permanently. In backward disassemblage analysis, the form traveler moving must be related to a change of the structural system.

The backward analysis procedure can establish the necessary data for the erection at each stage such as the elevations of deck, the cable forces, the deformations of structure, and the stresses at critical sections of deck and pylon.

One of the disadvantages of backward analysis is that creep effects are not able to be estimated; therefore, forward and backward simulations should be used alternately to determine the initial tension and the length of stay cables.

58.5 Construction Control

58.5.1 Objectives and Control Means

Obviously, the objective of construction control is to build a bridge that achieves the design aim with acceptable error. During the construction of a cable-stayed bridge, some discrepancies may occur between the actual state and the state of design expectation [14,15]. The discrepancies may arise from elevation error in laying forms, errors in stressing cable stays by jacks, errors of the first part of the dead load, i.e., the self-weights of the girder segments, and the second part of the dead load, i.e., the self-weights of the surfacing, curbs and fencing, etc. On the other hand, a system error may occur in measuring the deflection of the girder and the pylons. It is impossible to eliminate all the errors. Actually, there are two basic requirements for the completed structure [12]: (1) the geometric profile matches the designed shape well and (2) the internal forces are within the designed

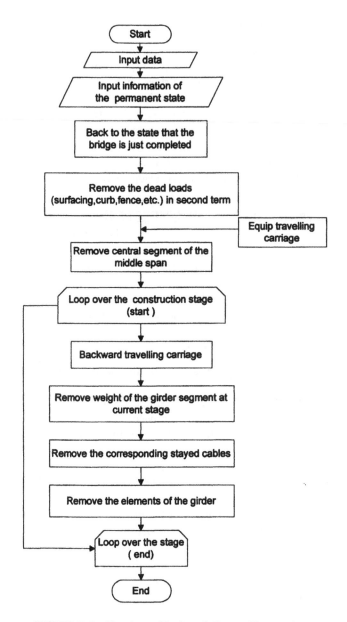

FIGURE 58.7   Flowchart of backward disassemblage analysis.

envelope values; specifically, the bending moments of the girder and the pylons are small and evenly distributed.

Since the internal forces of the girder and the pylons are closely related to the cable forces, the basic method of construction control is to adjust the girder elevation and the cable forces. If the error of the girder elevation deviated from the design value is small, such error can be reduced or eliminated by adjusting the elevation of the segment without inducing an angle between two adjacent deck segments. In this way we only change the geometric position of the girder without changing the internal force state of the structure. When the errors are not small, it is necessary to adjust the cable forces. In this case, both the geometric position change and the changes of internal forces occur in the structure.

Nevertheless, cable force adjustment are not preferable because they may take a lot of time and money. The general exercise at each stage is to find out the correct length of the cables and set the elevation of the segment appropriately. Cable tensioning is performed for the new stays only. Generally, a comprehensive adjustment of all the cables is only applied before connecting the two cantilever ends [21]. In case a group of cables needs to be adjusted, careful planning for the adjustment based on a detailed analysis is absolutely necessary.

58.5.2 Construction Control System

To guarantee structural safety and to reach the design aim, a monitoring and controlling system is important [13,15,19]. A typical construction controlling system consists of four subsystems: measuring subsystem, error and sensitivity analysis subsystem, control/prediction subsystem, and new design value calculation subsystem. An example of a construction control system for a PC cable-stayed bridge [18,20] is shown in Figure 58.8.

1. Measuring subsystem — The measuring items mainly include the elevation/deflection of the girder, the cable forces, the horizontal displacement of the pylon(s), the stresses of sections in the girder and the pylon(s), the modulus of elasticity and mass density of concrete, the creep and shrinkage of concrete, the temperature/temperature gradient in the structure.
2. Error and sensitivity analysis subsystem — In this subsystem, the temperature effects are first determined and removed. Then the sensitivity of structural parameter such as the elasticity modulus of concrete, self-weight, stiffness of the girder segment or pylon, etc., are analyzed. Through the analysis, the causes of errors can be found so that the corresponding adjustment steps are utilized.
3. Control/prediction subsystem — Compare the measuring values with those of the design expectation; if the differences are lower than the prescribed limits, then go to the next stage. And the elevation is determined appropriately. Otherwise, it is necessary to find out the reasons, then to eliminate or reduce those errors through proper measures. The magnitude of cable tension adjustment can be determined by a linear programming model.
4. New design value calculation subsystem — Since the structural parameters for the completed part have deviated from the designed values, the design expectation must be updated with the changed state of the structure. And the sequential construction follows new design values so that the final state of structure can be achieved optimally.

58.6 An Engineering Example

The general view of a PC cable-stayed bridge is shown in Figure 58.1. The cable-stayed portion of the bridge has a total length of 702 m. The main span between towers is 380 m and the side anchor span is 161 m. The side anchor spans consist of two spans of 90 and 71 m with an auxiliary pier. The main girder is composed with two edge girders and a deck plate. The edge girders are laterally stiffened by a T-shaped PC girder with 6-m spacing. The edge girder is a solid section whose height is 2.2 m and width varies from 2.6 m at intersection of girder and pylon to 2.0 m at middle span. The deck plate is 28 cm thick. The width of the deck is 37.80 m out-to-out with eight traffic lanes. Spatial 244 stay cables are arranged in a semifan configuration. The pylon is shaped like a diamond with an extension mast (see Figure 58.1). All the cable stays are anchored in the mast part of the pylon. The stay cables are attached to the edge girders at 6 m spacing.

At the side anchor span an auxiliary pier is arranged to increase the stiffness of the bridge. And an anchorage segment of deck is set up to balance the lifting forces from anchorage cables.

58.6.1 Construction Process

The bridge deck structure is erected by the balance cantilevering method utilizing cable-supported form travelers. The construction process is briefly described as follows.

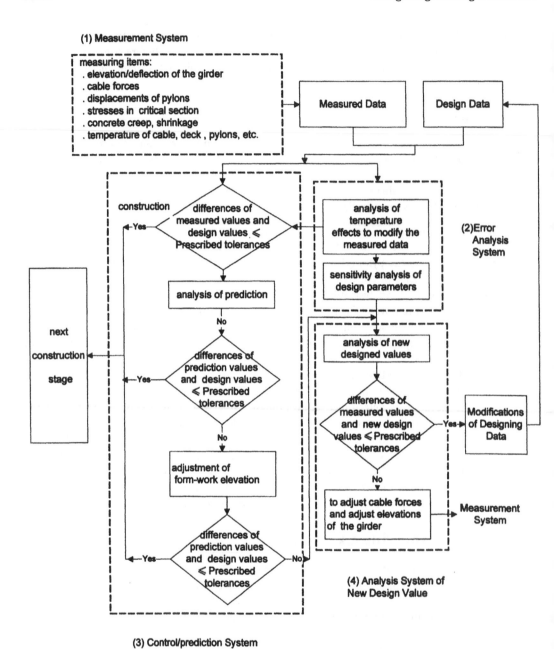

FIGURE 58.8 A typical construction control system for a PC cable-stayed bridge.

- Build the towers.
- Cast in place the first segment on timbering support.
- Erect the No. 1 cable and stress to its final length.
- Hoist the traveling carriages and positioning.
- Erect the girder segments one by one on the two sides of the pylons.
- Connect the cantilever ends of the side span with the anchorage parts.
- Continue to erect the girder segments in the center span.

- Connect the cantilever ends of the center span.
- Remove traveling carriages and temporary supports.
- Connect the girder with the auxiliary piers.
- Cast pavement and set up fence, etc.

A typical erection stage of one segment is described as follows:

- Move the traveling carriage forward and set up the form at proper levels.
- Erect and partially stress the stay cables attached to the traveler.
- Place reinforcement, post-tensioning bars and couple the stressed bars with those of the previously completed deck segment.
- Cast in place the deck concrete.
- Stress the stay cables to adjust the girder segments to proper levels.
- Cure deck panel and stress the longitudinal and lateral bars and strands.
- Loosen the connection between the stay cable and traveling carriage.
- Stress cable stays to the required value.

The above erection steps are repeated until the bridge is closed at the middle span.

58.6.2 Construction Simulation

The above construction procedure can be simulated stage by stage as illustrated in Section 58.4.2. Since creep and shrinkage occur and the second part of the dead weight will be loaded on the bridge girder after completion of the structure, a downward displacement is induced. Therefore, as the erection is just finished the elevation of the girder profile should be set higher than that of the design profile and the pylons should be leaning toward the side spans. In this example, the maximum value which is set higher than the designed profile in the middle of the bridge is about 35.0 cm, while the displacement of pylon top leaning to anchorage span is about 9.0 cm. The initial cable forces are listed in Table 58.1 to show the effects of creep. As can be seen, considering the long-term effects of concrete creep, the initial cable forces are a little greater than those without including the time-dependent effects.

58.6.3 Construction Control System

In the construction practice of this PC cable-stayed bridge, a construction control system is employed to control the cable forces and the elevation of the girder. Before starting concrete casting the reactions of the cable-supported form traveler are measured by strain gauge equipment. Thus the weights of the four travelers used in this bridge are known.

At each stage the mass density of concrete and the elasticity modulus of concrete are tested in the laboratory *in situ*. The calculation of construction is carried out with the measured parameters. In several sections of the deck and pylon, strain gauges are embedded to measure the strains of the structure during the whole construction period, thus the stress of the structure can be monitored.

In this example the main flowchart of the construction control system for a typical erecting segment is shown in Figure 58.9.

TABLE 58.1 Predicted Initial Cable Forces (kN)

No.	S1	S2	S3	S4	S5	S6	S7	S8	S9	S10
NCS	7380	3733	4928	5130	5157	5373	5560	5774	5969	6125
CS	7745	3854	5101	5331	5348	5549	5694	5873	6050	6185

No.	S11	S12	S13	S14	S15	S16	S17	S18	S19	S20
NCS	6320	6870	7193	7283	7326	7519	7317	7478	7766	8035
CS	6367	6908	7209	7331	7467	7639	7429	7580	7856	8114

No.	S21	S22	S23	S24	S25	S26	S27	S28	S29	S30
NCS	8366	8169	7693	8207	9579	9649	9278	9197	9401	12820
CS	8442	8246	7796	8354	9778	9791	9509	9296	9602	12930

No.	M1	M2	M3	M4	M5	M6	M7	M8	M9	M10
NCS	7334	3662	4782	4970	5035	5426	5479	5628	5778	6108
CS	7097	3433	4591	4877	5006	5477	5567	5736	5904	6231

No.	M11	M12	M13	M14	M15	M16	M17	M18	M19	M20
NCS	6139	6572	6627	6918	6973	7353	7540	7742	7882	7978
CS	6263	6706	6756	7031	7088	7422	7624	7813	7942	8060

No.	M21	M22	M23	M24	M25	M26	M27	M28	M29	M30
NCS	8384	8479	8617	8833	9175	9359	9394	9480	9641	13440
CS	8452	8561	8694	8931	9212	9413	9459	9570	9716	13570

No.: Cable number; M: middle span; S: side span; CS: with the effects of creep and shrinkage of concrete; NCS: without the effects of creep and shrinkage of concrete.

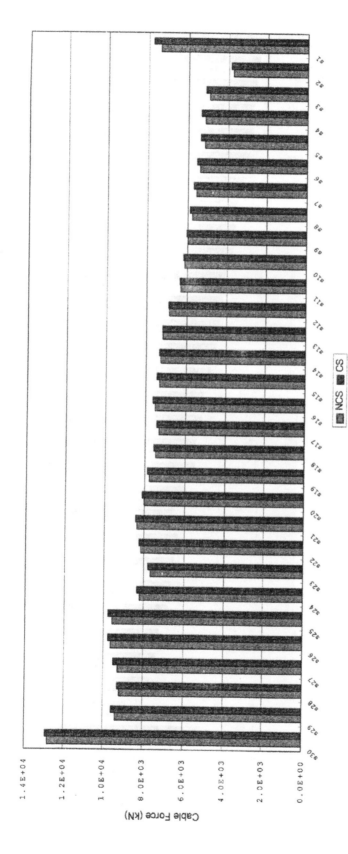

FIGURE 58.9 (a) Initial cable forces in side span determined by simulation analysis. (b) Initial cable forces in middle span determined by simulation analysis.

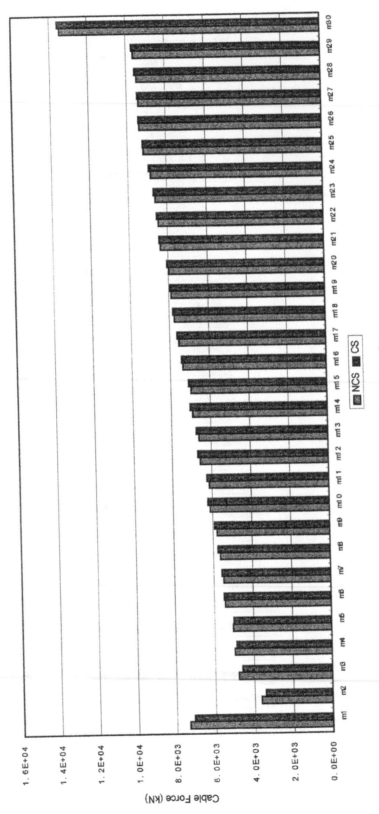

FIGURE 58.9 (continued)

References

1. Tang, M.C. The 40-year evolution of cable-stayed bridges, in *International Symposium on Cable-Stayed Bridges,* Lin Yuanpei et al., Eds., Shanghai, 1994, 30–11.
2. Leonhardt, F. and Zellner, W., Past, present and future of cable-stayed bridges, in *Cable-Stayed Bridges, Recent Developments and Their Future,* M. Ito et al., Eds., Elsevier Science Publishers, New York, 1991.
3. Podolny, W. and Scalmi, J., *Construction and Design of Cable-Stayed Bridges,* John Wiley & Sons, New York, 1983.
4. Walther, R., Houriet, B., Lsler, W., and Moia, P., *Cable-Stayed Bridges,* Thomas Telford, London, 1988.
5. Gimsing, N. J., *Cable Supported Bridges, Concept and Design,* John Wiley & Sons, New York, 1983.
6. Kasuga, A., Arai, H., Breen, J. E., and Furukawa, K., Optimum cable-force adjustment in concrete cable-stayed bridges, *J. Struct. Eng.,* ASCE, 121(4), 685–694, 1995.
7. Ma, W. T., Cable Force Adjustment and Construction Control of PC Cable-Stayed Bridges, Ph.D. dissertation of Department of Civil Engineering, South China University of Technology, 1997 [in Chinese].
8. Wang, X. W. et al., A study of determination of cable tension under dead loads, *Bridge Constr.,* 4, 1–5, 1996 [in Chinese].
9. Yan, D. H. et al., Simulation analysis of Tongling Cable-Stayed Bridge for construction control, in *National Symposium on Highway Bridge,* Dai Jing, Ed., Beijing Renming Jiaotong Press, Guangzhou, 347–355, 1995. [in Chinese].
10. Zhou, L. X. et al. Prestressed Concrete Cable-Stayed Bridges, Beijing Renming Jiaotong Press, 1989 [in Chinese].
11. Xiao, R. C., Ling, P. Application of computational structural mechanics in construction design and control of bridge structures, *Comput. Struct. Mech. Appl.,* 10(1) 92–98, 1993 [in Chinese].
12. Fang, Z. and Liu, G. D., A Study of Construction Control System of Cable-Stayed Bridges, Research Report of Department of Civil Engineering, Hunan University, 1995 [in Chinese].
13. Chen, D. W., Xiang, H. F., and Zheng, X. G., Construction control of PC cable-stayed bridge, J. Civil Eng., 26(1) 1–11, 1993 [in Chinese].
14. Yoshimura, M., Ueki, Y., and Imai, Y., Design and construction of a prestressed concrete cable-stayed bridge: the Tsukuhara Ohashi Bridge, *J. Jpn. Prestressed Concrete Eng. Assoc.,* Tokyo, Japan, 29(1) 1987 [in Japanese].
15. Fujisawa, N. and Tomo, H., Computer-aided cable adjustment of cable-stayed bridges, *IABSE Proc.,* P-92/85, 1985.
16. Furukawa, K., Inoue, K., Nakkkayama, H., and Ishido, K., Studies on the management system of cable-stayed bridges under construction using multi-objective programming method. *Proc. JSCE,* Tokyo, Japan, 374(6), 1986 [in Japanese].
17. Furuta, H. et al., Application on fuzzy mathematical programming to cable tension adjustment of cable-stayed bridges, in *International Symposium on Cable-Stayed Bridges,* Lin Yuanpei et al. Eds., Shanghai, 1994, 584–595.
18. Takuwa, I. et al., Prestressed concrete cable-stayed bridge constructed on an expressway — the Tomei Ashigra Bridge, in *Cable-Stayed Bridges, Recent Developments and Their Future.* M. Ito et al., Eds., Elsevier Science Publishers, New York, 1991.
19. Yasuhiro, K. et al., Construction of Tokachi Ohashi Bridge Superstructure (PC cable-stayed bridge), *Bridge Found.,* 1, 7–15, 1995 [in Japanese].
20. Hidemi, O. et al., Construction of Ikara Bridge superstructure (PC cable-stayed bridge), *Bridge Found.,* No. 11, 7–14, 1995 [in Japanese].
21. Fushimi, T., et al., Erection of the Tsurumi Fairway Bridge superstructure, *Bridge Found.,* 10, 2–10, 1994 [in Japanese].

59

Active Control in Bridge Engineering

Zaiguang Wu
*California Department of
Transportation*

59.1   Introduction

In bridge engineering, one of the constant challenges is to find new and better means to design new bridges or to strengthen existing ones against destructive natural effects. One avenue, as a traditional way, is to design bridges based on strength theory. This approach, however, can sometimes be untenable both economically and technologically. Other alternatives, as shown in Chapter 41, include installing isolators to isolate seismic ground motions or adding passive energy dissipation devices to dissipate vibration energy and reduce dynamic responses. The successful application of these new design strategies in bridge structures has offered great promise [11]. In comparison with passive energy dissipation, research, development, and implementation of active control technology has a more recent origin. Since an active control system can provide more control authority and adaptivity than a passive system, the possibility of using active control systems in bridge engineering has received considerable attention in recent years.

Structural control systems can be classified as the following four categories [6]:

- **Passive Control** — A control system that does not require an external power source. Passive control devices impart forces in response to the motion of the structure. The energy in a passively controlled structural system cannot be increased by the passive control devices.

0-8493-7434-0/00/$0.00+$.50
© 2000 by CRC Press LLC

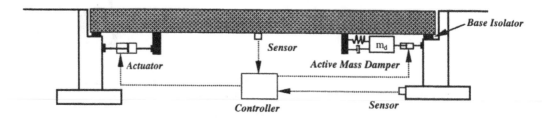

FIGURE 59.1 Base-isolated bridge with added active control system.

- **Active Control** — A control system that does require an external power source for control actuator(s) to apply forces to the structure in a prescribed manner. These controlled forces can be used both to add and to dissipate energy in the structure. In an active feedback control system, the signals sent to the control actuators are a function of the response of the system measured with physical sensors (optical, mechanical, electrical, chemical, etc.).
- **Hybrid Control** — A control system that uses a combination of active and passive control systems. For example, a structure equipped with distributed viscoelastic damping supplemented with an active mass damper on or near the top of the structure, or a base-isolated structure with actuators actively controlled to enhance performance.
- **Semiactive Control** — A control system for which the external energy requirements are an order of magnitude smaller than typical active control systems. Typically, semiactive control devices do not add mechanical energy to the structural system (including the structure and the control actuators); therefore, bounded-input and bounded-output stability is guaranteed. Semiactive control devices are often viewed as controlled passive devices.

Figure 59.1 shows an active bracing control system and an active mass damper installed on each of the abutments of a seismically isolated concrete box-girder bridge [8]. As we know, base isolation systems can increase the chances of the bridge surviving a seismic event by reducing the effects of seismic vibrations on the bridge. These systems have the advantages of simplicity, proven reliability, and no need for external power for operation. The isolation systems, however, may have difficulties in limiting lateral displacement and they impose severe constraints on the construction of expansion joints. Instead of using base isolation, passive energy dissipation devices, such as viscous fluid dampers, viscoelastic dampers, or friction dampers, can also be employed to reduce the dynamic responses and improve the seismic performance of the bridge. The disadvantage of passive control devices, on the other hand, is that they only respond passively to structural systems based on their designed behaviors.

The new developed active systems, a typical example as shown in Figure 59.1, have unique advantages. Based on the changes of structural responses and external excitations, these intelligent systems can actively adapt their properties and controlling forces to maximize the effectiveness of the isolation system, increase the life span of the bridge, and allow it to withstand extreme loading effects. Unfortunately, in an active control system, the large forces required from the force generator and the necessary power to generate these forces pose implementation difficulties. Furthermore, a purely active control system may not have proven reliability. It is natural, therefore, to combine the active control systems (Figure 59.1) with abutment base isolators, which results in the so-called hybrid control. A hybrid control system is more reliable than a purely active system, since the passive devices can still protect the bridge from serious damage if the active portion fails during the extreme earthquake events. But the installation and maintenance of the two different systems are the major shortcoming in a hybrid system. Finally, if the sliding bearings are installed at the bridge abutments and if the pressure or friction coefficient between two sliding surfaces can be adjusted actively based on the measured bridge responses, this kind of controlled bearing will then be known as semiactive control devices. The required power supply essential for signal processing and mechanical operation

TABLE 59.1   Bridge Control Systems

Systems	Typical Devices	Advantages	Disadvantages
Passive	Elastomeric bearings Lead rubber bearings Metallic dampers Friction dampers Viscoelastic dampers Tuned mass dampers Tuned liquid dampers	Simple Cheap Easy to install Easy to maintain No external energy Inherently stable	Large displacement Unchanged properties
Active	Active tendon Active bracing Active mass damper	Smart system	Need external energy May destabilize system Complicated system
Hybrid	Active mass damper + bearing Active bracing + bearing Active mass damper + VE damper	Smart and reliable	Two sets of systems
Semiactive	Controllable sliding bearings Controllable friction dampers Controllable fluid dampers	Inherently stable Small energy required Easy to install	Two sets of systems

is very small in a semiactive control system. A portable battery may have sufficient capacity to store the necessary energy before an earthquake event. This feature thus enables the control system to remain effective regardless of a major power supply failure. Therefore, the semiactive control systems seem quite feasible and reliable.

The various control systems with their advantages and disadvantages are summarized in Table 59.1.

Passive control technologies, including base isolation and energy dissipation, are discussed in Chapter 41. The focus of this chapter is on active, hybrid, and semiactive control systems. The relationships among different stages during the development of various intelligent control technologies are organized in Figure 59.2. Typical control configurations and control mechanisms are described first in Section 59.2. Then, the general control strategies and typical control algorithms are presented in Section 59.3, along with discussions of practical concerns in actual bridge applications of active control strategies. The analytical development and numerical simulation of various control systems applied on different types of bridge structures are shown as case studies in Section 59.4. Remarks and conclusions are given in Section 59.5.

59.2   Typical Control Configurations and Systems

As mentioned above, various control systems have been developed for bridge vibration control. In this section, more details of these systems are presented. The emphasis is placed on the motivations behind the development of special control systems to control bridge vibrations.

59.2.1   Active Bracing Control

Figure 59.3 shows a steel truss bridge with several actively braced members [1]. Correspondingly, the block diagram of the above control system is illustrated in Figure 59.4. An active control system generally consists of three parts. First, *sensors*, like human eyes, nose, hands, etc., are attached to the bridge components to measure either external excitations or bridge response variables. Second, *controllers*, like the human brain, process the measured information and compute necessary actions needed based on a given control algorithm. Third, *actuators*, usually powered by external sources, produce the required control forces to keep bridge vibrations under the designed safety range.

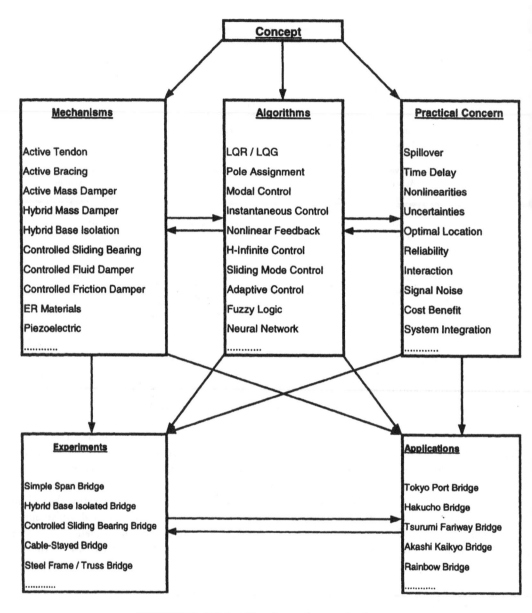

FIGURE 59.2   Relationship of control system development.

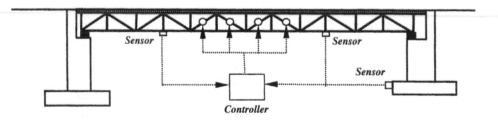

FIGURE 59.3   Active bracing control for steel truss bridge.

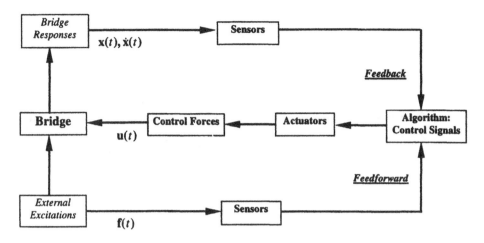

FIGURE 59.4 Block diagram of active control system.

Based on the information measured, in general, an active control system may be classified as three different control configurations. When only the bridge response variables are measured, the control configuration is referred to as *feedback control* since the bridge response is continually monitored and this information is used to make continuous corrections to the applied control forces. On the other hand, when only external excitations, such as earthquake accelerations, are measured and used to regulate the control actions, the control system is called *feedforward control*. Of course, if the information on both the response quantities and excitation are utilized for control design, combining the previous two terms, we get a new term, *feedback/feedforward control*. A bridge equipped with an active control system can adapt its properties based on different external excitations and self-responses. This kind of self-adaptive ability makes the bridge more effective in resisting extraordinary loading and relatively insensitive to site conditions and ground motions. Furthermore, an active control system can be used in multihazard mitigation situations, for example, to control the vibrations induced by wind as well as earthquakes.

59.2.2 Active Tendon Control

The second active control configuration, as shown in Figure 59.5, is an active tendon control system controlling the vibrations of a cable-stayed bridge [17,18]. Cable-stayed bridges, as typical flexible bridge structures, are particularly vulnerable to strong wind gusts. When the mean wind velocity reaches a critical level, referred to as the flutter speed, a cable-stayed bridge may exhibit vibrations with large amplitude, and it may become unstable due to bridge flutter. The mechanism of flutter is attributed to "vortex-type" excitations, which, coupled with the bridge motion, generate motion-dependent aerodynamic forces. If the resulting aerodynamic forces enlarge the motion associated with them, a self-excited oscillation (flutter) may develop. Cable-stayed bridges may also fail as a result of excessively large responses such as displacement or member stresses induced by strong earthquakes or heavy traffic loading. The traditional methods to strength the capacities of cable-stayed bridges usually yield a conservative and expensive design. Active control devices, as an alternative solution, may be feasible to be employed to control vibrations of cable-stayed bridges. Actuators can be installed at the anchorage of several cables. The control loop also includes sensors, controller, and actuators. The vibrations of the bridge girder induced by strong wind, traffic, or earthquakes are monitored by various sensors placed at optimal locations on the bridge. Based on the measured amplitudes of bridge vibrations, the controller will make decisions and, if necessary, require the actuators to increase or decrease the cable tension forces through hydraulic servomechanisms. Active tendon control seems ideal for the suppression of vibrations in a cable-stayed bridge since the existing stay cables can serve as active tendons.

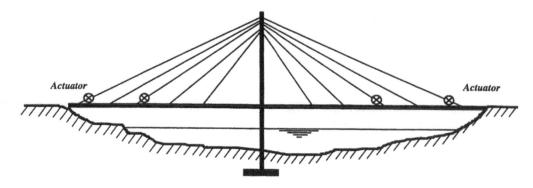

FIGURE 59.5   Active tendon control for cable-stayed bridge.

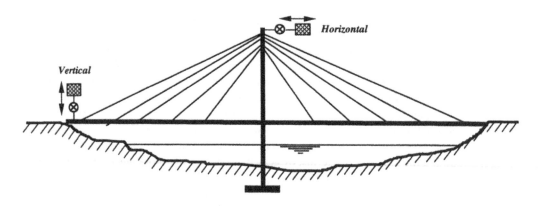

FIGURE 59.6   Active mass damper on cable-stayed bridge.

59.2.3   Active Mass Damper

Active mass damper, which is a popular control mechanism in the structural control of buildings, can be the third active control configuration for bridge structures. Figure 59.6 shows the application of this system in a cable-stayed bridge [12]. Active mass dampers are very useful to control the wind-induced vibrations of the bridge tower or deck during the construction of a cable-stayed bridge. Since cable-stayed bridges are usually constructed using the cantilever erection method, the bridge under construction is a relatively unstable structure supported only by a single tower. There are certain instances, therefore, where special attention is required to safeguard against the external dynamic forces such as strong wind or earthquake loads. Active mass dampers can be especially useful for controlling this kind of high tower structure. The active mass damper is the extension of the passive tuned mass damper by installing the actuators into the system. Tuned mass dampers (Chapter 41) are in general tuned to the first fundamental period of the bridge structure, and thus are only effective for bridge control when the first mode is the dominant vibration mode. For bridges under seismic excitations, however, this may not be always the case since the vibrational energy of an earthquake is spread over a wider frequency band. By providing the active control forces through the actuators, multimodal control can be achieved, and the control efficiency and robustness will be increased in an active mass damper system.

59.2.4   Seismic Isolated Bridge with Control Actuator

An active control system may be added to a passive control system to supplement and improve the performance and effectiveness of the passive control. Alternatively, passive devices may be installed in an active control scheme to decrease its energy requirements. As combinations of active and passive systems, hybrid control systems can sometimes alleviate some of the limitations and restrictions that exist in either an active or a passive control system acting alone. Base isolators are finding more and more applications in bridge engineering. However, their shortcomings are also becoming clearer. These include (1) the relative displacement of the base isolator may be too large to satisfy the design requirements, (2) the fundamental frequency of the base-isolated bridge cannot vary to respond favorably to different types of earthquakes with different intensities and frequency contents, and (3) when bridges are on a relatively soft ground, the effectiveness of the base isolator is limited. The active control systems, on the other hand, are capable of varying both the fundamental frequency and the damping coefficient of the bridge instantly in order to respond favorably to different types of earthquakes. Furthermore, the active control systems are independent of the ground or foundation conditions and are adaptive to external ground excitations. Therefore, it is natural to add the active control systems to the existing base-isolated bridges to overcome the above shortcomings of base isolators. A typical setup of seismic isolators with a control actuator is illustrated at the left abutment of the bridge in Figure 59.1 [8,19].

59.2.5 Seismic Isolated Bridge with Active Mass Damper

Another hybrid control system that combines isolators with active mass dampers is installed on the right abutment of the bridge in Figure 59.1 [8,19]. In general, either base isolators or tuned mass dampers are only effective when the responses of the bridge are dominated by its fundamental mode. Adding an actuator to this system will give the freedom to adjust the controllable frequencies based on different types of earthquakes. This hybrid system utilizes the advantages of both the passive and active systems to extend the range of applicability of both control systems to ensure integrity of the bridge structure.

59.2.6 Friction-Controllable Sliding Bearing

Currently, two classes of seismic base isolation systems have been implemented in bridge engineering: elastomeric bearing system and sliding bearing system. The elastomeric bearing, with its horizontal flexibility, can protect a bridge against strong earthquakes by shifting the fundamental frequency of the bridge to a much lower value and away from the frequency range where the most energy of the earthquake ground motion exists. For the bridge supported by sliding bearings, the maximum forces transferred through the bearings to the bridge are always limited by the friction force at the sliding surface, regardless of the intensity and frequency contents of the earthquake excitation. The vibrational energy of the bridge will be dissipated by the interface friction. Since the friction force is just the product of the friction coefficient and the normal pressure between two sliding surfaces, these two parameters are the critical design parameters of a sliding bearing. The smaller the friction coefficient or normal pressure, the better the isolation performance, due to the correspondingly small rate of transmission of earthquake acceleration to the bridge. In some cases, however, the bridge may suffer from an unacceptably large displacement, especially the residual displacement, between its base and ground. On the other hand, if the friction coefficient or normal pressure is too large, the bridge will be isolated only under correspondingly large earthquakes and the sliding system will not be activated under small to moderate earthquakes that occur more often. In order to substantially alleviate these shortcomings, therefore, the ideal design of a sliding system should vary its friction coefficient or normal pressure based on measured earthquake intensities and bridge responses. To this purpose, a friction-controllable sliding bearing has been developed, and Figure 59.7 illustrates one of its applications in bridge engineering [4,5]. It can be seen from Figure 59.7 that the friction forces in the sliding bearings are actively controlled by adjusting the fluid pressure in the fluid chamber located inside the bearings.

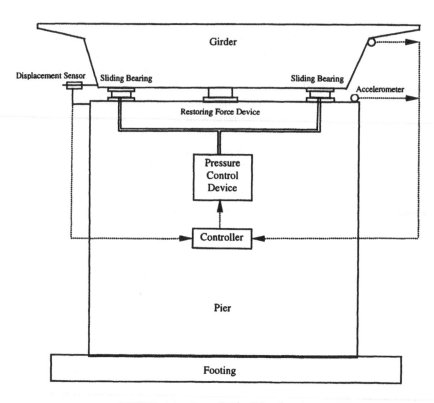

FIGURE 59.7 Controllable sliding bearing.

59.2.7 Controllable Fluid Damper

Dampers are very effective in reducing the seismic responses of bridges. Various dampers, as discussed in Chapter 41, have been developed for bridge vibration control. One of them is fluid damper, which dissipates vibrational energy by moving the piston in the cylinder filled with viscous material (oil). Depending on the different function provided by the dampers, different damping coefficients may be required. For example, one may set up a large damping coefficient to prevent small deck vibrations due to braking loads of vehicles or wind effects. However, when bridge deck responses under strong earthquake excitations exceed a certain threshold value, the damping coefficient may need to be reduced in order to maximize energy dissipation. Further, if excessive deck responses are reached, the damping coefficient needs to be set back to a large value, and the damper will function as a stopper. As we know, it is hard to change the damping coefficient after a passive damper is designed and installed on a bridge. The multifunction requirements for a damper have motivated the development of semiactive strategy. Figure 59.8 shows an example of a semiactive controlled fluid damper. The damping coefficient of this damper can be controlled by varying the amount of viscous flow through the bypass based on the bridge responses. The new damper will function as a damper stopper at small deck displacement, a passive energy dissipator at intermediate deck displacement, and a stopper with shock absorber for excessive deck displacement.

59.2.8 Controllable Friction Damper

Friction dampers, utilizing the interface friction to dissipate vibrational energy of a dynamic system, have been widely employed in building structures. A few feasibility studies have also been performed to exploit their capacity in controlling bridge vibrations. One example is shown in Figure 59.9, which has been utilized to control the vibration of a cable-stayed bridge [20]. The interface pressure

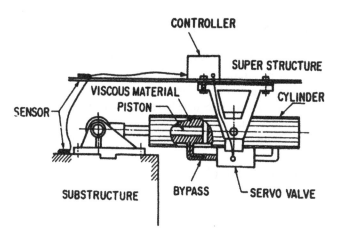

FIGURE 59.8 Controllable fluid damper. (*Source: Proceedings of the Second US–Japan Workshop on Earthquake Protective Systems for Bridges.* p. 481, 1992. With permission.)

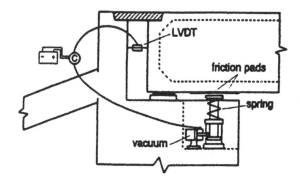

FIGURE 59.9 Controllable friction damper.

of this damper can be actively adjusted through a prestressed spring, a vacuum cylinder, and a battery-operated valve. Since a cable-stayed bridge is a typical flexible structure with relatively low vibration frequencies, its acceleration responses are small due to the isolation effect of flexibility, and short-duration earthquakes do not have enough time to generate large structural displacement responses. In order to take full advantage of the isolation effect of flexibility, it is better not to impose damping force in this case since the increase of large damping force will also increase bridge effective stiffness. On the other hand, if the earthquake excitation is sufficiently long and strong, the displacement of this flexible structure may be quite large. Under this condition, it is necessary to impose large friction forces to dissipate vibrational energy and reduce the moment demand at the bottom of the towers. Therefore, a desirable control system design will be a multistage control system having friction forces imposed at different levels to meet different needs of response control.

The most attractive advantage of the above semiactive control devices is their lower power requirement. In fact, many can be operated on battery power, which is most suitable during seismic events when the main power source to the bridge may fail. Another significant characteristic of semiactive control, in contrast to pure active control, is that it does not destabilize (in the bounded input/bounded output sense) the bridge structural system since no mechanical energy is injected into the controlled bridge system (i.e., , including the bridge and control devices) by the semiactive control devices. Semiactive control devices appear to combine the best features of both passive and active control systems. That is the reason this type of control system offers the greatest likelihood of acceptance in the near future of control technology as a viable means of protecting civil engineering structural systems against natural forces.

59.3   General Control Strategies and Typical Control Algorithms

In this section, the general control strategies, including linear and nonlinear controllers, are introduced first. Then, the linear quadratic regulator (LQR) controlling a simple single-degree-of-freedom (SDOF) bridge system is presented. Further, an extension is made to the multi-degree-of-freedom (MDOF) system that is more adequate to represent an actual bridge structure. The specific characteristics of hybrid and semiactive control systems are also discussed. Finally, the practical concerns about implementation of various control systems in bridge engineering are addressed.

59.3.1   General Control Strategies

Theoretically, a real bridge structure can be modeled as an MDOF dynamic system and the equations of motion of the bridge without and with control are, respectively, expressed as

$$\mathbf{M}\ddot{\mathbf{x}}(t) + \mathbf{C}\dot{\mathbf{x}}(t) + \mathbf{K}\mathbf{x}(t) = \mathbf{E}\mathbf{f}(t) \tag{59.1}$$

$$\mathbf{M}\ddot{\mathbf{x}}(t) + \mathbf{C}\dot{\mathbf{x}}(t) + \mathbf{K}\mathbf{x}(t) = \mathbf{D}\mathbf{u}(t) + \mathbf{E}\mathbf{f}(t) \tag{59.2}$$

where $\mathbf{M}$, $\mathbf{C}$, and $\mathbf{K}$ are the mass, damping, and stiffness matrices, respectively, $\mathbf{x}(t)$ is the displacement vector, $\mathbf{f}(t)$ represents the applied load or external excitation, and $\mathbf{u}(t)$ is the applied control force vector. The matrices $\mathbf{D}$ and $\mathbf{E}$ define the locations of the control force vector and the excitation, respectively.

Assuming the feedback/feedforward configuration is utilized in the above controlled system and the control force is a linear function of the measured displacements and velocities, i.e.,

$$\mathbf{u}(t) = \mathbf{G}_x \mathbf{x}(t) + \mathbf{G}_{\dot{x}} \dot{\mathbf{x}}(t) + \mathbf{G}_f \mathbf{f}(t) \tag{59.3}$$

where $\mathbf{G}_x$, $G_{\dot{x}}$, and $\mathbf{G}_f$ are known as control gain matrices.

Substituting Eq. (59.3) into Eq. (59.2), we obtain

$$\mathbf{M}\ddot{\mathbf{x}}(t) + (\mathbf{C} - \mathbf{D}\mathbf{G}_{\dot{x}})\dot{\mathbf{x}}(t) + (\mathbf{K} - \mathbf{D}\mathbf{G}_x)\mathbf{x}(t) = (\mathbf{E} + \mathbf{D}\mathbf{G}_f)\mathbf{f}(t) \tag{59.4}$$

Alternatively, it can be written as

$$\mathbf{M}\ddot{\mathbf{x}}(t) + \mathbf{C}_c(t)\dot{\mathbf{x}}(t) + \mathbf{K}_c(t)\mathbf{x}(t) = \mathbf{E}_c(t)\mathbf{f}(t) \tag{59.5}$$

Comparing Eq. (59.5) with Eq. (59.1), it is clear that the result of applying a control action to a bridge is to modify the bridge properties and to reduce the external input forces. Also this modification, unlike passive control, is real-time adaptive, which makes the bridge respond more favorably to the external excitation.

It should be mentioned that the above control effect is just an ideal situation: linear bridge structure with linear controller. Actually, physical structure/control systems, such as a hybrid base-isolated bridge, are inherently nonlinear. Thus, all control systems are nonlinear to a certain extent. However, if the operating range of a control system is small and the involved nonlinearities are smooth, then the control system may be reasonably approximated by a linearized system, whose dynamics is described by a set of linear differential equations, for instance, Eq. (59.5).

In general, nonlinearities can be classified as *inherent* (natural) and *intentional* (artificial). Inherent nonlinearities are those that naturally come with the bridge structure system itself. Examples

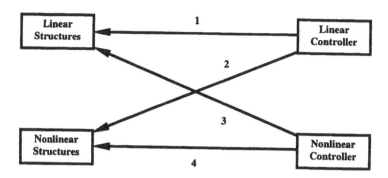

FIGURE 59.10 General control strategies.

of inherent nonlinearities include inelastic deformation of bridge components, seismic isolators, friction dampers, etc. Intentional nonlinearities, on the other hand, are artificially introduced into bridge structural systems by the designer [14,16]. Nonlinear control laws, such as optimal bang–bang control, sliding mode control, and adaptive control, are typical examples of intentional nonlinearities.

According to the properties of the bridge itself and properties of the controller selected, general control strategies may be classified into the following four categories, shown in Figure 59.10 [13].

- **Inherent linear control strategy**: A linear controller controlling a linear bridge structure. This is a simple and popular control strategy, such as LQR/LQG control, pole assignment/mode space control, etc. The implication of this kind of control law is based on the assumption that a controlled bridge will remain in the linear range. Thus, designing a linear controller is the simplest yet reasonable solution. The advantages of linear control laws are well understood and easy to design and implement in actual bridge control applications.

- **Intentional linearization strategy**: A linear controller controlling a nonlinear structure. This belongs to the second category of control strategy, as shown in Figure 59.10. Typical examples of this kind of control laws include instantaneous optimal control, feedback linearization, and gain scheduling, etc. This control strategy retains the advantages of the linear controller, such as simplicity in design and implementation. However, since linear control laws rely on the key assumption of small-range operation, when the required operational range becomes large, a linear controller is likely to perform poorly or sometimes become unstable, because nonlinearities in the system cannot be properly compensated.

- **Intentional nonlinearization strategy**: A nonlinear controller controlling a linear structure. Basically, if undesirable performance of a linear system can be improved by introducing a nonlinear controller intentionally, instead of using a linear controller, the nonlinear one may be preferable. This is the basic motivation for developing intentional nonlinearization strategy, such as optimal bang–bang control, sliding mode control, and adaptive control.

- **Inherent nonlinear control strategy**: A nonlinear controller controlling a nonlinear structure. It is reasonable to control a nonlinear structure by using a nonlinear controller, which can handle nonlinearities in large-range operations directly. Sometimes a good nonlinear control design may be simple and more intuitive than its linear counterparts since nonlinear control designs are often deeply rooted in the physics of the structural nonlinearities. However, since nonlinear systems can have much richer and more complex behaviors than linear systems, there are no systematic tools for predicting the behaviors of nonlinear systems, nor are there systematic procedures for designing nonlinear control systems. Therefore, how to identify and describe structural nonlinearities accurately and then design a suitable nonlinear controller based on those specified nonlinearities is a difficult and challenging task in current nonlinear bridge control applications.

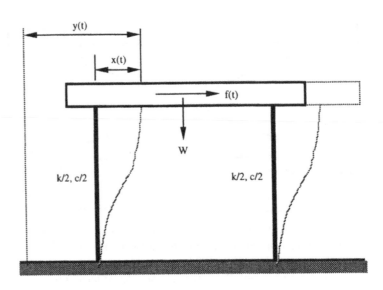

FIGURE 59.11 Simplified bridge model — SDOF system.

59.3.2 Single-Degree-of-Freedom Bridge System

Figure 59.11 shows a simplified bridge model represented by an SDOF system. The equation of motion for this SDOF system can be expressed as

$$m\ddot{x}(t) + c\dot{x}(t) + kx(t) = f(t) \tag{59.6}$$

where m represents the total mass of the bridge, k and c are the linear elastic stiffness and viscous damping provided by the bridge columns and abutments, $f(t)$ is an external disturbance, and $x(t)$ denotes the lateral movement of the bridge. For a specified disturbance, $f(t)$, and with known structural parameters, the responses of this SDOF system can be readily obtained by any step-by-step integration method.

In the above, $f(t)$ represents an arbitrary environmental disturbance such as earthquake, traffic, or wind. In the case of an earthquake load,

$$f(t) = -m\ddot{x}_g(t) \tag{59.7}$$

where $\ddot{x}_g(t)$ is earthquake ground acceleration. Then Eq. (59.6) can be alternatively written as

$$\ddot{x}(t) + 2\xi \, \omega \, \dot{x}(t) + \omega^2 x(t) = -\ddot{x}_0(t) \tag{59.8}$$

in which ω and ξ are the natural frequency and damping ratio of the bridge, respectively.

If an active control system is now added to the SDOF system, as indicated in Figure 59.12, the equation of motion of the extended SDOF system becomes

$$\ddot{x}(t) + 2\xi \, \omega \, \dot{x}(t) + \omega^2 x(t) = u(t) - \ddot{x}_0(t) \tag{59.9}$$

where $u(t)$ is the normalized control force per unit mass. The central topic of control system design is to find an optimal control force $u(t)$ to minimize the bridge responses. Various control strategies,

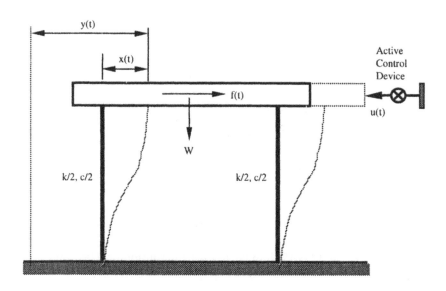

FIGURE 59.12 Simplified bridge with active control system.

as discussed before, have been proposed and implemented to control different structures under different disturbances. Among them, the LQR is the simplest and most widely used control algorithm [10,13].

In LQR, the control force $u(t)$ is designed to be a linear function of measured bridge displacement, $x(t)$, and measured bridge velocity, $\dot{x}(t)$,

$$u(t) = g_x x(t) + g_{\dot{x}} \dot{x}(t) \tag{59.10}$$

where g_x and $g_{\dot{x}}$ are two constant feedback gains which can be found by minimizing a performance index:

$$J = \frac{1}{2} \int_0^\infty [q_x x^2(t) + q_{\dot{x}} \dot{x}^2(t) + r u^2(t)] dt \tag{59.11}$$

where q_x, $q_{\dot{x}}$, and r are called weighting factors. In Eq. (59.11), the first term represents bridge vibration strain energy, the second term is the kinetic energy of the bridge, and the third term is the control energy input by external source powers. Minimizing Eq. (59.11) means that the total bridge vibration energy will be minimized by using minimum input control energy, which is an ideal optimal solution.

The role of weighting factors in Eq. (59.11) is to apply different penalties on the controlled responses and control forces. The assignment of large values to the weight factors q_x and $q_{\dot{x}}$ implies that a priority is given to response reductions. On the other hand, the assignment of a large value to weighting factor r means that the control force requirement is the designer's major concern. By varying the relative magnitudes of q_x, $q_{\dot{x}}$, and r, one can synthesize the controllers to achieve a proper trade-off between control effectiveness and control energy consumption. The effects of these weighting factors on the control responses of bridge structures will be investigated in the next section of case studies.

It has been found [15] that analytical solutions of the feedback constant gains, g_x and $g_{\dot{x}}$, are

$$g_x = -\omega^2(s_x - 1) \tag{59.12}$$

$$g_{\dot{x}} = -2\xi\,\omega\,(s_{\dot{x}} - 1) \tag{59.13}$$

where the coefficients s_x and $s_{\dot{x}}$ are derived as

$$s_x = \sqrt{1 + \frac{(q_x/r)}{\omega^4}} \tag{59.14}$$

$$s_{\dot{x}} = \sqrt{1 + \frac{(q_{\dot{x}}/r)}{4\xi^2\omega^2} + \frac{(s_x - 1)}{2\xi^2}} \tag{59.15}$$

Substituting Eqs. (59.14) and (59.15) into Eq. (59.10), the control force becomes

$$u(t) = -\omega^2(s_x - 1)x(t) - 2\xi\,\omega\,(s_{\dot{x}} - 1)\dot{x}(t) \tag{59.16}$$

Inserting the above control force into Eq. (59.9), one obtains the equation of motion of the controlled system as

$$\ddot{x}(t) + 2\xi\,\omega\,s_{\dot{x}}\dot{x}(t) + \omega^2 s_x x(t) = -\ddot{x}_0(t) \tag{59.17}$$

It is interesting to compare Eq. (59.8), which is an uncontrolled system equation, with Eq. (59.17), which is a controlled system equation. It can be seen that the coefficient s_x reflects a shift of the natural frequency caused by applying the control force, and the coefficient $s_{\dot{x}}$ indicates a change in the damping ratio due to control force action.

The concept of active control is clearly exhibited by Eq. (59.17). On the one hand, an active control system is capable of modifying properties of a bridge in such a way as to react to external excitations in the most favorable manner. On the other hand, direct reduction of the level of excitation transmitted to the bridge is also possible through an active control if a feedforward strategy is utilized in the control algorithm.

Major steps to design an SDOF control system based on LQR are

- Calculate the responses of the uncontrolled system from Eq. (59.8) by response spectrum method or the step-by-step integration, and decide whether a control action is necessary or not.
- If a control system is needed, then assign the values to the weighting factors q_x, $q_{\dot{x}}$, and r, and evaluate the adjusting coefficients s_x and $s_{\dot{x}}$ from Eqs. (59.14) and (59.15) directly.
- Find the responses of the controlled system and control force requirement from Eq. (59.17) and Eq. (59.16), respectively.
- Make the final trade-off decision based on concern about the response reduction or control energy consumption and, if necessary, start the next iterative process.

59.3.3 Multi-Degree-of-Freedom Bridge System

An actual bridge structure is much more complicated than the simplified model shown in Figure 59.11, and it is hard to model as an SDOF system. Therefore, a MDOF system will be introduced next to handle multispan or multimember bridges. The equation of motion for an MDOF system without and with control has been given in Eqs. (59.1) and (59.2), respectively. In the control system design, Eq. (59.2) is generally transformed into the following state equation for convenience of derivation and expression:

$$\dot{z}(t) = Az(t) + Bu(t) + Wf(t) \tag{59.18}$$

where

$$z(t) = \begin{bmatrix} x(t) \\ \dot{x}(t) \end{bmatrix}; \quad A = \begin{bmatrix} 0 & 1 \\ -M^{-1}K & -M^{-1}C \end{bmatrix}; \quad B = \begin{bmatrix} 0 \\ M^{-1}D \end{bmatrix}; \quad W = \begin{bmatrix} 0 \\ M^{-1}E \end{bmatrix} \tag{59.19}$$

Similar to SDOF system design, the control force vector $u(t)$ is related to the measured state vector $z(t)$ as the following linear function:

$$u(t) = Gz(t) \tag{59.20}$$

in which G is a control gain matrix which can be found by minimizing the performance index [10]:

$$J = \frac{1}{2} \int_0^\infty [z^T(t)Qz(t) + u^T(t)Ru(t)]dt \tag{59.21}$$

where Q and R are the weighting matrices and have to be assigned by the designer. Unlike an SDOF system, an analytical solution of control gain matrix G in Eq. (59.21) is currently not available. However, the matrix numerical solution is easy to find in general control program packages. Theoretically, designing a linear controller to control an MDOF system based on LQR principle is easy to accomplish. But the implementation of a real bridge control is not so straightforward and many challenging issues still remain and need to be addressed. This will be the last topic of this section.

59.3.4 Hybrid and Semiactive Control System

It should be noted from the previous section that most of the hybrid or semiactive control systems are intrinsically nonlinear systems. Development of control strategies that are practically implementable and can fully utilize the capacities of these systems is an important and challenging task. Various nonlinear control strategies have been developed to take advantage of the particular characteristics of these systems, such as optimal instantaneous control, bang–bang control, sliding mode control, etc. Since different hybrid or semiactive control systems have different unique features, it is impossible to develop a universal control law, like LQR, to handle all these nonlinear systems. The particular control strategy for a particular nonlinear control system will be discussed as a case study in the next section.

59.3.5 Practical Considerations

Although extensive theoretical developments of various control strategies have shown encouraging results, it should be noted that these developments are largely based on idealized system descriptions. From theoretical development to practical application, engineers will face a number of important issues; some of these issues are listed in Figure 59.2 and are discussed in this section.

59.3.5.1 Control Single Time Delay

As shown in Figure 59.1, from the measurement of vibration signal by the sensor to the application of a control action by the actuator, time has to be consumed in processing measured information, in performing online computation, and in executing the control forces as required. However, most of the current control algorithms do not incorporate this time delay into the programs and assume that all operations can be performed instantaneously. It is well understood that missing time delay may render the control ineffective and, most seriously, may cause instability of the system. One example is discussed here. Suppose:

1. The time periods consumed in processing measurement, computation, and force action are 0.01, 0.2, and 0.3 s, respectively;
2. The bridge vibration follows a harmonic motion with a period of 1.02 s; and
3. The sensor picks up a positive peak response of the bridge vibration at 5.0 s.

After the control system finishes all processes and applies a large control force onto the bridge, the time is 5.51 s. At this time, the bridge vibration has already changed its phase and reached the negative peak response. It is evident that the control force actually is not controlling the bridge but exciting the bridge. This kind of excitation action is very dangerous and may lead to an unstable situation. Therefore, the time delay must be compensated for in the control system implementation. Various techniques have been developed to compensate for control system time delay. The details can be found in Reference [10].

59.3.5.2 Control and Observation Spillover

Although actual-bridge structures are distributed parameter systems, in general, they are modeled as a large number of degrees of freedom discretized system, referred as the full-order system, during the analytical and simulation process. Further, it is difficult to design a control system based on the full-order bridge model due to online computation process and full state measurement. Hence, the full-order model is further reduced to a small number of degrees of freedom system, referred as a reduced-order system. Then, the control design is performed based on the reduced-order bridge model. After finishing the design, however, the implementation of the designed control system is applied on the actual distributed parameter bridge. Two problems may result. First, the designed control action can only control the reduced-order modes and may not be effective with the residual (uncontrolled) modes, and sometimes even worse to excite the residual modes. This kind of action is called *control spillover*, i.e., the control actions spill over to the uncontrolled modes and enhance the bridge vibration. Second, the control design is based on information observed from the reduced-order model. But, in reality, it is impossible to isolate the vibration signals from residual modes and the measured information must be contaminated by the residual modes. After the contaminated information is fed back into the control system, the control action, originally based on the "pure" measurements, may change, and the control performance may be degraded seriously. This is the so-called *observation spillover*. Again, all spillover effects must be compensated for in the control system implementation [10].

59.3.5.3 Optimal Actuator and Sensor Locations

Because a large number of degrees of freedom are usually involved in the bridge structure, it is impractical to install sensors on each degree-of-freedom location and measure all state variables. Also, in general, only fewer (often just one) control actuators are installed at the critical control locations. Two problems are (1) How many sensors and actuators are required for a bridge to be completely observable and controllable? (2) Where are the optimal locations to install these sensors and actuators in order to measure vibration signals and exert control forces most effectively? Actually, the vibrational control, property identification, health monitoring, and damage detection are closely related in the development of optimal locations. Various techniques and schemes have been successfully developed to find optimal sensor and actuator locations. Reference [10] provides more details about this topic.

59.3.5.4 Control–Structure Interaction

Like bridge structures, control actuators themselves are dynamic systems with inherent dynamic properties. When an actuator applies control forces to the bridge structure, the structure is in turn applying the reaction forces on the actuator, exciting the dynamics of the actuator. This is the so-called *control–structure interaction*. Analytical simulations and experimental verifications have indicated that disregarding the control–structure interaction may significantly reduce both the achievable control performance and the robustness of the control system. It is important to model the dynamics of the actuator properly and to account for the interaction between the structure and the actuator [3,9].

59.3.5.5 Parameter Uncertainty

Parameter identification is a very important part in the loop of structural control design. However, due to limitations in modeling and system identification theory, the exact identification of structural parameters is virtually impossible, and the parameter values used in control system design may deviate significantly from their actual values. This type of *parameter uncertainty* may also degrade the control performance. The sensitivity analysis and robust control design are effective means to deal with the parameter uncertainty and other modeling errors [9].

The above discussions only deal with a few topics of practical considerations in real bridge control implementation. Some other issues that must be investigated in the design of control system include the stability of the control, the noise in the digitized instrumentation signals, the dynamics of filters required to attenuate the signal noise, the potential for actuator saturation, any system nonlinearities, control system reliability, and cost-effectiveness of the control system. More-detailed discussions of these topics are beyond the scope of this chapter. A recent state-of-the-art paper is a very useful resource that deals with all the above topics [6].

59.4 Case Studies

59.4.1 Concrete Box-Girder Bridge

59.4.1.1 Active Control for a Three-Span Bridge

The first example of case studies is a three-span concrete box-girder bridge located in a seismically active zone. Figure 59.13a shows the elevation view of this bridge. The bridge has the span lengths of 38, 38, and 45 m, respectively. The width of the bridge is 32 m and the depth is 2.1 m. The column heights are 15 and 16 m at Bents 2 and 3, respectively. Each span has four oblong-shaped columns with 1.67×2.51 m cross section. The columns are monolithically connected with bent cap at the top and pinned with footing at the bottom. The bridge has a total weight of 81,442 kN or a total mass of 8,302,000 kg. The longitudinal stiffness, including abutments and columns, is 82.66 kN/mm. Two servo-hydraulic actuators are installed on the bridge abutments and controlled by the same controller to keep both actuators in the same phase during the control operation. The objective of using the active control system is to reduce the bridge vibrations induced by strong earthquake excitations. Only longitudinal movement will be controlled.

The analysis model of this bridge is illustrated in Figure 59.13b, and a simplified SDOF model is shown in Figure 59.13c. The natural frequency of the SDOF system $\omega = 19.83$ rad/s, and damping ratio $\xi = 5\%$. Without loss of the generality, the earthquake ground motion, $\ddot{x}_0(t)$, is described as a stationary random process. The well-known Kanai–Tajimi spectrum is utilized to represent the power spectrum density of the input earthquake, i.e.,

$$G_{\ddot{x}_0}(\omega) = \frac{G_0[1 + 4\xi_g{}^2(\omega/\omega_g)^2]}{[1 - (\omega/\omega_g)^2]^2 + 4\xi_g{}^2(\omega/\omega_g)^2} \tag{59.22}$$

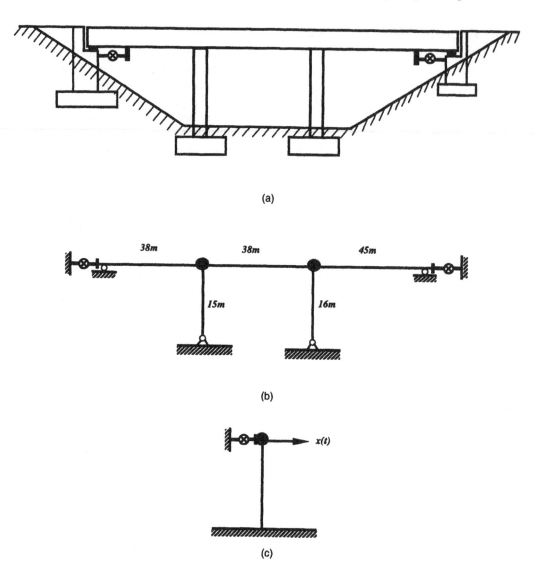

FIGURE 59.13 Three-span bridge with active control system. (a) Actual bridge; (b) bridge model for analysis; (c) SDOF system controlled by actuator.

where ω_g and ξ_g are, respectively, the frequency and damping ratio of the soil, whose values are taken as $\omega_g = 22.9$ rad/s and $\xi_g = 0.34$ for average soil condition. The parameter G_0 is the spectral density related to the maximum earthquake acceleration a_{max} [15]. At this bridge site, the maximum ground acceleration $a_{max} = 0.4\,g$.

The maximum response of an SDOF system with natural frequency ω and damping ratio ξ under $\ddot{x}_0(t)$ excitation can be estimated as

$$x_{max}(\omega,\xi) = \gamma_p \sigma_x \tag{59.23}$$

in which γ_p is a peak factor and σ_x is the root-mean-square response which can be determined by random vibration theory [2].

From Eq. (59.17), it is known that the frequency and damping ratio of a controlled system are

$$\omega_c = \sqrt{s_x}\,\omega; \quad \xi_c = (s_\dot{x}\,/\sqrt{s_x}\,)\xi \tag{59.24}$$

TABLE 59.2 Summary of Three-Span Bridge Control

			ω_c	ξ	d_{max}		a_{max}		u_{max}	
r	s_x	$s_{\dot{x}}$	(rad/s)	(%)	(cm)	Redu (%)	(g)	Redu (%)	(kN)	Weight (%)
1E+07	1.000	1.001	19.83	0.05	3.15	0	1.23	0	10	0
100,000	1.000	1.103	19.83	0.06	2.86	9	1.12	9	1046	1
10,000	1.000	1.777	19.83	0.09	2.30	27	0.90	27	7894	10
5,000	1.001	2.305	19.83	0.12	1.97	37	0.77	37	13258	16
1,000	1.003	4.750	19.85	0.24	1.30	59	0.51	59	38097	47
500	1.005	6.643	19.88	0.33	0.72	77	0.28	77	57329	70

where s_x and $s_{\dot{x}}$ can be found from Eq. (59.14) and Eq. (59.15), respectively, once the weighting factors q_x, $q_{\dot{x}}$, and r are assigned by the designer. The maximum response of the controlled system is obtained from Eq. (59.23).

In this case study, the weighting factors are assigned as $q_x = 100m$ and $q_{\dot{x}} = k$. Through varying the weight factor r, one can obtain different control efficiencies by applying different control forces. Table 59.2 lists the control coefficients, controlled frequencies, damping ratios, maximum bridge responses, and maximum control force requirements based on various assignments of the weight factor r.

It can be seen from Table 59.2 that no matter how small the weighting factor r is, the coefficient s_x is always close to 1, which means that the structural natural frequency is hard to shift by LQR algorithm. However, the coefficient $s_{\dot{x}}$ increases significantly with decrease of the weighting factor r, which means that the major effect of LQR algorithm is to modify structural damping. This is just what we wanted. In fact, extensive simulation results have shown the same trend as indicated in Table 59.2 [13]. The maximum acceleration of the bridge deck is 1.23 g without control. If the control force is applied on the bridge with maximum value of 13,258 kN (16% bridge weight), the maximum acceleration response reduces to 0.77 g, the reduction factor is 37%. The larger the applied control force, the larger the response reduction. But, in reality, current servo-hydraulic actuators may not generate such a large control force.

59.4.1.2 Hybrid Control for a Simple-Span Bridge

The second example of the case studies, as shown in Figure 59.14, is a simple-span bridge equipped with rubber bearings and active control actuators between the bridge girder and columns [19]. The bridge has a span length of 30 m and column height of 22 m. The bridge is modeled as a nine-degree-of-freedom system, as shown in Figure 59.14b. Duo to symmetry, it is further reduced to a four-degree-of-freedom system, as shown in Figure 59.14c. The mass, stiffness, and damping properties of this bridge can be found in Reference [19].

The bridge structure is considered to be linear elastic except the rubber bearings. The inelastic stiffness restoring force of the rubber bearing is expressed as

$$F_s = \alpha k x(t) + (1 - \alpha) k D_y v \qquad (59.25)$$

in which $x(t)$ is the deformation of the rubber bearing, k is the elastic stiffness, α is the ratio of the postyielding to preyielding stiffness, D_y is the yield deformation, and v is the hysteretic variable with $|v| \leq 1$, where

$$\dot{v} = D_y^{-1} \{ A\dot{x} - \beta |\dot{x}| |v|^{n-1} v - \gamma \dot{x} |v|^n \} \qquad (59.26)$$

In Eq. (59.26), the parameters A, β, γ, and n govern the scale, general shape, and smoothness of the hysteretic loop. It can be seen from Eq. (59.25) that if $\alpha = 1.0$, then the rubber bearing has a linear stiffness, i.e., $F_s = kx(t)$.

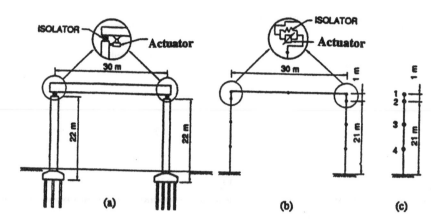

FIGURE 59.14 Simple-span bridge with hybrid control system. (a) Actual bridge; (b) lumped mass system; (c) four-degree-of-freedom system. (*Source: Proceedings of the Second U.S.–Japan Workshop on Earthquake Protective Systems for Bridges*, p. 482, 1992. With permission.)

TABLE 59.3 Summary of Simple-Span Bridge Control

Control System	d_{1max} (cm)	d_{2max} (cm)	d_{3max} (cm)	d_{4max} (cm)	a_{1max} (g)	V_{bmax} (kN)	u_{max} (% W_1)
Passive	24.70	3.96	3.07	1.25	1.31	1648	0
Hybrid	5.53	1.46	1.14	0.46	0.48	628	41

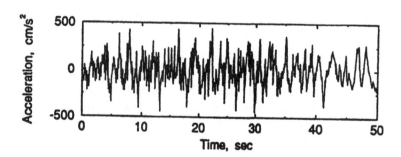

FIGURE 59.15 Simulated earthquake ground acceleration.

The LQR algorithm is incapable of handling the nonlinear structure control problem, as indicated in Eq. (59.25). Therefore, the sliding mode control (SMC) is employed to develop a suitable control law in this example. The details of SMC can be found from Reference [19].

The input earthquake excitation is shown in Figure 59.15, which is simulated such that the response spectra match the target spectra specified in the Japanese design specification for highway bridges. The maximum deformations (d_{1max}, d_{2max}, d_{3max}, and d_{4max}), maximum acceleration (a_{1max}), maximum base shear of the column (V_{bmax}), and maximum actuator control force (u_{max}) are listed in Table 59.3. It is clear that adding an active control system can significantly improve the performance and effectiveness of the passive control. Comparing with passive control alone, the reductions of displacement and acceleration at the bridge deck can reach 78 and 63%, respectively. The base shear of the column can be reduced to 38%. The cost is that each actuator has to provide the maximum control force up to 20% of the deck weight.

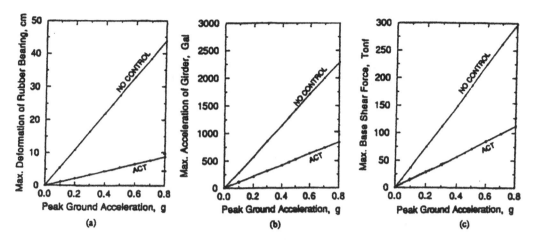

FIGURE 59.16 Bridge maximum responses. (a) deformation of rubber bearing; (b) acceleration of girder; (c) base shear force of pier.

In order to evaluate and compare the effectiveness of a hybrid control system over a wide range of earthquake intensities, the design earthquake shown in Figure 59.15 is scaled uniformly to different peak ground acceleration to be used as the input excitations. The peak response quantities for the deformation of rubber bearing, the acceleration of the bridge deck, and the base shear of the column are presented as functions of the peak ground acceleration in Figure 59.16. In this figure, "no control" means passive control alone, and "act" denotes hybrid control. Obviously, the hybrid control is much more effective over passive control alone within a wide range of earthquake intensities.

59.4.2 Cable-Stayed Bridge

59.4.2.1 Active Control for a Cable-Stayed Bridge

Cable-supported bridges, as typical flexible bridge structures, are particularly vulnerable to strong wind gusts. Extensive analytical and experimental investigations have been performed to increase the "critical wind speed" since wind speeds higher than the critical will cause aerodynamic instability in the bridge. One of these studies is to install an active control system to enhance the performance of the bridge under strong wind gusts [17,18].

Figure 59.17 shows the analytical model of the Sitka Harbor Bridge, Sitka, Alaska. The midspan length of the bridge is 137.16 m. Only two cables are supported by each tower and connected to the bridge deck at distance $a = l/3 = 45.72m$. The two-degree-of-freedom system is used to describe the vibrations of the bridge deck. The fundamental frequency in flexure $\omega_g = 5.083$ rad/s, and the fundamental frequency in torsion $\omega_f = 8.589$ rad/s. In this case study, the four existing cables, which are designed to carry the dead load, are also used as active tendons to which the active feedback control systems (hydraulic servomechanisms) are attached. The vibrational signals of the bridge are measured by the sensors installed at the anchorage of each cable, and then transmitted into the feedback control system. The sensed motion, in the form of electric voltage, is used to regulate the motion of hydraulic rams in the servomechanisms, thus generating the required control force in each cable.

Suppose that the accelerometer is used to measure the bridge vibration. Then the feedback voltage $v(t)$ is proportional to the bridge acceleration $\ddot{w}(t)$:

$$v(t) = p\ddot{w}(t) \qquad (59.27)$$

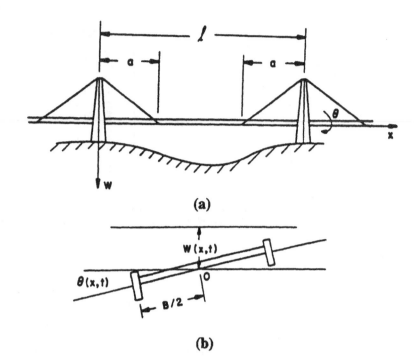

FIGURE 59.17 Cable-stayed bridge with active tendon control. (a) Side view with coordinate system; (b) two-degree-of-freedom model. (*Source:* Yang, J.N. and Giannopolous, F., *J. Eng. Mech. ASCE*, 105(5), 798–810, 1979. With permission.)

where p is the proportionality constant associated with each sensor. For active tendon configuration, the displacement $s(t)$ of hydraulic ram, which is equal to the additional elongation of the tendon (cable) due to active control action, is related to the feedback voltage $v(t)$ through the first-order differential equation:

$$\dot{s}(t) + R_1 s(t) = \frac{R_1}{R} v(t) \tag{59.28}$$

in which R_1 is the loop gain and R is the feedback gain of the servomechanism. The cable control force generated by moving the hydraulic ram is

$$u(t) = ks(t) \tag{59.29}$$

where k is the cable stiffness.

Combining Eq. (59.27) and Eq. (59.29), we have

$$u(t) = g(R_1, R)\ddot{w}(t) \tag{59.30}$$

It is obvious that Eq. (59.30) represents an acceleration feedback control and the control gain $g(R_1, R)$ depends on the control parameters R_1 and R which will be assigned by the designer. Further, two nondimensional parameters ε and τ are introduced to replace R_1 and R

$$\varepsilon = \frac{R_1}{\omega_f} \quad \text{and} \quad \tau = \frac{p\omega_f^2}{R} \tag{59.31}$$

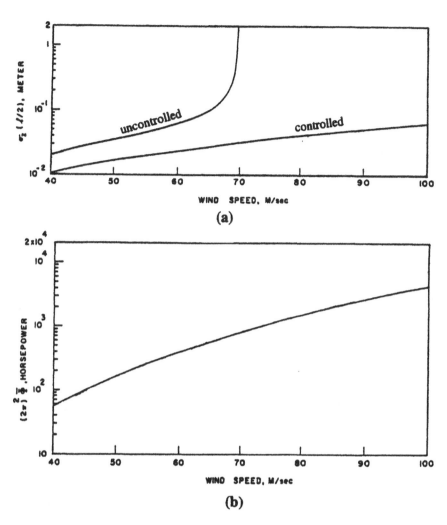

FIGURE 59.18 Root-mean-square displacement and average power requirement. (a) Root-mean-square displacement of bridge deck; (b) average power requirement. (*Source*: Yang, J.N., and Giannnopolous, F., *J. Eng. Mech., ASCE*, 105(5) 798-810, 1979. With permission.)

Finally, the critical wind speed and the control power requirement are all related to the control parameters ε and τ.

Figure 59.18a shows the root-mean-square displacement response of the bridge deck without and with control. In the control case the parameter $\varepsilon = 0.1$, $\tau = 10$. Correspondingly, the average power requirement to accomplish active control is illustrated in Figure 59.18b. It can be seen that the bridge response is reduced significantly (up to 80% of the uncontrolled case) with a small power requirement by the active devices. In terms of critical wind speed, the value without control is 69.52 m/s, while with control it can be raised to any desirable level provided that the required control forces are realizable. Based on the studies, it appears that the active feedback control is feasible for applications to cable-stayed bridges.

59.4.2.2 Active Mass Damper for a Cable-Stayed Bridge under Construction

Figure 59.19 shows a cable-stayed bridge during construction using the cantilever erection method. It can be seen that not only the bridge weight but also the heavy equipment weights are all supported by a single tower. Under this condition, the bridge is a relatively unstable structure, and special

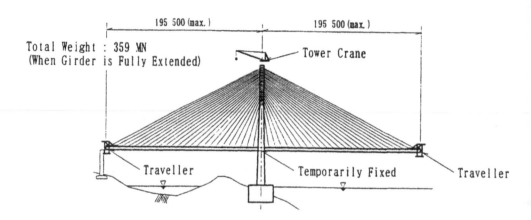

FIGURE 59.19 Construction by cantilever erection method. (*Source:* Tsunomoto, M., et al. *Proceedings of Fourth U.S.–Japan Workshop on Earthquake Protective Systems for Bridges*, 115–129, 1996. With permission.)

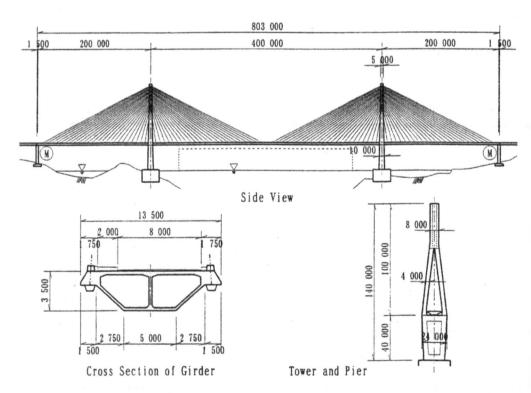

FIGURE 59.20 General view of cable-stayed bridge studied. (*Source:* Tsunomoto, M., et al. *Proceedings of Fourth U.S.–Japan Workshop on Earthquake Protective Systems for Bridges*, 115–129, 1996. With permission.)

attention is required to safeguard against dynamic external forces such as earthquake and wind loads. Since movable sections are temporarily fixed during the construction, the seismic isolation systems that will be adopted after the completion of the construction are usually ineffective for the bridge under construction. Active tendon control by using the bridge cable is also difficult to install on the bridge at this period. However, active mass dampers, as shown in Figure 59.6, have proved to be effective control devices in reducing the dynamic responses of the bridge under construction [12].

The bridge in this case study is a three-span continuous prestressed concrete cable-stayed bridge with a central span length of 400 m, as shown in Figure 59.20. When the girder is fully extended,

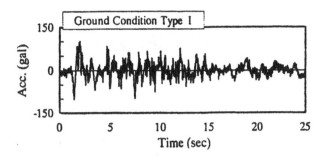

FIGURE 59.21 Input earthquake ground motion. (*Source:* Tsunomoto, M., et al. *Proceedings of Fourth U.S.–Japan Workshop on Earthquake Protective Systems for Bridges*, 115–129, 1996. With permission.)

the total weight is 359 MN, including bridge self-weight, traveler weight (1.37 MN at each end of the girder), and crane weight (0.78 MN at the top of the tower). The damping ratio for dynamic analysis is 1%. The ground input acceleration is shown in Figure 59.21. Since the connections between pier and footing and between girder and pier are fixed during construction, the moments at pier bottom and at tower bottom are the critical response parameters to evaluate the safety of the bridge at this period. Two control cases are investigated. In the first case, the active mass damper (AMD) is installed at the tower top and operates in the longitudinal direction. In the second case, the AMD is installed at the cantilever girder end and operates in the vertical direction. The AMD is controlled by the direct velocity feedback algorithm, in which the control force is only related to the measured velocity response at the location of the AMD. By changing the control gain, the maximum control force is adjusted to around 3.5 MN, which is about 1% of the total weight of the bridge.

Figures 59.22 and 59.23 show the time histories of the bending moments and control forces in Case 1 and Case 2, respectively. In Case 1, the maximum bending moment at the pier bottom is reduced by about 15%, but the maximum bending moment at the tower bottom is reduced only about 5%. In Case 2, the reduction of the bending moment at the tower bottom is the same as that in Case 1, but the reduction of the bending moment at the pier bottom is about 35%, i.e., 20% higher than the reduction in Case 1. The results indicate that the AMD is an effective control device to reduce the dynamic responses of the bridge under construction. Installing an AMD at the girder end of the bridge is more effective than installing it at the tower top. The response control in reducing the bending moment at the tower bottom is less effective than that at the pier bottom.

59.5 Remarks and Conclusions

Various structural protective systems have been developed and implemented for vibration control of buildings and bridges in recent years. These modern technologies have generated a strong impact in the traditional structural design and construction fields. The entire structural engineering discipline is undergoing a major change. It now seems desirable to encourage structural engineers and architects to seriously consider exploiting the capabilities of structural control systems for retrofitting existing structures and also enhancing the performance of prospective new structures.

The basic concepts of various control systems are introduced in this chapter. The emphasis is put on active control, hybrid control, and semiactive control for bridge structures. The different bridge control configurations are presented. The general control strategies and typical control algorithms are discussed. Through several case studies, it is shown that the active, hybrid, and semiactive control systems are quite effective in reducing bridge vibrations induced by earthquake, wind, or traffic.

It is important to recognize that although significant progress has been made in the field of active response control to bridge structures, we are now still in the study-and-development stage and await coming applications. There are many topics related to the active control of bridge structures that need research and resolution before the promise of smart bridge structures is fully realized. These topics are

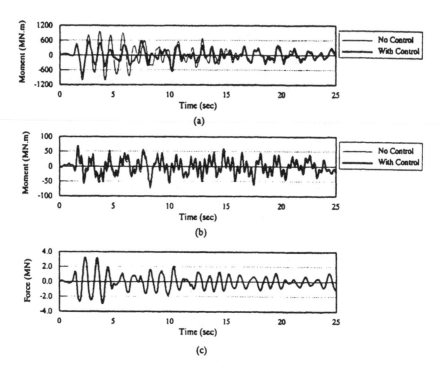

FIGURE 59.22 Bridge responses and control force with AMD at tower top. (a) moment at pier bottom; (b) moment at tower bottom; (c) control force of AMD. (*Source:* Tsunomoto, M., et al. *Proceedings of Fourth U.S.–Japan Workshop on Earthquake Protective Systems for Bridges,* 115–129, 1996. With permission.)

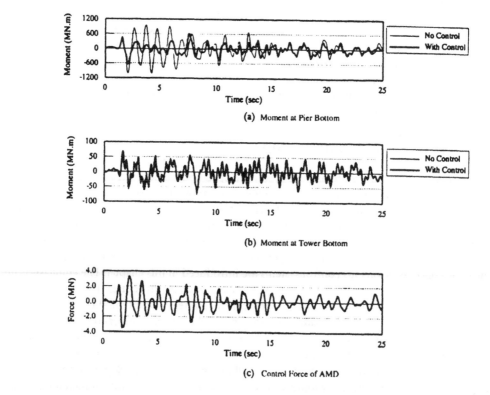

FIGURE 59.23 Bridge responses and control force with AMD at girder end. (a) moment at pier bottom; (b) moment at tower bottom; (c) control force of AMD. (*Source:* Tsunomoto, M., et al. *Proceedings of Fourth U.S.–Japan Workshop on Earthquake Protective Systems for Bridges,* 115–129, 1996. With permission.)

- Algorithms for active, hybrid, and semiactive control of nonlinear bridge structures;
- Devices with energy-efficient features able to handle strong inputs;
- Integration of control devices into complex bridge structures;
- Identification and modeling of nonlinear properties of bridge structures;
- Standardized performance evaluation and experimental verification;
- Development of design guidelines and specifications;
- Implementation on actual bridge structures.

References

1. Adeli, H. and Saleh, A., Optimal control of adaptive/smart structures, *J. Struct. Eng. ASCE*, 123(2), 218–226, 1997.
2. Ben-Haim, Y., Chen, G., and Soong, T. T., Maximum structural response using convex models, *J. Eng. Mech. ASCE*, 122(4), 325–333, 1996.
3. Dyke, S. J., Spencer, B. F., Quast, P., and Sain, M. K., The role of control-structure interaction in protective system design, *J. Eng. Mech. ASCE*, 121(2), 322–338, 1995.
4. Feng, M. Q., Shinozuka, M., and Fujii, S., Friction-controllable sliding isolation system, *J. Eng. Mech. ASCE*, 119(9), 1845–1864, 1993.
5. Feng, M. Q., Seismic response variability of hybrid-controlled bridges, *Probabilistic Eng. Mech.*, 9, 195–201, 1994.
6. Housner, G. W., Bergman, L. A., Caughey, T. K., Chassiakos, A. G., Claus, R. O., Masri, S. F., Skeleton, R. E., Soong, T. T., Spencer, B. F. and Yao, J. T. P., Structural control: past, present, and future, *J. Eng. Mech. ASCE*, 123(9), 897–971, 1997.
7. Kawashima, K. and Unjoh, S., Seismic response control of bridges by variable dampers, *J. Struct. Eng. ASCE*, 120(9), 2583–2601, 1994.
8. Reinhorn, A. M. and Riley, M., Control of bridge vibrations with hybrid devices. Proceedings of First World Conference on Structural Control, II, TA2, 1994, 50–59.
9. Riley, M., Reinhorn, A. M., and Nagarajaiah, S., Implementation issues and testing of a hybrid sliding isolation system, *Eng. Struct.*, 20(3), 144–154, 1998.
10. Soong, T. T., *Active Control: Theory and Practice*, Longman Scientific and Technical, Essex, England, and Wiley, New York, 1990.
11. Soong, T. T. and Dargush, G. F., *Passive Energy Dissipation Systems in Structural Engineering*, John Wiley & Sons, London, 1997.
12. Tsunomoto, M., Otsuka, H., Unjoh, S., and Nagaya, K., Seismic response control of PC cable-stayed bridge under construction by active mass damper, in *Proceedings of Fourth U.S. — Japan Workshop on Earthquake Protective Systems for Bridges*, 115–129, 1996.
13. Wu, Z., Nonlinear Feedback Strategies in Active Structural Control, Ph.D. dissertation, State University of New York at Buffalo, Buffalo, 1995.
14. Wu, Z., Lin. R. C., and Soong, T. T., Nonlinear feedback control for improved response reduction, *Smart Mat. Struct.*, 4(1), A140–A148, 1995.
15. Wu, Z. and Soong, T. T., Design spectra for actively controlled structures based on convex models, *Eng. Struct.*, 18(5), 341–350, 1996.
16. Wu, Z. and Soong, T. T., Modified bang-bang control law for structural control implementation, *J. Struct. Eng. ASCE*, 122(8), 771–777, 1996.
17. Yang, J. N. and Giannopolous, F., Active control and stability of cable-stayed bridge, *J. Eng. Mech. ASCE*, 105(4), 677–694, 1979.
18. Yang, J. N. and Giannopolous, F., Active control of two-cable-stayed bridge, *J. Eng. Mech. ASCE*, 105(5), 795–810, 1979.
19. Yang, J. N., Wu, J. C., Kawashima, K., and Unjoh, S., Hybrid control of seismic-excited bridge structures, *Earthquake Eng. Struct. Dyn.*, 24, 1437–1451, 1995.
20. Yang, C. and Lu, L. W., Seismic response control of cable-stayed bridges by semiactive friction damping, in *Proceedings of Fifth U.S. National Conference on Earthquake Engineering*, Vol. I, 1994, 911–920.

60

Vessel Collision Design of Bridges

Michael Knott
Moffatt & Nichol Engineers

Zolan Prucz
Modjeski and Masters, Inc.

Notations

The following symbols are used in this chapter. The section number in parentheses after definition of a symbol refers to the section or figure number where the symbol first appears or is identified.

AF annual frequency of bridge element collapse (Section 60.5.2)
B_M beam (width) of vessel (Figure 60.2)
B_P width of bridge pier (Figure 60.2)
DWT size of vessel based on deadweight tonnage (one tonne = 2205 lbs = 9.80 kN) (Section 60.4.1)
H ultimate bridge element strength (Section 60.5.2)
N number of one-way vessel passages through the bridge (Section 60.5.2)
P vessel collision impact force (Section 60.5.2)
P_{BH} ship collision impact force for head-on collision between ship bow and a rigid object (Section 60.6.1)
P_{DH} ship collision impact force between ship deckhouse and a rigid superstructure (Section 60.6.1)

0-8493-7434-0/00/$0.00+$.50
© 2000 by CRC Press LLC

P_{MT} ship collision impact force between ship mast and a rigid superstructure (Section 60.6.1)
P_S ship collision impact force for head-on collision between ship bow and a rigid object (Section 60.6.1)
PA probability of vessel aberrancy (Section 60.5.2)
PC probability of bridge collapse (Section 60.5.2)
PG geometric probability of vessel collision with bridge element (Section 60.5.2)
R_{BH} ratio of exposed superstructure depth to the total ship bow depth (Section 60.6.1)
R_{DH} reduction factor for ship deckhouse collision force (Section 60.6.1)
V design impact speed of vessel (Section 60.6.1)
x distance to bridge element from the centerline of vessel transit path (Figure 60.2)
ϕ angle between channel and bridge centerlines (Figure 60.2)

60.1 Introduction

60.1.1 Background

It was only after a marked increase in the frequency and severity of vessel collisions with bridges that studies of the vessel collision problem have been initiated in recent years. In the period from 1960 to 1998, there have been 30 major bridge collapses worldwide due to ship or barge collision, with a total loss of life of 321 people. The greatest loss of life occurred in 1983 when a passenger ship collided with a railroad bridge on the Volga River, Russia; 176 were killed when the aberrant vessel attempted to transit through a side span of the massive bridge. Most of the deaths occurred when a packed movie theater on the top deck of the passenger ship was sheared off by the low vertical clearance of the bridge superstructure.

Of the bridge catastrophes mentioned above, 15 have occurred in the United States, including the 1980 collapse of the Sunshine Skyway Bridge crossing Tampa Bay, Florida, in which 396 m of the main span collapsed and 35 lives were lost as a result of the collision by an empty 35,000 DWT (deadweight tonnage) bulk carrier (Figure 60.1).

One of the more publicized tragedies in the United States involved the 1993 collapse of a CSX Railroad Bridge across Bayou Canot near Mobile, Alabama. During dense fog, a barge tow became lost and entered a side channel of the Mobile River where it struck a railroad bridge causing a large displacement of the structure. The bridge collapsed a few minutes later when a fully loaded Amtrak passenger train attempted to cross the damaged structure; 47 fatalities occurred as a result of the collapse and the train derailment.

It should be noted that there are numerous vessel collision accidents with bridges which cause significant damage, but do not necessarily result in collapse of the structure. A study of river towboat collisions with bridges located on the U.S. inland waterway system during the short period from 1970 to 1974 revealed that there were 811 accidents with bridges costing $23 million in damages and 14 fatalities. On the average, some 35 vessel collision incidents are reported every day to U.S. Coast Guard Headquarters in Washington, D.C.

A recent accident on a major waterway bridge occurred in Portland, Maine in September 1996 when a loaded tanker ship (171 m in length and 25.9 m wide) rammed the guide pile fender system of the existing Million Dollar Bridge over the Fore River. A large portion of the fender was destroyed; the flair of the ship's bow caused significant damage to one of the bascule leafs of the movable structure (causing closure of the bridge until repairs were made); and 170,000 gallons of fuel oil were spilled in the river due to a 9-m hole ripped in the vessel hull by an underwater protrusion of the concrete support pier (a small step in the footing). Although the main cause of the accident was attributed to pilot error, a contributing factor was certainly the limited horizontal clearance of the navigation opening through the bridge (only 29 m).

The 1980 collapse of the Sunshine Skyway Bridge was a major turning point in awareness and increased concern for the safety of bridges crossing navigable waterways. Important steps in the development of modern ship collision design principles and specifications include:

FIGURE 60.1 Sunshine Skyway Bridge, May 9, 1980 after being struck by the M/V *Summit Venture.*

- In 1983, a "Committee on Ship/Barge Collision," appointed by the Marine Board of the National Research Council in Washington, D.C., completed a study on the risk and consequences of ship collisions with bridges crossing navigable coastal waters in the United States [1].

- In June 1983, a colloquium on "Ship Collision with Bridges and Offshore Structures" was held in Copenhagen, Denmark under the auspices of the International Association for Bridge and Structural Engineering (IABSE), to bring together and disseminate the latest developments on the subject [2].

- In 1984, the Louisiana Department of Transportation and Development incorporated criteria for the design of bridge piers with respect to vessel collision for structures crossing waterways in the state of Louisiana [3,4].

- In 1988, a pooled-fund research project was sponsored by 11 states and the Federal Highway Administration to develop vessel collision design provisions applicable to all of the United States. The final report of this project [5] was adopted by AASHTO as a Vessel Collision Design Guide Specification in February, 1991 [6].

- In 1993, the International Association for Bridge and Structural Engineering (IABSE) published a comprehensive document that included a review of past and recent developments in the study of ship collisions and the interaction between vessel traffic and bridges [7].

- In 1994, AASHTO adopted the recently developed LRFD bridge design specifications [8], which incorporate the vessel collision provisions developed in Reference [6] as an integral part of the bridge design criteria.

- In December 1996, the Federal Highway Administration sponsored a conference on "The Design of Bridges for Extreme Events" in Atlanta, Georgia to discuss developments in design loads (vessel collision, earthquake, and scour) and issues related to the load combinations of extreme events [9].

- In May 1998, an international symposium on "Advances in Bridge Aerodynamics, Ship Collision Analysis, and Operation & Maintenance" was held in Copenhagen, Denmark in conjunction with the opening of the record-setting Great Belt Bridge to disseminate the latest developments on the vessel collision subject [10].

Current highway bridge design practices in the United States follow the AASHTO specifications [6,8]. The design of railroad bridge protection systems against vessel collision is addressed in the American Railway Engineering and Maintenance-of-Way Association (AREMA) Manual for Railway Engineering [11]. Research and development work in the area of vessel collision with bridges continues. Several aspects, such as the magnitude of the collision loads to be used in design, and the appropriate combination of extreme events (such as collision plus scour) are not yet well established and understood. As further research results become available, appropriate code changes and updates can be expected.

60.1.2 Basic Concepts

The vulnerability of a bridge to vessel collision is affected by a variety of factors, including:

- Waterway geometry, water stage fluctuations, current speeds, and weather conditions;
- Vessel characteristics and navigation conditions, including vessel types and size distributions, speed and loading conditions, navigation procedures, and hazards to navigation;
- Bridge size, location, horizontal and vertical geometry, resistance to vessel impact, structural redundancy, and effectiveness of existing bridge protection systems;
- Serious vessel collisions with bridges are extreme events associated with a great amount of uncertainty, especially with respect to the impact loads involved. Since designing for the worst-case scenario could be overly conservative and economically undesirable, a certain amount of risk must be considered as acceptable. The commonly accepted design objective is to minimize (in a cost-effective manner) the risk of catastrophic failure of a bridge component, and at the same time reduce the risk of vessel damage and environmental pollution.

The intent of vessel collision provisions is to provide bridge components with a "reasonable" resistance capacity against ship and barge collisions. In navigable waterway areas where collision by merchant vessels may be anticipated, bridge structures should be designed to prevent collapse of the superstructure by considering the size and type of vessel, available water depth, vessel speed, structure response, the risk of collision, and the importance classification of the bridge. It should be noted that damage to the bridge (even failure of secondary structural members) is usually permitted as long as the bridge deck carrying motorist traffic does not collapse (i.e., sufficient redundancy and alternate load paths exist in the remaining structure to prevent collapse of the superstructure).

60.1.3 Application

The vessel collision design recommendations provided in this chapter are consistent with the AASHTO specifications [6,8] and they apply to all bridge components in navigable waterways with water depths over 2.0 ft (0.6 m). The vessels considered include merchant ships larger than 1000 DWT and typical inland barges.

60.2   Initial Planning

It is very important to consider vessel collision aspects as early as possible in the planning process for a new bridge, since they can have a significant effect on the total cost of the bridge. Decisions related to the bridge type, location, and layout should take into account the waterway geometry, the navigation channel layout, and the vessel traffic characteristics.

60.2.1 Selection of Bridge Site

The location of a bridge structure over a waterway is usually predetermined based on a variety of other considerations, such as environmental impacts, right-of-way, costs, roadway geometry, and political considerations. However, to the extent possible, the following vessel collision guidelines should be followed:

- Bridges should be located away from turns in the channel. The distance to the bridge should be such that vessels can line up before passing the bridge, usually at least eight times the length of the vessel. An even larger distance is preferable when high currents and winds are likely to occur at the site.
- Bridges should be designed to cross the navigation channel at right angles and should be symmetrical with respect to the channel.
- An adequate distance should exist between bridge locations and areas with congested navigation, port facilities, vessel berthing maneuvers, or other navigation problems.
- Locations where the waterway is shallow or narrow so that bridge piers could be located out of vessel reach are preferable.

60.2.2 Selection of Bridge Type, Configuration, and Layout

The selection of the type and configuration of a bridge crossing should consider the characteristics of the waterway and the vessel traffic, so that the bridge would not be an unnecessary hazard to navigation. The layout of the bridge should maximize the horizontal and vertical clearances for navigation, and the bridge piers should be placed away from the reach of vessels. Finding the optimum bridge configuration and layout for different bridge types and degrees of protection is an iterative process which weighs the costs involved in risk reduction, including political and social aspects.

60.2.3 Horizontal and Vertical Clearance

The horizontal clearance of the navigation span can have a significant impact on the risk of vessel collision with the main piers. Analysis of past collision accidents has shown that bridges with a main span less than two to three times the design vessel length or less than two times the channel width are particularly vulnerable to vessel collision.

The vertical clearance provided in the navigation span is usually based on the highest vessel that uses the waterway in a ballasted condition and during periods of high water level. The vertical clearance requirements need to consider site-specific data on actual and projected vessels, and must be coordinated with the Coast Guard in the United States. General data on vessel height characteristics are included in References [6,7].

60.2.4 Approach Spans

The initial planning of the bridge layout should also consider the vulnerability of the approach spans to vessel collision. Historical vessel collisions have shown that bridge approach spans were damaged in over 60% of the total number of accidents. Therefore, the number of approach piers exposed to vessel collision should be minimized, and horizontal and vertical clearance considerations should also be applied to the approach spans.

60.2.5 Protection Systems

Bridge protection alternatives should be considered during the initial planning phase, since the cost of bridge protection systems can be a significant portion of the total bridge cost. Bridge protection systems include fender systems, dolphins, protective islands, or other structures designed to redirect, withstand, or absorb the impact force and energy, as described in Section 60.8.

60.3 Waterway Characteristics

The characteristics of the waterway in the vicinity of the bridge site such as the width and depth of the navigation channel, the current speed and direction, the channel alignment and cross section, the water elevation, and the hydraulic conditions, have a great influence on the risk of vessel collision and must be taken into account.

60.3.1 Channel Layout and Geometry

The channel layout and geometry can affect the navigation conditions, the largest vessel size that can use the waterway, and the loading condition and speed of vessels.

The presence of bends and intersections with other waterways near the bridge increases the probability of vessels losing control and become aberrant. The navigation of downstream barge tows through bends is especially difficult.

The vessel transit paths in the waterway in relation to the navigation channel and the bridge piers can affect the risk of aberrant vessels hitting the substructure.

60.3.2 Water Depth and Fluctuations

The design water depth for the channel limits the size and draft of vessels using the waterway. In addition, the water depth plays a critical role in the accessibility of vessels to piers outside the navigation channel. The vessel collision analysis must include the possibility of ships and barges transiting ballasted or empty in the waterway. For example, a loaded barge with a 6 m draft would run aground before it could strike a pier in 4 m of water, but the same barge empty with a 1 m draft could potentially strike the pier.

The water level along with the loading condition of vessels influences the location on the pier where vessel impact loads are applied, and the susceptibility of the superstructure to vessel hits. The annual mean high water elevation is usually the minimum water level used in design. In waterways with large water stage fluctuations, the water level used can have a significant effect on the structural requirements for the pier and/or pier protection design. In these cases, a closer review of the water stage statistics at the bridge site is necessary in order to select an appropriate design water level.

60.3.3 Current Speed and Direction

Water currents at the location of the bridge can have a significant effect on navigation and on the probability of vessel aberrancy. The design water currents commonly used represent annual average values rather than the occasional extreme values that occur only a few times per year, and during which vessel traffic restrictions may also apply.

60.4 Vessel Traffic Characteristics

60.4.1 Physical and Operating Characteristics

General knowledge on the operation of vessels and their characteristics is essential for safe bridge design. The types of commercial vessels encountered in navigable waterways may be divided into ships and barge tows.

60.4.1.1 Ships

Ships are self-propelled vessels using deep-draft waterways. Their size may be determined based on the DWT. The DWT is the weight in metric tonnes (1 tonne = 2205 lbs = 9.80 kN) of cargo, stores, fuel, passenger, and crew carried by the ship when fully loaded. There are three main classes of merchant ships: bulk carriers, product carriers/tankers, and freighter/containers. General information on ship profiles, dimensions, and sizes as a function of the class of ship and its DWT is provided in References [6,7]. The dimensions given in References [6,7] are typical values, and due to the large variety of existing vessels, they should be regarded as general approximations.

The steering of ships in coastal waterways is a difficult process. It involves constant communications between the shipmaster, the helmsman, and the engine room. There is a time delay before a ship starts responding to an order to change speed or course, and the response of the ship itself is relatively slow. Therefore, the shipmaster has to be familiar with the waterway and be aware of obstructions and navigation and weather conditions in advance. Very often local pilots are used to navigate the ships through a given portion of a coastal waterway. When the navigation conditions are difficult, tugboats are used to assist ships in making turns. Ships need speed to be able to steer and maintain rudder control. A minimum vessel speed of about 5 knots (8 km/h) is usually needed to maintain steering. Fully loaded ships are more maneuverable, and in deep water they are directionally stable and can make turns with a radius equal to one to two times the length of the ship. However, as the underkeel clearance decreases to less than half the draft of the ship, many ships tend to become directionally unstable, which means that they require constant steering to keep them traveling in a straight line. In the coastal waterways of the United States, the underkeel clearance of many laden ships may be far less than this limit, in some cases as small as 5% of the draft of the ship. Ships riding in ballast with shallow draft are less maneuverable than loaded ships, and, in addition, they can be greatly affected by winds and currents. Historical accident data indicate that most bridge accidents involve empty or ballasted vessels.

60.4.1.2 Barge Tows

Barge tows use both deep-draft and shallow-draft waterways. The majority of the existing bridges cross shallow draft waterways where the vessel fleet comprises barge tows only. The size of barges in the United States is usually defined in terms of the cargo-carrying capacity in short tons (1 ton = 2000 lbs = 8.90 kN). The types of inland barges include open and covered hoppers, tank barges, and deck barges. They are rectangular in shape and their dimensions are quite standard so they can travel in tows. The number of barges per tow can vary from one to over 20, and their configuration, is affected by the conditions of the waterway. In most cases barges are pushed by a towboat. Information on barge dimensions and capacity, as well as on barge tow configurations is included in References [6,7]. A statistical analysis of barge tow types, configurations, and dimensions, which utilizes barge traffic data from the Ohio River, is reported in Reference [12].

It is very difficult to control and steer barge tows, especially in waterways with high stream velocities and cross currents. Taking a turn in a fast waterway with high current is a serious undertaking. In maneuvering a bend, tows experience a sliding effect in a direction opposite to the direction of the turn, due to inertial forces, which are often coupled with the current flow. Sometimes, bridge piers and fenders are used to line up the tow before the turn. Bridges located in a high-velocity waterway near a bend in the channel will probably be hit by barges numerous times during their lifetime. In general, there is a high likelihood that any bridge element that can be reached by a barge will be hit during the life of the bridge.

60.4.2 Vessel Fleet Characteristics

The vessel data required for bridge design include types of vessels and size distributions, transit frequencies, typical vessel speeds, and loading conditions. In order to determine the vessel size distribution at the bridge site, detailed information on both present and projected future vessel traffic is needed. Collecting data on the vessel fleet characteristics for the waterway is an important and often time-consuming process.

Some of the sources in the United States for collecting vessel traffic data are listed below:

- U.S. Army Corps of Engineers, District Offices
- Port authorities and industries along the waterway

- Local pilot associations and merchant marine organizations
- U.S. Coast Guard, Marine Safety & Bridge Administration Offices
- U.S. Army Corps of Engineers, "Products and Services Available to the Public," Water Resources Support Center, Navigation Data Center, Fort Belvoir, Virginia, NDC Report 89-N-1, August 1989
- U.S. Army Corps of Engineers, "Waterborne Commerce of the United States (WCUS), Parts 1 thru 5," Water Resources Support Center (WRSC), Fort Belvoir, Virginia
- U.S. Army Corps of Engineers, "Lock Performance Monitoring (LPM) Reports," Water Resources Support Center (WRSC), Fort Belvoir, Virginia
- Shipping registers (American Bureau of Shipping Register, New York; and Lloyd's Register of Shipping, London)
- Bridge tender reports for movable bridges

Projections for anticipated vessel traffic during the service life of the bridge should address both changes in the volume of traffic and in the size of vessels. Factors that need to be considered include:

- Changes in regional economics;
- Plans for deepening or widening the navigation channel;
- Planned changes in alternate waterway routes and in navigation patterns;
- Plans for increasing the size and capacity of locks leading to the bridge;
- Port development plans.

Vessel traffic projections that are made by the Maritime Administration of the U.S. Department of Transportation, Port Authorities, and U.S. Army Corps of Engineers in conjunction with planned channel-deepening projects or lock replacements are also good sources of information for bridge design. Since a very large number of factors can affect the vessel traffic in the future, it is important to review and update the projected traffic during the life of the bridge.

60.5 Collision Risk Analysis

60.5.1 Risk Acceptance Criteria

Bridge components exposed to vessel collision could be subjected to a very wide range of impact loads. Due to economic and structural constraints bridge design for vessel collision is not based on the worst-case scenario, and a certain amount of risk is considered acceptable.

The risk acceptance criteria consider both the probability of occurrence of a vessel collision and the consequences of the collision. The probability of occurrence of a vessel collision is affected by factors related to the waterway, vessel traffic, and bridge characteristics. The consequences of a collision depend on the magnitude of the collision loads and the bridge strength, ductility, and redundancy characteristics. In addition to the potential for loss of life, the consequences of a collision can include damage to the bridge, disruption of motorist and marine traffic, damage to the vessel and cargo, regional economic losses, and environmental pollution.

Acceptable risk levels have been established by various codes and for individual bridge projects [2–10]. The acceptable annual frequencies of bridge collapse values used generally range from 0.001 to 0.0001. These values were usually determined in conjunction with the risk analysis procedure recommended, and should be used accordingly.

The AASHTO provisions [6,8] specify an annual frequency of bridge collapse of 0.0001 for critical bridges and an annual frequency of bridge collapse of 0.001 for regular bridges. These annual frequencies correspond to return periods of bridge collapse equal to 1 in 10,000 years, and 1 in 1000 years, respectively. Critical bridges are defined as those bridges that are expected to continue to function after a major impact, because of social/survival or security/defense requirements.

60.5.2 Collision Risk Models

60.5.2.1 General Approach

Various collision risk models have been developed to achieve design acceptance criteria [2–10]. In general, the occurrence of a collision is separated into three events: (1) a vessel approaching the bridge becomes aberrant, (2) the aberrant vessel hits a bridge element, and (3) the bridge element that is hit fails. Collision risk models consider the effects of the vessel traffic, the navigation conditions, the bridge geometry with respect to the waterway, and the bridge element strength with respect to the impact loads. They are commonly expressed in the following form [6,8]:

$$AF = (N) \, (PA) \, (PG) \, (PC) \tag{60.1}$$

where *AF* is the annual frequency of collapse of a bridge element; *N* is the annual number of vessel transits (classified by type, size, and loading condition) which can strike a bridge element; *PA* is the probability of vessel aberrancy; *PG* is the geometric probability of a collision between an aberrant vessel and a bridge pier or span; *PC* is the probability of bridge collapse due to a collision with an aberrant vessel.

60.5.2.2 Vessel Traffic Distribution, *N*

The number of vessels, *N*, passing the bridge based on size, type, and loading condition and available water depth has to be developed for each pier and span component to be evaluated. All vessels of a given type and loading condition have to be divided into discrete groupings of vessel size by DWT to determine the contribution of each group to the annual frequency of bridge element collapse. Once the vessels are grouped and their frequency distribution is established, information on typical vessel characteristics may be obtained from site-specific data, or from published general data such as References [6,7].

60.5.2.3 Probability of Aberrancy, *PA*

The probability of vessel aberrancy reflects the likelihood that a vessel is out of control in the vicinity of a bridge. Loss of control may occur as a result of pilot error, mechanical failure, or adverse environmental conditions. The probability of aberrancy is mainly related to the navigation conditions at the bridge site. Vessel traffic regulations, vessel traffic management systems, and aids to navigation can improve the navigation conditions and reduce the probability of aberrancy.

The probability of vessel aberrancy may be evaluated based on site-specific information that includes historical data on vessel collisions, rammings, and groundings in the waterway, vessel traffic, navigation conditions, and bridge/waterway geometry. This has been done for various bridge design provisions and specific bridge projects worldwide [2,3,7,9,12]. The probability of aberrancy values determined range from 0.5×10^{-4} to over 7.0×10^{-4}.

As an alternative, the AASHTO provisions [6,8] recommend base rates for the probability of vessel aberrancy that are multiplied by correction factors for bridge location relative to bends in the waterway, currents acting parallel to vessel transit path, crosscurrents acting perpendicular to vessel transit path, and the traffic density of vessels using the waterway. The recommended base rates are 0.6×10^{-4} for ships, and 1.2×10^{-4} for barges.

60.5.2.4 Geometric Probability, *PG*

The geometric probability is the probability that a vessel will hit a particular bridge pier given that it has lost control (i.e., is aberrant) in the vicinity of the bridge. It is mainly a function of the geometry of the bridge in relation to the waterway. Other factors that can affect the likelihood that an aberrant vessel will strike a bridge element include the original vessel transit path, course, rudder position, velocity at the time of failure, vessel type, size, draft and maneuvering characteristics, and the hydraulic and environmental conditions at the bridge site. Various geometric probability models, some based on simulation studies, have been recommended and used on different bridge projects

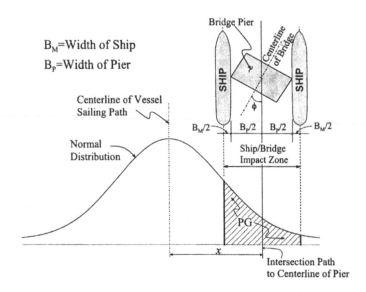

B$_M$=Width of Ship
B$_P$=Width of Pier

Centerline of Vessel
Sailing Path

Normal
Distribution

FIGURE 60.2 Geometric probability of pier collision.

[2,3,7]. The AASHTO provisions [6,8] use a normal probability density function about the centerline of the vessel transit path for estimating the likelihood of an aberrant vessel being within a certain impact zone along the bridge axis. Using a normal distribution accounts for the fact that aberrant vessels are more likely to pass under the bridge closer to the navigation channel than farther away from it. The standard deviation of the distribution equals the length of the design vessel considered. The probability that an aberrant vessel is located within a certain zone is the area under the normal probability density function within that zone (Figure 60.2).

Bridge elements beyond three times the standard deviation from the centerline of vessel transit path are designed for specified minimum impact load requirements, which are usually associated with an empty vessel drifting with the current.

60.5.2.5 Probability of Collapse, *PC*

The probability of collapse, *PC*, is a function of many variables, including vessel size, type, forepeak ballast and shape, speed, direction of impact, and mass. It is also dependent on the ultimate lateral load strength of the bridge pier (particularly the local portion of the pier impacted by the bow of the vessel). Based on collision damages observed from numerous ship–ship collision accidents which have been correlated to the bridge–ship collision situation [2], an empirical relationship has been developed based on the ratio of the ultimate pier strength, *H*, to the vessel impact force, *P*. As shown in Figure 60.3, for *H/P* ratios less than 0.1, *PC* varies linearly from 0.1 at *H/P* = 0.1 to 1.0 at *H/P* = 0.0. For *H/P* ratios greater than 0.1, *PC* varies linearly from 0.1 at *H/P* = 0.1 to 0.0 at *H/P* = 1.0.

60.6 Vessel Impact Loads

60.6.1 Ship Impact

The estimation of the load on a bridge pier during a ship collision is a very complex problem. The actual force is time dependent, and varies depending on the type, size, and construction of the vessel; its velocity; the degree of water ballast in the forepeak of the bow; the geometry of the collision; and the geometry and strength characteristics of the bridge. There is a very large scatter among the collision force values recommended in various vessel collision guidelines or used in various bridge projects [2–10].

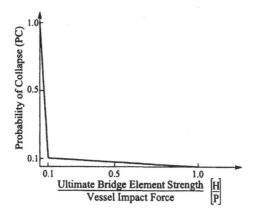

FIGURE 60.3 Probability of collapse distribution.

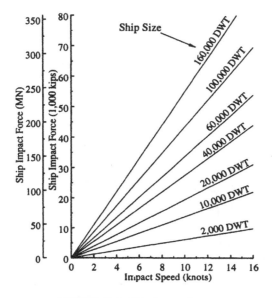

FIGURE 60.4 Ship impact force.

Ship collision forces are commonly applied as equivalent static loads. Procedures for evaluating dynamic effects when the vessel force indentation behavior is known are included in References [3,4,10,13,14]. The AASHTO provisions [6,8] use the following formula for estimating the static head-on ship collision force, P_S, on a rigid pier:

$$P_s = 0.98(DWT)^{\frac{1}{2}}(V/16) \tag{60.2}$$

where P_S is the equivalent static vessel impact force (MN); DWT is the ship deadweight tonnage in tonnes; and V is the vessel impact velocity in knots (Figure 60.4). This formulation was primarily developed from research conducted by Woisin in West Germany during 1967 to 1976 on physical ship models to generate data for protecting the reactors of nuclear power ships from collisions with other ships. A schematic representation of a typical impact force time history is shown in Figure 60.6 based on Woisin's test data. The scatter in the results of these tests is of the order of ±50%. The formula recommended (Eq. 60.2) uses a 70% fractile of an assumed triangular distribution with zero values at 0% and 100% and a maximum value at the 50% level (Figure 60.7).

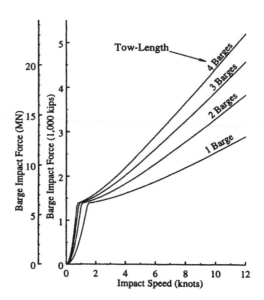

FIGURE 60.5   Barge impact force.

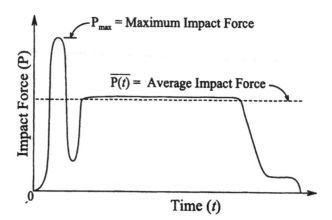

FIGURE 60.6   Typical ship impact force time history by Woisin.

Formulas for computing design ship collision loads on a bridge superstructure are given in the AASHTO provisions [6,8] as a function of the design ship impact force, P_S, as follows:

- Ship Bow Impact Force, P_{BH}:

$$P_{BH} = (R_{BH}) (P_S) \qquad (60.3)$$

where R_{BH} is a reduction coefficient equal to the ratio of exposed superstructure depth to the total bow depth.

- Ship Deckhouse Impact Force, P_{DH}:

$$P_{DH} = (R_{DH}) (P_S) \qquad (60.4)$$

where R_{DH} is a reduction coefficient equal to 0.10 for ships larger than 100,000 DWT, and

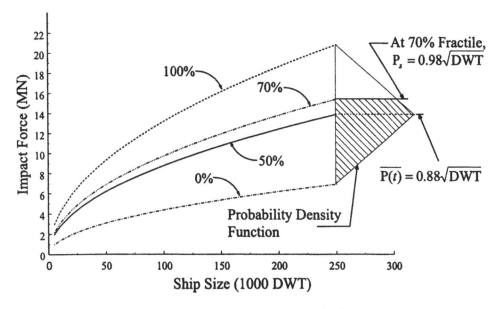

FIGURE 60.7 Probability density function of ship impact force.

$$0.2 - \frac{DWT}{100,000}(0.10)$$

for ships under 100,000 DWT.

- Ship Mast Impact Force, P_{MT}:

$$P_{MT} = 0.10\, P_{DH} \qquad\qquad (60.5)$$

where P_{DH} is the ship deckhouse impact force.

The magnitude of the impact loads computed for ship bow and deckhouse collisions are quite high relative to the strength of most bridge superstructure designs. Also, there is great uncertainty associated with predicting ship collision loads on superstructures because of the limited data available and the ship–superstructure load interaction effects. It is therefore suggested that superstructures, and also weak or slender parts of the substructure, be located out of the reach of a ship's hull or bow.

60.6.2 Barge Impact

The barge collision loads recommended by AASHTO for the design of piers are shown in Figure 60.5 as a function of the tow length and the impact speed. Numerical formulations for deriving these relationships may be found in References [6,8].

The loads in Figure 60.5 were computed using a standard 59.5 × 10.7 m hopper barge. The impact force recommended for barges larger than the standard hopper barge is determined by increasing the standard barge impact force by the ratio of the width of the wider barge to the width of the standard hopper barge.

60.6.3 Application of Impact Forces

Collision forces on bridge substructures are commonly applied as follows:

- 100% of the design impact force in a direction parallel to the navigation channel (i.e., head-on);
- 50% of the design impact force in the direction normal to the channel (but not simultaneous with the head-on force);
- For overall stability, the design impact force is applied as a concentrated force at the mean high water level;
- For local collision forces, the design impact force is applied as a vertical line load equally distributed along the ship's bow depth for ships, and along head log depth for barges;
- For superstructure design the impact forces are applied transversely to the superstructure component in a direction parallel to the navigation channel.

When determining the bridge components exposed to physical contact by any portion of the hull or bow of the vessel considered, the bow overhang, rake, or flair distance of vessels have to be taken into account. The bow overhang of ships and barges is particularly dangerous for bridge columns and for movable bridges with relatively small navigation clearances.

60.7 Bridge Analysis and Design

Vessel collisions are extreme events with a very low probability of occurrence; therefore the limit state considered is usually structural survival. Depending on the importance of the bridge, various degrees of damage are allowed — provided that the structure maintains its integrity, hazards to traffic are minimized, and repairs can be made in a relatively short period of time. When the design is based on more frequent but less severe collisions, structural damage and traffic interruptions are not allowed.

Designing for vessel collision is commonly based on equivalent static loads that include global forces for checking overall capacity and local forces for checking local strength of bridge components. A clear load path from the location of the vessel impact to the bridge foundation needs to be established and the components and connections within the load path must be adequately designed and detailed. The design of individual bridge components is based on strength and stability criteria. Overall stability, redundancy, and ductility are important criteria for structural survival.

The contribution of the superstructure to the transfer of loads to adjacent substructure units depends on the capacity of the connection of the superstructure to substructure and the relative stiffness of the substructure at the location of the impact. Analysis guidelines for determining the distribution of collision loads to adjacent piers are included in Reference [15]. To find out how much of the transverse impact force is taken by the pier and how much is transferred to the superstructure, two analytical models are typically used. One is a two-dimensional or a three-dimensional model of the complete pier, and the other is a two-dimensional model of the super-structure projected on a horizontal plane. The projected superstructure may be modeled as a beam with the moment of inertia referred to a vertical axis through the center of the roadway, and with hinges at expansion joint locations. The beam is supported at pier locations by elastic horizontal springs representing the flexibility of each pier. The flexibility of the piers is obtained from pier models using virtual forces. The superstructure model is loaded with a transverse virtual force acting at the place where the pier under consideration is located. The spring in the model at that place is omitted to obtain a flexibility coefficient of the superstructure at the location of the top of the pier under consideration. Thus, the horizontal displacement of the top of the pier due to the impact force on the pier (usually applied at mean high water level) is equal to the true displacement of the superstructure due to the transmitted part of the impact force. The magnitude of the force transmitted to the superstructure is obtained by equating the total true displacement of the top of the pier from the pier model to the displacement of the superstructure. However, in order to consider partial transfer of lateral forces to the superstructure, positive steel or concrete connections of superstructure to substructure, such as shear keys must be provided. Similarly, for partial transfer to the superstructure of the longitudinal component of the impact force the shear capacity of the bearings must be adequate. When elastomeric bearings are used their longitudinal flexibility may be added to the longitudinal flexibility of the piers. If the ultimate capacity of the bearings is exceeded, then the pier must take the total longitudinal force and be treated as a cantilever.

The modeling of pile foundations could vary from the simple assumption of a point of fixity to nonlinear soil–structure interaction models, depending on the limit state considered and the sensitivity of the response to the soil conditions. Lateral load capacity analysis methods for pile groups that include nonlinear behavior are recommended in References [15,16] and the features of a finite-element analysis computer program developed for bridge piers composed of pier columns and cap supported on a pile cap and nonlinear piles and soil are presented in Reference [17]. Transient foundation uplift or rocking involving separation from the subsoil of an end bearing foundation pile group or the contact area of a foundation footing could be allowed under impact loading provided sufficient consideration is given to the structural stability of the substructure.

60.8 Bridge Protection Measures

The cost associated with protecting a bridge from catastrophic vessel collision can be a significant portion of the total bridge cost, and must be included as one of the key planning elements in establishing a bridge's type, location, and geometry. The alternatives listed below are usually evaluated in order to develop a cost-effective solution for a new bridge project:

- Design the bridge piers, foundations, and superstructure to withstand directly the vessel collision forces and impact energies;
- Design a pier fender system to reduce the impact loads to a level below the capacity of the pier and foundation;
- Increase span lengths and locate piers in shallow water out of reach of large vessels in order to reduce the impact design loads; and
- Protect piers from vessel collision by means of physical protection systems.

60.8.1 Physical Protection Systems

Piers exposed to vessel collision can be protected by special structures designed to absorb the impact loads (forces or energies), or redirect the aberrant vessel away from the pier. Because of the large forces and energies involved in a vessel collision, protection structures are usually designed for plastic deformation under impact (i.e., they are essentially destroyed during the head-on design collision and must be replaced). General types of physical protection systems include:

Fender Systems. These usually consist of timber, rubber, steel, or concrete elements attached to a pier to fully, or partially, absorb vessel impact loads. The load and energy absorbing characteristics of such fenders is relatively low compared with typical vessel impact design loads.

Pile-Supported Systems. These usually consist of pile groups connected by either flexible or rigid caps to absorb vessel impact forces. The piles may be vertical (plumb) or battered depending on the design approach followed, and may incorporate relatively large-diameter steel pipe or concrete pile sizes. The pile-supported protection structure may be either freestanding away from the pier, or attached to the pier itself. Fender systems may be attached to the pile structure to help resist a portion of the impact loads.

Dolphin Protection Systems. These usually consist of large-diameter circular cells constructed of driven steel sheet piles, filled with rock or sand, and topped by a thick concrete cap. Vessel collision loads are absorbed by rotation and lateral deformation of the cell during impact.

Island Protection Systems. These usually consist of protective islands built of a sand or quarry-run rock core and protected by outer layers of heavy rock riprap for wave, current, and ice protection. The island geometry is developed to stop an aberrant vessel from hitting a pier by forcing it to run aground. Although extremely effective as protection systems, islands are often difficult to use due to adverse environmental impacts on river bottoms (dredge and fill permits) and river currents (increase due to blockage), as well as impacts due to settlement and downdrag forces on the bridge piers.

Floating Protection Systems. These usually consist of cable net systems suspended across the waterway to engage and capture the bow of an aberrant vessel, or floating pontoons anchored in front of the piers. Floating protection systems have a number of serious drawbacks (environmental, effectiveness, maintenance, cost, etc.) and are usually only considered for extremely deep water situations where other protection options are not practicable.

The AASHTO Guide Specification [6] provides examples and contains a relatively extensive discussion of various types of physical protection systems, such as fenders, pile-supported structures, dolphins, protective islands, and floating structures. However, the code does not include specific procedures and recommendations on the actual design of such protection structures. Further research is needed to establish consistent analysis and design methodologies for protection structures, particularly since these structures undergo large plastic deformations during the collision.

60.8.2 Aids to Navigation Alternatives

Since 60 to 85% of all vessel collisions are caused by pilot error, it is important that all aspects of the bridge design, siting, and aids to navigation with respect to the navigation channel be carefully evaluated with the purpose of improving or maintaining safe navigation in the waterway near the bridge. Traditional aids include buoys, range markers, navigation lighting, and radar reflectors, as well as standard operating procedures and regulations specifically developed for the waterway by government agencies and pilot associations. Modern aids include advanced vessel traffic control systems (VTS) using shore-based radar surveillance and radio-telephone communication systems; special electronic transmitters known as Raycon devices mounted to bridge spans for improved radar images indicating the centerline of the channel; and advanced navigation positioning systems based on shipboard global positioning satellite (GPS) receivers using differential signal techniques to improve location accuracy.

Studies have indicated that improvements in the aids to navigation near a bridge can provide extremely cost-effective solutions to reducing the risk of collisions to acceptable levels. The cost of such aid to navigation improvements and shipboard electronic navigation systems is usually a fraction of the cost associated with expensive physical protection alternatives. However, few electronic navigation systems have ever been implemented (worldwide) due to legal complications arising from liability concerns; impacts on international laws governing trade on the high seas; and resistance by maritime users.

It should be noted that the traditional isolation of the maritime community must come to an end. In addition to the bridge costs, motorist inconvenience, and loss of life associated with a catastrophic vessel collision, significant environmental damage can also occur due to spilled hazardous or noxious cargoes in the waterway. The days when the primary losses associated with an accident rested with the vessel and her crew are over. The $13 million value of the *M/V Summit Venture* was far below the $250 million replacement cost of the Sunshine Skyway Bridge which the vessel destroyed. The losses associated with the 11 million gallons of crude oil spilled from the *M/V Exxon Valdez* accident off the coast of Alaska in 1989 are over $3.5 billion. Both of these accidents could have been prevented using shipboard advanced electronic navigation systems.

60.9 Conclusions

Experience to date has shown that the use of the vessel impact and bridge protection requirements (such as the AASHTO specifications [6,8]) for planning and design of new bridges has resulted in a significant change in proposed structure types over navigable waterways. Incorporation of the risk of vessel collision and cost of protection in the total bridge cost has almost always resulted in longer-span bridges being more economical than traditional shorter span structures, since the design goal for developing the bridge pier and span layout is the least cost of the total structure (including the protection costs). Typical costs for incorporating vessel collision and protection issues in the planning stages of a new bridge have ranged from 5% to 50% of the basic structure cost without protection.

Experience has also shown that it is less expensive to include the cost of protection in the planning stages of a proposed bridge, than to add it after the basic span configuration has been established without considering vessel collision concerns. Typical costs for adding protection, or for retrofitting an existing bridge for vessel collision, have ranged from 25% to over 100% of the existing bridge costs.

It is recognized that vessel collision is but one of a multitude of factors involved in the planning process for a new bridge. The designer must balance a variety of needs including political, social, and economic in arriving at an optimal bridge solution for a proposed highway crossing. Because of the relatively high bridge costs associated with vessel collision design for most waterway crossings, it is important that additional research be conducted to improve our understanding of vessel impact mechanics, the response of the structure, and the development of cost-effective protection systems.

References

1. National Research Council, *Ship Collisions with Bridges — The Nature of the Accidents, Their Prevention and Mitigation,* National Academy Press, Washington, D.C., 1983.
2. IABSE, *Ship Collision with Bridges and Offshore Structures,* International Association for Bridge and Structural Engineering, Colloquium Proceedings, Copenhagen, Denmark, 3 vols. (Introductory, Preliminary, and Final Reports), 1983.
3. Modjeski and Masters, Criteria for the Design of Bridge Piers with Respect to Vessel Collision in Louisiana Waterways, Report prepared for Louisiana Department of Transportation and Development and the Federal Highway Administration, 1984.
4. Prucz, Z. and Conway, W. B., Design of bridge piers against ship collision, in *Bridges and Transmission Line Structures,* L. Tall, Ed., ASCE, New York, 1987, 209–223.
5. Knott, M. A. and Larsen, O. D. 1990. *Guide Specification and Commentary for Vessel Collision Design of Highway Bridges,* U.S. Department of Transportation, Federal Highway Administration, Report No. FHWA-RD-91-006.
6. AASHTO, *Guide Specification and Commentary for Vessel Collision Design of Highway Bridges,* American Association of State Highway and Transportation Officials, Washington, D.C., 1991.
7. Larsen, O. D., Ship Collision with Bridges: The Interaction between Vessel Traffic and Bridge Structures, IABSE Structural Engineering Document 4, IABSE-AIPC-IVBH, Zürich, Switzerland, 1993.
8. AASHTO, *LRFD Bridge Design Specifications and Commentary,* American Association of State Highway and Transportation Officials, Washington, D.C., 1994.
9. FHWA, *The Design of Bridges for Extreme Events,* Proceedings of Conference in Atlanta, Georgia, December 3–6, 1996.
10. *International Symposium on Advances in Bridge Aerodynamics, Ship Collision Analysis, and Operation & Maintenance,* Copenhagen, Denmark, May 10–13, Balkema Publishers, Rotterdam, Netherlands, 1998.
11. AREMA, *Manual for Railway Engineering,* Chapter 8, Part 23, American Railway Engineering Association, Washington, D.C., 1999.
12. Whitney, M. W., Harik, I. E., Griffin, J. J., and Allen, D. L. Barge collision design of highway bridges, *J. Bridge Eng. ASCE,* 1(2), 47–58, 1996.
13. Prucz, Z. and Conway, W. B., Ship Collision with Bridge Piers — Dynamic Effects, Transportation Research Board Paper 890712, Transportation Research Board, Washington, D.C., 1989.

14. Grob, B. and Hajdin, N., Ship impact on inland waterways, *Struct. Eng. Int.*, IABSE, Zürich, Switzerland, 4, 230–235, 1996.
15. Kuzmanovic, B. O. and Sanchez, M. R., Design of bridge pier pile foundations for ship impact, *J. Struct. Eng. ASCE*, 118(8), 2151–2167, 1992.
16. Brown, D. A. and Bollmann, H. T. Pile supported bridge foundations designed for impact loading, Transportation Research Record 1331, TRB, National Research Council, Washington, D.C., 87–91, 1992.
17. Hoit, M., McVay, M., and Hays, C., Florida Pier Computer Program for Bridge Substructure Analysis: Models and Methods, Conference Proceedings, Design of Bridges for Extreme Events, FHWA, Washington, D.C., 1996.

61

Bridge Hydraulics

Jim Springer
*California Department
of Transportation*

Ke Zhou
*California Department
of Transportation*

61.1 Introduction

This chapter presents bridge engineers basic concepts, methods, and procedures used in bridge hydraulic analysis and design. It involves hydrology study, hydraulic analysis, on-site drainage design, and bridge scour evaluation.

Hydrology study for bridge design mainly deals with the properties, distribution, and circulation of water on and above the land surface. The primary objective is to determine either the peak discharge or the flood hydrograph, in some cases both, at the highway stream crossings. Hydraulic analysis provides essential methods to determine runoff discharges, water profiles, and velocity distribution. The on-site drainage design part of this chapter is presented with the basic procedures and references for bridge engineers to design bridge drainage.

Bridge scour is a big part of this chapter. Bridge engineers are systematically introduced to concepts of various scour types, presented with procedures and methodology to calculate and evaluate bridge scour depths, provided with guidelines to conduct bridge scour investigation and to design scour preventive measures.

61.2 Bridge Hydrology and Hydraulics

61.2.1 Hydrology

61.2.1.1 Collection of Data

Hydraulic data for the hydrology study may be obtained from the following sources: as-built plans, site investigations and field surveys, bridge maintenance books, hydraulic files from experienced report writers, files of government agencies such as the U.S. Corps of Engineers studies, U.S. Geological Survey (USGS), Soil Conservation Service, and FEMA studies, rainfall data from local water agencies, stream gauge data, USGS and state water agency reservoir regulation, aerial photographs, and floodways, etc.

Site investigations should always be conducted except in the simplest cases. Field surveys are very important because they can reveal conditions that are not readily apparent from maps, aerial

0-8493-7434-0/00/$0.00+$.50
© 2000 by CRC Press LLC

photographs and previous studies. The typical data collected during a field survey include high water marks, scour potential, stream stability, nearby drainage structures, changes in land use not indicated on maps, debris potential, and nearby physical features. See HEC-19, Attachment D [16] for a typical Survey Data Report Form.

61.2.1.2 Drainage Basin

The area of the drainage basin above a given point on a stream is a major contributing factor to the amount of flow past that point. For given conditions, the peak flow at the proposed site is approximately proportional to the drainage area.

The shape of a basin affects the peak discharge. Long, narrow basins generally give lower peak discharges than pear-shaped basins. The slope of the basin is a major factor in the calculation of the time of concentration of a basin. Steep slopes tend to result in shorter times of concentration and flatter slopes tend to increase the time of concentration. The mean elevation of a drainage basin is an important characteristic affecting runoff. Higher elevation basins can receive a significant amount of precipitation as snow. A basin orientation with respect to the direction of storm movement can affect peak discharge. Storms moving upstream tend to produce lower peaks than those moving downstream.

61.2.1.3 Discharge

There are several hydrologic methods to determine discharge. Most of the methods for estimating flood flows are based on statistical analyses of rainfall and runoff records and involve preliminary or trial selections of alternative designs that are judged to meet the site conditions and to accommodate the flood flows selected for analysis.

Flood flow frequencies are usually calculated for discharges of 2.33 years through the overtopping flood. The frequency flow of 2.33 years is considered to be the mean annual discharge. The base flood is the 100-year discharge (1% frequency). The design discharge is the 50-year discharge (2% frequency) or the greatest of record, if practical. Many times, the historical flood is so large that a structure to handle the flow becomes uneconomical and is not warranted. It is the engineer's responsibility to determine the design discharge. The overtopping discharge is calculated at the site, but may overtop the roadway some distance away from the site.

Changes in land use can increase the surface water runoff. Future land-use changes that can be reasonably anticipated to occur in the design life should be used in the hydrology study. The type of surface soil is a major factor in the peak discharge calculation. Rock formations underlying the surface and other geophysical characteristics such as volcanic, glacial, and river deposits can have a significant effect on runoff. In the United States, the major source of soil information is the Soil Conservation Service (SCS). Detention storage can have a significant effect on reducing the peak discharge from a basin, depending upon its size and location in the basin.

The most commonly used methods to determine discharges are

1. Rational method
2. Statistical Gauge Analysis Methods
3. Discharge comparison of adjacent basins from gauge analysis
4. Regional flood-frequency equations
5. Design hydrograph

The results from various methods of determining discharge should be compared, not averaged.

61.2.1.3.1 Rational Method

The rational method is one of the oldest flood calculation methods and was first employed in Ireland in urban engineering in 1847. This method is based on the following assumptions:

TABLE 61.1 Runoff Coefficients for Developed Areas

Type of Drainage Area	Runoff Coefficient
Business	
Downtown areas	0.70–0.95
Neighborhood areas	0.50–0.70
Residential areas	
Single-family areas	0.30–0.50
Multiunits, detached	0.40–0.60
Multiunits, attached	0.60–0.75
Suburban	0.25–0.40
Apartment dwelling areas	0.50–0.70
Industrial	
Light areas	0.50–0.80
Heavy areas	0.60–0.90
Parks, cemeteries	0.10–0.25
Playgrounds	0.20–0.40
Railroad yard areas	0.20–0.40
Unimproved areas	0.10–0.30
Lawns	
Sandy soil, flat, 2%	0.05–0.10
Sandy soil, average, 2–7%	0.10–0.15
Sandy soil, steep, 7%	0.15–0.20
Heavy soil, flat, 2%	0.13–0.17
Heavy soil, average, 2–7%	0.18–0.25
Heavy soil, steep, 7%	0.25–0.35
Streets	
Asphaltic	0.70–0.95
Concrete	0.80–0.95
Brick	0.70–0.85
Drives and walks	0.75–0.85
Roofs	0.75–0.95

1. Drainage area is smaller than 300 acres.
2. Peak flow occurs when all of the watershed is contributing.
3. The rainfall intensity is uniform over a duration equal to or greater than the time of concentration, T_c.
4. The frequency of the peak flow is equal to the frequency of the rainfall intensity.

$$Q = CiA \tag{61.1}$$

where
Q = discharge, in cubic foot per second
C = runoff coefficient (in %) can be determined in the field and from Tables 61.1 and 61.2 [5,16] or a weighted C value is used when the basin has varying amounts of different cover. The weighted C value is determined as follows:

$$C = \frac{\sum C_j A_j}{\sum A_j} \tag{61.2}$$

i = rainfall intensity (in inches per hour) can be determined from either regional IDF maps or individual IDF curves
A = drainage basin area (in acres) is determined from topographic map

(*Note:* 1 sq. mile = 640 acres = 0.386 sq. kilometer)

TABLE 61.2   Runoff Coefficients for Undeveloped Area Watershed Types

Soil	0.12–0.16	0.08–0.12	0.06–0.08	0.04–0.06
	No effective soil cover, either rock or thin soil mantle of negligible infiltration capacity	Slow to take up water, clay or shallow loam soils of low infiltration capacity, imperfectly or poorly drained	Normal, well-drained light or medium-textured soils, sandy loams, silt and silt loams	High, deep sand or other soil that takes up water readily, very light well-drained soils
Vegetal Cover	0.12–0.16	0.08–0.12	0.06–0.08	0.04–0.06
	No effective plant cover, bare or very sparse cover	Poor to fair; clean cultivation crops, or poor natural cover, less than 20% of drainage area over good cover	Fair to good; about 50% of area in good grassland or woodland, not more than 50% of area in cultivated crops	Good to excellent; about 90% of drainage area in good grassland, woodland or equivalent cover
Surface Storage	0.10–0.12	0.08–0.10	0.06–0.08	0.04–0.06
	Negligible surface depression few and shallow, drainageways steep and small, no marshes	Low, well-defined system of small drainageways; no ponds or marshes	Normal; considerable surface depression storage; lakes and pond marshes	High; surface storage, high; drainage system not sharply defined; large floodplain storage or large number of ponds or marshes

The time of concentration for a pear-shaped drainage basin can be determined using a combined overland and channel flow equation, the Kirpich equation:

$$T_c = 0.0195(L / S^{0.5})^{0.77} \tag{61.3}$$

where
T_c = Time of concentration in minutes
L = Horizontally projected length of watershed in meters
$S = H/L$ (H = difference in elevation between the most remote point in the basin and the outlet in meters)

61.2.1.3.2 Statistical Gauge Analysis Methods
The following two methods are the major statistical analysis methods which are used with stream gauge records in the hydrological analysis.

1. Log Pearson Type III method
2. Gumbel extreme value method

The use of stream gauge records is a preferred method of estimating discharge/frequencies since they reflect actual climatology and runoff. Discharge records, if available, may be obtained from a state department of water resources in the United States. A good record set should contain at least 25 years of continuous records.

It is important, however, to review each individual stream gauge record carefully to ensure that the database is consistent with good statistical analysis practice. For example, a drainage basin with a large storage facility will result in a skewed or inconsistent database since smaller basin discharges will be influenced to a much greater extent than large discharges.

The most current published stream gauge description page should be reviewed to obtain a complete idea of the background for that record. A note should be given to changes in basin area over time, diversions, revisions, etc. All reliable historical data outside of the recorded period should be included. The adjacent gauge records for supplemental information should be checked and utilized to extend the record if it is possible. Natural runoff data should be separated from later controlled data. It is known that high-altitude basin snowmelt discharges are not compatible with rain flood discharges. The zero years must also be accounted for by adjusting the final plot positions, not by inclusion as minor flows. The generalized skew number can be obtained from the chart in Bulletin No.17 B [8].

Quite often the database requires modification for use in a Log Pearson III analysis. Occasionally, a high outlier, but more often low outliers, will need to be removed from the database to avoid skewing results. This need is determined for high outliers by using $Q_H = \bar{Q}_H + K S_H$, and low outliers by using $Q_L = \bar{Q}_L + K S_L$, where K is a factor determined by the sample size, $\bar{Q}_H$ and $\bar{Q}_L$ are the high and low mean logarithm of systematic peaks, Q_H and Q_L are the high and low outlier thresholds in log units, S_H and S_L are the high and low standard deviations of the logarithmic distribution. Refer to FHWA HEC-19, Hydrology [16] or USGS Bulletin 17B [8] for this method and to find the values of K.

The data to be plotted are "PEAK DISCHARGE, Q (CFS)" vs. "PROBABILITY, Pr" as shown in the example in Figure 61.1. This plot usually results in a very flat curve with a reasonably straight center portion. An extension of this center portion gives a line for interpolation of the various needed discharges and frequencies.

The engineer should use an adjusted skew, which is calculated from the generalized and station skews. Generalized skews should be developed from at least 40 stations with each station having at least 25 years of record.

The equation for the adjusted skew is

$$G_w = \frac{MSE_{G_S}(G_L) + MSE_{G_L}(G_S)}{MSE_{G_S} + MSE_{G_L}} \tag{61.4}$$

where

G_w = weighted skew coefficient
G_S = station skew
G_L = generalized skew
MSE_{G_S} = mean square error of station skew
MSE_{G_L} = mean square error of generalized skew

The entire Log Pearson type III procedure is covered by Bulletin No. 17B, "Guidelines for Determining Flood Flow Frequency" [8].

The Gumbel extreme value method, sometimes called the double-exponential distribution of extreme values, has also been used to describe the distribution of hydrological variables, especially the peak discharges. It is based on the assumption that the cumulative frequency distribution of the largest values of samples drawn from a large population can be described by the following equation:

$$f(Q) = e^{-e^{a(Q-b)}} \tag{61.5}$$

where

$a = \dfrac{1.281}{S}$

$b = \bar{Q} - 0.450 S$

S = standard deviation
$\bar{Q}$ = mean annual flow

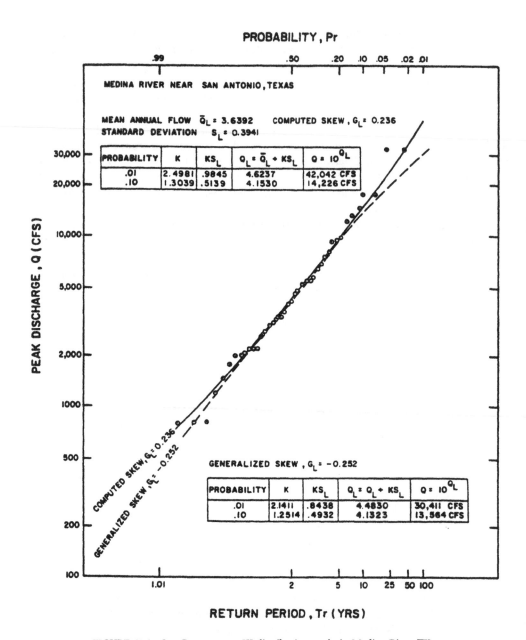

FIGURE 61.1 Log Pearson type III distribution analysis, Medina River, TX.

Values of this distribution function can be computed from Eq. (61.5). Characteristics of the Gumbel extreme value distribution are that the mean flow, $\overline{Q}$, occurs at the return period of $T_r = 2.33$ years and that it is skewed toward the high flows or extreme values as shown in the example of Figure 61.2. Even though it does not account directly for the computed skew of the data, it does predict the high flows reasonably well. For this method and additional techniques, please refer to USGS Water Supply Paper 1543-A, Flood-Frequency Analysis, and Manual of Hydrology Part 3.

The Gumbel extreme value distribution is given in "Statistics of Extremes" by E.J. Gumbel and is also found in HEC-19, p.73. Results from this method should be plotted on special Gumbel paper as shown in Figure 61.2.

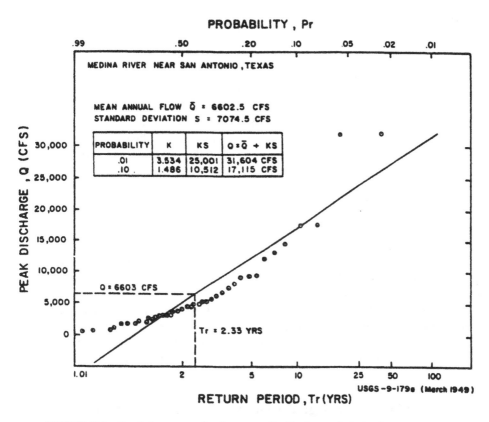

FIGURE 61.2 Gumbel extreme value frequency distribution analysis, Medina River, TX.

61.2.1.3.3 Discharge Comparison of Adjacent Basins

HEC 19, Appendix D [16] contains a list of reports for various states in the United States that have discharges at gauges that have been determined for frequencies from 2-year through 100-year frequencies. The discharges were determined by the Log Pearson III method. The discharge frequency at the gauges should be updated by the engineer using Log Pearson III and the Gumbel extreme value method.

The gauge data can be used directly as equivalent if the drainage areas are about the same (within less than 5%). Otherwise, the discharge determination can be obtained by the formula:

$$Q_u = Q_g (A_u / A_g)^b \tag{61.6}$$

where
Q_u = discharge at ungauged site

Q_g = discharge at gauged site

A_u = area of ungauged site

A_g = area of gauged site

b = exponent of drainage area

61.2.1.3.4 Regional Flood-Frequency Equations

If no gauged site is reasonably nearby, or if the record for the gauge is too short, then the discharge can be computed using the applicable regional flood-frequency equations. Statewide regional regression equations have been established in the United States. These equations permit peak flows to be estimated for return periods varying between 2 and 100 years. The discharges were determined by the Log Pearson III method. See HEC-19, Appendix D [16] for references to the studies that were conducted for the various states.

61.2.1.3.5   Design Hydrographs

Design hydrographs [9] give a complete time history of the passage of a flood at a particular site. This would include the peak flow. A runoff hydrograph is a plot of the response of a watershed to a particular rainfall event. A unit hydrograph is defined as the direct runoff hydrograph resulting from a rainfall event that lasts for a unit duration of time. The ordinates of the unit hydrograph are such that the volume of direct runoff represented by the area under the hydrograph is equal to 1 in. of runoff from the drainage area. Data on low water discharges and dates should be given as it will control methods and procedures of pier excavation and construction. The low water discharges and dates can be found in the USGS Water Resources Data Reports published each year. One procedure is to review the past 5 or 6 years of records to determine this.

61.2.1.4   Remarks

Before arriving at a final discharge, the existing channel capacity should be checked using the velocity as calculated times the channel waterway area. It may be that a portion of the discharge overflows the banks and never reaches the site.

The proposed design discharge should also be checked to see that it is reasonable and practicable. As a rule of thumb, the unit runoff should be 300 to 600 s-ft per square mile for small basins (to 20 square miles), 100 to 300 s-ft per square mile for median areas (to 50 square miles) and 25 to 150 s-ft for large basins (above 50 square miles). The best results will depend on rational engineering judgment.

61.2.2   Bridge Deck Drainage Design (On-Site Drainage Design)

61.2.2.1   Runoff and Capacity Analysis

The preferred on-site hydrology method is the rational method. The rational method, as discussed in Section 61.2.1.3.1, for on-site hydrology has a minimum time of concentration of 10 min. Many times, the time of concentration for the contributing on-site pavement runoff is less than 10 min. The initial time of concentration can be determined using an *overland flow* method until the runoff is concentrated in a curbed section. Channel flow using the roadway-curb cross section should be used to determine velocity and subsequently the time of flow to the first inlet. The channel flow velocity and flooded width is calculated using Manning's formula:

$$V = \frac{1.486}{n} A R^{2/3} S_f^{1/2} \qquad (61.7)$$

where
V = velocity
A = cross-sectional area of flow
R = hydraulic radius
S_f = slope of channel
n = Manning's roughness value [11]

The intercepted flow is subtracted from the initial flow and the bypass is combined with runoff from the subsequent drainage area to determine the placement of the next inlet. The placement of inlets is determined by the allowable flooded width on the roadway.

Oftentimes, bridges are in sump areas, or the lowest spot on the roadway profile. This necessitates the interception of most of the flow before reaching the bridge deck. Two overland flow equations are as follows.

1. **Kinematic Wave Equation:**

$$t_o = \frac{6.92 L^{0.6} n^{0.6}}{i^{0.4} S^{0.3}}$$ (61.8)

2. **Overland Equation:**

$$t_o = \frac{3.3(1.1-C)(L)^{1/2}}{(100\,S)^{1/3}}$$ (61.9)

where

t_o = overland flow travel time in minutes
L = length of overland flow path in meters
S = slope of overland flow in meters
n = manning's roughness coefficient [12]
i = design storm rainfall intensity in mm/h
C = runoff coefficient (Tables 61.1 and 61.2)

61.2.2.2 Select and Size Drainage Facilities

The selection of inlets is based upon the allowable flooded width. The allowable flooded width is usually outside the traveled way. The type of inlet leading up to the bridge deck can vary depending upon the flooded width and the velocity. Grate inlets are very common and, in areas with curbs, curb opening inlets are another alternative. There are various monographs associated with the type of grate and curb opening inlet. These monographs are used to determine interception and therefore the bypass [5].

61.2.3 Stage Hydraulics

High water (HW) stage is a very important item in the control of the bridge design. All available information should be obtained from the field and the Bridge Hydrology Report regarding HW marks, HW on upstream and downstream sides of the existing bridges, high drift profiles, and possible backwater due to existing or proposed construction.

Remember, observed high drift and HW marks are not always what they seem. Drift in trees and brush that could have been bent down by the flow of the water will be extremely higher than the actual conditions. In addition, drift may be pushed up on objects or slopes above actual HW elevation by the velocity of the water or wave action. Painted HW marks on the bridge should be searched carefully. Some flood insurance rate maps and flood insurance study reports may show stages for various discharges. Backwater stages caused by other structures should be included or streams should be noted.

Duration of high stages should be given, along with the base flood stage and HW for the design discharge. It should be calculated for existing and proposed conditions that may restrict the channel producing a higher stage. Elevation and season of low water should be given, as this may control design of tremie seals for foundations and other possible methods of construction. Elevation of overtopping flow and its location should be given. Normally, overtopping occurs at the bridge site, but overtopping may occur at a low sag in the roadway away from the bridge site.

61.2.3.1 Waterway Analysis

When determining the required waterway at the proposed bridge, the engineers must consider all adjacent bridges if these bridges are reasonably close. The waterway section of these bridges should be tied into the stream profile of the proposed structure. Structures that are upstream or downstream of the proposed bridge may have an impact on the water surface profile. When calculating the effective waterway area, adjustments must be made for the skew and piers and bents. The required waterway should be below the 50-year design HW stage.

If stream velocities, scour, and erosive forces are high, then abutments with wingwall construction may be necessary. Drift will affect the horizontal clearance and the minimum vertical clearance line of the proposed structure. Field surveys should note the size and type of drift found in the channel. Designs based on the 50-year design discharge will require drift clearance. On major streams and rivers, drift clearance of 2 to 5 m above the 50-year discharge is needed. On smaller streams 0.3 to 1 m may be adequate. A formula for calculating freeboard is

$$\text{Freeboard} = 0.1Q^{0.3} + 0.008V^2 \tag{61.10}$$

where
Q = discharge
V = velocity

61.2.3.2 Water Surface Profile Calculation

There are three prominent water surface profile calculation programs available [1,2]. The first one is HEC-2 which takes stream cross sections perpendicular to the flow. WSPRO is similar to HEC-2 with some improvements. SMS is a new program that uses finite-element analysis for its calculations. SMS can utilize digital elevation models to represent the streambeds.

61.2.2.3 Flow Velocity and Distribution

Mean channel, overflow velocities at peak stage, and localized velocity at obstructions such as piers should be calculated or estimated for anticipated high stages. Mean velocities may be calculated from known stream discharges at known channel section areas or known waterway areas of bridge, using the correct high water stage.

Surface water velocities should be measured roughly, by use of floats, during field surveys for sites where the stream is flowing. Stream velocities may be calculated along a uniform section of the channel using Manning's formula Eq. (61.7) if the slope, channel section (area and wetted perimeter), and roughness coefficient (n) are known.

At least three profiles should be obtained, when surveying for the channel slope, if possible. These three slopes are bottom of the channel, the existing water surface, and the HW surface based on drift or HW marks. The top of low bank, if overflow is allowed, should also be obtained. In addition, note some tops of high banks to prove flows fall within the channel. These profiles should be plotted showing existing and proposed bridges or other obstruction in the channel, the change of HW slope due to these obstructions and possible backwater slopes.

The channel section used in calculating stream velocities should be typical for a relatively long section of uniform channel. Since this theoretical condition is not always available, however, the nearest to uniform conditions should be used with any necessary adjustments made for irregularities.

Velocities may be calculated from PC programs, or calculator programs, if the hydraulic radius, roughness factor, and slope of the channel are known for a section of channel, either natural or artificial, where uniform stream flow conditions exists. The hydraulic radius is the waterway area divided by the wetted perimeter of an average section of the uniform channel. A section under a bridge whose piers, abutments, or approach fills obstruct the uniformity of the channel cannot be used as there will not be uniform flow under the structure. If no part of the bridge structure seriously obstructs or restricts the channel, however, the section at the bridge could be used in the above uniform flow calculations.

The roughness coefficient n for the channel will vary along the length of the channel for various locations and conditions. Various values for n can be found in the References [1,5,12,17].

At the time of a field survey the party chief should estimate the value of n to be used for the channel section under consideration. Experience is required for field determination of a relatively close to actual *n* value. In general, values for natural streams will vary between 0.030 and 0.070. Consider both low and HW *n* value. The water surface slope should be used in this plot and the slope should be adjusted for obstructions such as bridges, check dams, falls, turbulence, etc.

The results as obtained from this plot may be inaccurate unless considerable thought is given to the various values of slope, hydraulic radius, and *n*. High velocities between 15 and 20 ft/s (4.57 and 6.10 m/s through a bridge opening may be undesirable and may require special design considerations. Velocities over 20/ 6.10 m/s should not be used unless special design features are incorporated or if the stream is mostly confined in rock or an artificial channel.

61.3 Bridge Scour

61.3.1 Bridge Scour Analysis

61.3.1.1 Basic Scour Concepts

Scour is the result of the erosive action of flowing water, excavating and carrying away material from the bed and banks of streams. Determining the magnitude of scour is complicated by the cyclic nature of the scour process. Designers and inspectors need to study site-specific subsurface information carefully in evaluating scour potential at bridges. In this section, we present bridge engineers with the basic procedures and methods to analyze scour at bridges.

Scour should be investigated closely in the field when designing a bridge. The designer usually places the top of footings at or below the total potential scour depth; therefore, determining the depth of scour is very important. The total potential scour at a highway crossing usually comprises the following components [11]: aggradation and degradation, stream contraction scour, local scour, and sometimes with lateral stream migration.

61.3.1.1.1 Long-Term Aggradation and Degradation

When natural or human activities cause streambed elevation changes over a long period of time, aggradation or degradation occurs. Aggradation involves the deposition of material eroded from the channel or watershed upstream of the bridge, whereas degradation involves the lowering or scouring of the streambed due to a deficit in sediment supply from upstream.

Long-term streambed elevation changes may be caused by the changing natural trend of the stream or may be the result of some anthropogenic modification to the stream or watershed. Factors that affect long-term bed elevation changes are dams and reservoirs up- or downstream of the bridge, changes in watershed land use, channelization, cutoffs of meandering river bends, changes in the downstream channel base level, gravel mining from the streambed, diversion of water into or out of the stream, natural lowering of the fluvial system, movement of a bend, bridge location with respect to stream planform, and stream movement in relation to the crossing. Tidal ebb and flood may degrade a coastal stream, whereas littoral drift may cause aggradation. The problem for the bridge engineer is to estimate the long-term bed elevation changes that will occur during the lifetime of the bridge.

61.3.1.1.2 Stream Contraction Scour

Contraction scour usually occurs when the flow area of a stream at flood stage is reduced, either by a natural contraction or an anthropogenic contraction (like a bridge). It can also be caused by the overbank flow which is forced back by structural embankments at the approaches to a bridge. There are some other causes that can lead to a contraction scour at a bridge crossing [11]. The decreased flow area causes an increase in average velocity in the stream and bed shear stress through the contraction reach. This in turn triggers an increase in erosive forces in the contraction. Hence, more bed material is removed from the contracted reach than is transported into the reach. The natural streambed elevation is lowered by this contraction phenomenon until relative equilibrium is reached in the contracted stream reach.

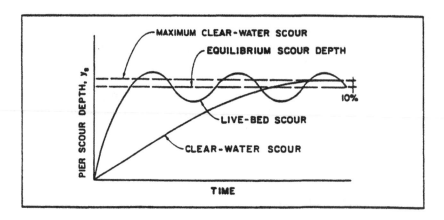

FIGURE 61.3   Illustrative pier scour depth in a sand-bed stream as a function of time.

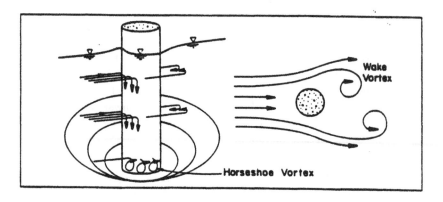

FIGURE 61.4   Schematic representation of local scour at a cylindrical pier.

There are two forms of contraction scour: live-bed and clear-water scours. Live-bed scour occurs when there is sediment being transported into the contracted reach from upstream. In this case, the equilibrium state is reached when the transported bed material out of the scour hole is equal to that transported into the scour hole from upstream. Clear-water scour occurs when the bed sediment transport in the uncontracted approach flow is negligible or the material being transported in the upstream reach is transported through the downstream at less than the capacity of the flow. The equilibrium state of the scour is reached when the average bed shear stress is less than that required for incipient motion of the bed material in this case (Figure 61.3).

61.3.1.1.3 Local Scour

When upstream flow is obstructed by obstruction such as piers, abutments, spurs, and embankments, flow vortices are formed at their base as shown in Figure 61.4 (known as horseshoe vortex). This vortex action removes bed material from around the base of the obstruction. A scour hole eventually develops around the base. Local scour can also be either clear-water or live-bed scour. In considering local scour, a bridge engineer needs to look into the following factors: flow velocity, flow depth, flow attack angle to the obstruction, obstruction width and shape, projected length of the obstruction, bed material characteristics, bed configuration of the stream channel, and also potential ice and debris effects [11, 13].

61.3.1.1.4 Lateral Stream Migration

Streams are dynamic. The lateral migration of the main channel within a floodplain may increase pier scour, embankment or approach road erosion, or change the total scour depth by altering the flow angle of attack at piers. Lateral stream movements are affected mainly by the geomorphology of the stream, location of the crossing on the stream, flood characteristics, and the characteristics of the bed and bank materials [11,13].

61.3.1.2 Designing Bridges to Resist Scour

It is obvious that all scour problems cannot be covered in this special topic section of bridge scour. A more-detailed study can be found in HEC-18, "Evaluating Scour at Bridges" and HEC-20, "Stream Stability at Highway Structures" [11,18]. As described above, the three most important components of bridge scour are long-term aggradation or degradation, contraction scour, and local scour. The total potential scour is a combination of the three components. To design a bridge to resist scour, a bridge engineer needs to follow the following observation and investigation steps in the design process.

1. **Field Observation** — Main purposes of field observation are as follows:
    - Observe conditions around piers, columns, and abutments (Is the hydraulic skew correct?),
    - Observe scour holes at bends in the stream,
    - Determine streambed material,
    - Estimate depth of scour, and
    - Complete geomorphic factor analysis.

    There is usually no fail-safe method to protect bridges from scour except possibly keeping piers and abutments out of the HW area; however, proper hydraulic bridge design can minimize bridge scour and its potential negative impacts.
2. **Historic Scour Investigation** — Structures that have experienced scour in the past are likely to continue displaying scour problems in the future. The bridges that we are most concerned with include those currently experiencing scour problems and exhibiting a history of local scour problems.
3. **Problem Location Investigation** — Problem locations include "unsteady stream" locations, such as near the confluence of two streams, at the crossing of stream bends, and at alluvial fan deposits.
4. **Problem Stream Investigation** — Problem streams are those that have the following characteristics of aggressive tendencies: indication of active degradation or aggradation; migration of the stream or lateral channel movement; streams with a steep lateral slope and/or high velocity; current, past, or potential in-stream aggregate mining operations; and loss of bank protection in the areas adjacent to the structure.
5. **Design Feature Considerations** — The following features, which increase the susceptibility to local scour, should be considered:
    - Inadequate waterway opening leads to inadequate clearance to pass large drift during heavy runoff.
    - Debris/drift problem: Light drift or debris may cause significant scour problems, moderate drift or debris may cause significant scour but will not create severe lateral forces on the structure, and heavy drift can cause strong lateral forces or impact damage as well as severe scour.
    - Lack of overtopping relief: Water may rise above deck level. This may not cause scour problems but does increase vulnerability to severe damage from impact by heavy drift.
    - Incorrect pier skew: When the bridge pier does not match the channel alignment, it may cause scour at bridge piers and abutments.
6. **Traffic Considerations** — The amount of traffic such as average daily traffic (ADT), type of traffic, the length of detour, the importance of crossings, and availability of other crossings should be taken into consideration.

7. **Potential for Unacceptable Damage** — Potential for collapse during flood, safety of traveling public and neighbors, effect on regional transportation system, and safety of other facilities (other bridges, properties) need to be evaluated.
8. **Susceptibility of Combined Hazard of Scour and Seismic** — The earthquake prioritization list and the scour-critical list are usually combined for bridge design use.

61.3.1.3 Scour Rating

In the engineering practice of the California Department of Transportation, the rating of each structure is based upon the following:

1. **Letter grading** — The letter grade is related to the potential for scour-related problems at this location.
2. **Numerical grading** — The numerical rating associated with each structure is a determination of the severity for the potential scour:

   A-1   No problem anticipated
   A-2   No problem anticipated/new bridge — no history
   A-3   Very remote possibility of problems
   B-1   Slight possibility of problems
   B-2   Moderate possibility of problems
   B-3   Strong possibility of problems
   C-1   Some probability of problems
   C-2   Moderate probability of problems
   C-3   Very strong probability of problems

Scour effect of storms is usually greater than design frequency, say, 500-year frequency. FHWA specifies 500-year frequency as 1.7 times 100-year frequency. Most calculations indicate 500-year frequency is 1.25 to 1.33 times greater than the 100-year frequency [3,8]; the 1.7 multiplier should be a maximum. Consider the amount of scour that would occur at overtopping stages and also pressure flows. Be aware that storms of lesser frequency may cause larger scour stress on the bridge.

61.3.2 Bridge Scour Calculation

All the equations for estimating contraction and local scour are based on laboratory experiments with limited field verification [11]. However, the equations recommended in this section are considered to be the most applicable for estimating scour depths. Designers also need to give different considerations to clear-water scour and live-bed scour at highway crossings and encroachments.

Prior to applying the bridge scour estimating methods, it is necessary to (1) obtain the fixed-bed channel hydraulics, (2) determine the long-term impact of degradation or aggradation on the bed profile, (3) adjust the fixed-bed hydraulics to reflect either degradation or aggradation impact, and (4) compute the bridge hydraulics accordingly.

61.3.2.1 Specific Design Approach

Following are the recommended steps for determining scour depth at bridges:

Step 1: Analyze long-term bed elevation change.
Step 2: Compute the magnitude of contraction scour.
Step 3: Compute the magnitude of local scour at abutments.
Step 4: Compute the magnitude of local scour at piers.
Step 5: Estimate and evaluate the total potential scour depths.

The bridge engineers should evaluate if the individual estimates of contraction and local scour depths from Step 2 to 4 are reasonable and evaluate the total scour derived from Step 5.

61.3.2.2 Detailed Procedures

1. **Analyze Long-Term Bed Elevation Change** — The face of bridge sections showing bed elevation are available in the maintenance bridge books, old preliminary reports, and sometimes in FEMA studies and U.S. Corps of Engineers studies. Use this information to estimate aggradation or degradation.

2. **Compute the Magnitude of Contraction Scour** — It is best to keep the bridge out of the normal channel width. However, if any of the following conditions are present, calculate contraction scour.

   a. Structure over channel in floodplain where the flows are forced through the structure due to bridge approaches

   b. Structure over channel where river width becomes narrow

   c. Relief structure in overbank area with little or no bed material transport

   d. Relief structure in overbank area with bed material transport

   The general equation for determining contraction scour is

$$y_s = y_2 - y_1 \tag{61.11}$$

   where

   y_s = depth of scour
   y_1 = average water depth in the main channel
   y_2 = average water depth in the contracted section

   Other contraction scour formulas are given in the November 1995 HEC-18 publication — also refer to the workbook or HEC-18 for the various conditions listed above [11]. The detailed scour calculation procedures can be referenced from this circular for either live-bed or clear-water contraction scour.

3. **Compute the Magnitude of Local Scour at Abutments** — Again, it is best to keep the abutments out of the main channel flow. Refer to publication HEC-18 from FHWA [13]. The scour formulas in the publication tend to give excessive scour depths.

4. **Compute the Magnitude of Local Scour at Piers** — The pier alignment is the most critical factor in determining scour depth. Piers should align with stream flow. When flow direction changes with stages, cylindrical piers or some variation may be the best alternative. Be cautious, since large-diameter cylindrical piers can cause considerable scour. Pier width and pier nose are also critical elements in causing excessive scour depth.

Assuming a sand bed channel, an acceptable method to determine the maximum possible scour depth for both live-bed and clear-water channel proposed by the Colorado State University [11] is as follows:

$$\frac{y_s}{y_1} = 2.0\, K_1 K_2 K_3 \left(\frac{a}{y_1}\right)^{0.65} F_r^{0.43} l \tag{61.12}$$

where

y_s = scour depth
y_1 = flow depth just upstream of the pier
K_1 = correction for pier shape from Figure 61.5 and Table 61.3
K_2 = correction for angle of attack of flow from Table 61.4
K_3 = correction for bed condition from Table 61.5
a = pier width
l = pier length
F_r = Froude number = $\dfrac{V}{(gy)}$ (just upstream from bridge)

Drift retention should be considered when calculating pier width/type.

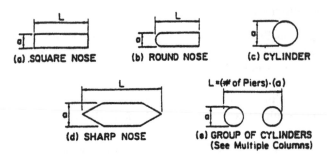

FIGURE 61.5   Common pier shapes.

TABLE 61.3   Correction Factor, K_1, for Pier Nose Shape

Shape of Pier Nose	K_1
Square nose	1.1
Round nose	1.0
Circular cylinder	1.0
Sharp nose	0.9
Group of cylinders	1.0

TABLE 61.4   Correction Factor, K_2, for Flow Angle of Attack

Angle	$L/a = 4$	$L/a = 8$	$L/a = 12$
0	1.0	1.0	1.0
15	1.5	2.0	2.5
30	2.0	2.75	3.5
45	2.3	3.3	4.3
90	2.5	3.9	5

TABLE 61.5   Increase in Equilibrium Pier Scour Depths K_3 for Bed Conditions

Bed Conditions	Dune Height H, ft	K_3
Clear-water scour	N/A	1.1
Plane bed and antidune flow	N/A	1.1
Small dunes	$10 > H > 2$	1.1
Medium dunes	$30 > H > 10$	1.1–1.2
Large dunes	$H > 30$	1.3

61.3.2.3   Estimate and Evaluate Total Potential Scour Depths

Total potential scour depths is usually the sum of long-term bed elevation change (only degradation is usually considered in scour computation), contraction scour, and local scour. Historical scour depths and depths of scourable material are determined by geology. When estimated depths from the above methods are in conflict with geology, the conflict should be resolved by the hydraulic engineer and the geotechnical engineer; based on economics and experience, it is best to provide for maximum anticipated problems.

61.3.3 Bridge Scour Investigation and Prevention

61.3.3.1 Steps to Evaluate Bridge Scour

It is recommended that an interdisciplinary team of hydraulic, geotechnical, and bridge engineers should conduct the evaluation of bridge scour. The following approach is recommended for evaluating the vulnerability of existing bridges to scour [11]:

Step 1. Screen all bridges over waterways into five categories: (1) low risk, (2) scour-susceptible, (3) scour-critical, (4) unknown foundations, or (5) tidal. Bridges that are particularly vulnerable to scour failure should be identified immediately and the associated scour problem addressed. These particularly vulnerable bridges are

1. Bridges currently experiencing scour or that have a history of scour problems during past floods as identified from maintenance records, experience, and bridge inspection records
2. Bridges over erodible streambeds with design features that make them vulnerable to scour
3. Bridges on aggressive streams and waterways
4. Bridges located on stream reaches with adverse flow characteristics

Step 2. Prioritize the scour-susceptible bridges and bridges with unknown foundations by conducting a preliminary office and field examination of the list of structures compiled in Step 1 using the following factors as a guide:

1. The potential for bridge collapse or for damage to the bridge in the event of a major flood
2. The functional classification of the highway on which the bridge is located.
3. The effect of a bridge collapse on the safety of the traveling public and on the operation of the overall transportation system for the area or region

Step 3. Conduct office and field scour evaluations of the bridges on the prioritized list in Step 2 using an interdisciplinary team of hydraulic, geotechnical, and bridge engineers:

1. In the United States, FHWA recommends using 500-year flood or a flow 1.7 times the 100-year flood where the 500-year flood is unknown to estimate scour [3,6]. Then analyze the foundations for vertical and lateral stability for this condition of scour. The maximum scour depths that the existing foundation can withstand are compared with the total scour depth estimated. An engineering assessment must be then made whether the bridge should be classified as a scour-critical bridge.
2. Enter the results of the evaluation study in the inventory in accordance with the instructions in the FHWA "Bridge Recording and Coding Guide" [7].

Step 4. For bridges identified as scour critical from the office and field review in Steps 2 and 3, determine a plan of action for correcting the scour problem (see Section 61.3.3.3).

61.3.3.2 Introduction to Bridge Scour Inspection

The bridge scour inspection is one of the most important parts of preventing bridge scour from endangering bridges. Two main objectives to be accomplished in inspecting bridges for scour are

1. To record the present condition of the bridge and the stream accurately; and
2. To identify conditions that are indicative of potential problems with scour and stream stability for further review and evaluation by other experts.

In this section, the bridge inspection practice recommended by U.S. FHWA [6,10] is presented for engineers to follow as guidance.

61.3.3.2.1 Office Review

It is highly recommended that an office review of bridge plans and previous inspection reports be conducted prior to making the bridge inspection. Information obtained from the office review provides a better foundation for inspecting the bridge and the stream. The following questions should be answered in the office review:

- Has an engineering scour evaluation been conducted? If so, is the bridge scour critical?
- If the bridge is scour-critical, has a plan of action been made for monitoring the bridge and/or installing scour prevention measures?
- What do comparisons of streambed cross sections taken during successive inspections reveal about the stream bed? Is it stable? Degrading? Aggrading? Moving laterally? Are there scour holes around piers and abutments?
- What equipment is needed to obtain stream-bed cross sections?
- Are there sketches and aerial photographs to indicate the planform locations of the stream and whether the main channel is changing direction at the bridge?
- What type of bridge foundation was constructed? Do the foundations appear to be vulnerable to scour?
- Do special conditions exist requiring particular methods and equipment for underwater inspections?
- Are there special items that should be looked at including damaged riprap, stream channel at adverse angle of flow, problems with debris, etc.?

61.3.3.2.2 Bridge Scour Inspection Guidance

The condition of the bridge waterway opening, substructure, channel protection, and scour prevention measures should be evaluated along with the condition of the stream during the bridge inspection. The following approaches are presented for inspecting and evaluating the present condition of the bridge foundation for scour and the overall scour potential at the bridge.

Substructure is the key item for rating the bridge foundations for vulnerability to scour damage. Both existing and potential problems with scour should be reported so that an interdisciplinary team can make a scour evaluation when a bridge inspection finds that a scour problem has already occurred. If the bridge is determined to be scour critical, the rating of the substructures should be evaluated to ensure that existing scour problems have been considered. The following items should be considered in inspecting the present condition of bridge foundations:

- Evidence of movement of piers and abutments such as rotational movement and settlement;
- Damage to scour countermeasures protecting the foundations such as riprap, guide banks, sheet piling, sills, etc.;
- Changes in streambed elevation at foundations, such as undermining of footings, exposure of piles; and
- Changes in streambed cross section at the bridge, including location and depth of scour holes.

In order to evaluate the conditions of the foundations, the inspectors should take cross sections of the stream and measure scour holes at piers and abutments. If equipment or conditions do not permit measurement of the stream bottom, it should be noted for further investigation.

To take and plot measurement of stream bottom elevations in relation to the bridge foundations is considered the single most important aspect of inspecting the bridge for actual or potential damage from scour. When the stream bottom cannot be accurately measured by conventional means, there are other special measures that need to be taken to determine the condition of the substructures or foundations such as using divers and using electronic scour detection equipment. For the purposes of evaluating resistance to scour of the substructures, the questions remain essentially the same for foundations in deep water as for foundations in shallow water [7] as follows:

- How does the stream cross section look at the bridge?
- Have there been any changes as compared with previous cross section measurements? If so, does this indicate that (1) the stream is aggrading or degrading or (2) is local or contraction scour occurring around piers and abutments?
- What are the shapes and depths of scour holes?
- Is the foundation footing, pile cap, or the piling exposed to the stream flow, and, if so, what is the extent and probable consequences of this condition?
- Has riprap around a pier been moved or removed?

Any condition that a bridge inspector considers to be an emergency or of a potentially hazardous nature should be reported immediately. This information as well as other conditions, which do not pose an immediate hazard but still warrant further investigation, should be conveyed to the interdisciplinary team for further review.

61.3.3.3 Introduction to Bridge Scour Prevention

Scour prevention measures are generally incorporated after the initial construction of a bridge to make it less vulnerable to damage or failure from scour. A plan of preventive action usually has three major components [11]:

1. Timely installation of temporary scour prevention measures;
2. Development and implementation of a monitoring program;
3. A schedule for timely design and construction of permanent scour prevention measures.

For new bridges [11], the following is a summary of the best solutions for minimizing scour damage:

1. Locating the bridge to avoid adverse flood flow patterns;
2. Streamlining bridge elements to minimize obstructions to the flow;
3. Designing foundations safe from scour;
4. Founding bridge pier foundations sufficiently deep to not require riprap or other prevention measures; and
5. Founding abutment foundations above the estimated local scour depth when the abutment is protected by well-designed riprap or other suitable measures.

For existing bridges, the available scour prevention alternatives are summarized as follows:

1. Monitoring scour depths and closing the bridge if excessive bridge scour exists;
2. Providing riprap at piers and/or abutments and monitoring the scour conditions;
3. Constructing guide banks or spur dikes;
4. Constructing channel improvements;
5. Strengthening the bridge foundations;
6. Constructing sills or drop structures; and
7. Constructing relief bridges or lengthening existing bridges.

These scour prevention measures should be evaluated using sound hydraulic engineering practice. For detailed bridge scour prevention measures and types of prevention measures, refer to Chapter 7 of "Evaluating Scour at Bridges" from U.S. FHWA. [10,11,18,19].

References

1. AASHTO, *Model Drainage Manual,* American Association of State Highway and Transportation Officials, Washington, D.C., 1991.
2. AASHTO, *Highway Drainage Guidelines,* American Association of State Highway and Transportation Officials, Washington, D.C., 1992.
3. California State Department of Transportation, *Bridge Hydraulics Guidelines,* Caltrans, Sacramento
4. California State Department of Transportation, *Highway Design Manual,* Caltrans, Sacramento,
5. Kings, *Handbook of Hydraulics,* Chapter 7 (*n* factors).
6. U.S. Department of the Interior, Geological Survey (USGS), Magnitude and Frequency of Floods in California, Water-Resources Investigation 77–21.
7. U.S. Department of Transportation, Recording and Coding Guide for the Structure Inventory and Appraisal of the Nation's Bridges, FHWA, Washington D.C., 1988.
8. U.S. Geological Survey, Bulletin No. 17B, Guidelines for Determining Flood Flow Frequency.
9. U.S. Federal Highway Administration, Debris-Control Structures, Hydraulic Engineering Circular No. 9, 1971.
10. U.S. Federal Highway Administration, Design of Riprap Revetments, Hydraulic Engineering Circular No. 11, 1989.
11. U.S. Federal Highway Administration, Evaluating Scour at Bridges, Hydraulic Engineering Circular No. 18, Nov. 1995.
12. U.S. Federal Highway Administration, Guide for Selecting Manning's Roughness Coefficient (*n* factors) for Natural Channels and Flood Plains, Implementation Report, 1984.
13. U.S. Federal Highway Administration, Highways in the River Environment, Hydraulic and Environmental Design Considerations, Training & Design Manual, May 1975.
14. U.S. Federal Highway Administration, Hydraulics in the River Environment, Spur Dikes, Sect. VI-13, May 1975.
15. U.S. Federal Highway Administration, Hydraulics of Bridge Waterways, Highway Design Series No. 1, 1978.
16. U.S. Federal Highway Administration, Hydrology, Hydraulic Engineering Circular No. 19, 1984.
17. U.S. Federal Highway Administration, Local Design Storm, Vol. I–IV (*n* factor) by Yen and Chow.
18. U.S. Federal Highway Administration, Stream Stability at Highway Structures, Hydraulic Engineering Circular No. 20, Nov. 1990.
19. U.S. Federal Highway Administration, Use of Riprap for Bank Protection, Implementation Report, 1986.

62

Sound Walls and Railings

Farzin Lackpour
Parsons Brinckerhoff-FG, Inc.

Fuat S. Guzaltan
Parsons Brinckerhoff-FG, Inc.

62.1   Sound Walls

62.1.1   Introduction

62.1.1.1   Need for Sound Walls

Population growth experienced during past decades in metropolitan areas has prompted the expansion and improvement of highway systems. As a direct result of these improvements, currently 90 million people in the United States live close to high-volume, high-speed highways. Rush-hour traffic on a typical high-volume, high-speed urban highway generates noise levels in the 80 to 90 dBA range. Within 50 to 100 yd (45 to 90 m) from the highway, due to absorption by the ground cover, the noise level dissipates to about 70 to 80 dBA. This ambient noise level, in comparison with a 50 to 55 dBA noise level in an average quiet house, is very intrusive to the majority of people, and should be further reduced to at least 60 to 70 dBA level by implementing noise abatement measures.

62.1.1.2   Design Noise Levels

In 1982, the Federal Highway Administration (FHWA) published the "Procedures for Abatement of Highway Traffic Noise and Construction Noise" in the Federal Aid Highway Program Manual, and therein established the acceptable noise levels at the location of the receivers (houses, schools, etc.) after the installation of the sound walls. This publication regulates the average allowable noise levels, $L_{eq}(h)$, and the peak allowable noise levels, $L_{10}(h)$ (the noise level that is exceeded more than 10% of the given period of time used to measure the allowable noise level) (Table 62.1)[1].

0-8493-7434-0/00/$0.00+$.50
© 2000 by CRC Press LLC

TABLE 62.1   Noise Abatement Design Criteria

Activity Category	L_{eq}(h), dBA	L_{10}(h), dBA	Land Use Category
A	57	60 (exterior)	Tracts of lands in which serenity and quiet are of extraordinary significance and serve an important public need and where the preservation of those quantities is essential if the area is to continue to serve its intended purpose; such areas could include amphitheaters, particular parks or portions of parks, or open spaces which are dedicated or recognized by appropriate local officials for activities requiring special quantities of serenity and quiet
B	67	70 (exterior)	Residences, motels, hotels, public meeting rooms, schools, churches, libraries, hospitals, picnic areas, playgrounds, active sports areas, and parks
C	72	75 (exterior)	Developed lands, properties, or activities not included in categories A and B above
D			Undeveloped lands; for requirements see paragraphs 5.a (5) and (6) of Publication PPM 90-2
E		55 (interior)	Residences, motels, hotels, public meeting rooms, schools, churches, libraries, hospitals, and auditoriums

62.1.2 Selection of Sound Walls

62.1.2.1 Sound Wall Materials

When a sound barrier is inserted in the line of sight between a noise source and a receiver, the intensity of the noise diminishes on the receiver side of the wall. This reduction in the noise intensity is referred to as insertion loss. The main factors that contribute to the insertion loss are the diffraction and reflection of the noise by the sound wall, and transmission loss as noise travels through the wall material. The amount of diffraction and reflection can be controlled by varying the height and inclination of the wall, installing specially shaped closure pieces at the top of the wall, or coating the wall surface with a sound absorbent material. The transmission loss can be controlled by varying the thickness and density of the wall material. The transmission loss levels for several common construction materials are given in Table 62.2 [2].

Earth berms, concrete, timber, and to a certain extent steel have been the traditional choices of material for sound walls. Other materials such as composite plaster panels, concrete blocks, bricks, and plywood panels have also been successfully utilized in smaller quantities in comparison to the traditional materials. In recent years, the awareness and need have risen to recycle materials rather than bury them in landfills. This trend has led to the use of recycled tires, glass, and plastics as sound wall material.

Given the variety of materials available for use in sound wall construction, selection of an appropriate type of sound wall becomes a difficult task. An intelligent decision can only be made after investigating the major factors contributing to successful implementation of a sound wall project such as cost, aesthetics, durability/life cycle, constructibility, etc.

62.1.2.2 Decision Matrix

A decision matrix is a convenient way of comparing the performance of different sound wall alternatives. The first step in building a decision matrix is to determine the parameters that will be the basis of the evaluation and selection process. The most important parameters are cost, aesthetics, durability/life cycle, and constructibility. The cost of the sound wall should include the cost of the surface finish or treatment, landscaping, utility relocation, drainage system, right-of-way, environmental mitigation, maintenance, and future replacement. Better durability and longer life cycle almost always translate into higher initial construction cost and lower maintenance cost. Aesthetic treatment of the sound wall should provide visual compatibility with the surrounding environment. Restrictions such as right-of-way limitations and presence of nearby residential areas may affect the constructibility of certain sound wall types. Other parameters such as construction access, and

TABLE 62.2   Sound Wall Materials

Materials	Thickness, in. (mm)	Transmission Loss (TL)[a], dBA
	Woods[b]	
Fir	½ (13)	17
	1 (25)	20
	2 (50)	24
Pine	½ (13)	16
	1 (25)	19
	2 (50)	23
Redwood	½ (13)	16
	1 (25)	19
	2 (50)	23
Cedar	½ (13)	15
	1 (25)	18
	2 (50)	22
Plywood	½ (13)	20
	1 (25)	23
Particle Board[c]	½ (13)	20
	Metals[d]	
Aluminum	1/16 (1.6)	23
	1/8 (3)	25
	1/4 (6)	27
Steel	24 ga (0.6)	18
	20 ga (0.9)	22
	16 ga (15)	25
Lead	1/16 (1.6)	28
	Concrete, Masonry, etc.	
Light concrete	4 (100)	36
	6 (150)	39
Dense concrete	4 (100)	40
Concrete block	4 (100)	32
	Composites	
Aluminum-faced plywood[e]	¾ (20)	21–23
Aluminum-faced particle board[e]	¾ (20)	21–23
Plastic lamina on plywood	¾ (20)	21–23
Plastic lamina on particle board	¾ (20)	21–23
	Miscellaneous	
Glass (safety glass)	¼ (6)	22
Plexiglas (shatterproof)	—	22–25
Masonite	½ (13)	20
Fiber glass/resin	¼ (6)	20
Stucco on metal lath	1 (25)	32
Polyester with aggregate surface[f]	3 (75)	20–30

[a] A weighted TL based on generalized truck spectrum.
[b] Tongue-and-groove boards recommended to avoid leaks (for fir, pine, redwood, and cedar).
[c] Should be treated for water resistance.
[d] May require treatment to reduce glare (for aluminum and steel).
[e] Aluminum is 0.01 in. thick. Special care is necessary to avoid delamination (for all composites).
[f] TL depends on surface density of the aggregate.

impacts on residences, parks, utilities, drainage systems, traffic, and environment should also be considered. In this step, the crucial issue is the identification of the relevant parameters in collaboration with the project owner. If the project owner is willing, receiving input from the local governments, residents, and the traveling public can be an invaluable asset in the success and acceptance of the project.

The second step in the process is assigning a percent weight to each parameter that is considered to be relevant in the first step (Table 62.3).

The third step involves assigning a rating ranging from 1 to 10 to each parameter. A rating of 10 represents the most desirable case, and a rating of 1 represents the least desirable case. For parameters that are associated with costs (Items 1, 3, 5, 7, 8, and 9 in Table 62.3), the rating can be based on the following formula once these costs are determined for each sound wall alternative:

$$\frac{\text{Cost of Least Expensive Alternative} \times 10}{\text{Cost of Alternative Considered}}$$

For the rating of less quantitative items such as aesthetics, constructibility, and construction access, the best approach is to define the factors which will give satisfactory results for the parameter in question. For instance, if a sound wall is considered atop an existing retaining wall, we may select balance (a pleasing proportion between the heights of the proposed sound wall and existing retaining wall), integration (presence of a fully integrated appearance between the proposed sound wall and existing retaining wall), and tonal value (uniformity of color and a pleasing contrast in textures between the proposed sound wall and existing retaining wall) as desirable parameters. We can assign a 10 rating to a sound wall alternative that displays all three factors, a 9 rating to the alternative that displays any two of the three factors, and an 8 rating to the alternative that satisfies only one of the factors.

The next step is to sum up the scores for each sound wall alternative, and rank them from the highest score to the lowest score (Table 62.3). The alternative with the highest score should be selected and recommended for design and construction.

62.1.3 Design Considerations

AASHTO *Guide Specifications for Structural Design of Sound Barriers* [3] is currently the main reference for the design loads, load combinations, and design criteria for concrete, steel, and masonry sound walls.

62.1.3.1 Design Loads

The loads that should be considered in the design of the sound barriers are dead load, wind load, seismic load, earth pressure, traffic impact, and ice and snow loads.

Dead Loads

The weight of all the components making up the sound wall are to be applied at the center of gravity of each component.

Wind Loads

Wind loads are to be applied perpendicular to the wall surface and at the centroid of the exposed surface. Minimum wind pressure is to be computed by the following formula [3]:

$$P = 0.00256 \ (1.3V)^2 \ C_d \ C_c \quad (0.0000473 \ (1.3V)^2 \ C_d \ C_c) \tag{62.1}$$

where

P = wind pressure in pounds per square foot (kilopascals)
V = wind speed in miles per hour (km/h) based on 50-year mean recurrence interval

TABLE 62.3   Decision Matrix for Sound Wall Alternatives

| Rating Parameters | Relative weight, % | Generic Post and Panel Concrete Sound Wall | | | | Proprietary Concrete Panel Sound Wall | | | |
| | | Alternative I (Sound wall at the top of a retaining wall) | | Alternative II (Sound wall in front of a retaining wall) | | Alternative I (Sound wall at the top of a retaining wall) | | Alternative II (Sound wall in front of a retaining wall) | |
		Rating	Score	Rating	Score	Rating	Score	Rating	Score
1. Initial construction cost	40	7.8	3.12	10.00	4.00	6.20	2.48	7.30	2.92
2. Aesthetics	15	10.0	1.50	9.00	1.35	8.00	1.20	7.00	1.05
3. Right-of-way impact	10	10.0	1.00	10.0	1.00	10.0	1.00	10.0	1.00
4. Constructibility	10	7.00	0.70	10.0	1.00	5.00	0.50	8.00	0.80
5. Drainage impact	5	9.00	0.45	9.00	0.45	9.00	0.45	9.00	0.45
6. Construction access	5	9.00	0.45	8.00	0.40	9.00	0.45	8.00	0.40
7. Utility impact	5	10.00	0.50	9.00	0.45	10.00	0.50	9.00	0.45
8. Maintenance cost	5	10.00	0.50	8.00	0.40	10.00	0.50	8.00	0.40
9. Maintenance and protection of traffic	5	10.00	0.50	8.00	0.40	10.00	0.50	8.00	0.40
Total Score			8.72		9.45		7.58		7.87
Ranking			2		1		4		3

TABLE 62.4 Coefficient C_c

Exposure Category	Height Zone[a]		
	0 < H Û 14 (4)	14 (4) < H Û 29 (9)	Over 29 (9)
Exposure Bl — Urban and suburban areas with numerous closely spaced obstructions having the size of single-family dwellings or larger that prevail in the upwind direction from the sound wall for a distance of at least 1500 ft. (450 m); for sound walls not located on structures	0.37	0.50	0.59
Exposure B2 — Urban and suburban areas with more open terrain not meeting the requirements of Exposure Bl; for sound walls not located on structures.	0.59	0.75	0.85
Exposure C — Open terrain with scattered obstructions; this category includes flat, open country and grasslands; this exposure is to be used for sound walls located on bridge structures, retaining walls, or traffic barriers	0.80	1.00	1.10

[a] Given as the distance from average level of adjacent ground surface to centroid of loaded area in ft (m).

$(1.3V)$ = gust speed, 30% increase in design wind velocity
C_d = drag coefficient (1.2 for sound barriers)
C_c = combined height, exposure, and location coefficient

The three exposure categories and related C_c values shown in Table 62.4 are to be considered for determining the wind pressure.

Seismic Loads
The following load applies to sound walls if the structures in the same area are designed for seismic loads:

$$\text{Seismic Load} = EQD = A \times f \times D \qquad (62.2)$$

where
EQD = seismic dead load
D = dead load of sound wall
A = acceleration coefficient
f = dead-load coefficient (use 0.75 for dead load, except on bridges; 2.50 for dead load on bridges; 8.0 for dead load for connections of non-cast-in-place walls to bridges; 5.0 for dead loads for connections of non-cast-in-place walls to retaining walls)

The product of A and f is not to be taken as less than 0.10.

Earth Loads
Earth loads that are applied to any portion of the sound wall and its foundations should conform to AASHTO *Standard Specifications for Highway Bridges*, Section 3.20 — Earth Pressure, except that live-load surcharge is not to be combined with seismic loads.

Traffic Loads
It will not be necessary to apply traffic impact loads to sound walls unless they are combined with concrete traffic barriers. The foundation systems for those sound wall and traffic barrier combinations that are located adjacent to roadway side slopes are not to be less than that required for the traffic impact load alone.

When a sound wall and traffic barrier combination is supported on a bridge superstructure, the design of the traffic barrier attachment details are based on the group loads that apply or the traffic load as given in AASHTO *Standard Specifications for Highway Bridges*, whichever controls.

TABLE 62.5   Load Combinations

Working Stress Design (WSD)		Allowable Over-stress as % of	Load Factor Design (LFD)	
Load Groups		Unit Stress	Load Groups	
Group I:	D + E + SC	100%	Group I:	$\beta \times D + 1.7E + 1.7SC$
Group II:	D + W + E + SC	133%	Group II:	$\beta \times D + 1.7E + 1.3W + 1.3I$
Group III:	D + EQD + E	133%	Group III:	$\beta \times D + 1.3\ E + 1.3\ EQE$
Group IV:	D + W + E +I	133%	Group IV:	$\beta \times D + 1.3\ E + 1.3\ EQD$
			Group V:	$\beta \times D + 1.1\ E + 1.1\ (EQE + EQD)$

$\beta = 1.0$ or 1.3, whichever controls the design; D = dead load; E = lateral earth pressure; SC = live-load surcharge; W = wind load; EQD = seismic dead load; EQE = seismic earth load; I = ice and snow loads

Ice and Snow Loads

Where snow drifts are encountered, their effects need to be considered.

Bridge Loads

When a sound wall is supported by a bridge superstructure, the wind or seismic load to be transferred to the superstructure and substructure of the bridge is to be as specified above under Wind Loads and Seismic Loads. Additional reinforcement may be required in traffic barriers and deck overhangs to resist the loads transferred by the sound wall.

62.1.3.2 Load Combinations

The groups in Table 62.5 represent various combinations of loads to which the sound wall structure may be subjected. Each part of the wall and its foundation is to be designed for these load groups.

62.1.3.3 Functional Requirements

The basic functional requirements for sound walls are as follows:

- To prevent vehicular impacts the sound walls should be located as far away as possible from the roadway clear zone. At locations where right-of-way is limited, a guide rail or concrete barrier curb should be utilized in front of the sound wall.
- A sound wall, especially along curved alignments, should not block the line of sight of the driver, and therefore reduce the driver's sight distance to less than the distance required for safe stopping.
- To avoid undesirable visual impacts on the aesthetic features of the surrounding area, the minimum sound wall height should not be less than the height of the right-of-way fence, and walls higher than 15 ft (4.5 m) should be avoided.
- To prevent icing on the roadway, the sound walls should not be located within a distance of less than one and a half times the height to the traveled roadway.
- To prevent saturation of the sloped embankments and avoid unstable soil conditions, transverse and longitudinal drainage facilities should be provided along the sound wall.
- To control fire or chemical spills on the highway, fire hose connections should be provided through the sound wall to the fire hydrants on the opposite side.

62.1.3.4 Maintenance Considerations

Sound walls should be placed as close as possible to the right-of-way line to avoid creating a strip of land behind the sound wall and adjacent to the right-of-way line. If this is not practical, then consideration should be given to accommodating independent maintenance and landscaping functions behind the wall. In cases where the access to the right-of-way side of the sound wall is not possible via local streets, then access through the sound wall should be provided at set intervals along the wall by using a solid door or overlapping two parallel sound walls. Parallel sound walls

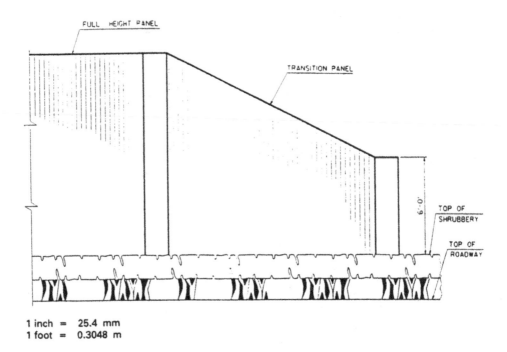

1 inch = 25.4 mm
1 foot = 0.3048 m

FIGURE 62.1 Typical transition at wall ends.

concealing an access opening should be overlapped a minimum of four times the offset distance in order to maintain the integrity of the noise attenuation of the main sound wall.

In urban settings, sound walls may be targets for graffiti. As a deterrence, the surface texture on the residential side of the wall should be selected rough and uneven so as to make the placement of graffiti difficult, or very smooth to facilitate easy removal of the graffiti. Sound walls with rough textures and dark colors are known to discourage graffiti.

62.1.3.5 Aesthetic Considerations

The selected sound wall alternative should address two aesthetic requirements: visual quality of the sound wall as a dynamic whole viewed from a vehicle in motion, and as a stationary form and texture as seen by the residents [4,5]. The appearance of the sound wall should avoid being monotonous to drivers; neither should it be too distractive. There are several ways of achieving a pleasing dynamic balance:

- Using discrete but balanced drops — 1 to 2 ft (0.3 to 0.6 m) — at the top of the walls to break the linear monotony.
- Implementing a gradual transition from the ground to the top of the sound wall by utilizing low-level slow-growing shrubbery in front of the wall and tapering wall panels at the ends of the wall (see Figure 62.1).
- Creating landmarks to give a sense of distance and location to the drivers. This can be achieved by utilizing distinct landscaping features with trees and plantings, or creating gateways with distinct architectural features such as planter boxes, wall niches, and terraces (Figure 62.2), or special surface finishes or textures using form liners (Figure 62.3). Gateways can also be used to delineate the limits of the individual communities along the sound wall by using a unique gateway design for each community.

As for the stationary view of the noise barrier, as seen by the residents, surface texturing and coloring are the most commonly used tools to gain acceptance by the public. By using a textured

1 inch = 25.4 mm
1 foot = 0.3048 m

FIGURE 62.2 Gateway.

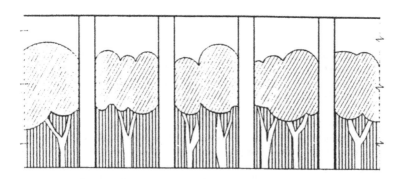

FIGURE 62.3 Architectural finish using form liners.

finish, it is possible to obtain different levels of light reflection on the wall surfaces and to evoke a sense of a third dimension. The most commonly used texturing method is raking the exposed face of the panel after concrete is placed in the formwork. Stamping a pattern in the fresh concrete surface is another texturing method. Coloring of concrete can be achieved by adding pigments to the concrete mix (internal coloring) or coating the surface of the panel with a water-based stain (external coloring). Although internal coloring would require less maintenance during the life cycle of the wall, achieving color consistency among panels is extremely difficult due to variations in the cement color and pigment dispersion rates. Staining offers uniformity in color. However, restaining of panels may be required after 10 to 15 years of service.

62.1.4 Ground-Mounted Sound Walls

62.1.4.1 Generic Sound Wall Systems

Generic sound wall systems (Table 62.6) make use of common construction materials such as earth, concrete, brick, masonry blocks, metal, and wood. With the exception of earth, all other materials are usually fabricated into post and panel systems in the shop, and installed on precast or cast-in-place concrete foundations at the site.

62.1.4.2 Proprietary Sound Wall Systems

In the late 1970s and early 1980s the regulatory actions by the Congress, Environmental Protection Agency (EPA), and Federal Highway Administration (FHWA) effectively launched a new industry. Ever since, a number of proprietary sound wall systems (Table 62.7) have been introduced and have become successful.

62.1.4.3 Foundation Types

The following foundation types are commonly used for sound walls:

TABLE 62.6   Generic Sound Wall Systems

Type	Features
Concrete walls	Approximately 45% of all existing sound walls are made of concrete; durability, ease of construction, and low construction and maintenance costs make concrete the most favored material in sound wall construction; precast posts and panels are usually used in combination with cast-in-place footings
Earth berms	Earth berms, alone or in combination with other types of sound walls, make up about 25% of existing sound walls; ease of construction, low cost, and availability for landscaping make earth berms the first choice of sound wall construction material wherever sufficient right-of-way is present
Timber walls	Timber is the choice of construction material for 15% of existing sound walls; timber is a flexible construction material, and can be used in a variety of ways in sound wall construction; timber posts can be solid sawn, glue laminated, or round pole type; the posts can be driven or embedded in concrete footings, and timber planks or plywood panels can be nailed or bolted to the posts at the site
Brick and concrete masonry block walls	These types of walls account for about 10% of the total sound wall construction; brick and masonry blocks can be preassembled into panels off site or can be mortared at the site; depending on the height of the wall, horizontal and/or vertical reinforcement may need to be used
Metal walls	These make up about 5% of sound wall construction; metal posts can be driven into the ground, embedded into concrete foundations, or attached to the top of the foundations; generally, metal panels that are made up of corrugated pans are connected to each other to form a solid surface, and then the entire panel assembly is bolted to the posts
Combination walls	It is sometimes advisable to combine two or more construction materials in a sound wall construction to take advantage of the superior characteristics of each material; a commonly encountered combination is the use of earth berms with other types of walls at locations where construction of a full-height earth berm is not feasible due to right-of-way limitations; in this case, an earth berm can be constructed to the edge of the right-of-way line, and the remainder of the height required for a full-level sound abatement can be provided by using a concrete, timber, or metal wall; another most commonly practiced combination in the field is the use of steel posts with concrete, timber, or composite wall panels or planks; the inherent advantage in this combination is the ability to make quick-bolted or welded connections between the steel posts and foundations.

Pile Foundations

Timber, steel, and concrete piles can be driven into the ground to act as a foundation, and also as a post for sound walls. One shortcoming of this foundation system is the problem of controlling the plumbness and location of driven posts. Also, damage to the pile end may often require trimming and/or repairs. On the other hand, the advantages of pile foundations are the ease and economy of installation into almost any kind of soil except rock, and completion of foundation and post installation in a single-step process.

Caisson (Bored) Foundations

Caissons are the most frequently used type of foundation in sound wall construction. It involves excavating a round hole using augering equipment (use of a metal casing may be required to prevent the collapse of the hole walls), installing a reinforcement cage, inserting the post or, alternatively, installing anchor bolt assemblies for the post connection, and finally placement of concrete. The advantages of caisson foundations are the ease of installation into any kind of soil (except soils containing large boulders), the convenience of performing the construction in tight spaces with a minor amount of disturbance to the surrounding environment, and ability to locate posts accurately in the horizontal plane and plumb in the vertical plane. The disadvantages are the high possibility of water intrusion into the excavated holes in areas with a high water table, interference of boulders with the augering process, and the presence of concurrent construction activities, such as augering, placement of reinforcement, insertion of post, and placement of concrete. In a caisson construction, all these tasks should be carefully orchestrated to achieve a cost-effective and expedited operation.

Spread Footings

Spread footings are the ideal choice where the construction site is suitable for a continuous trench-type excavation using heavy excavation equipment. Once the trench excavation is completed to the bottom of the proposed foundation, cast-in-place footings can be constructed on the ground, or

TABLE 62.7 Proprietary Sound Wall Systems

Type	Features
Siera Wall	This sound wall system consists of precast wall panels and cast-in-place foundations; each panel is precast integrally with a pilaster post along one edge, and attached to the adjacent panels with a tongue-and-groove connection; the connection between the wall panels and foundation is secured by welding the steel plates embedded at the base of the posts to the steel plates mounted on top of the foundations
Port-O-Wall	This is another sound wall system which utilizes precast panels and cast-in-place foundations; each precast concrete wall panel is secured to the adjacent panel with a tongue-and-groove connection; the panels are also mechanically connected by horizontal tie bars at the top and bottom; a rectangular hole at the bottom of each panel allows the installation of the transverse reinforcement for the construction of a continuous cast-in-place concrete footing
Fan Wall	This precast concrete wall system is a castellated freestanding wall that does not require a concrete footing; a rotatable and interlocking joint system allows the joining of panels at any angle; the joint along the sides of each panel features mating concave and convex edges and stainless steel aircraft-type cable connector assemblies
Sound Zero	This lightweight (8 lb/sf) (380 Pa) panel system is fabricated to the full wall height, and installed on top of cast-in-place footings or bridge parapets using concrete or steel posts
Carsonite© Sound Barrier	This panel system features tongue-and-groove modular sections made from a fiberglass-reinforced polymer composite shell that is filled with ground, recycled tires; the lightweight (7.5 lb/sf) (360 Pa) panels are preassembled off site and installed between posts anchored into cast-in-place footings, traffic barriers, or bridge parapets
Contech Noise Walls	This wall system consists of hot-rolled steel posts and cold-formed interlocking steel panels; all wall components are galvanized, and the panels are additionally protected by a choice of colored coating systems
Evergreen Noise Abatement Walls	This wall system consists of precast concrete units, and is supported on individual or continuous cast-in-place concrete foundations; the precast units are stacked up to form a freestanding wall; a select granular material is placed in each wall unit and compacted prior to the installation of the next higher unit; the trays of the wall units are filled with topsoil to support the planting and growth of evergreen and deciduous plants
Maccaferri Gabion Sound Walls	This sound wall system utilizes stacked-up gabion baskets to form a freestanding sound wall; zinc or PVC-coated wire baskets are filled with rock to blend with the natural environment; the cores of the baskets can also be filled with topsoil to allow planting of vegetation

precast footings can be erected. If precast footings are used, roughening of the bottom of footing in combination with placement of a crushed stone layer with a thin cement grout topping under the footing may be necessary to achieve a desired safety factor against sliding.

Tie-Down Foundations

At locations where rock is at or close to the surface, augering for caissons or excavating for spread footings may be too costly and time-consuming. A more practical solution in this case is the use of concrete pedestals anchored into rock with post-tensioned tie-downs. A pedestal detail similar to the stem of the spread footing can be constructed to allow insertion of the post into a recess in the stem, or installation of anchorage assemblies for the connection of the post to the top of the stem.

62.1.4.4 Typical Details

The sketches shown in this chapter depict a concrete sound wall system consisting of precast posts and panels supported on spread footings [6].

Precast panels (Figure 62.4) are 5 in. (130 mm) thick and reinforced with wire mesh fabric to sustain the wind loading. Additional mild reinforcement is placed along the periphery of the panels to carry the vertical loads. Two cast-in-place inserts are provided at the top of the panels for handling and erection. The panels are 4 ft (1.2 m) high and span between the posts spaced at 15 ft (4.6 m) on centers. Overlapping joints between the panels provide a leakproof surface. The roadway side of the panels is finished smooth and the residential side of the panel displays a raked finish to deter graffiti.

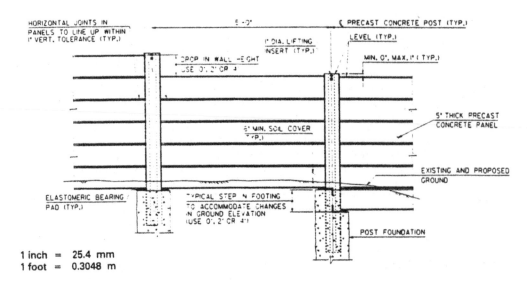

FIGURE 62.4 Precast concrete post and panel sound walls. (*Source:* Guzaltan, F., *PCI J.*, 27(4), 60, 1992. With permission.)

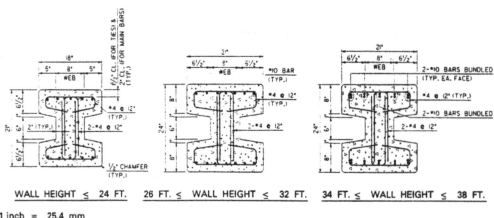

FIGURE 62.5 Concrete precast post details. (*Source:* Guzaltan, F., *PCI J.*, 27(4), 61, 1992. With permission.)

The precast concrete posts (Figure 62.5) are H-shaped to accept the insertion of panels and allow approximately a 10° angle change in the orientation of panels. Both flanges are reinforced with mild steel reinforcement to carry the wind loads that are transmitted from the panels. For posts longer than 34 ft (10 m), the use of prestressing strands may need to be considered to prevent reinforcement congestion in the flanges. Shear reinforcement wrapping the outline of the web and flanges is also provided.

Four types of foundations for use in different soil and site conditions are featured: caisson foundations (Figure 62.6) for sites without boulders; cast-in-place and precast spread footings (Figure 62.6) at sites where large boulders are frequent; and tie-down foundations (Figure 62.7) at locations where rock is close to the surface.

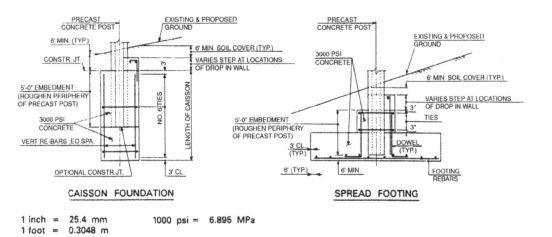

FIGURE 62.6 Caisson foundation and spread footing. (*Source*: Guzaltan, F., *PCI J.*, 27(4), 61, 1992. With permission.)

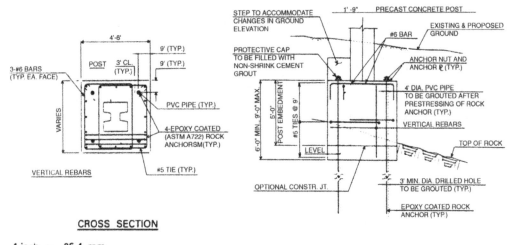

FIGURE 62.7 Tie-down foundation. (*Source*: Guzaltan, F., *PCI J.*, 27(4), 62, 1992. With permission.)

62.1.5 Bridge-Mounted Sound Walls

62.1.5.1 Assessment of an Existing Bridge to Carry a Sound Wall

The capacity of the existing superstructure components (barrier curb, deck slab, girders, and diaphragms) should be checked prior to deciding to attach sound barriers to an existing bridge. Furthermore, the capacity of the existing bearings and piers may also be investigated due to increases in the girder reactions. The main forces to be considered are the dead load of the sound wall, wind and ice loads on the panels and posts, and torsion created by the eccentricity of these forces. Quite often, the existing deck slabs, beams, girders, and floor beams tend to have excess flexural capacity to carry additional loads. However, they may not have any spare capacity to carry additional torsion.

The transmission of forces from the sound wall to the bridge superstructure can be best determined by using a three-dimensional grid or a finite-element model. The structural model should consider the stiffness of the deck slab, girders, and diaphragms, as well as the constraints introduced by the bearings at the girder ends. In any case, significant torsional impact should be expected in the deck slab, diaphragms, fascia girder, and first and second interior girders in multigirder superstructures.

62.1.5.2 Strengthening an Existing Bridge Superstructure to Carry a Sound Wall

Multisteel girder superstructures are relatively easy to strengthen to carry the superimposed moments, shear, and torsion due to the installation of a sound wall. Bolting an additional cover plate to the bottom flange may be sufficient to strengthen a steel girder. However, strengthening a precast concrete I-beam poses a greater challenge. Adding post-tensioned strands, longitudinal and shear reinforcement, and encasing the strands and reinforcing bars in a high-strength concrete mass may be a feasible way of strengthening an existing I-beam. Nevertheless, the time and cost involved in strengthening concrete I-beams should always be compared with the cost of replacing the superstructure before a strengthening alternative is seriously pursued.

Even if the deck slab and girders of an existing bridge are found to be adequate or amenable to strengthening, there may be still a need to replace at least the bridge parapets in order to be able to anchor the posts of the sound wall.

62.1.5.3 Typical Details

The typical retrofit sound wall details (Figure 62.8) on an existing bridge feature steel posts anchored into a New Jersey barrier curb. The top of the new barrier curb is oversized to accept a base plate for the sound wall post. The lightweight concrete precast panels are used to lessen the impacts of dead load.

The bridge incorporates several strengthening features. The deck slab is thickened over the new fascia girder and two adjacent girders. All new girders as well as the new bearings and diaphragms are designed to carry additional sound wall loads. At the existing hammerhead pier, the length of the pier cap cantilever is reduced by adding an auxiliary column. This measure is intended to counteract the effects of the increased girder reactions.

62.1.6 Independent Sound Wall Structures

62.1.6.1 Need for an Independent Sound Wall Structure

When it is not feasible to strengthen or replace the components of an existing bridge to carry a proposed sound wall, the alternative is to provide a freestanding structure to support the sound wall along an existing bridge.

In the design of an independent sound wall structure the following guidelines may be followed:

- There should be no noise leakage between the bridge and the independent structure. A closure device should be provided between two structures to block the noise leakage and, at the same time, permit thermal movements and differential settlement of structures without adversely affecting each other.
- In order to maintain consistency in the appearance of the sound wall, the panels on the independent structure should match the panels of the adjacent ground-mounted sound wall at least in texture and appearance.
- The independent structure should mimic the appearance of the existing bridge as much as possible and follow the span arrangement of the existing bridge. The substructure components of the independent structure should also utilize the same material and architectural treatment used on the adjacent bridge.

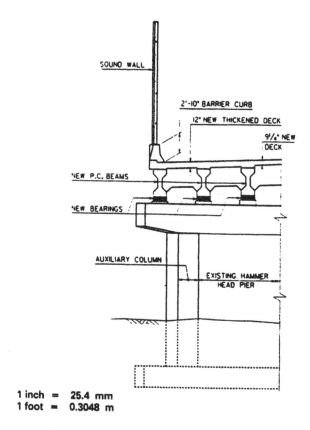

SOUND WALL

2'-10" BARRIER CURB

12" NEW THICKENED DECK

9¼" NEW DECK

NEW P.C. BEAMS

NEW BEARINGS

AUXILIARY COLUMN

EXISTING HAMMER HEAD PIER

1 inch = 25.4 mm
1 foot = 0.3048 m

FIGURE 62.8 Bridge-mounted sound wall.

- The superstructure of the independent structure that carries the sound panels should match the horizontal lines of the bridge and be capable of carrying the dead load of the panels and the wind load acting on the panels.
- Effects of fatigue, due to the reversal of wind direction, should be considered in the design of the structural steel components of the independent structure as well as their connections.
- The independent sound barrier structure should not infringe on the lateral and vertical clearance envelope of the existing bridge.
- Since an independent sound wall structure usually carries small vertical loads (dead and ice loads) and highly eccentric transverse loads (wind loads), it should either be founded on piles with high tensile capacity or be connected with dowels to the abutments and piers of the existing bridge, providing that these bridge components display adequate capacity.

62.1.6.2 Typical Details

The following are the typical features of an independent sound wall structure (Figure 62.9) [6]:

- Cast-in-place reinforced concrete footings and pedestals that are cast against and doweled into the existing bridge abutments and piers in order to prevent the rocking of the independent sound wall structure in the transverse direction under large wind loads.
- A steel tower made of a pair of structural steel tubes is supported on the concrete pedestal and carries a twin Vierendeel truss.
- A Vierendeel truss is made of structural steel tubing, and consists of continuous top and bottom chords and groove welded struts spanning between the chords. A twin truss is formed

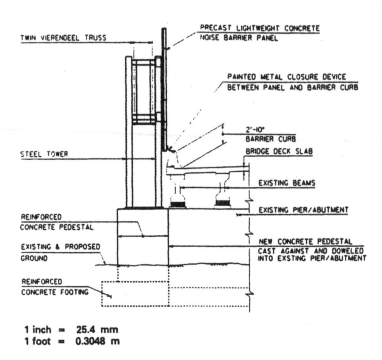

1 inch = 25.4 mm
1 foot = 0.3048 m

FIGURE 62.9 Independent sound wall structure. (*Source*: Guzaltan, F., *PCI J.*, 27(4), 62, 1992. With permission.)

by connecting each top and bottom chord member to the adjacent ones using groove-welded horizontal struts.

- The sound wall consists of 5-in. (130-mm)-thick lightweight reinforced concrete panels bolted to vertical posts (W shaped) spaced at 5-ft (1.5-m) intervals. The vertical posts are bolted to two spacers (W shaped) at the level of the top and bottom chords. These spacers are in turn shop-welded to the vertical struts of the truss.
- The gap between the wall panels and bridge parapet is closed using an L-shaped bent steel plate. The horizontal leg of the steel plate is bolted to the top of the parapet. The vertical leg is positioned upward and overlaps the wall panel approximately 10 in. (250 mm). A 2-in. (50-mm) gap is provided between the wall panels and upright leg of the bent plate to allow the independent movement of the bridge and sound wall.

62.2 Bridge Railings

62.2.1 Introduction

Railings are provided along the edges of structures in order to protect pedestrians, bicyclists, and vehicular traffic. Depending on the function they are designed to serve, bridge railings are classified as: pedestrian railing, bicycle railing, traffic railing, and combination railing. For each category, the AASHTO *Standard Specifications for Highway Bridges*, 16th ed. [7], has defined specific geometric requirements and the loads to be applied at various elements of the railings. An alternative approach for determining the geometries and the loads of unique or new railings is presented in the AASHTO *Guide Specifications for Bridge Railings* [8].

In the United States most states have their own standards for bridge railing geometry and design criteria. These standards generally follow the AASHTO Standard or Guide Specifications. In instances where a special design must be provided for a new railing or an existing railing must be upgraded, the following discussions are presented to help in understanding the AASHTO Standard and Guide Specifications and their differences.

By necessity, throughout the following section there are statements that paraphrase the contents of these references. Similarly, the figures and tables are those of AASHTO presented in a slightly different format to better serve the text. This treatise is not intended to substitute for the AASHTO Standard and Guide Specifications; rather it is meant to facilitate the AASHTO codes and their intention, as the author sees them.

As related to traffic railings and combination traffic and pedestrian railings, AASHTO Standard Specifications specifies only one set of loads applicable to all classes of vehicles and all types of roadways and traffic; it does not recognize variations in the vehicle type, percent of truck usage, design speeds, average daily traffic (ADT), etc. These variables are addressed in the AASHTO Guide Specifications which establishes guidelines for crash testing and evaluation of bridge railings based on three performance levels. Each performance level is based on the ADT of the roadway for which the railing is being considered. But other variables, such as design speed, percent truck use, rail offset to travel lane, and the type of highway (divided, undivided, and one-way), are also considered in the selection of the performance levels.

For each performance level the Guide Specifications establishes a certain crash test procedure and evaluation criteria. The testing procedure includes the type, weight, size, and geometry of the test vehicle as well as its speed and impact angle. Based on the crash test results, the railing is evaluated in accordance with the crash test evaluation criteria, which establishes the pertinent performance level and the type of traffic railing to suit the functional needs of a site.

Recognizing that crash tests are expensive and time-consuming, the Guide Specifications provide a table (Table 62.8), based on the results of actual crash tests conducted by the National Cooperative Highway Research Program (NCHRP). In this table design loads and traffic rail geometries are given for the three standard performance levels and two optional levels. These optional levels relate to heavier trucks and higher railings. The loads and dimensions provided in this table are applicable to four conceptual traffic rails that are representative of traffic rails or combination rails in the United States.

Table 62.8 is generally used to design the prototype railings that are to be crash tested. It is also used for the design of one-of-a-kind railings where the cost of crash testing cannot be justified. The highway agencies are encouraged to conduct crash test programs when specific traffic/combination railing is being considered for the first time. Railings that meet the crash test criteria for a desired performance level are exempt from the requirements of AASHTO Standard Specifications.

For bicycle and pedestrian railings the AASHTO Standard Specifications and Guide Specifications provide similar design requirements with minor differences in geometry. It is important to note that where pedestrian or bicycle traffic is expected, this traffic must be separated from travel lanes by a traffic railing or barrier. The height of this railing above the sidewalk or bikeway surface should be no less than 24 in. (610 mm). Also the face of the railing should have a smooth surface to prevent any snag potential. Where it is desirable to raise the height of the traffic barrier to prevent the bicycles from falling over the railing onto the roadway, or to improve the level of comfort, a traffic railing or a modified combination railing may be used.

Where a raised sidewalk curb with a width greater than 3.5 ft (1.067 m) is provided, the Guide Specifications would consider it acceptable to use only a crash-tested combination railing along the edge of the bridge. The curb must be included in the crash test that is used for determining the combination railing design. The Standard Specifications, however, makes no specific reference to raised sidewalks. Where the roadway curb projects more than 9 in. (229 mm) from traffic face of railing (construed to be a safetywalk or a sidewalk) the Standard Specifications allow the use of a combination traffic and pedestrian railing along the edge of the structure. On urban expressways where the curb projects less than 9 in. (229 mm), the Standard Specifications call for a combination railing to separate pedestrian walkways from the adjacent roadway. On rural expressways it calls for a traffic railing or barrier to separate pedestrians from vehicular traffic. A pedestrian railing must be provided along the edge of the structure where the pedestrian walkway is separated from the roadway.

TABLE 62.8　Bridge Railing Design Information — Bridge Railing Loads, Load Distribution, and Location

Quantity Designations	Railing Performance Level				
	PL-1	PL-2	PL-3	Optional PL-4	Optional PL-4T
			Group I[a] Loads (Body and Wheels)		
F_{BWH}	30 (133) kips	80 (356) kips	140 (623) kips	200 (890) kips	200 (890) kips
F_{BWL}	±9 (±40) kips	±24 (±107) kips	±42 (±187) kips	±60 (267) kips	±60 (±267) kips
F_{BWV}	±12 (±53) kips (down) -4 (-18) kips (up)	15 (67) kips (down) -5 (-22) kips (up)	+18 (80) kips (down) -6 (-27) kips (up)	+18 (80) kips (down) -6 (-27) kips (up)	+18 (80) kips (down) -6 (-27) kips (up)
			Group II[a] Loads (Trailer Floor)		
F_{FH}	—	—	—	240 (890) kips	200 (890) kips
F_{FL}	—	—	—	±60 (267) kips	±50 (±222) kips
F_{FV}	—	—	—	-18 (80) kips (down) -6 (-27) kips (up)	+18 (80) kips (down) -6 (-27) kips (up)
			Group III[a] Loads (Tank Trailer)		
F_{TH}	—	—	—	—	200 (890) kips
F_{TL}	—	—	—	—	±50 (±222) kips
F_{TV}	—	—	—	—	+18 (80) kips (down) -6 (-27) kips (up)

Load Distribution Pattern Dimensions

a	24 in. (610)	28 in. (711)	32 in. (813)	36 in. (914)	36 in. (914)
b	12 in. (305)	14 in. (356)	16 in. (406)	18 in. (457)	18 in. (457)
c	—	—	—	12 in. (305)	12 in. (305)
d	—	—	—	6 in. (152)	6 in. (152)
e	—	—	—	—	36 in. (914)
f	—	—	—	—	8 in. (203)

Load Locations

h_{BW}	16 in. thru (H-6 in.) [(406) thru (H-152)]	17 in. thru (H-7 in.) [(432) thru (H-178)]	18 in. thru (H-8 in.) [(457) thru (H-203)]	19 in. thru (H_{BW}-9 in.) [(483) thru (H_{BW}-229)]	19 in. thru (H_{BW}-9 in.) [(483) thru (H_{BW}-229)]
h_F	—	—	—	51 in. (1295)	51 in. (1295)
h_T	—	—	—	—	74 in. (min) 84 in. (max) (1.880 m) (2.134 m)

Railing Geometry Dimensions

H	27 in. (686) (min)	32 in. (813) (min)	42 in. (1067) (min)	54 in. (1372) (min)	78 in. (1981) (min)
H_A	10 in. (254) (max)	10 in. (254) (max)	10 in. (1254) (max)	10 in. (254) (max)	10 in. (254) (max)
H_{BW}	27 in. (686) (min)	32 in. (813) (min)	42 in. (1067) (min)	32 in. to 42 in. (813 to 1067)	32 in. to 42 in. (813 to 1067)
H_{BWR}	12 in. (305) (min)	12 in. (305) (min)	12 in. (305) (min)	12 in. (305) (min)	12 in. (305) (min)
H_F	—	—	—	54 in. (1372) (min)	54 in. (1372) (min)
H_{FR}	—	—	—	6 in. (152) (min)	6 in. (152) (min)
H_T	—	—	—	—	78 in. (1981) (min)
H_{TR}	—	—	—	—	8 in. (152) (min)

Note: PL = Performance Level. Where kips are indicated values in ( ) indicate metric equivalent in kilonewtons; where inches are indicated values in ( ) indicate metric equivalent in millimeters.

a Each set of Group Loads to be applied separately.

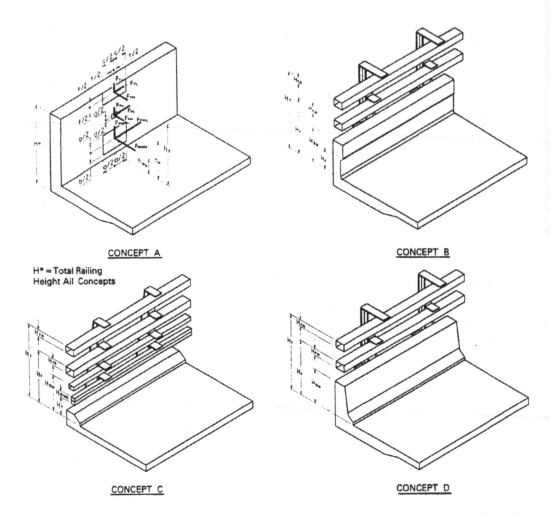

CONCEPT A

CONCEPT B

H* = Total Railing
Height All Concepts

CONCEPT C

CONCEPT D

FIGURE 62.10   Bridge railing concepts — configuration and loading patterns. (*Source*: AASHTO, *Guide Specifications for Bridge Railings*, American Association of State Highway and Transportation Officials, Washington, D.C., 1989. With permission.)

62.2.2   Vehicular Railing

62.2.2.1   Geometry

Bridge railings are primarily provided to contain the vehicles using the bridge, but they are also required to (1) protect the occupants of an errant vehicle in a collision with the railing; (2) protect other vehicles near the collision; (3) protect the people and properties on the roadways or other areas underneath the structure; and (4) have the appearance and freedom of view from passing vehicles.

Figure 62.10 shows four conceptual railing configurations identified in the Guide Specifications. The railing dimensions and the magnitude of design loads to be applied at the points of load application are given in Table 62.8. All three standard performance levels and two optional performance levels are represented in Figure 62.10. Railings for each performance level can be constructed from this figure by assuming that railing elements for which no dimension is given in Table 62.8 do not exist.

The dimensions and configurations shown for these railings are designed to provide a smooth and continuous surface for the traffic side of the railings. The geometry of the rail system should

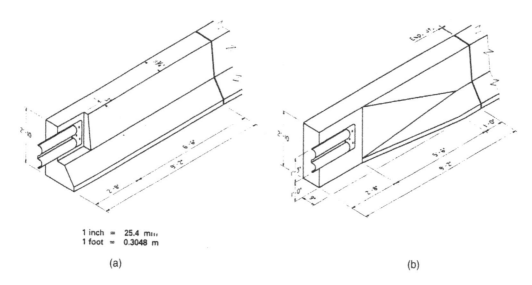

1 inch = 25.4 mm
1 foot = 0.3048 m

(a) (b)

FIGURE 62.11 Guide rail attachment at end of bridge.

be such as to preclude any potential contact with the posts by major vehicle parts, should there be a penetration or opening through the railing. The Guide Specifications requires a minimum of 10 in. (254 mm) between the face of the railing and the face of the posts, where snagging is an obvious possibility.

At bridge ends where an open-face railing (guide rail) meets the bridge parapet or barrier, a transition is normally required to provide a smooth flow of traffic while eliminating any snag potential. The Guide Specifications requires that the close face railing (a parapet or a barrier) be flared a maximum of 3.5 longitudinal to one lateral, starting a minimum of 10 in. (254 mm) back of the open-face railing (a guide rail).

Figure 62.11 represents the standard details used in New Jersey for the attachment of guide rails to concrete barriers. These details can be applied at the ends of a parallel wingwall, the ends of a pylon created to provide the transition from guide rail to barrier, or at the ends of a bridge parapet where the thermal movements of the bridge can be accommodated by slotted holes in the guide rail, i.e., small movements.

Where posts are used in a bridge traffic railing, post spacing should not exceed 10 ft 0 in. (3.05 m).

The height of traffic railing is measured from the top of the highest rail or parapet to the top of the roadway, the top of the future overlay, or the top of a raised sidewalk. A raised sidewalk can be defined as a raised roadway curb located along the edge of the bridge, sufficiently wide to accommodate passage of two pedestrians shoulder to shoulder. While 4 ft (1.219 m) is the nominal minimum width, the Guide Specifications allows an absolute minimum width of 3 ft 6 in. (1.067 m).

In the past, safetywalks were used as a means of providing access for infrequent pedestrians or maintenance personnel along a long bridge where a sidewalk could not be justified. These safetywalks, varying in width from 1 ft 6 in. to 3 ft (457 mm to 914 mm), are no longer used since they do not provide sufficient protection for the pedestrians and traffic. Raised curbs supporting a railing, however, are still in use. These curbs generally have a horizontal projection of 9 in. (229 mm) or less from the railing face.

The minimum height of traffic railings or the traffic portion of combination railings is 2 ft 3 in. (686 mm). Where a parapet has a sloping face, intended to allow redirecting of vehicles back to the roadway, the minimum height allowed by the Standard Specifications is 2 ft 8 in. (813 mm). The Guide Specifications, however, lists a 2 ft 3-in. (686-mm) minimum dimension for the same height for Performance Level 1 (see Table 62.8).

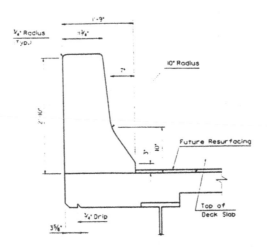

1 inch = 25.4 mm
1 foot = 0.3048 m

FIGURE 62.12 Standard New Jersey barrier bridge parapet to be used where there is no walkway on the bridge.

PL = Performance Level. Where kips are indicated values in () indicate metric equivalent in kilonewtons; where inches are indicated values in () indicate metric equivalent in millimeters.

The minimum height of a traffic barrier in some states exceeds the AASHTO requirements. For instance, in New Jersey, where the barrier is along the edge of the bridge, the minimum height is 2 ft 10 in. (864 mm), as shown in Figure 62.12. Where the barrier is used to separate pedestrians from highway traffic, the minimum height is 2 ft 8 in. (813 mm).

In a combination railing where the lower element is a parapet, or in a traffic railing, the height of the lower element should be no less than 1 ft 6 in. (457 mm). If a rail is the lower element, its height, measured from its center to reference surface, should be between 15 and 20 in. (381 and 508 mm).

The Guide Specifications requires that the clear distance between the bottom rail and the reference surface be no less than 10 in. (254 mm). The maximum clearance for the same dimension is given as 17 in. (432 mm) by the Standard Specifications. The maximum distance between adjacent rails should not exceed 15 in. (381 mm). Additionally, the traffic face of all rails should be within 1 in. (25 mm) of a vertical plane through the traffic face of the rail closest to traffic.

Thermal movements of the rails are normally provided for by the use of joints in sleeves in the rails. For short-length bridges these joints can be located anywhere along the bridge length. But for long-span bridges, where thermal movements are expected to exceed 2 to 3 in. (51 to 76 mm), it is prudent to place the splices at the expansion joint locations. To eliminate snag potential at these joints, sleeves for pipes (or steel hoods for concrete parapets) must be provided. The projection or depression of the rails at rail joints or steel hoods should not have a depth (thickness) greater than the rail wall thickness or ⅜ in. (10 mm), whichever is less.

62.2.2.3 Loads

Method 1 — Guide Specifications

The Guide Specifications provides criteria for the selection of performance levels of various sites based on the ADT, design speed, percent truck use, bridge rail offset from traffic, and the type of highway under consideration. Upon the selection of the performance level, Table 62.8 can be used to determine the magnitude of loads to be applied to the railing at locations shown in Figure 62.11. While the performance levels PL-1, PL-2, and PL-3 represent small automobiles, pickup trucks, medium-size single unit trucks, and van-type tractor-trailers, the optional performance levels PL-4 and PL-4T are used where heavier and larger trucks at higher volumes are likely to use the roadway. The optional level PL-4 is given for 54 in. (1.372 m) high and higher railings, and where truck volumes and truck type, size, and weight would be greater than PL-3. The optimal level, PL-4T, represents railings that have a minimum height of 6 ft 6 in. (1.981 m), and where truck volumes and highway alignment and use would justify such high railings, e.g., closed face barrier curbs over electrified tracks in high-speed, high-volume curved alignments.

The loads to be applied at the lowest level of the railing, F_{BWH}, F_{BWL}, and F_{BWW} represent the impact loads of the body and wheels of an errant vehicle; they are given under Group 1 Loads for all five performance levels. The loads to be applied at the midrail of the railing, F_{FH}, F_{FL}, and F_{FV}, represent the impact loads of a trailer floor; they are given under Group II for optional performance levels PL-4 and PL-4T. The loads to be applied at the top level of performance level PL-T4, representing the potential impact of a tank trailer, are given under Group III Loading. While all three loads in each group should be applied simultaneously, and distributed over the designated area, only one group of loading should be applied at a time. All three loads in each group should be distributed evenly over the loaded areas. Dimensions of the load areas over which the forces for each loading category must be evenly distributed are given in Table 62.8.

Where the railing width is less than the related load area, the entire load should be distributed over the available width. Where a load area bears on more than one rail, the load applied to each rail should be prorated based on the distance from each rail to the reference surface. It is to be noted that Load F_{BWH} must be applied over a range of heights as shown in Table 62.8.

The loads on posts are transmitted through the longitudinal rail elements. These loads are to be distributed to no more than three posts.

As stated previously, the loading criteria outlined above are to be used for the design of prototype railings that are to be crash tested and for the design of one-of-a-kind railings where the cost of crash testing cannot be justified. Otherwise, the best way to ensure suitability of a particular railing for a given site is to subject it to applicable crash tests and evaluate the results in accordance with the performance criteria for a desired performance level. The crash test procedures are described in the NCHRP Report 350, Recommended Procedures for the Safety Performance Evaluation of Highway Features [9]. The criteria for evaluation of crash test results and the procedure for the selection of performance levels are provided in the *Guide Specifications for Bridge Railings*.

Since performance level selection is based on the assumption that a railing will be near its ultimate strength when subjected to its specific maximum containment load, it is recommended that ultimate strength approach be used in the analysis and design of the railings, posts, and supporting deck slab.

Where a railing is selected and successfully crash tested in accordance with the provisions of the Guide Specifications and NCHRP 350, it does not need to meet the requirements of AASHTO Standard Specifications, as described below.

Method 2 — Standard Specifications

The nominal transverse load $P = 10$ kips (44.5 kN) is to be applied and distributed as shown in Figure 62.13.

- Where the height of the top traffic rail exceeds 2 ft 9 in. (838 mm), the rail and the post is to be designed for a transverse load of CP, where

$$C = 1 + \frac{h-33}{18} > 1 \left(C = \frac{h-381}{457.2} \right)$$

  and h = height of top rail from reference surface in inches (millimeters). However, the maximum load applied to any element is not to exceed P.

- Where the rail face is more than 1 in. (25.4 mm) behind a vertical plane through the face of traffic rail closest to traffic, or where the rail center is less than 1 ft 3 in. (38 mm) from the reference surface, the rail should be designed for a $P/2$ or what is applied to an adjacent traffic rail, whichever is less.

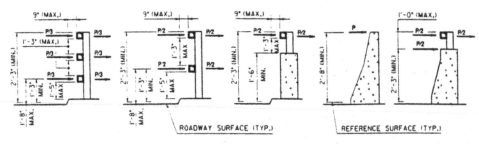

TRAFFIC RAILING

1 inch = 25.4 mm   **To be used where there is no curb or curb projects 9"**
1 foot = 0.3048 m   **or less from the traffic face of railing**

FIGURE 62.13 Traffic railing to be used where there is no curb or curb projects 9 ft or less from the traffic face of the railing. (*Source*: AASHTO, *Standard Specifications for Highway Bridges*, 16th ed., American Association of State Highway and Transportation Officials, Washington, D.C., 1996. With permission.)

- The posts are to be designed for P or CP, as shown in Figure 62.13. Simultaneous with the transverse loads a longitudinal load equal to $P/2$ or $CP/2$ is to be applied, divided among a maximum of four posts, assuming that railing is continuous. Also, posts are to be designed for an inward load, equal to ¼ of the outward loads.

- The rail attachment to the post is to be designed for a vertical load of ¼ of transverse loading, to be applied either upward or downward. This attachment should be also designed for an inward load of ¼ of transverse load.

- The rail members are to be designed for a moment of $P'L/6$, at the center of the panel and at the posts, where L is the post spacing and P' is P, $P/2$, or $P/3$, as modified by the factor C, where required. The handrail members of combination railings are to be designed for a moment of $0.1\,wL^2$ at the center and at the posts, where w is the pedestrian loading per unit length of rail; $w = 50$ lb/ft (729 N/m).

Where a concrete parapet or barrier curb is used, the transverse load of P or CP should be spread over a 5-ft (1.5-m) length of the parapet. Since AASHTO does not specify the location of load application it can be construed that even where the parapet/barrier curb ends the 5-ft (1.5-m) distribution length applies.

Where possible, bridge railings should preferably provide continuity for moment and shear throughout their length. To meet this goal, continuity transfer splices and expansion devices (capable of handling moments) and end anchorages (for transferring shear) must be provided for beam-and-post railings. Providing continuity transfer sleeves for concrete parapets may be more difficult because of the frequency of transverse joints in the parapets. These joints are normally provided at 10 to 20 ft (3 to 6 m) intervals to arrest potential temperature and shrinkage cracks that may otherwise develop, and to prevent parapet participation in the composite behavior of the fascia girders.

In designing the deck slab and distributing the loads from the posts to the deck slab, it is highly desirable to design the deck slab such that it would not sustain any damage due to a potential destruction of the post. This may be accomplished by providing additional reinforcement in the deck slab in order to distribute the concentrated loads from the posts over a larger area of the deck slab.

62.2.3 Bicycle Railing

62.2.3.1 Geometry

Bicycle railings are provided along the edges of the structure to contain the bicyclist where bicycle use is anticipated. Such railings are to be designed to provide safety while meeting the aesthetic

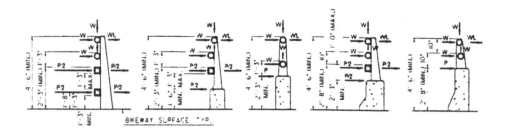

COMBINATION TRAFFIC AND BICYCLE RAILING

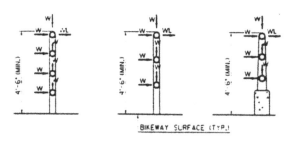

BIKEWAY SURFACE (TYP.)

1 inch = 25.4 mm
1 foot = 0.3048 m

BICYCLE RAILING

NOTES:

Loads on left are applied to rails
Loads on right are applied to posts
W = Pedestrian loading per unit length of rail
L = Post Spacing

FIGURE 62.14 Bicycle railings. (*Source:* AASHTO, *Standard Specifications for Highway Bridges,* 16th ed., American Association of State Highway and Transportation Officials, Washington, D.C., 1996. With permission.)

requirements of the bridge owner. Where it is used in conjunction with vehicular traffic, freedom of view from passing vehicles is to be maintained.

As shown in Figure 62.14, the minimum height of a bicycle railing is 4 ft 6 in. (1372 mm), measured from the top of the riding surface to the top of the top rail.

The Standard Specifications requires that from the bikeway surface to 2 ft 3 in. (686 mm) above it, all railing elements be spaced such that a 6-in. (152-mm) sphere cannot pass through the railing. From 2 ft 3 in. to 4 ft 6 in. (686 to 1372 mm) from the bikeway surface an 8-in. (203 mm) sphere should not be able to pass through the railing. Where the railing is made up of horizontal and vertical elements, the spacing requirements apply to one or the other, but not to both.

The Guide Specifications requires that from the bikeway surface to 4 ft 6 in. (1372 mm) above it, horizontal elements of the railing have a clear spacing of 1 ft 3 in. (381 mm) and vertical elements a clear spacing of 8 in. (203 mm). Where the railing has both horizontal and vertical elements, the spacing requirements will apply to one or the other, but not to both.

62.2.3.2 Loads

As shown on Figure 62.14, the horizontal elements (rails) in a bicycle railing are to be designed for a minimum design loading of $w = 50$ lb/ft (729 N/m) acting laterally and vertically at the same time. The vertical elements (posts) should be designed for wL (where L is the post spacing) acting at the center of the upper rail, but at a height not greater than 4 ft 6 in. (1372 mm). Where a rail is located more than this limit, the design load will be determined by the designer.

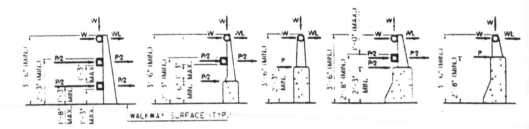

COMBINATION TRAFFIC AND PEDESTRIAN RAILING

To be used when curb projects more then
9" from the traffic face of railing

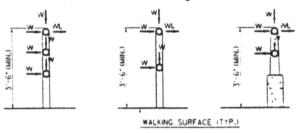

WALKING SURFACE (TYP.)

PEDESTRIAN RAILING

1 inch = 25.4 mm   To be used on the outer edge of a sidewalk
1 foot = 0.3048 m   when highway traffic is separeted from
                     pedestrian traffic by a traffic railing.

NOTES:

Loads on left are applied to rails
Loads on right are applied to posts
W = Pedestrian loading per unit length of rail
L = Post Spacing

FIGURE 62.15 Pedestrian railings. (*Source*: AASHTO, *Standard Specification for Highway Bridges*, 16th ed., American Association of State Highway and Transportation Officials, Washington, D.C., 1996. With permission).

Where vehicular traffic and bicycles are contained by a single combination railing, the Standard Specifications provides the geometry and loading requirements for five railing options, as reproduced in Figure 62.14.

62.2.4 Pedestrian Railing

62.2.4.1 Geometry

Pedestrian railings are provided along the outer edge of a sidewalk to contain pedestrians. Where the sidewalk is not raised, a traffic railing (a concrete barrier or a guide rail) must separate the pedestrians from highway traffic. In an urban setting, where there is a raised sidewalk, a combination traffic and pedestrian railing, along the outer edge of the sidewalk will be required. Such railings will be designed to provide safety while meeting the aesthetic requirements of the bridge owner. Consideration should also be given to freedom of view from passing vehicles.

As shown in Figure 62.15, the minimum height of a pedestrian or a combination railing is to be 3 ft 6 in. (1067 mm), measured from the top of the walkway to the top of the upper rail.

The Standard Specifications requires that, from the walkway surface to 27 in. (686 mm) above it, all railing elements be spaced such that a 6-in. (152 mm) sphere cannot pass through any opening in the railing. From 2 ft 3 in. to3 ft 6 in. (686 to 1067 mm) from the walkway surface, an 8-in. (203-mm) sphere should not be able to pass through the railing.

The Guide Specifications requires that, within the 3 ft 6-in. (1067 mm) height of the railing, the horizontal elements have a maximum clear spacing of 1 ft 3 in. (381 mm) while the vertical elements have a maximum clear spacing of 8 in. (203 mm). Where the railing has both horizontal and vertical elements, the spacing requirements will apply to one or the other, but not to both.

62.2.4.2 Loads

As shown on Figure 62.15, the horizontal elements (rails) in a pedestrian railing are to be designed for a minimum design loading of $w = 50$ lb/ft (729 N/m), acting laterally and vertically at the same time. Rail members located more than 5 ft (1.524 m) above the walkway are excluded from this requirement. The vertical elements (posts) are to be designed for wL (where L is the post spacing) acting at the center of the upper rail.

Where combination traffic and pedestrian railing is to be provided, the geometry and the loads may be obtained from one of the five options shown in the Standard Specifications and reproduced in Figure 62.15.

62.2.5 Structural Specifications and Guidelines for Bicycle and Pedestrian Railings

Bicycle and pedestrian railings are to be designed by the elastic method to the allowable stresses for the appropriate material. The following requirements are those specified in the AASHTO Standard Specifications, used by permission.

For aluminum alloys the design stresses given in the *Specifications for Aluminum Structures* 5th ed., December 1986, published by the Aluminum Association, Inc., for "Bridge and Similar Type Structures" apply. For alloys 6061-T6 (Table A.6), 6351-T5 (Table A.6), and 6063-T6 (Table A.8) apply, and for cast aluminum alloys the design stresses given for alloys A444.0-T4 (Table A.9), A356.0-T61 (Table A.9), and A356.0-T6 (Table A.9) apply.

For fabrication and welding of aluminum railing see Article 11.5 of the AASHTO *Standard Specifications for Highway Bridges.*

The allowable unit stresses for steel are as given in Article 10.32 of the AASHTO Standard Specifications, except as modified below.

For steels not generally covered by the Standard Specifications but having a guaranteed yield strength, F_y the allowable unit stress, is derived by applying the general formulas as given in the Standard Specifications under "Unit Stresses" except as indicated below.

The allowable unit stress for shear is $F_v = 0.33 F_y$

Round or oval steel tubes may be proportioned using an allowable bending stress, $F_b = 0.66 F_y$, provided the R/t ratio (radius/thickness) is less than or equal to 40.

Square and rectangular steel tubes and steel W and I sections in bending with tension and compression on extreme fibers of laterally supported compact sections having an axis of symmetry in the plane of loading may be designed for an allowable stress $F_b = 0.60 F_y$

The requirements for a compact section are as follows:

In the above formulas b, t, and ℓ are in inches (millimeters) and f_a, F_a, and F_y are in psi (Mpa).

	English Units	S.I Units	
1. The width-to-thickness ratio of projecting elements of the compression flange of W and I sections not to exceed:	$\dfrac{b}{t} \le \dfrac{1600}{\sqrt{F_y}}$	$\left(\dfrac{b}{t} \le \dfrac{133}{\sqrt{Fy}} \right)$	(62.3)
2. The width-to-thickness ratio of the compression flange of square or rectangular tubes is not to exceed:	$\dfrac{b}{t} \le \dfrac{6000}{\sqrt{F_y}}$	$\left(\dfrac{b}{t} \le \dfrac{499}{\sqrt{Fy}} \right)$	(62.4)
3. The D/t ratio of webs is not to exceed:	$\dfrac{D}{t} \le \dfrac{13{,}300}{\sqrt{F_y}}$	$\left(\dfrac{D}{t} \le \dfrac{1106}{\sqrt{Fy}} \right)$	(62.5)
4. If subject to combined axial force and bending, the D/t ratio of webs is not to exceed:	$\dfrac{D}{t} < \dfrac{13{,}300\left[1 - 1.43\left(\dfrac{f_a}{F_a}\right)\right]}{\sqrt{F_y}}$	$\left(\dfrac{1106\left[1 - 1.43\dfrac{fa}{Fa}\right]}{\sqrt{Fy}} \right)$	(62.6)
but need not be less than:	$\dfrac{D}{t} < \dfrac{7000}{\sqrt{F_y}}$	$\left(\dfrac{D}{t} < \dfrac{581}{\sqrt{Fy}} \right)$	(62.7)
5. The distance between lateral supports in inches of W or I sections is not to exceed:	$\ell \le \dfrac{2400b}{\sqrt{F_y}}$	$\left(\le \dfrac{199.26}{\sqrt{F_y}} \right)$	(62.8)
or:	$\ell \le \dfrac{20{,}000{,}000\, A_f}{dF_y}$	$\left(\ell \le \dfrac{137{,}640\, A_f}{dF_y} \right)$	(62.9)

References

1. Procedures for Abatement of Highway Traffic Noise and Construction Noise, Federal Aid Highway Program Manual, Vol. 7, Chap. 7, Section 3, 1982.
2. Simpson, M. A., Noise Barrier Design Handbook, Publication No. FHWA-RD-76-58, U.S. Department of Transportation, 1976.
3. AASHTO, *Guide Specifications for Structural Design of Sound Barriers*, American Association of State Highway and Transportation Officials, Washington, D.C., 1989.
4. Blum, R. F. A Guide to Visual Quality in Noise Barrier Design, Publication No. FHWA-HI-94-039, U.S. Department of Transportation, 1976.
5. Highway Noise Barriers, National Cooperative Highway Research Program, Synthesis of Highway Practice 87, 1981.
6. Guzaltan, F., Precast concrete noise barrier walls for New Jersey Interstate Route 80, *PCI J.*, 27(4), 1992.
7. AASHTO, *Standard Specifications for Highway Bridges*, 16th ed., American Association of State Highway and Transportation Officials, Washington, D.C., 1996.
8. AASHTO, *Guide Specifications for Bridge Railings*, American Association of State Highway and Transportation Officials, Washington, D.C., 1989.
9. Recommended Procedures for Safety Evaluation of Highway Features, National Cooperative Highway Research Program Report 350.

Section VII
Worldwide Practice

63

Design Practice in China

Guohao Li
Tongji University, China

Rucheng Xiao
Tongji University, China

63.1 Introduction

63.1.1 Historical Evolution

With a recorded history of about 5000 years, China has a vast territory, topographically higher in the northwest and lower in the southeast. Networked with rivers, China has the well-known valleys of the Yangtze River, the Yellow River, and the Pearl River, which are the cradle of the Chinese nation and culture. Throughout history, the Chinese nation erected thousands of bridges, which form an important part of Chinese culture.

0-8493-7434-0/00/$0.00+$.50
© 2000 by CRC Press LLC

FIGURE 63.1   Anji Bridge.

Ancient Chinese bridges are universally acknowledged and have enjoyed high prestige in world bridge history. They can be classified into four categories: beam, arch, cable suspension, and pontoon bridges.

The earliest reference to the beam bridge in Chinese history is the Ju Bridge dating from the Shang Dynasty (16th to 11th century B.C.). During the Song Dynasty (A.D. 960 to 1279), a large number of stone pier and stone-beam bridges were constructed. In Quanzhou alone, as recorded in ancient books, 110 bridges were erected during the two centuries, including 10 well-known ones. For example, the 362-span Anping Bridge was known for its length of 2223 m, a national record for over 700 years. To elongate the span, either the timber beams or the stone ones were placed horizontally on top of each other, the upper layer cantilevering over the lower one, thus supporting the simple beam in the middle. The extant single-span timber cantilever bridge, the Yinping Bridge built in Qing Dynasty (A.D. 1644 to 1911) has a span of more than 60 m with a covered housing on it.

The oldest arch bridge in China, which still survives and is well preserved, is the Anji Bridge, also known as the Zhaozhou Bridge, at Zhouxian, Hebei Province, built in the Sui Dynasty (Figure 63.1). It is a single segmental stone arch, composed of 28 individual arches bonded transversely, 37.02 m in span and rising 7.23 m above the chord line. Narrower in the upper part and wider in the lower, the bridge averages 9 m in width. The main arch ring is 1.03 m thick with protective arch stones on it. Each of its spandrels is perforated by two small arches, 3.8 and 2.85 m in clear span, respectively, so that flood can be drained and the bridge weight is lightened as well. The Anji Bridge has a segmental deck and the parapets are engraved with dragons and other animals. Its construction started in the 15th year of the reign of Kaithuang (A.D. 595) and was completed in the first year of Day's reign (A.D. 605) of the Sui Dynasty. To date, it has survived for 1393 years. The bridge, exquisite in workmanship, unique in structure, well proportioned and graceful in shape, with its meticulous yet lively engraving, has been regarded as one of the greatest achievements in China. Great attention has been paid to its preservation through successive dynasties. In 1991, the Anji Bridge was named among the world cultural relics.

Stone arches in China vary in accordance with different land transport and different natures between the north and south waterways. In the north, what prevails is the flat-deck bridge with solid spandrels, thick piers, and arch rings, whereas in the south crisscrossed with rivers, the hump-shaped bridge with thin piers and shell arches prevails.

In the southeastern part of China, Jiangsu and Zhejiang Provinces, networked with navigable rivers, boats were the main means of transportation. As bridges were to be built over tidal waters and their foundations laid in soft soil, even the stone arch bridge had to be built with thin piers and shell arches in order that its weight could be reduced as much as possible. The thinnest arch

FIGURE 63.2   Suzhou Baodai Bridge.

ring is merely ⅟₆₇ of the span, whereas for an average the depth of the arch ring is ⅟₂₀ of the span. The longest surviving combined multispan bridge with shell arches and thin piers is the Baodai Bridge (Figure 63.2) in Suzhou, Jiangsu Province. Built in the Tang Dynasty (A.D. 618 to 907) and having undergone a series of renovations in successive dynasties, the bridge is now 316.8 m long, 4.1 m wide, with 53 spans in all, the three central arches being higher than the rest for boats to pass through. Both ends of the bridge are ornamented with lions or pavilions and towers, all of stone.

Cable suspension bridges vary in kind according to the material of which the cables are made: rattan, bamboo, leather, and iron chain. According to historical records, 285 B.C. saw the Zha Bridge (bamboo cable bridge). Li Bin of the Qin State, who guarded Shu (256 to 251 B.C.), superintended the establishment of seven bridges in Gaizhou (now Chengdu, Sichuan Province), one of which was built of bamboo cables. The Jihong Bridge at Yongping County, Yunnan Province, is the oldest and broadest bridge with the mostly iron chains in China today. Spanning the Lanchang River, it is 113.4 m long, 4.1 m wide, and 57.3 m in clear span. There are 16 bottom chains and a handrail chain on each side. The bridge is situated on the ancient road leading to India and Burma.

The Luding Iron-Chain Bridge (Figure 63.3) in Sichuan Province, the most exquisite of the extant bridges of the same type, spans the Dadu River and has served as an important link between Sichuan Province and Tibet. It is 104 m in clear span, 2.8 m in width, with boards laid on the bottom chains. There are nine bottom chains, each about 128 m long, and 2 handrail chains on each side. On each bank, there is a stone abutment, whose deadweight balances the pulling force of the iron chains. Its erection began in 1705 and was completed in the following year.

According to historical records, a great number of pontoon bridges were built at nine and five different places over the Yangtze and the Yellow Rivers, respectively, in ancient times. In 1989 unearthed in Yongji, Shanxi Province, were four iron oxen, weighing over 10 tons each, and four life-size iron men, all with lively charm, exquisitely cast. They were intended to anchor the iron chains on the east bank of the Pujing Floating Bridge in the Tang Dynasty.

Ancient Chinese bridges, with various structures, exquisite workmanship, and reasonable details are the fruit of practical experience. Calculations and analyses by modern means prove that the great majority is in conformity with scientific principles. Ancient Chinese bridges are of great artistic and scientific value and have made remarkable achievements, from which we can assimilate rich nourishment to give birth to new and future bridges.

Comparatively speaking, the construction of modern bridges in China started late. Before the 1950s, many bridges were invested, designed, and constructed by foreigners. Most highway bridges were made up of wood. After the 1950s, China's bridge construction entered a new era. In 1956, the first prestressed concrete highway bridge was constructed. After 1 year, Wuhan Yangtze River Bridge was erected, which ended the history of the Yangtze River having no bridges. Nanjing Yangtze

FIGURE 63.3 Luding Iron-Chain Bridge.

River bridge was completed in 1969. In the 1960s, China began to adopt cantilever construction technology to construct T-type rigid frame bridges. During the 1970s, more prestressed concrete continuous bridges were constructed. China also began to practice new construction technology such as the lift-push launching method, the traveling formwork method, the span-by-span erecting method, etc. Two reinforced concrete cable-stayed bridges were constructed in 1975, which signified the start of cable-stayed bridge construction in China. Since 1980, China began to develop long-span bridges. One after another, many long-span bridges such as Humen Bridge (prestressed concrete continous rigid frame) in Guangdong Province with a main span of 270 m, Wanxian Yangtze River Bridge (arch reinforced concrete) in Shichuan Province with a main span of 420 m, Yangpu Bridge (cable-stayed) in Shanghai City with a main span of 602 m, etc. have been completed. The Jiangying Yangtze River (suspension) Bridge with a main span of 1385 m is under construction. The first two bridges mentioned above have the longest spans of their respective types in the world. Today, five large-scale and across-sea projects for high-class road arteries along the coast are under planning by the Ministry of Communications of China. From north to south, the road arteries cut across Bohai Strait, Yangtze Seaport, Hangzhou Bay, Pearl Seaport, Lingdingyang Ocean, and Qiongzhou strait. A large number of long-span bridges have to be constructed in these projects. The Lingdingyang long-span bridge project across Pearl Seaport has started.

63.1.2 Bridge Design Techniques

63.1.2.1 Design Specifications and Codes

There are two series of bridge design specifications and codes in China. One is for highway bridges [3] and the other for railway bridges [4]. In addition, there are design guides such as the wind-resistant guide for bridges [6]. Design Specifications for Highway Bridges are mainly for concrete bridges, which are widely constructed in China. Here only these specifications are presented because of space limitations.

The current Design Specifications for Highway Bridges [3], which were issued by the Ministry of Communications of the People's Republic of China in 1989, include six parts. They are the General Design Specification for Bridges, the Design Specification for Masonry Bridges, the Design Specification for Reinforced and Prestressed Concrete Bridges, the Design Specification for Footing and Foundations of Bridges, the Design Specification for Steel and Timber Members of Bridges, and the Seismic Design Specification for Bridges. The design philosophies and loads are provided in the General Design Specification.

In the specifications, two design philosophies are adopted: load and resistance factor design (RFD) theory for reinforced prestressed concrete members and allowable stress design (ASD) theory for steel and timber members.

Three basic requirements for strength, rigidity, and durability need to be checked for all bridge members. For a bridge member that may be subjected to bending, axial tension, or compression, combined bending and axial forces etc. should be checked in accordance with its loading states. To ensure its strength requirement, the rigidity of a bridge is evaluated according to the displacement range at the midspan or cantilever end. By checking the widths of cracks and taking some measurements, the durability of structures may be ensured.

63.1.2.2 Analysis Theories and Methods

The analysis of a bridge structure in terms of service is based on the assumption of linear elastic theory and general mechanics of materials. According to design requirements, the enveloping curves of internal forces and displacements of members of a bridge are calculated. Then, checking for strength, rigidity, and durability is done carefully in accordance with the design specifications. For simple structures, they are usually simplified as plane structures but they can also be analyzed more accurately by 3D-FEM.

For example, simply supported girder bridges are usually simplified in the following way. According to the cross section shape and the construction method, the bridge may be divided into several longitudinal basic members such as T-girders or hollow plate girders or box girders. The internal forces of the basic members caused by dead loads are calculated under an assumption of every basic member carrying the same loads. In order to consider the effect of space structure under live loads, the influence surfaces of internal forces and displacements are approximately simplified as two univariant curves; one is the influence line of internal forces or displacements of a basic member and another is the influence line of the transverse load distribution.

To prove the feasibility and reliability of the approximate method, extensive tests and theoretical studies have been conducted. Several methods to determine the influence lines of transverse load distribution for different structures and construction methods have been developed [5]. In the current practice, the transversely hinge-connected slab (or beam) method, rigid-connected beam method, rigid cross beam method, and lever principle method are used according to structures and construction methods. They may satisfy the design requirement for a lot of bridges. With computer programs, these simplified analysis methods have become very easy.

However, some bridges, such as irregular skewed bridges, curved bridges, and composite bridges, cannot be divided into several longitudinal girders that mainly have behaviors of vertical plane structures. They are not suited to the simplified analysis methods mentioned above. For those complex space structures, the influence surfaces of internal forces and displacements due to dead load are obtained by the static finite-element method and the maximal impact responses of internal forces and displacements caused by live loads can be obtained using dynamic analysis proceedures.

63.1.2.3 Theories and Methods for Long-Span Bridges

Long-span bridges are usually expensive to construct and are flexible in structural nature. In view of the economic and functional requirements, the problems of structural optimization, nonlinear analysis, stability analysis, and construction control become especially important to long-span bridges. Chinese bridge experts who participate in the study and design of China's long-span bridges have put forward many theories and methods to solve the problems mentioned above. In respect to the nonlinear analysis of long-span bridges, they developed an influence area method for geometric nonlinear analysis of live loads, nonlinear adjustment calculation method, and nonlinear construction simulation calculation method, for construction control [8]. Using finite displacement theory, a three-dimensional nonlinear analysis system considering dead load, live load, and construction stage and methods was developed [9]. Stability problems of truss, frame, and arch bridge have been studied extensively [1]. A stability analysis approach was developed for the wind effect on long-span bridges. Optimization theory and techniques have been applied to all kinds of bridges successfully. The accuracy and efficiency of those methods developed have been verified by practical application.

63.1.2.4 Bridge CAD Techniques

Since the late 1970s, computer technologies have been widely employed for structural analysis in bridge design practice in China. Many special-purpose structural analysis programs for bridge design were developed. With full concern for the special feature of bridge design, for example, the Synthetical Bridge Program [9], provided the capability of construction stage transferring, concrete creep and shrinkage analysis, prestress calculation, etc. To a certain extent, widespread adoption of this program reflected the application status of computational technology in the field of highway bridge design in China during the years from the late 1970s to the early 1980s.

Since the 1980s, the popularization of computer graphics devices, such as the rolled drafting plotter and digitizer, have brought computational application from merely structural analyzing to aided design including both structural analysis and detail drafting. With the development of the highway system, standardized simply supported bridges have spread over China. Based on the microcomputer platform, many researchers and engineers began to develop automated CAD systems integrating structural analysis and detail drafting. The "Automated Medium and Short-Span Bridge CAD System on Micro-computer" cooperatively developed by the membership of China Highway Computer Application Association, for example, has the capabilities to accomplish all processes of simply supported T-beam and plate bridge design. With the aid of this system, only a few primary pieces of information are required to be input, and the computer will automatically produce a set of design documents including both specifications and drawings in a short time. The design efficiency is excellent compared with the traditional manner. Many design institutes and firms employed this system to design medium- and short-span bridges.

During the7th Five Year Plan of China (1985 to 1990), to develop a new highway bridge system, a special task group consisting of more than 40 practical bridge engineers and scholars was formed and organized by the Ministry of Communications. As a national key scientific research project, the allied group invested $2 million of RMB to research and develop the CAD techniques applied in the construction of highway bridges. In 1991, the "Highway Bridge CAD System (JT-HBCADS)" was successfully developed. More than 10 large highway bridge design institutes have installed this system and fulfilled the design of about 10 large bridges such as Nanpu Bridge, Yangpu Bridge, etc.

During the years from 1991 to 1995, the increase in personal computer (PC) hardware performance and software technology has issued a critical challenge to the development of research and application of bridge CAD techniques. Many advanced software development techniques, such as kernel database accessing, object-oriented programming, application visualizing, and rapid application developing, were entirely developed and made available for the personal computer, which brought forth lots of chances that had never appeared before in developing the new generation of integrated and intelligent bridge CAD systems.

With full regard to, and on the basis of, experience and acquaintance with the development of JT-HBCADS and many newly available support software technologies, the developing ideas of integrated bridge CAD system (BICADS) has been brought up, and the new generation BICADS was successfully developed thoroughly under the guidance of this thought. Taking the Windows NT operating system as the platform, the system architectural design of BICADS entirely adopted the kernel database accessing techniques to avoid the difficulties of system maintenance and upgrading the innate and unavoidable weakness caused by the traditional file system. The first version of BICADS consists of five subsystems including the Design Documentation, Pre-Processing of Bridge FEM, Bridge FEM Kernel, Post-Processing of Bridge FEM, and the Preliminary Design of Box Girder Bridges. Several detailed design subsystems of other commonly used bridges can be included by employing a good integrating and expanding mechanism in the main system. Additionally, the research of some fundamental problems in the field of bridge intelligent CAD techniques and the development of bridge experts system tools with graphics processing abilities have already yielded considerable promise. It is predicted that, motivated by the rapid development of computer technologies by the end of this century, a new generation in China's bridge CAD techniques application and research is being opened.

63.1.3 Experimental Research of Dynamic and Seismic Loads

Model Tests for Bridges
To establish the dynamic behavior base line for health monitoring bridge structures, the model tests are usually done just after construction of bridges. Experimental procedures that have been used in the past include (1) impact tests and (2) ambient vibrations. For large bridges, such as Shanghai Yangpu Bridge (cable-stayed bridge) and Shanghai Fengpu Bridge (continuous box-girder bridge), the method of using test vehicles (controlled traffic) for exciting bridges was successfully verified.

Shaking Table Test of Bridge Models
The tests of a simply supported beam and a continuous girder bridge model were performed on the shaking table (made by the MTS Co.). These tests were to evaluate the effect of ductility and seismic isolation on bridges, in which the viaduct of Shanghai Inner Ring Road was regarded as the background of the continuous girder bridge model; meanwhile, the analytical models of bridges and elements were verified.

Ductility Performance and Seismic Retrofitting Techniques for Bridge Piers
Recently, high-strength concrete with cylindrical compressive strength up to 100 MPa or higher can be made with locally obtainable materials, such as ordinary cement, sand, crushed stone, a water-reducing superplasticizer, standard mixing methods, and careful quality control in production. There are many characteristics for high-strength concrete that are beneficial in civil engineering, but, on the other hand, there are some shortcomings to the increasing use of high-strength concrete. For instance, brittle features and less postpeak deformability may cause brittle failure during earthquakes or under other conditions. Much work, theoretical and experimental, has been done by Chinese researchers for ductility design and improving design code of bridges. Through the tests and analyses, some important conclusions may be summarized briefly as follows:

1. Test results indicate that for high-strength concrete columns, very large ductility could be achieved by using lateral confining reinforcement.
2. All retrofitted piers using steel jackets, steel fiber concrete, expoxy concrete, and fiberglass-expoxy performed extremely satisfactorily. Good ductility, energy-dissipation capacity, and stable-deformation behavior were achieved.

Dynamic Behavior Test of Isolation Devices

To meet the requirements of earthquake resistance design of bridge, seismic design of isolated bridge and optimization have been widely used in China. The dynamic properties of elastomeric pad bearings (EP bearings) has been evaluated, including the shear modulus, hysteretic behavior, and sliding friction coefficient of EP bearings and Teflon plate-coated sliding bearings (TPCS bearings). The tests were done on an electro-hydraulic fatigue machine (made by INSTRON Co.) with an auxiliary clamping apparatus. These results may be summarized as follows:

1. At constant shear strain amplitude, the shear modulus of EP bearings increases with the increase in frequency. At constant frequency, the shear modulus obviously decreases with the increase in shear strain amplitude. Sizes and compression have no obvious effect on dynamic shear modulus.
2. At constant compression and sliding displacement amplitude, the hysteretic energy of TPCS bearings increases with the increase in frequency. At constant sliding displacement amplitude and frequency, the increased compression results in an increase in the hysteretic energy of TPCS bearings.
3. The friction coefficient of TPCS bearing decreases with the increase in compression.

Based on experimental research of rubber bearings and steel damping, a system of seismic isolation and energy absorption, composed of curved steel-strip energy absorbers and TPCS bearings, was developed, and then a seismic rubber bearing with curved mild-steel strip, was invented.

Recently, some kinds of improved seismic bearings have come out. A great number of dynamic experiments show that these types of bearings have better hysteretic characteristics than elastomeric laminated bearings. To avoid span failures of bridges upon impact, restricting blocks are usually placed at the end of beams. To compare the behavior of the blocks, three kinds of blocks [4] have been manufactured and an experiment has been conducted on these blocks: (1) " T-type" rubber blocks, (b) "bowl-type" rubber blocks, and (3) cubic reinforced concrete blocks. During the tests, the impact hammer freely fell from a given height and contact forces between the block and high-strength concrete hammer were recorded. The test results show it is very obvious that T-type rubber blocks have the best energy absorption capacity and the impact force of T-type rubber blocks is much lower than that of concrete blocks.

63.1.4 Wind Tunnel Test Techniques

Since the 1980s, with the building of long-span cable-stayed and suspension bridges, China has made great progress in wind engineering. For example, there are three boundary-layer wind tunnels in the National Key Laboratory for Disaster Reduction in Civil Engineering at Tongji University. TJ-1, TJ-2, and TJ-3 BLWTs, which have been put into service only for several years, have working sections of 1.2 m (width), 1.8 m (height); 3 m (width), 2.5 m (height); and 15 m (width), 2 m (height), respectively. The maximum wind speeds of these are 32, 17, and 65 m/s, respectively. Until now, about 30 model tests have been carried out in these wind tunnels. Wind-resistant researches on about 40 cable-stayed bridges and suspension bridges have been carried out mainly at Tongji University, Shanghai, China. More than 10 full-scale aeroelastic bridge model tests have been performed. To meet the requirements of the wind-resistant design of highway bridges with increasing spans, a Chinese Wind Resistant Design Guideline of Highway Bridge was compiled. Some achievements of flutter analysis, buffeting analysis, and wind-induced vibration control have been made and are introduced in the following.

Flutter Analysis

As is well known, the critical flutter velocity is the first factor that controls the design for a long-span bridge, especially located in typhoon areas. Precision of torsional frequency in the calculation is very important. The traditional single-beam model test of bridge deck usually gives estimates of torsional frequencies lower than the actual ones and may make a lower critical flutter velocity estimation. A three-beam model of a bridge deck which was developed by Xiang et al., [6] has been proved to be efficient in improving the precision of torsional frequency to a great extent.

The state-space method for flutter analysis overcomes the shortcomings of Scanlan's method for flutter analysis in which only one vertical mode and one torsional mode can be considered. A multimode flutter phenomenon was found. Participation of more than two modes in flutter make the critical velocity higher than that from Scanlan's method.

Buffeting Analysis

With the increase in span length, bridge structures tend to become more flexible. Excessive buffeting in near-ground turbulent wind, although not destructive, may cause fatigue problems due to high frequency of occurrence and traffic discomfort. Davenport and Scanlan et al., proposed buffeting analysis methods in the 1960s and 1970s, respectively. Since then, refinement studies on these methods have been made. It is possible to establish practical methods for buffeting response spectrum and buffeting-based selection.

Aerodynamic selection of deck cross section shape is important in the preliminary design stage of a long-span bridge. In the past year, this selection aimed mainly at flutter-based selection. The concept of "buffeting-based selection" and the corresponding method were used in the wind-resistant design of the Jiangying Yangtze River Bridge and the Humen Bridge, a suspension bridge with a main span of 888 m.

To investigate the nonlinear response characteristics of long-span bridges, a nonlinear buffeting analysis method in the time domain has been used to analyze the Jiangying Yangtze River Bridge and the Shantou Bay Bridge, etc. Analysis results show that for long-span suspension bridges the aerodynamic and structural nonlinear effects on the buffeting response should be considered.

Wind-Induced Vibration Control

In practice today, the increment of critical flutter velocity of a long-span bridge is usually achieved using aerodynamic measures. The theoretical analysis and experiments indicate that passive TMD may also be an effective device for flutter control. A couple of TMDs with proper parameters can increase the critical flutter velocity of the Humen-Gate Bridge with wind screens on the deck (for improving vehicle moving condition) by 50%, although the efficiency, duration, and reliability of the device for long-time-period use still have some problems to be solved.

The buffeting response increases with wind speed, and may become very strong at high wind speed. Two new methods were proposed for determination of optimal parameters of the TMD system for controlling buffeting response with only the vertical mode and with coupling the vertical and torsional modes, respectively.

63.1.5 Bridge Construction Techniques

63.1.5.1 Constructional Materials

According to the design specifications for bridges in China, the maximum strength of concrete is 60 MPa; the prestressing tendons include hard-drawn steel bars, high-strength steel wires, and high-strength strands, the strengths of which are from 750 to 1860 MPa; the general reinforcement bars are made of A3, 16Mn, etc.; the steel plate is made of A3 or 16Mn or 15MnVN, etc. In normal designs, the concrete used in prestressed bridges should have a strength higher than 40 MPa; the prestressing tendons used in pretensioned slab girders are hard-drawn 45 SiMnV bars with the strength of 750 MPa or steel strands with strength of 1860 MPa; the high tensile strength and low relaxation strands are widely used in post-tensioned concrete bridges. Now a viaduct usually has a lower depth of girders so high-strength concrete over 50 MPa is often adopted. Concrete having a strength of over 60 MPa and tensile wires and strands will be used in bridges in the future.

63.1.5.2 Prestressing Techniques

Prestressing techniques including internal and external prestressing have been used for about 40 years in China. Not only were the full and partial prestressed bridges constructed speedily, but also the preflex prestressed girders and double-prestressed girders have been used in viaducts and separation structures. The high tensile strength and low relaxation strands, the reliable anchorages, such as the OVM system, and the high-tonnage jacks have been widely used in many bridges including continuous girder bridges, T-frame bridges, cable-stayed bridges, and suspension bridges. The design and construction of prestressed concrete structures is a normal process in China. The external prestressing tendons, including unbonded tendons, have been used in new bridges and in the strengthening of many old bridges. Now, several external prestressed long-span composite bridges are being built in China.

63.1.5.3 Precast Techniques of Concrete and Steel Girders

Most simply supported girder bridges are made with fabricated methods in China, and factory production is usually adopted. When the span is shorter than about 22 m, the pretensioned, prestressed voided slab girder is often the best choice, and the high-strength and low-relaxation strands are used as the prestressed reinforcement. When the span is over about 25 m, the post-tensioned T-girder may be used, in which the strands are arranged with curved profiles. In the construction of some bridges and urban viaducts and in precasting yards, steam curing is often used to increase the strength of concrete early and to raise the working efficiency. Usually the weight and length of a precast girder are limited to below about 1200 kN and 50 m to ease transport and erection.

Segmental bridges are usually built using the cantilever casting method, or other casting methods; nevertheless, only a few segmental bridges are constructed with the cantilever erection method. We usually cast in place because it is noticed that the rusting of prestressing strands at the segment joints may cut down the service life of bridges. The high anticorrosive external prestressing tendon or strand cable is not widely adopted yet in post-tensioned segmental bridges.

In China, complete riveting techniques have been replaced by welding and high-strength bolting techniques. Complete welded box and composite girders have been used in urban viaducts, separation structures, and cable-stayed bridges; techniques adopted in shipbuilding, such as computer layout and precision cutting, are being introduced.

63.1.5.4 Cable Fabrication Techniques

About 10 to 20 years ago, the stay cable in China was fabricated mainly on the construction site and consisted of 5-mm-diameter or 7-mm-diameter parallel galvanized steel wires. It was protected with PE casing pipe grouted with cement, or with corrosion paint and three layers of glass fibers coated by epoxy resin. A lot of cable-stayed bridges have been built in the last decade and the cable fabrication techniques have developed rapidly. With the construction of Shanghai Nanpu Bridge in 1988, the first factory, which mechanically produced long-lay spiral parallel wire cables with a hot-extruded PE or PE and PU sheath, was established. Since then, the quality of stay cables has greatly improved, especially in resistance to corrosion. Now the maximum working tension of stay cables is over 10,000 kN and high-quality anchorage has been developed. In recent years, the parallel and spiral strand cables of factory production with maximum working tensions at over 10,000 kN have been frequently used in cable-stayed bridges.

At the same time, the main cables of Santo and Humen (suspension) Bridges were successfully fabricated in China; the parallel wire strand consisted of 127ϕ 5.2-mm zinc-coated steel wires and had a length of over 1600 m; the mean square root error in the length of wires was lower than 1/36,000. Now, Jiangyin Yangtze River Bridge, having the longest span, close to 1400 m, in China, is under construction; its main cables will also be prefabricated.

63.1.5.5 Construction Techniques of Large-Diameter Piles

In China, bored piles are usually adopted for large bridges. When the ground is poor or the rock formation is near the Earth's surface or riverbed, piles have to be built in the rock and they become the bearing piles. Normally, the diameter of bearing piles is about 0.8 to 2.5 m. A large-diameter pile can be adopted to replace the pile group in order to reduce material construction time. Usually this large-diameter pile has a diameter of 2.5 to 7 m, is hollow, and consists of two or three segments. The first segment of the pile is a double-wall steel and concrete composite drive pipe which is driven into a weathered layer as a cofferdam; the second segment is a hollow concrete bearing pile which has a smaller diameter than the first segment, and the pier shaft is connected on the top of this segment; the last segment has a minimum diameter or, similar to the second, it is built in the rock. As a result, construction is easy, and no platform or hollow pile uses up a lot of concrete and steel.

63.1.5.6 Advanced Construction Techniques

With the development of transportation in China, more and more large bridges have been built and new construction techniques have been developed. Continuous curved bridges have been built with the incremental launching method, and the speed of the cantilever casting construction method is about 5 or 6 days per segment. The cable-stayed composite bridges, whose composite girders are composed of prefabricated, wholly welded steel girders and precast reinforced concrete deck slabs, were constructed with the cantilever erection method — for example, the 602-m Shanghai Yangpu Bridge, built in 1993. For prestressed concrete cable-stayed bridges, the tensions of stay cables and alignment of girder can easily achieve their best states by using computer-automated control techniques. The construction method of modern long-span suspension bridges was a new technique in China several years ago, most using PWS (prefabricated parallel wire strand) methods.

The improvement of construction techniques is not only in continuous girder bridges, rigid frame bridges, cable-stayed bridges, and suspension bridges. In a deep valley or flood river, the stiff reinforcement skeleton consisting of steel pipes is used as the reinforcement of a long-span concrete arch ring; after the stiff reinforcement skeleton is erected and closed up at midspan, the concrete is pumped into the steel pipes; then, by using the traveling form, which is supported on the stiff reinforcement skeleton, the concrete is cast and the reinforced concrete box arch ring is formed. Another construction method used in long-span composite arch bridges is the swing method. The two halves of the arch are separately erected on each side of river embankments or hillsides; then, by using jacks, they are rotated around their supports under arch seats and closed at midspan; finally, the concrete is pumped into the pipe arch. In order to keep the balance of a half arch, water containers are usually used as the ballast weights.

The progress of construction techniques has not only been made for superstructures but also for substructures. The height of reinforced, prestressed hollow piers and precast piers used in deep valleys has reached over 80 m. Large-diameter hollow piles and large concrete and steel caissons and double-wall steel and concrete composite cofferdams are adopted in river or sea depths over 50 m.

63.2 Beam Bridges

63.2.1 General Description

Simple in structure, convenient to fabricate and erect, easy to maintain, and with less construction time and low cost, beam structures have found wide application in short- to medium-span bridges. In 1937, over the Qiantang River, in the city of Hangzhou, a railway-highway bipurpose bridge was erected, with a total length of 1453 m, the longest span being 67 m. When completed, it was a remarkable milestone of the beam bridge designed and built by Chinese engineers themselves.

Reinforced concrete beam structures are most commonly used for short- to medium-span bridges. A representative masterpiece is the Rong River Bridge completed in 1964 in the city of Nanning, the capital of Guangxi Zhuangzu Autonomous Region. The bridge, with a main span of 55 m and a cross section of a thin-walled box with continuous cells, designed in accordance with closed thin-walled member theory, is the first of its kind in China.

Prestressed concrete beam bridges are a new type of structure. China began to research and develop their construction in the 1950s. In early 1956, a simply supported prestressed concrete beam railway bridge with a main span of 23.9 m was erected over the Xinyi River along the Longhai Railway. Completed at the same time, the first prestressed concrete highway bridge was the Jingzhou Highway Bridge. The longest simply supported prestressed concrete beam which reaches 62 m in span is the Feiyun River Bridge in Ruan'an, Zhejiang Province, built in 1988. Another example is the 4475.09-m Yellow River Bridge, built in the city of Kaifeng, Henan Province in 1989. Its 77 spans are 50-m simply supported prestressed concrete beams and its continuous deck extends to 450 m. It is also noticeable that the Kaifeng Yellow River Bridge is designed on the basis of partially prestressed concrete theory. Representative of prestressed concrete continuous girder railway bridges, the second Qiantang River Bridge (completed in 1991) boasts its large span and its great length, its main span being 80 m long and continuous over 18 spans. Its erection was an arduous task as the piers were subjected to a wave height of 1.96 m and a tidal pressure of 32 kPa when under construction. The extensive construction of continuous beam bridges has led to the application of the incremental launching method especially to straight and plane curved bridges. In addition, large capacity (500-t) floating crane installation and movable slip forms as well as span erection schemes have also attained remarkable advancement.

Beam bridges are also used widely in overcrossings. In the 1980s, with the growth of urban construction and the development of highway transportation, numerous elevated freeways were built, which provide great traffic capacity and allow high vehicle speed, for instance, Beijing's Second and Third Freeway and East City Freeway, the Intermediate and Outer Freeway in Tianjin, and Guangzhou's Inner and Outer Freeway and viaduct. In Shanghai, the elevated inner beltway was completed in 1996. Subsequently, there has appeared an upsurge of erecting different-sized grade separation structures on urban main streets and express highways. Uutil now, in Beijing alone, 80-odd large overcrossings have been erected, which makes the city rank the first in the whole country in number and scale.

To optimize the bridge configuration, to reduce the peak moment value at supports, and to minimize the constructional depth of girders, V-shaped or Y-shaped piers are developed for prestressed concrete continuous beam, cantilever, or rigid frame bridges. The prominent examples are the Zhongxiao Bridge (1981) in Taiwan Province and the Lijiang Bridge (1987) at Zhishan in the city of Guilin.

63.2.2 Examples of Beam Bridges

Kaifeng Yellow River Bridge

Kaifeng Yellow River Bridge (Figure 63.4) is an extra large highway bridge, located at the northwest part of Kaifeng City, Henan Province. It consists of 108 spans (77 × 50 + 31 × 20) m, its total length reaching 4475.09 m.

Simply supported prestressed concrete T-girders are adopted for its superstructure. The deck is 18.5 m wide, including 12.3 m for motor vehicle traffic and two sidewalks 3.1 m wide each on both sides. Substructure applies single-row double-column piers, which rest on 2200-mm large-diameter bored pile foundations.

The bridge is of the same type as those built earlier over the Yellow River in Luoyang and Zhengzhou. Kaifeng Bridge has obtained an optimized design scheme, with its construction cost reduced and schedule shortened. The main characteristics of the bridge are as follows:

FIGURE 63.4   Kaifeng Yellow River Bridge.

1. Adoption of partial prestress concrete in the design of T-girder;
2. Modification of the beams over central piers as prestressed concrete structure;
3. Increase in the continuous length of the deck reaching 450 m.

The bridge was designed by Highway Planning, Survey and Design Institute of Henan Province, and constructed by Highway Engineering Bureau of Henan Province. It was opened in 1989.

Xuzhuangzi Overcrossing

Xuzhuangzi Overcrossing (Figure 63.5), a long bell-mouth interchange grade crossing on the freeway connecting Beijing-Tianjin and Tangshan, is a main entrance to the city of Tianjin.

The overcrossing has a total length of 4264 m. The superstructure consists of simply supported prestressed concrete T-griders and multispan continuous box girders. The 1.5-m-diameter bored piles and invested trapezoidal piers are adopted for the substructure.

The bridge was designed by the first Highway Survey and Design Institute, Tianjin Municipal Engineering Co. and constructed by the first Highway Co., Ministry of Communications, Kumagai Co., Ltd, Japan. It was opened to traffic in 1992.

Liuku Nu River Bridge

Liuku Nu River Bridge (Figure 63.6), the longest prestressed concrete continuous bridge in China at present, is located in the Nu River Lisu Autonomous Prefecture, Yunnan Province. It has three spans of length (85 + 154 + 85) m. The superstructure is a single-box single-cell girder with two 2.5-m-wide overhangs on both sides. The beam depth at the support is 8.5 m, i.e., 1/18 of the span, while at the midspan it is only 2.8 m, i.e., 1/55 of the span. The whole bridge has only two diaphragms at the hammer-headed block.

FIGURE 63.5 Xuzhuangzi Overcrossing.

FIGURE 63.6 Liuku Nu River Bridge.

Three-way prestress is employed. A large tonnage strand group anchorage system is applied. With tendons installed only in the top and bottom slabs, no bent-up or bent-down tendon is needed and the widening of the web is avoided, which makes the construction very convenient. Vertical prestress is provided by Grade 4 high-strength rolled screwed rebars with diameter of 32 mm, which also served as the rear anchorage devices of the form traveler during cantilever casting. For the substructure hollow piers supported by bored piles foundation on rock stratum were adopted.

The bridge was completed in 1993, designed by Highway Survey and Design Institute of Yunnan Province and constructed by Chongqing Bridge Engineering Co.

FIGURE 63.7   Second Qiantang River Bridge.

The Second Qiantang River Bridge

The second Qiantang River Bridge, located on Sibao in Hangzhou, Zhejiang Province, is a parallel and separate highway–railway bipurpose bridge (Figure 63.7). The 11.4-m-wide railway bridge carries two tracks, with a total length of 2861.4 m. The highway bridge, which was designed according to freeway standard, is 20 m wide and 1792.8 m long, carrying four-lane traffic. Both main bridges are of prestressed concrete continuous box girders, and the continuous beams reach a total length of 1340 m, i.e., 45 + 65 + 14 × 80 + 65 + 45 m, the longest in China at present.

To obtain the 506 mm expansion magnitude of the main bridge, composite expansion joints were applied in the highway bridge, whereas transition beams and expansion rails were used for the railway bridge. Pot neoprene bearings were specially designed to accommodate the large displacement and to offer sufficient vertical resistance.

Three-way prestress was introduced to the box girder. Strands and group anchorage system were adopted longitudinally, with the maximum stretching force in excess of 2000 kN. The cantilever casting method was used for the main construction of the bridge, while the bored piles foundation was constructed at river sections of rare strong tidal surge with a height of 1.96 m and a pressure reaching 32 kPa. The bridge was designed and constructed by Major Bridge Engineering Bureau, Ministry of Railway. It was completed in November 1991.

63.3   Arch Bridges

63.3.1   General Description

Of all types of bridges in China, the arch bridge takes the leading role in variety and magnitude. Statistics from all the sources available show that close to 60% of highway bridges are arch bridges. China is renowned for its mountains with an abundant supply of stone. Stone has been used as the main construction material for arch bridges. The Wuchao River Bridge in Hunan Province, for instance, with a span of 120 m is the longest stone arch bridge in the world. However, reinforced concrete arch bridges are also widely used in various forms and styles.

Most of the arches used in China fall into the following categories: box arch, two-way curved arch, ribbed arch, trussed arch, and rigid framed arch. The majority of these structures are deck bridges with wide clearance, and it costs less to build such bridges. The box arch is especially suitable for long-span bridges. The longest stone arch ever built in China is the Wu River Bridge in Beiling, Sichuan Province, whose span is as long as 120 m. The Wanxian Yangtze River Bridge in Wanxian, Sichuan Province with a spectacular span of 420 m set a world record in the concrete arch literature. A unique and successful improvement of the reinforced concrete arch, the two-way curved arch structure, which originated in Wuxi, Jiangsu Province, has found wide application all over the country, because of its advantages of saving labor and falsework. The largest span of this type goes to the 150-m-span Qianhe River Bridge in Henan Province, built in 1969. This trussed arch with light deadweight performs effectively on soft subsoil foundations. It has been adopted to improve the composite action between the rib and the spandrel. On the basis of the truss theory, a light and congruous reinforced concrete arch bridge has been gradually developed for short and medium spans. Through prestressing and with the application of cantilevering erection process, a special type of bridge known as a "cantilever composite trussed arch bridge" has come into use. An example of this type is the 330 m-span Jiangjie River Bridge in Guizhou Province. The Yong River Bridge, located in Yunnan Province, is a half-through ribbed arch bridge with a span of 312 m, the longest of its kind. With a simplified spandrel construction, the rigid framed arch bridge has a much better stress distribution on the main rib by means of inclined struts, which transfer to the springing point the force induced by the live load on the critical position. In the city of Wuxi, Jiangsu Province, three such bridges with a span of 100 m each were erected in succession across the Great Canal. Many bridges, quite a number of which are ribbed arch bridges, have been built either with tied-arches or with Langer's girders. The recently completed Wangcang Bridge in Sichuan Province and the Gaoming Bridge in Guangdong Province are both steel pipe arch bridges. The former has a 115-m prestressed tied-arch, while the latter has a 110-m half-through fixed rib arch. A few steel arch bridges and slant-legged rigid frame bridges have also been constructed.

In building arch bridges of short and medium spans, precast ribs are used to serve as temporary falsework. And sometimes a cantilever paving process is used. Large-span arch bridges are segmented transversely and longitudinally. With precast ribs, a bridge can be erected without scaffolding, its components being assembled complemented by cast-in-place concrete. Also, successful experience has been accumulated on arch bridge erection, particularly erection by the method of overall rotation without any auxiliary falsework or support.

Along with the construction of reinforced concrete arch bridges, research on the following topics has been carried out: optimum arch axis locus, redistribution of internal forces between concrete and reinforcement caused by concrete creep, analytical approach to continuous arch, and lateral distribution of load between arch ribs.

63.3.2 Examples of Masonry Arch Bridge

Longmen Bridge

Longmen Bridge (Figure 63.8), 12 km south of Luoyang City, Henan Province, is an entrance of the Longmen Grottoes over Yihe River. It is a 60 + 90 + 60 m three-span stone arch bridge, with a width of 12.6 m. A catenary of 1:8 rise-to-span ratio was chosen as the arch axis. The main arch ring has a constant cross section, with a depth of 1.1 m. Two stone arches of 6 m long each were arranged on either bank providing under crossing traffic. The bridge was constructed on steel truss falsework supported by temporary piers. It was designed and constructed by Highway Engineering Bureau, Communications Department of Henan Province and completed in 1961.

FIGURE 63.8 Longmen Bridge.

Wuchao River Bridge

Wuchao River Bridge (Figure 63.9), a structure on Fenghuang County Highway Route, Hunan Province, spans the valley of the Wuchao River with a total length of 241 m. To use local materials, a masonry arch bridge scheme was adopted. On the basis of the experience accumulated in the last 20 years of construction of masonry arch bridges in China, the bridge has a main span of 120 m, which is a world record for this type of bridge.

The bridge is 8 m wide. There are nine spandrel spans of 13 m each over the main spans; three spans of 13 m each for the south approach; a single span of 15 m for the north approach. The main arch ring is a structure of twin separated arch ribs, connected by eight reinforced concrete floor beams. A catenary of $m = 1.543$ was chosen as the arch axis, with a rise-to-span ratio of 1:5. The arch rib has a variable width and a uniform depth of 1.6 m. It is made up of block stone with a strength of 100 kPa and ballast concrete of 20 Mpa.

The lateral stability of the bridge was checked. Because the masonry volume of its superstructure is only 1.36 m^3/m^2, the structure achieves a slim and graceful aesthetic effect. The bridge was designed and constructed by Communication Bureau of Fenghuang County, Hunan Province. It was completed in 1990.

Heyuan DongRiver Bridge

Heyuan DongRiver Bridge (Figure 63.10) is on the Provincial Route near Heyuan County. It is a 6×50 m multispan masonry arch bridge with a width of $7 + 2 \times 1$ m and a total length of 420.06 m. The rise-to-span ratio of the arch ring is 1:6.

A transversely cantilevered setting method was applied for its arch ring construction. The arch ring was divided into several arch ribs, and each rib was longitudinally divided into several precast

FIGURE 63.9   Wuchao River Bridge.

FIGURE 63.10   Heyuan DongRiver Bridge.

concrete hollow blocks. Side ribs were erected by transversely setting with the support of the erected central rib. The bridge was designed by Highway Survey and Design Institute of Guangdong Province and constructed by Highway Engineering Department of Guangdong Province. It was completed in 1972.

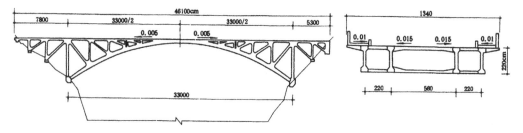

FIGURE 63.11   Jiangjie River Bridge.

FIGURE 63.12   Jinkui Grand Canal Bridge.

63.3.3   Examples of Prestressed Concrete, Reinforced Concrete, and Arch Bridges

Jiangjie River Bridge

Jiangjie River Bridge (Figure 63.11), located in Weng'an County, Guizhou Province, is a prestressed concrete truss arch bridge crossing Wujiang Valley at a height of 270 m above normal water level. It has a record-breaking main span of 330 m in China. Its side truss spans, 30 + 20 m on one side and 30 + 25 + 20 m for the other, are arranged along the mountain slopes. The total length of the bridge is 461 m.

The most obvious characteristics of the bridge are the use of batholite as the lower chords of the side spans and the anchoring of the prestress bars in tensile diagonals on the batholite. The deck is 13.4 m wide with 9 m for lanes and two pedestrian walkways of 1.5 m each. The arch depth is 2.7 m, L/122, and its width is 10.56 m, L/31.3, with a rise-to-span ratio being 1:6. The bridge was constructed by cantilever assembling. A derrick mast with a hoisting duty of 1200 kN was used. The bridge was designed by Communication Department of Guizhou Province and constructed by Bridge Engineering Co. of Guizhou Province.

Jinkui Grand Canal Bridge

Jinkui Grand Canal Bridge (Figure 63.12), with a main span reaching 100 m, is one of the longest rigid-framed prestressed concrete arch bridges on soft-soil foundation. It crosses the Grand Canal in Wuxi County, Jiangsu Province.

The bridge has a rise-to-span ratio of 1:10. The arch rib is of the I type with a constant cross section, while the solid spandrel segment has a variable cross section. Only two inclined braces are

FIGURE 63.13   Taibai Bridge.

arranged on either side to get a aesthetic effect. In order to reduce the deadweight, ribbed slabs are employed for the deck. The substructure includes combined-type thin-wall abutments, which are designed to resist the horizontal thrusts from superstructure by boring piles and slide-resistant slabs working jointly. The bridge was designed by Shanghai Urban Construction College and constructed by Bridge Engineering Co. of Wuxi County. It was completed in 1980.

Taibai Bridge

Taibai Bridge (Figure 63.13), a rigid-framed reinforced concrete arch highway bridge with a span of 130 m, is located in Dexi copper mining area, Jiangxi Province. The bridge was constructed by the swing method. After assembling steel bar skeletons and casting 100 mm bottom slab on simple scaffoldings, 42 25-mm tensile bars were stretched to get the structure separate from the scaffoldings. The whole swing system, with a total weight of 18,100 kN, was supported by a reinforced concrete spherical hinge on abutment foundation. The bridge was designed by Nanchang Non-ferrous Metallurgical Design Institute and constructed by Huachang Engineering Co. It was completed in March 1993.

Wanxian Yangtze River Bridge

The bridge located in Huangniu Kong, 7 km upstream from Wanxian, is an important structure on the No. 318 national highway (Figure 63.14). It is 864.12 m long. A reinforced concrete box arch with a rise-to-span ratio of 1:5 offers a single span of 420 m. Steel pipes are used to form stiffening arch skeletons before the erection of the main arch; there are 14 spans of 30 m prestressed concrete. Simply supported T-girders make up the spandrel structure, while 13 spans of the same girders are for the approaches. The continuous deck is 24 m wide, providing 2×7.75 m lanes for motor vehicle traffic and two sidewalks of 3.0 m each. A longitudinal slope of 1% is arranged from the midspan to either side with a radius of vertical curve being 5000 m, while the cross slope is 2%. The bridge was designed by Highway Survey and Design Institute of Sichuan Province and constructed by Highway Engineering Company of Sichuan Province. It was completed in 1997.

63.4   T-Type and Continuous Rigid Frame Bridges

63.4.1   General Description

The prestressed concrete rigid T-frame bridge was primarily developed and built in China in the 1960s. This kind of structure is most suitable to be erected by balanced cantilever construction

FIGURE 63.14 Wanxian Yangtze River Bridge.

process, either by cantilever segmental concreting with suspended formwork or by cantilever erection with segments of precast concrete. The first example of cantilever erection is the Wei River Bridge (completed in 1964) in Wuling, Henan Province, while the Liu River Bridge (completed in 1967) in Liuzhou in Guangxi Zhuangzu Autonomous Region is the first by cantilever casting. The Yangtze River Highway Bridge at Chongqing (completed in 1980), having a main span of 174 m, is regarded as the largest of this kind at present.

From prestressed concrete rigid T-frame bridges were developed multiple prestressed concrete continuous beam and continuous rigid frame bridges, which can have longer spans and offer better traffic conditions. Among others, the Luoxi Bridge in Guangzhou, Guangdong Province (completed in 1988) features a 180-m main span. The Huangshi Yangtze River Bridge in Hubei Province has a main span of 245 m. And the Humen Continuous Rigid Frame Bridge in Guangdong Province (completed in 1997), which has a 270-m main span, is regarded as the largest of this kind in the world.

63.4.2 Examples of T-Type Rigid Frame Bridges

Qingtongxia Yellow River Highway Bridge

Qingtongxia Highway Bridge (Figure 63.15) is 80 km south of Yinchuan, Ningxia. It is 743 m long and 14 m wide. The spans arrangement is $4 \times 30 + 60 + 3 \times 90 + 60 + 6 \times 30 + 20$ m. Prestressed concrete T-girders were adopted for the three main spans, while prestressed concrete simply supported beams were used for approaches. The T-frame is a two-cell single-box thin-wall structure, which was built by cantilever casting. The substructure consists of thin-wall hollow box piers, resting on elevated bored pile foundations, the piles having a diameter of 1.5 m. The bridge was designed by Highway Survey and Design Institute of Ningxia Province and constructed by Highway Engineering Bureau of Ningxia Province. It was completed in October 1991.

Huanglingji Bridge

Huanglingji Bridge (Figure 63.16), located in Hanyang County, Hubei Province, is a prestressed concrete truss T-frame highway bridge. It has spans of $7 \times 20 + 53 + 90 + 53 + 2 \times 20$ m, with a

FIGURE 63.15   Qingtongxia Yellow River Highway Bridge.

FIGURE 63.16   Huanglingji Bridge.

total length of 380.19 m. The 90-m-long main span is composed of two cantilever arms of 37 m each and a 16-m-long suspended span.

Caisson foundations and box piers were adopted for the substructure. Its superstructure consists of two trusses, with prestressed concrete simply supported slab on the top, which serves as upper bracing after transverse prestressing has been introduced. Prestressing tendons are used for tensile members, while common rebars for compressive members. Longitudinal prestress tendons are arranged in open channels which makes stretching convenient. The cantilever assembling method was employed. The bridge type featuresa slim configuration and saves construction materials. The bridge was designed at Tongji University in cooperation with Highway Engineering Bureau of Hubei Province. The construction unit was Road and Bridge Co. of Hubei Province. It was completed in 1979.

Hongtang Bridge

Hongtang Bridge (Figure 63.17), the longest highway bridge over Min River, is west of Fuzhou City, Fujian Province. It is 1843 m long and 12 m wide. The main span is a three-hinge connected lower chord supported prestressed concrete truss T-frame, which synthesizes the virtues of cable-stayed bridges, truss bridges, and T-frame bridges.

The bridge was erected by cantilever assembling with cable cranes. On the side shoal 31 spans are of prestressed concrete continuous girders erected by adopting nonglued segmental assembling span by span, a new technology first applied in Chinese bridge construction. Spans on the banks are of simply supported prestressed concrete beams.

The substructure of the bridge is prestressed concrete V-type hollow piers on bored piles foundation for the main span and dual-column bored piles foundation for the side spans. The bridge

FIGURE 63.17 Hongtang Bridge.

was designed by Communication Planning and Design Institute of Fujian Province and constructed by the Second Highway Engineering Co. of Fujian Province. It was completed in December of 1990.

63.4.3 Examples of Continuous Rigid Frame Bridges

Luoxi Bridge

Luoxi bridge (Figure 63.18), the longest prestressed concrete continuous rigid frame bridge in China, spans Pearl River in Guangzhou, Guangdong Province. It is 1916.04 m long and 15.5 m wide. The main bridge has spans of 65 + 125 + 180 + 110 m, providing a navigation clearance of 34 × 120 m.

The single-cell box beam has a variable depth, 10 m (i.e., ⅛ of the main span) at root and 3 m (i.e., ⅙ of the main span) at midspan. Three-way prestresses were introduced. A great tonnage group anchorage system with a post-tension force of 4275 kN for each group, which set a record in China, was employed longitudinally, with the tendons reaching 190 m long.

The superstructure was erected by cantilever casting. As the thickness is only 500 mm, the dual-wall hollow box piers of the main span have rather small thrust-resistant rigidity. Artificial islands were constructed around the piers to safeguard against the collision of passing vessels. The top diameter of each island is 23 and 28 m at bottom, with a height of 20 m. Two types of spans, 16 and 32 m, were chosen for the 1376.24-m-long approach mainly based on economical consideration, thus achieving a rather low construction cost. The bridge was designed by Highway Survey and Design Institute of Guangdong Province. It was constructed by Highway Engineering Department of Guangdong Province and completed in August 1988.

Huangshi Yangtze River Bridge

Huangshi Yangtze River Bridge (Figure 63.19) is located in Huangshi, Hubei Province, with its total length reaching 2580.08 m. A 162.5 + 3 × 245 + 162.5 m prestressed concrete continuous box girder rigid frame bridge was designed for the main bridge. The deck is 20 m wide, providing 15 m for motor vehicle traffic and 2.5 m on both sides for non-motor-vehicle traffic.

The approach along the Huangshi bank is 840.7 m long, consisting of continuous bridges and simply supported T-girder bridges with continuous decks, while the approach along the Xishui bank is 679.21 m, being single supported T-girder bridges with continuous decks.

FIGURE 63.18   Luoxi Bridge.

FIGURE 63.19   Huangshi Yangtze River Bridge.

FIGURE 63.20   Humen Bridge (continuous rigid frame).

A 28-m-diameter double-wall steel cofferdam with 16 3-m-diameter bored piles foundation was employed for piers of the main span, which provided enough capacity to resist impact force of ships. The navigation clearance of the bridge is 200 × 24 m, which allows the navigation of a vessel of 5000 tons. The bridge was designed by Highway Planning and Design Institute affiliated with the Ministry of Communications. It was constructed by China Road and Bridge Corporation and completed in 1996.

Humen Bridge

The Humen Bridge (Figure 63.20), an extra major highway bridge over Pearl River, is on the freeway connecting Guangzhou, Zhuhai and Shenzhen. It is composed of bridges of different types.

A rigid frame bridge (150 + 270 + 150 m) is arranged over the auxiliary navigation channel, with its main span reaching 270 m, a world record of the same type. The superstructure of the bridge consists of two separate bridges, each a single-box, single-cell prestressed concrete continuous rigid frame. The 24-m-wide deck provides 214.25 m for motor vehicle traffic. Adoption of 15.24 mm VSL prestress system makes thinner top slabs and no bottom slabs of the box girder possible, the single-box single-cell thin-wall section offers a greater moment of inertia per unit area, and the depth of main girder at the supports is 14.8 m ($\frac{1}{18}$ of the main span) and 5 m ($\frac{1}{54}$ of the main span) in midspan. The substructure consists of double thin-wall piers resting upon group piles foundations. The symmetrical cantilever casting method is employed for the erection of the superstructure. The bridge was designed by GuangDong Highway Planning and Design Institute and constructed by Highway Engineering Construction Ltd, Guangdong Province. It was opened to traffic in July 1997.

63.5 Steel Bridges

63.5.1 General Description

Steel structures are employed primarily for railway and railway–highway bipurpose bridges. In 1957, in the city of Wuhan, a railway–highway bipurpose bridge was erected over the Yangtze River, another milestone in China's bridge construction history. The bridge has continuous steel trusses with 128 m main spans. The rivet-connected truss is made of grade No. 3 steel. A newly developed cylinder shaft 1.55 m in diameter was initially used in the deep foundation. (Later in 1962, a 5.8-m cylinder shaft foundation was laid in the Ganjiang South Bridge in Nanchang, Jiangxi Province.) In 1968, another such bridge over the Yangtze River — the Nanjing Yangtze River Bridge — came into being. The whole project, including its material, design, and installation, was completed through the

Chinese own efforts. It is a rivet-connected continuous truss bridge with 160-m main spans. The material used is high-quality steel of 16 Mn. In construction, a deep-water foundation was developed. Open caissons were sunk to a depth of 54.87 m, and pretensioned concrete cylinder shafts 3.6 m in diameter were laid, thus forming a new type of compound foundation. And underwater cleaning was performed in a depth of 65 m. China's longest steel highway bridge is the Beige Yellow River Bridge in Shandong Province (1972), its main span being 113 m long. It has a continuous truss of bolt-connected welded members. The foundation is composed of 1.5-m-diameter concrete bored piles, whose penetration depth into subsoil reaches 107 m, the deepest pile ever drilled in China. A new structure of field-bolting welded box girder paved with orthotropic steel deck was first introduced in the North River Highway Bridge at Mafang, Guangdong Province, which was completed in 1980.

Another attractive and gigantic structure standing over the Yangtze River is the Jiujiang railway–highway Bridge completed in 1992. Chinese-made 15 MnVN steel was used and shop-welded steel plates 56 mm thick were bolted on site. The main span reaches 216 m. The continuous steel truss is reinforced by flexible stiffening arch ribs. In laying the foundation, a double-walled sheet piling cofferdam was built, in which a concrete bored pile was cast in place. When erecting the steel beams, double suspended cable frame took the place of a single one, which is another innovation.

Since the 1980s, stell girder or composite girder bridges have been adopted in the construction of long-span or complex-structure city bridges in China. For example, they were applied in Guangzhong Road Flyover and East Yanan Road Viaduct in Shanghai.

63.5.2 Examples of Steel Bridges

Nanjing Yangtze River Bridge

Nanjing Yangtze River Bridge (Figure 63.21) is a highway and railway double-deck continuous steel truss bridge in Nanjing, Jiangsu Province. On the upper deck there are four lanes of highway traffic, which are 15 m wide, plus two sidewalks of 2.25 m wide each, and on the lower deck two tracks for railway. The main bridge is 1576 m long. If approaches are taken into account, the length of the railway bridge reaches 6772 m and the highway bridge is 4588 m long.

Ten spans were arranged for the main bridge, including a side span of 128 m being simply supported steel truss and nine spans of 160 m each being continuous steel trusses, continued every three spans. The main truss is a parallel chord rhombic truss with reinforcing bottom chord. It was erected by cantilever assembling.

Considering the complex geologic conditions at the bridge site, different types of foundations were used: heavy concrete caissons with a depth of penetration reaching 54.87 m for areas with shallow water and deep coverings; a floating-type steel caisson combined with pipe column foundations was used for the first time at sites of deep water. The bridge was designed and constructed by the Major Bridge Engineering Bureau, Ministry of Railways. It was completed in December 1968.

Jiujiang Yangtze River Bridge

Jiujiang Bridge (Figure 63.22), on the border of Hubei Province and Jiangxi Province, is a double-deck highway and railway bipurpose bridge, with the longest truss span, 216 m, in China at present. The four-lane highway is on the upper deck, with a width of 14 m for motor vehicles and two sidewalks of 2 m each on both sides, while the double-track railway is carried on the lower deck.

The main bridge, divided into 11 spans, is all of steel. The three main spans (180 + 216 + 180 m) are combined truss-arch system, which consists of continuous steel truss beams and flexible steel stiffening arches. Two continuous steel truss beams of 3 × 162 m each are for the side span on the northern bank, while a 2 × 126 m continuous steel beam is over the south bank.

The main truss is of parallel chord triangular type with reinforcing bottom chord members, and its depth is 16 m and doubled at the supports. The stiffening arch over main span has a rise of 32 m and those over the side spans have a rise of 24 m. All steel structures are bolted and welded. The

FIGURE 63.21 Nanjing Yangtze River Bridge.

FIGURE 63.22 Jiujiang Yangtze River Bridge.

first-used 15 MnVN steel has a yield strength of 420 MPa. Double layers of suspenders were applied for assembling the truss beams.

There are different types of foundations employed in the bridge: circular reinforced concrete caisson for No. 1 pier in shoal, application of clay slurry lubricating jacket makes the depth of penetration reach 50 m; double-wall steel cofferdam and bored piles foundation in deep water with favorable rock conditions. The bridge was designed and constructed by the Major Bridge Engineering Bureau, Ministry of Railways. It was completed in May 1992.

63.6   Cable-Stayed Bridges

63.6.1   General Description

Cable-stayed bridges were first introduced into China in the early 1960s. Two trial bridges, the Xinwu Bridge with a main span of 54 m in Shanghai and the Tangxi Bridge with a span of 75.8 m in Yuyang, Sichuan Province — are both reinforced concrete cable-stayed bridges and were completed in 1975.

In 1977 the construction of long-span cable-stayed bridges began. The Jinan Bridge across the Yellow River with a main span of 220 m was completed in 1982. In the 1980s, the construction of cable-stayed bridges developed rapidly over a wide area in China. More than 30 bridges of various types were built in different provinces and municipalities. Among them, the Yong River Bridge in Tianjin has a main span of 260 m, and the Dongying Bridge in Shandong Province has span of 288 m, China's first steel cable-stayed bridge. In addition, the Haiying Bridge in Guangzhou has a 35-m-wide deck, single cable plane and double thin-walled pylon piers; the Jiujiang Bridge in Nanhai of Guangdong Province was erected by a floating crane with a capacity of 5000 kN; the Shimen Bridge in Chongqing has an asymmetrical single cable plane arrangement and a 230 m cantilever cast in place; and the attractive-looking Xiang River North Bridge in Changsha, Hunan Province, was completed in 1990 with light traveling formwork. All are representative of this period with their respective features.

At the beginning of the 1990s, with the completion of the Nanpu Bridge in Shanghai in 1991, a new high tide of construction of cable-stayed bridges began to surge in China. Now more then 20 cable-stayed bridges with a span of over 400 m have been completed, and a large number of long-span cable-stayed bridges are under design and construction. The most outstanding is the Yangpu Bridge with a main span of 602 m, a composite deck cable-stayed bridge in Shanghai.

63.6.2   Examples of Cable-Stayed Bridges

Laibin Hongshui River Railway Bridge

Laibin Hongshui River Bridge (Figure 63.23), a structure 398 m long over Hongshui River, is on the second Xianggui Railway Route. The main bridge, having three spans of 48 + 96 + 48 m, is the first prestressed concrete cable-stayed railway bridge built in China, with two pylons and an H-type cable configuration. The main girder, with a box section of two cells dimensioning 4.8 m (width) 3.2 m (depth), is prestressed longitudinally by 245 mm tendons whose $\sigma_p = 1600$ MPa.

The pylons are 29 m high, rigidly connected with main girder by strong box cross beams. Pot neoprene bearings were employed in the bridge. Three groups of parallel cables were installed on either side of each pylon, and each group consists of six bunches of 705-mm steel strands. To guarantee sufficient fatigue strength, key-grooved composite anchorage was specially designed, which also made adjustment and replacement of a cable possible.

The girders over side spans were cast on scaffoldings, while the middle span was constructed by cantilever casting. During the design and construction, special studies and tests were carried out to obtain the characteristics of the structure under railway loads. The bridge was designed by China Academy of Railway Sciences. It was constructed by Liuzhou Railway Bureau and completed in 1981.

Dongying Yellow River Bridge

Dongying Yellow River Bridge (Figure 63.24), the first steel cable-stayed bridge built in China, is on the highway route along the northern coast in Shandong Province. The total length of the bridge is 2817.46 m. The main bridge, a continuous steel cable-stayed bridge, has five spans of 60.5 + 136.5 + 288 + 136.5 + 60.5 m, while the 2135.46-m approaches are 71 spans of pretensioning prestressed concrete box girders, each being 30 m long.

FIGURE 63.23 Laibin Bridge.

FIGURE 63.24 Dongying Yellow River Bridge.

FIGURE 63.25 Shanghai Yangpu Bridge.

The deck is 19.5 m wide, among which 16 m are for vehicle traffic. Steel box girders with orthotropic plate deck form the main cable-stayed spans. It was erected by cantilever assembling. Each segment for assembling is about 12 m on average and consists of two side boxes, four plate decks, and cross beam. The H-type pylons are 69.7 m high. A fan-type cable configuration was adopted. Ten pairs of cables were installed on either side of each pylon, with an anchorage distance of 12 m on the deck. Each cable consists of 73/127ϕ7 galvanized steel wires with hot-squeezed sheath for protection. Bored piles foundation was employed in the substructure. The pylon rests on separate elevated pile caps which are supported by 22 piles with a diameter of 1.5 m and a length of 96.5 m.

The bridge was designed by Communication Planning and Design Institute of Shandong Province. It was constructed by Communication Engineering Co. of Shandong Province and completed in September 1987.

Yangpu Bridge

Shanghai Yangpu Bridge (Figure 63.25), located in Shanghai Yangpu District, is an important bridge in an urban district, which spans Huangpu River and connects Puxi old district with Pudong Development Zone. It is an essential component of the Inner Ring Elevated Viaduct. The bridge site is 11 km from the Nanpu Bridge, a cable-stayed bridge with 423 m main span completed in 1991.

The overall length of the bridge is 8354 m, including main spans, approach spans, and guide passage spans. The width of the bridge is 30.35 m. The main span is a dual-pylon, space dual-cable

FIGURE 63.26   Wuhan Yangtze River Highway Bridge.

plane, steel–concrete composite structure. As pylons rigidly connected with piers and separate from the girder, the superstructure is a suspended system longitudinally, with displacement-resistant and anti-seismic devices installed transversely.

The 200-m-high diamond-shaped pylons are of reinforced concrete, resting on steel pipe piles foundations. Columnar piers supported by precast concrete piles have found wide application in auxiliary piers, anchor piers, and side piers as well. Steel side box girders and I-type steel cross beams composite with precast reinforced concrete slabs make up the girder over the main span and two side spans. The center-to-center distance between the two main steel side box girders is 25 m, while that between steel cross beams is 4.5 m. For transitional spans, simply supported prestressed concrete T-girders are used. The stayed cables are 256 in number and there are 32 pairs of them on either side of each pylon. The maximum length of stays is 330 m, and 312ϕ7 high-strength parallel wires form the maximum cross section of the stays.

The bridge was designed by Shanghai Municipal Engineering Design Institute in cooperation with Tongji University, Shanghai Urban Construction Design Institute, and Shanghai Urban Construction College. The construction of the bridge was presided over by the Headquarters of Shanghai Huangpujiang Bridge Engineering Construction. It was completed and opened to traffic in 1993.

Wuhan Yangtze River Highway Bridge

The bridge (Figure 63.26) is a 4687.73-m-long structure in Wuhan, Hubei Province. Its cable-stayed bridge consists of prestressed concrete girder with t spans of 180 + 400 + 180 m. The cable-stayed bridge is a suspended system, with its longitudinal displacement restrained by the devices installed at the intersection parts of the girder and the pylons. The deck is 29.4 m wide, carrying six lanes 23 m wide and two pedestrian walks of 1.75 m each.

An open section with two side boxes is adopted, which is 3 m in depth and stiffened by cross beams every 4 m. The H-type reinforced concrete pylons are 94 m high, and a fan-type multicable configuration is adopted. There are a total of 392 cables, which are made up of 7-mm-diameter galvanized parallel steel wires and protected by hot-squeezed PE sheath. The maximum cable force reaches 5000 kN. Double-wall steel cofferdam bored piles foundations are employed. The bridge was designed and constructed by Major Bridge Engineering Bureau, Ministry of Railways. It was completed in 1995.

Huangshan Taiping Lake Bridge

The bridge (Figure 63.27) is a prestressed concrete cable-stayed bridge with a single pylon and a single cable plane. It has two spans of 190 m each. The pylon is 86.6 m high. A fan-type cable configuration with a cable distance of 6 m on the girder was adopted. Four lanes are arranged on

FIGURE 63.27 Huangshan Taiping Lake Bridge.

the deck, which is 18.2 m wide. The main girder is a three-cell prestressed concrete box girder with skew webs. It has a uniform depth of 3.5 m. The thickness of the top slab and the bottom slab are 220 and 200 mm, respectively. Three-way prestress was introduced. The bridge was designed by Lin & Li Consultants Shanghai Ltd., constructed by Major Bridge Engineering Bureau, Ministry of Railways, and opened to traffic in 1996.

63.7 Suspension Bridges

63.7.1 General Description

The construction of modern suspension bridges in China started in the 1960s. Some flexible suspension bridges with spans less than 200 m were built in the mountain areas of southwestern China, the Chaoyang Bridge in Chongqing, Sichuan Province, being the most famous one. However, the Dazi Bridge in Tibet completed in 1984 has a span of 500 m.

The upsurge of transportation engineering construction in the 1990s has led to a new stage of modern suspension bridges. The Shantou Bay Bridge in Shantou, Guangdong Province, was completed in 1995, having a 452 m concrete stiffening girder. The Humen Pearl River Bridge, a steel box girder suspension bridge with a main span of 888 m, was completed in 1997. The Jiangying Yangtze River Bridge with a main span of 1385 m, is now under construction.

63.7.2 Examples of Suspension Bridges

Chaoyang Bridge

Chaoyang Bridge (Figure 63.28), a highway bridge crossing Jialin River, is located in Beipei District, Chongqing. The bridge has three spans with a total length of 233.2 m. A double-chain reinforced girder suspension bridge, over 186 m long, is the main span, and two reinforced concrete slim curved beams are for the two side spans of 21.6 m each. The deck is 8.5 m wide, providing 7 m for motor vehicle traffic.

FIGURE 63.28   Chaoyang Bridge.

There are four cables in total and every two make a chain that is installed on either side. Each cable is made up of 19ϕ42 steel ropes; ϕ 42 steel pipe and ϕ42 steel rope are used as upper hanger and lower hanger, respectively. The stiffening girder, with depth of 2.0 m, is a single open-steel box composited with a reinforced concrete deck slab. The 63.8-m-high pylons are reinforced concrete portal frames. A tunnel-type anchorage system was adopted, and the tunnel length reached 15 m. The anchorage slabs are 1.8 m in depth and anchored in rock stratum.

The bridge was designed by Chongqing Communication Research Institute and Chongqing Communication Institute, and was built by Chongqing Bridge Engineering Corporation. The bridge was completed in 1969.

Dazi Bridge

Dazi Bridge (Figure 63.29), a 500-m suspension bridge crossing Lasa River, is located in Dazi, 25 km east of Lasa, Tibet. It is 4.5 m wide, providing only one lane for highway traffic. The main cables are made up of 5-mm-diameter parallel wires and its sag-to-span ratio is 1:5. The main girder is an open welded steel truss with a depth of 1.5 m. A simple orthotropic deck was formed by the composition of the longitudinal girder and steel plates. Four trucks of 200 kN each were used as design loads.

The bridge was designed by Highway Planning, Survey and Design Institute of Tibet Autonomous Region and constructed by Highway Bureau of Tibet Autonomous Region. It was completed in December 1984.

Shantou Bay Bridge

Shantou Bay Bridge (Figure 63.30), a 2420-m-long structure, crosses Shantou Bay in Shantou, Grangdong Province. The main bridge is a prestressed concrete suspension bridge, with a central span of 452 m and two side spans of 154 m each. Four spans of 25 m prestressed concrete T-girders

FIGURE 63.29   Dazi Bridge.

FIGURE 63.30   Shantou Bay Bridge.

connect the main bridge with the bank on either side. The width of the bridge varies from approaches of 27.3 m to the main bridge of 23.8 m.

The stiffening girder is a prestressed concrete flat box with three cells. It is prestressed longitudinally by tendons inside the top slab and those outside the bottom slab. Two prestressed concrete beams, which are specially designed to connect the central span and two side spans, provide a smooth transition of the deck. A pair of flexible steel piers was installed between each beam and the lower cross beam of a pylon, to get the stiffening girder restrained elastically and maintain the structural flexible characteristics.

The pylons are three-layer reinforced concrete rigid frame structures, each resting on two separate caisson foundations connected by a strong bracing beam on the top. The center-to-center distance

FIGURE 63.31 Hong Kong Tsing Ma Bridge.

between the two cable saddles on one pylon is 25.2 m. Roller bearings were employed during the erection of cables, which were fixed to the saddles later.

The main cables have a sag-to-span ratio of 1:10. Each cable consists of 110 bunches of $9\phi.5$-mm parallel wires, which makes its diameter reach approximately 550 to 630 mm. Every 6 m a hanger, which is composed of a pair of $\phi42$ steel ropes, is installed. The segment assembling scheme was adopted for the girder erection. A 5.7-m-long precast segment, weighing 16,000 kN, was connected with the completed girder by wet joint.

The bridge was designed and constructed by Major Bridge Engineering Bureau, Ministry of Railways. It was open to traffic in 1995.

Hong Kong Tsing Ma Bridge

Tsing Ma Bridge (Figure 63.31), a highway–railway bipurpose suspension bridge on the freeway between the new airport and the urban district, connecting the Tsing Yi Island and Mawan Island in Hong Kong, is the world's longest of its kind. The main span is 1377 m. The design of its main girder was mainly based on consideration of its aerodynamic stability and a truss type was finally adopted: The cross beams are of Vierendeel truss. The whole longitudinal girder can be treated as a composite structure.

There are six lanes of highway traffic on the upper deck and three passages on the lower deck, the central one on lower deck for railway while the other opened to highway traffic under severe weather condition. The main cable is of $80 \times 368 + 11 \times 360$ steel wires, each wire having a diameter of 5.35 mm. The construction of the bridge was started in 1992 and it was opened to traffic in 1997.

Jiangyin Yangtze River Bridge

Jiangyin Yangtze River Bridge (Figure 63.32) is located on the planned North–South Principal Highway System in the coastal area between Jiangyin and Jinjiang in Jiangsu Province. It is a large suspension bridge with a central span of 1385 m, and will be the first bridge with a span in excess of 1000 m designed and built by Chinese engineers. Its total length reaches nearly 3 km. The deck is designed to carry six-lane highway traffic, while median and emergency parking strips are also considered, with two 1.5-m-wide pedestrian walks on the central span.

Flat steel box girder with wind fairing is adopted, whose depth and width are 3 and 37.7 m, respectively. The two main cables, with sag-to-span ratio being 1:10.5, are composed of five galvanized high-strength wires, and will be erected by the PWS method. The bridge provides a navigation clearance 50 m high. The 190-m-high pylons are reinforced concrete structures. The northern pylon, located in the shallow water area outside the north bank, rests on piles foundation constructed by the sand island method, whereas the southern one is on rock stratum of the bank.

FIGURE 63.32   Jiangyin Yangtze River Bridge.

The south anchorage is of gravity type embedded on rock bed in comparison with the north anchorage gravity-friction type on soft-soil ground. The north anchorage is a massive concrete caisson, measuring 69 by 51 m in plan and 58 m in depth. It is the largest concrete caisson in the world. All the approaches are prestressed concrete beam bridges. A multispan prestressed concrete continuous rigid frame structure was chosen for the northern side span.

The project was designed by Highway Planning and Design Institute, Ministry of Communications in cooperation with Communication Design Institute of Jiangsu Province and Tongji University. It was scheduled to be completed in 1999.

References

1. Guohao Li, *Stability and Vibration of Bridge Structures*, 2nd ed., China Railway Publishing House, 1996.
2. Haifang Xiang, *Bridges in China*, Tongji University Press, A&U Publication Ltd., Hong Kong, 1993.
3. Chinese Design Code for Highway Bridges, People's Communication Press House of China, 1991.
4. Chinese Design Code for Railway Bridges, China Railway Publishing House, 1988.
5. Guohao Li, Computation of Transverse Load Distribution for Highway Bridges, People's Communication Press House of China, 1977.
6. Haifang Xiang, Wind Resistance Design Guideline for Highway Bridges, People's Communication Press House of China, 1996.

7. Yuan Wancheng and Fan Lichu, Study on energy absorption of a new aseismic rubber bearing for bridge, *J. Vibration Shock,* 14(3), 1995 [in Chinese].
8. Rucheng Xiao, Influence matrix method for structural adjustment calculation of internal forces and displacements of concern sections, *Comput. Struct. Mech. Appl.* 9(1), 1992 [in Chinese].
9. Rucheng Xiao, Synthetical bridge nonlinear analysis program system, *China Comp. Appl. Highway Eng.,* 4(1), 1994.

64

Design Practice in Europe

Jean M. Muller
Jean Muller International, France

64.1 Introduction

Europe is one of the birthplaces of bridge design and technology, beginning with masonry bridges and aqueducts built under the Roman Empire throughout Europe. The Middle Ages also produced many innovative bridges. The modern role of the engineer in bridge design appeared in France in the 18th century. The first bridge made of cast iron was built in England at the end of the same century. Prestressed concrete was born in France before extending throughout the world. Cantilever construction and incremental launching of concrete decks were devised in Germany, as well as modern cable-stayed bridges. The streamlined box-girder deck for long-span suspension bridges was born in England. The variety of bridges in Europe is enormous, from the point of view of both their age and their type.

Outstanding works of bridge history in Europe can be presented as follows.

0-8493-7434-0/00/$0.00+$.50
© 2000 by CRC Press LLC

Bridge	Year	Country	Designer	Comments
Unknown	600 B.C.	I	Etruscans	Probable use of vaults for bridge construction
Gardon River Bridge *	13 B.C.	F	Romans	Aqueduct 49 m high, with three rows of superposed arches
Céret Bridge over the River Tech	1339	F	Unknown	Masonry bridge spanning 42 m
Wettingen Bridge	1764	CH	Johann Ulrich Grubenmann	Biggest wooden bridge in Europe with a 61 m span
Coalbrookdale Bridge	1779	GB	Abraham Darby III	First metallic bridge: cast iron structure
Sunderland Bridge	1796	GB	Rowland Burdon	Six cast iron arches, each made up of 105 segments
Saint-Antoine Bridge	1823	CH	Guillaume Henri Dufour	First permanent suspension bridge with metallic cables in the world
Britannia Bridge	1850	GB	Robert Stephenson	First tubular straight girder, spanning 140 m, consisting of wrought iron sheets
Crumlin Viaduct	1857	GB	Charles Liddell	First metallic truss girder viaduct
Bridge over the River Isar	1857	D	Von Pauli, Gerber, Werder	Welded and bolted iron truss girder
Royal Albert Bridge	1859	GB	Isambard Kingdom Brunel	Metal truss girder, first of a whole modern generation of railway bridges
Maria Pia Bridge over the River Douro	1877	P	Gustave Eiffel	Arch spanning 160 m, made up of metal structure
Antoinette Bridge	1884	F	Paul Séjourné	Culmination of masonry bridges
Firth of Forth Bridge *	1890	GB	Sir John Fowler and Sir Benjamin Baker	First large steel bridge in the world — two main spans 520 m long
Alexandre III Bridge *	1900	F	Jean Résal	15 very slender arches composed of molded steel segments
Salginatobel Bridge	1930	CH	Robert Maillart	Arch marking the concrete box-girder birth
Albert Louppe Bridge *	1930	F	Eugène Freyssinet	Three reinforced concrete vaults, each spanning 188 m — wooden formwork spanning 170 m
Linz Bridge over the River Danube	1938	AUT	A. Sarlay and R. Riedl	First welded girder 250 m long — three spans
Luzancy Bridge	1946	F	Eugène Freyssinet	Concrete bridge prestressed in three directions, made up of precast segments
Cologne Deutz Bridge	1948	D	Fritz Leonhardt	Composite steel plate-concrete box-girder bridge spanning 184 m
Percha Bridge	1949	D	Dyckerhoff and Widmann	First reinforced concrete large span cantilever construction
Donzère Mondragon Bridge	1952	F	Albert Caquot	First cable-stayed bridge — 81 m long main span
Düsseldorf Northern Bridge	1957	D	Fritz Leonhardt	First modern cable-stayed metallic bridge
Bendorf Bridge *	1964	D	Ulrich Finsterwalder	Cast-in-place balanced cantilever girder bridge — 208 m long main span
Choisy Bridge	1965	F	Jean Muller	First prestressed concrete bridge consisting of precast segments with match-cast epoxy joints
First Severn Bridge *	1966	GB	William Brown	Decisive stage: deck aerodynamic study in a low- and high-speed wind tunnel
Weitingen Viaduct	1975	D	Fritz Leonhardt	Steel span world record: 263-m-long span
Saint-Nazaire Bridge	1975	F	Jean-Claude Foucriat	Steel cable-stayed bridge world record — 400-m-long main span
Brotonne Bridge	1977	F	Jean Muller	Prestressed concrete cable-stayed bridge world record — 320-m-long main span
Kirk Bridge	1980	Croatie	Ilija Stojadinovic	World record — prestressed concrete arch spanning 390 m
Ganter Bridge	1980	CH	Christian Menn	174-m-long cable stayed span — stay planes protected by concrete walls
Normandie Bridge *	1995	F	Michel Virlogeux	World record — cable-stayed bridge with a 856-m-long main span
Storebaelt Bridge *	1998	DK	Cowi Consult	6.6- and 6.8-km-long bridges including a suspension bridge with a 1624-m long central span
Tagus Bridge	1998	P	Campenon Bernard	13-km-long bridge including a cable-stayed bridge with a 420-m-long main span
Gibraltar Straight Bridge	Project	E	Not yet known	Suspension bridge: 3.5- to 5-km long spans
Messina Straight Bridge	Project	I	Not yet known	Suspension bridge: 3.3-km-long main span

* A brief description of these bridges are given later with a photograph.

FIGURE 64.1 Gard Bridge over the River Gardon. (*Source*: Leonhardt, F., Ponts/Puentes — 1986 Presses Polytechniques Romandes. With permission.)

If we could choose only eight outstanding bridges, they would be as follows.

1. *Gardon River Bridge* (*13 B.C.*) — The Gardon River Bridge, also named Gard bridge, located in France, is an aqueduct consisting of three rows of superposed arches, composed of big blocks of stone assembled without mortar. Its total length is 360 m, and its main arches are 23 m long between pillar axes. It fully symbolizes Roman engineering expertise from 50 B.C. to 50 A.D. (Figure 64.1). Built with large rectangular stones, the bridge surprises by its architectural simplicity. Repetitivity, symmetry, proportions, solidity reach perfection, although the overall impression is that this work is lacking spirit.

2. *Firth of Forth Bridge* (*1890*) — The Forth Railway Bridge, located in Scotland, Great Britain, was the first large steel bridge built in the world. Its gigantic girder span of 521 m, longer than the main span length of the greatest suspension bridges of the time, made this bridge a technical achievement (Figure 64.2). In all, 55,000 tons of steel and 6,500,000 rivets were necessary to build this structure costing more than 3 million sterling pounds. The very strong stiff structure, made of riveted tubes connected at nodes, consists of three balanced slanting elements and two suspended spans, with two approach spans formed of truss girders. The total bridge length is 2.5 km.

3. *Alexandre III Bridge* (*1900*) — This roadway bridge over the River Seine in Paris, France, designed by Jean Résal, bears on 15 parallel arches made up of molded steel segments assembled by bolts. These arches are rather shallow, the ratio is $\frac{1}{17}$, and so, massive abutments are necessary. The River Seine is crossed by a single span, 107 m long; the bridge deck is 40 m wide (Figure 64.3).

4. *Albert Louppe Bridge* (*1930*) — This bridge, located in France, is the most beautiful expression of Eugène Freyssinet's reinforced concrete works. The three arches, each spanning 186.40 m

FIGURE 64.2   Firth of Forth Bridge. (Courtesy of J. Arthur Dixon.)

FIGURE 64.3   Alexandre III Bridge. (Courtesy of SETRA.)

(Figure 64.4) crossed the River Elorn for half the cost of a conventional metal bridge. The arches are three cell box girders, 9.50 m wide and 5.00 m deep on average. The deck is a girder with reinforced concrete truss webs. The formwork used for casting the three vaults, moved on two 35 by 8 m reinforced concrete barges, was the greatest and the most daring wooden structure in construction history with its 10-m-wide huge vault spanning 170 m.

FIGURE 64.4 Albert Louppe Bridge. (Courtesy of Jean Muller International.)

5. *Bendorf Bridge (1964)* — Built in 1964 near Koblenz, Germany, this structure has a total length of 1029.7 m with a navigation span 208 m long over the River Rhine. Designed by Ulrich Finsterwalder, it is an early and outstanding example of the cast-in-place balanced cantilever bridge (Figure 64.5). The continuous seven-span main river structure consists of twin independent single-cell box girders. Total width of the bridge cross section is 30.86 m. Girder depth is 10.45 m at the pier and 4.4 m at midspan. The main navigation span has a hinge at midspan, and the superstructure is cast monolithically with the main piers. The structure is three-dimensionally prestressed.
6. *First Severn Bridge (1966)* — The suspension bridge over the River Severn, Wales, Great Britain, designed and constructed in 1966, marks a distinct change in suspension bridge shape during the second half of the 20th century (Figure 64.6). William Brown, the main design engineer, created a 988-m-long central span. The deck is a stiff and streamlined box girder. Its aerodynamic stability was improved in a wind tunnel, with high-speed wind tests under compressed airflow. Since the opening of the bridge, many designers have been drawn from afar to its shape, new at the time, but now looked upon as classical.
7. *Normandie Bridge (1995)* — The cable-stayed bridge, crossing the River Seine near its mouth, in northern France, is 2140 m long. Its 856-m-long main span constitutes a world record for this kind of structure, although the bridge in principle does not bring much innovation in comparison with the Brotonne bridge from which it is derived (Figure 64.7). The central 624 m of the main span is made of steel, whereas the rest of the deck is made of prestressed concrete. The deck is designed specially to reduce the impact of wind blowing at 180 km/h. Reversed Y-shaped pylons are 200 m high. The stays, whose lengths vary from 100 to 440 m, have been the subject of an advanced aerodynamic study because they represent 60% of the bridge area on which the wind is applied.

FIGURE 64.5   Bendorf Bridge. (*Source*: Leonhardt, F., Ponts/Puentes — 1986 Presses Polytechniques Romandes. With permission.)

FIGURE 64.6   First Severn Bridge. (*Source*: Leonhardt, F., Ponts/Puentes — 1986 Presses Polytechniques Romandes. With permission.)

FIGURE 64.7 Normandie Bridge. (Courtesy of Campenon Bernard.)

8. *Great Belt Strait Crossing (1998)* — The Storebælt suspension bridge, located in Denmark, has a central span of 1624 m. It is the main piece of a complex comprising a combined highway and railway bridge 6.6 km long, a twin tube tunnel 8 km long, and a 6.8-km-long highway bridge (Figure 64.8). This link is part of one of the most ambitious projects in Europe, to join Sweden and the Danish archipelago to the European Continent by a series of bridges, viaducts, and tunnels, which can accommodate highway and railway traffic.

64.2. Design

64.2.1 Philosophy

To allow for the single internal market setup, the European legislation includes two directive types:

1. Directives "products," whose purpose is to unify the national rules in order to remove the obstacles in the way of the free product movement.
2. Directives "public markets," aiming to avoid national or even local behaviors from owners or public buyers.

By experience, the only means of ensuring that a bid based on a calculation method practiced in another state is not dismissed is to have a common set of calculation rules. These rules do not necessarily require the same numerical values.

Consequently, the European Community Commission has undertaken to set up a complex of harmonized technical rules with regard to building and civil engineering design, to propose an alternative to different codes and standards used by the individual member states, and finally to replace them. These technical rules are commonly referred to as "Structural Eurocodes."

The Eurocodes, common rules for structural design and justification, are the result of technical opinion and competence harmonization. These norms have a great commercial significance. The

FIGURE 64.8 Storebælt Bridge. (Courtesy of Cowi Consult.)

Eurocodes preparation began in 1976, and drafts of the four first Eurocodes were proposed during the 1980s. In 1990 the European Economical Community put the European Normalization Committee in charge of developing, publishing, and maintaining the Eurocodes.

In general, the Eurocode refers to an Interpretative Document. This is a very general text which makes a technical statement. In the European Community countries the mechanical resistance and stability verifications are generally based on consideration of limit states and on format of partial safety factors, without excluding the possibility of defining safety levels using other methods, for example, probability theory of reliability.

From this document which heads them up, the Eurocodes deal with projects and work execution modes. Numerical data included are given for well-defined application fields. Therefore, the Eurocodes are not only frameworks that define a philosophy allowing the various countries the possibility to tailor the contents individually, they are something completely unique in the normalization field.

A norm defines tolerances, materials, products, performances. The Eurocodes are entirely different because they attempt to be design norms, i.e., norms that define what is right and what is wrong. That is a unique venture of its kind.

The transformation of the Eurocodes into European norms was begun in 1996 and will be reality in 2001 for the first ones. For about 5 years before their final adoption, both the Eurocodes and the national norms will stay applicable.

Of course, there exists a need for connection between Eurocodes and various national rules. Variable numerical values and the possibility of defining certain specifications differently allow this adaptation. From 2007 to 2008 national norms will be progressively withdrawn. Concerning bridges, from 2008 to 2009 only the Eurocodes will be applicable.

These texts are completely coherent, thus it is possible to go from one to the other with coherent combinations. This coherence expands to the building field where its importance is more significant. Moreover, these texts are merely a part of vast normative whole which refers to construction norms, product norms, and test norms.

The Eurocodes are written by teams constituted of experts from the main European Union countries, who work unselfishly for the benefit of future generations. For this reason they are the fruit of a synthesis of different technical cultures. They constitute an open whole. Texts have been written with a clear distinction between principles of inviolable nature and applications rules. The latter can be modulated within certain limits, so that they do not act as a brake upon innovation, and appear as a decisive progress factor. They allow, by constituting an efficient rule of the game, the establishment of competition on intelligent and indisputable grounds.

The Eurocodes applicable to bridge design are as follows.

Eurocode 1:	Basis of design and actions on structures [1]
Part 2	Loads: dead loads, water, snow, temperature, wind, fire, etc
Part 3	Traffic loads on bridges
Eurocode 2:	Concrete structure design [2]
Part 2:	Concrete bridges
Eurocode 3:	Steel structure design [3]
Eurocode 4:	Steel–concrete composite structure design and dimensioning [4]
Eurocode 5:	Wooden work design [5]
Eurocode 6:	Masonry structure design [6]
Eurocode 7:	Geotechnical design [7]
Eurocode 8:	Earthquake-resistant structure design [8]
Eurocode 9:	Aluminum alloy structure design [9]

64.2.2 Loads

The philosophy of Eurocode 1 is to realize a partial unification of concepts used to determine the representative values of the actions. In this way, most of the natural actions are based on a return period of 50 years. These actions are generally multiplied by a ULS (ultimate limit state) factor taken as 1.5. The return period depends on the reference duration of the action and the probability of exceeding it. This return period is generally 50 years for buildings and 100 years for bridges. This definition is rather conventional. At the moment, the Eurocode is a temporary norm. Consequently, the Eurocode 1 annex make it possible to use a formula which allows one to change the return period. With regard to traffic loads, Eurocodes constitute a completely new code, not inspired by another code. That means the elaboration was done as scientifically as possible.

The database of traffic loads consists of real traffic recordings. The highway section chosen is representative of European traffic in terms of vehicle distribution. On these real data, a certain number of mathematical processes are realized. But not all data were processed by mathematics and probability. Some situations allow definition of the characteristic load. These are obstruction situations, hold-up situations on one lane with a heavy but freely flowing traffic on the other lane, and so forth, i.e., realistic situations.

All these elements were mathematically extrapolated so that they correspond to a 1000-year return period, that is to say, a 10% probability of exceeding a certain level in 100 years. The axle distribution curve leads one to take into account a 1.35 ULS factor instead of 1.5 for a heavy axle. Concerning abnormal vehicles, the Eurocode gives a catalog from which the client chooses. The Eurocode defines as well, how an abnormal vehicle can use the bridge while traffic is kept on other lanes, which is rather realistic.

With regard to loads on railway bridges, the UIC models were revised in the Eurocode. Loads corresponding to a high-speed passenger train were also introduced in the Eurocode.

There are no military loads in Eurocodes. This type of loads is the client responsibility.

Concerning the wind, the speed measured at 10 m above the ground averaged over 10 min, with a 50-year return period, is taken into account. This return period seems to be somewhat conventional, because this speed is transformed into pressure by models and factors themselves including safety margin.

FIGURE 64.9 Oise Bridge. (Courtesy of Fred Boucher, SANEF.)

The most detailed studies show that the return period of the characteristic wind pressure value is rather contained by the interval between 100 and 200 years. After multiplication by the 1.5 ULS factor, this characteristic value has a return period indeed contained by the interval between 1000 and 10,000 years. The code also defines a dynamic amplification coefficient, which depends on the geometric characteristics of the element, its vibration period, and its structural and aerodynamic damping.

With regard to snow loads, the Eurocodes give maps for each European country. These maps show the characteristic depth of snow on the ground corresponding to a 50-year return period. Then this snow depth is transformed into snow weight taking into account additional details.

It is the same case for temperature. The characteristic value is the temperature corresponding to a 50-year return period. The characteristic value for earthquake loads, in Eurocode 8, corresponds as well to a 10% probability of exceeding the load in 50 years.

Therefore, the philosophy is rather clear with regard to loads. Some people wish to go toward greater unification, but it seems to be difficult to realize. Nevertheless, the load definition constitutes a comprehensible and homogeneous whole which is finally satisfactory.

64.3 Short- and Medium-Span Bridges

64.3.1. Steel and Composite Bridges

64.3.1.1 Oise River Bridge

In France, the Paris Boulogne highway link crosses the River Oise on a single steel concrete composite bridge (Figure 64.9). The bridge is 219 m long with a 105-m-long main span over the river and two symmetric side spans. The foundation of the bridge consists of 14 2.80-m-long, about 30-m-deep, diaphragm walls with variable thickness. Pier and abutment design is standard.

FIGURE 64.10 Roize Bridge. (Courtesy of Jean Muller International.)

The bridge deck is a composite structure, 2.50 m deep at midspan and on abutments, and 4.50 m deep on the piers. The steel main girders are spaced 11.40 m. The main girder bottom and top flange widths are constant, but their thicknesses vary continuously from 40 to 140 mm. The concrete slab has an effective width of 18 m. It is transversely prestressed with 4T15 cables, six units every 2.50 m.

The deck steel structure was assembled in halves, one behind each abutment on the embankment. Each half was launched over the river and welded together at midspan. The concrete deck slab was poured using two traveling formworks. The midspan area was poured first, followed by the pier areas.

Since 1994, the link has carried two traffic lanes, which will continue until the foreseen construction of a second parallel bridge.

64.3.1.2 Roize River Bridge

The Roize Bridge carries one of the French highway A49 link roads. Its deck was designed by Jean Muller (Figure 64.10). The choice made was a result of 10 years of studies on reducing the weight of medium-span bridge decks. Here the weight saving was obtained by replacing prestressed concrete cores by steel trusses constituting two triangulation planes (Warren-type) inclined and intersecting at the centerline of the bottom flange, by using a bottom flange formed of a welded-up hexagonal steel tube, and by reducing the thickness of the top slab by the use of high-strength concrete prestressed by bonded strands. The bridge was completed in 1990.

Indeed, innovation of this structure lies in its modular design. The steel structure is composed of tetrahedrons built in the factory, brought to site, and then assembled. The concrete slab also consists of prefabricated elements assembled *in situ.*

The deck is prestressed longitudinally by external tendons to keep a normal compression force in the upper slab on the piers, and to reduce the steel area of the bottom. It is also prestressed transversely.

FIGURE 64.11   Saint Pierre Bridge. (Courtesy of Albert Berenguier, Egis Group.)

The Roize Bridge structure has several advantages: light weight, low consumption of structural steel, industrialized fabrication, ease and speed of assembly, adaptability to complex geometric profile, durability. The basic characteristics are length = 112 m; width = 12.20 m; equivalent thickness of B80 concrete = 0.18 m; structural steel = 112 kg/m^2 of deck; pretensioned prestress = 17 kg/m^3; transverse prestress = 15 kg/m^3; longitudinal prestress = 32 kg/m^3.

64.3.1.3 Saint Pierre Bridge

This bridge is located in the historical center of Toulouse in the southwest of France. Its architecture is inspired by 19th century metal truss bridges with variable depth, while using modern technologies for the execution (Figure 64.11). The bridge is a 240 m long steel–concrete composite structure, partially prestressed. The span lengths are the following: 36.88 m, 3 × 55.00 m, 36.88 m.

It is founded on 1.80-m-diameter molded piles. Each pair of piles is linked by a reinforced concrete box girder. This structure supports a pier consisting of two elements. The deck rests on inclined elastomeric bearings so that the bridge works as a frame in longitudinal direction.

The longitudinal composite structure is made up of two lateral metal truss girders. These girders of variable depth are spaced 11.4 m apart with a cross-beam joining them every 14 m. Both main girders and cross-beams are connected to the concrete slab. The concrete slab is 25 cm thick on the central part bearing the traffic lanes. Toward the edges the slab is 27 cm thick and is placed 75 cm higher than the central part, accommodating the sidewalks.

The structure is prestressed longitudinally by 4K15 cables constituted by greased strands located toward the edges of the slab. Transversely, it is prestressed by greased monostrands located in the slab central part. The steel deck structure is erected from the piers supporting on temporary piling. The concrete slab is poured *in situ* with formwork supported by the now self-supporting steel structure.

This bridge is perfectly integrated into its environment of historic monuments, and opened to traffic in 1987.

FIGURE 64.12   A1 highway overpasses. (Courtesy of J. P. Houdry, Egis Group.)

64.3.2. Concrete Bridges

64.3.2.1   Channel Bridges: Overpasses over Highway A1

A new segmental design for overpasses was developed in France in 1992 to 1993, taking into account the necessity of standardization. The bridges have decks comprising a single transverse slab supported by two longitudinal lateral ribs (Figure 64.12).

This concept, suitable for a wide variety of bridge types with span lengths of between 15 and 35 m, is encompassed in the following ideas:

- The deck is built using precast segments, match-cast, and longitudinally prestressed.
- The segments are transversely prestressed using greased monostrands.
- The lateral ribs are used as barriers.

The main advantages of this type of concept are the possibility of building the overpass without disruption of traffic very quickly, with longer spans, thus fewer spans (two instead of four spans), than for the usual precast conventional overpasses.

64.3.2.2   Progressively Placed Segmental Bridges

Fontenoy Bridge

Fontenoy Bridge is 621 m long and open to traffic in 1979. It allows the crossing of the River Moselle in the north east of France with the following spans: 43.12 m, 10×52.70 m, 50.80 m. The foundations are either coarse aggregate concrete footings or bored piles, depending on the resisting substratum. On typical piers the bearings are of the elastomeric type, and on the abutments they are of the sliding type. The deck is a simply supported concrete box girder, 10.50 m wide, with two inclined webs and a constant depth of 2.75 m.

FIGURE 64.13 Fontenoy Bridge. (Courtesy of Campenon Bernard.)

The progressive placement method is used to build the deck, starting at one end of the structure, proceeding continuously to the other end (Figure 64.13). A movable temporary stay arrangement is used to limit the cantilever stresses during construction. The temporary tower is located over the preceding pier. All stays are continuous through the tower and anchored in the previously completed deck structure.

Precast segments are transported over the completed portion of the deck to the tip of the cantilever span under construction, where they are positioned by a swivel crane that proceeds from one segment to the next. The box girder is longitudinally prestressed by internal 12T13 units.

Les Neyrolles Bridge
Nantua and Neyrolles Viaducts allow the A40 highway to link Geneva, Switzerland, to Macon, France. The Neyrolles Viaducts have a total length of 985.5 m divided into three independent structures. It is composed of 20 spans of 51 m approximately, except for one span of 62 m which crosses the "Bief du Mont" stream (Figure 64.14). The deck is a concrete box girder approximately 11 m wide. The box girder was erected of precast match-cast segments.

The assembly was performed by asymmetric cantilevering by means of temporary stays and a deck-mounted swivel crane. The mast ensured the stability through the back stays carried by the previous span. The mast allowed erection of spans up to 60 m. The side spans at the abutments could not be assembled likewise because of the absence of a balancing span. Consequently, these span segments were placed on falsework and finally each span was prestressed and put on its definitive supports by means of jacks. The largest span (62 m) was assembled by both methods of construction mentioned.

The first phase consisted of assembly by stay-supported asymmetric cantilevering until the last stay available. The second phase consisted in erecting the last precast segments on falsework. The bridge was completed in 1995.

FIGURE 64.14 The second Neyrolles viaduct. (Courtesy of Campenon Bernard.)

64.3.2.3 Rotationally Constructed Bridges

Gilly Bridge
The Gilly Bridge, close to Albertville in France, consisting of two perpendicular decks was opened
to traffic in 1991. The main bridge crosses the river Isère and the access road to the Olympic site
resorts (Figure 64.15).

It is a prestressed concrete cable-stayed bridge, with two spans, 102 m long above the river and
60 m long above the road. The A-shaped pylon is tilted backward 20°. The other bridge supports
are a standard abutment on the left bank and a massive abutment acting as counterweight on the
right bank. Transversely, the 12-m concrete deck consists of two 1.90-m-deep and 1.10-m-wide
lateral ribs with cross-beams spaced 3.0 m supporting the top slab.

The A-shaped pylon was built vertically. It was tilted to its definite position by pivoting around
two temporary hinges located at its basis, the pylon being held back by two 19T15 cables. After
tilting, hinges were frozen by prestressing and concreting.

FIGURE 64.15   The Gilly Bridge. (Courtesy of Razel.)

The 162-m-long main bridge deck was concreted on a general formwork located on the right bank, parallel to the river. After concreting and cable-stay tensioning, the deck was placed in its definite position by a 90° rotation around a vertical axis. During the deck rotation the whole structure weighing 6000 t is supported on three points. Vertical reactions are measured continuously by electronic equipment to check dynamic effects.

Resorting to original construction methods has allowed realization of a bridge of high quality both structurally and aesthetically.

Ben Ahin Bridge

The Ben Ahin Bridge crossing the river Meuse in Belgium is a cable-stayed asymmetric bridge, 341 m in overall length (Figure 64.16), constructed in 1988. The reinforced concrete bridge deck, partially prestressed, is suspended by 40 cables anchored to a single tower structure. The central span is 168 m long. The deck girder has a box section, 21.80 m wide at the top fiber and 8.70 m at the bottom fiber. The depth, constant along the whole bridge, is 2.90 m.

The entire structure consisting of the tower structure, the stay cables, and the deck girder was constructed on the left bank of the river. After completion it was rotated by 70° relative to the tower axis, in order to swing the bridge around to its final definite position (Figure 64.17). Two pairs of jacks, each 500 ton force, located underneath the pylon sliding on Teflon, and four jacks each 300 ton force, located 45 m from the pylon underneath a stability metal frame, allowed the rotation of the 16,000 ton structure.

This method, already used in France for lighter bridges, was in this case designed to set a world record.

64.3.3. Truss Bridges

64.3.3.1 Sylans Bridge

The Sylans Viaduct runs through the French Jura Mountain complex. In this location, along the shores of a lake, difficulty lies in the uncertainty of the foundation soil since the route runs along a very steep slope whose 30-m-thick surface stratum comprises an eroded and fractured material of very doubtful stability.

FIGURE 64.16 Ben Ahin Bridge. (Courtesy of Daylight for Greisch.)

FIGURE 64.17 Ben Ahin Bridge during rotation. (Courtesy of Photo Studio 9 for Greisch.)

FIGURE 64.18 Sylans Bridge — two parallel decks. (Courtesy of Bouygues.)

The 1266-m-long viaduct comprises 21 60-m-long spans, each composed of two identical parallel decks 15 m apart and staggered 10 m in height (Figure 64.18), and was constructed in 1988. The deck is a prestressed concrete space truss structure 10.75 m wide and 4.17 m deep all along the bridge. It consists of 586 precast segments, i.e., 14 segments for each viaduct span.

Each typical concrete segment consists of two slabs linked by four inclined planes of diagonal prestressed concrete braces of 20 cm² cross section, assembled in pairs in the form of Xs. For every segment the diagonal braces are precast separately with a concrete of 65 MPa cylinder strength, and assembled with the segment-reinforcing cage. Then, the top and bottom slabs are poured with 50-MPa concrete. Finally, the diagonals are prestressed.

The deck segments are put in place by the cantilever method using a 135 m long launching girder. The deck prestressing consists of four families:

- Cantilever cables located below the top slab: 4T15 units;
- Strongly inclined cables from pier to withstand the shear force: 12T15 units;
- Horizontal continuity cables on and inside the bottom slab: 12T15 units;
- Horizontal cables in the top and the bottom slabs: respectively, 4T15 and 7T15 units.

The deck bears on its piers through reinforced elastomeric bearings.

Piers are supported by 6- to 35-m-tall, 4-m-diameter caissons. A circular concrete cap is cast on the caissons and anchored to the hard bedrock. In all, 3.5 years were necessary to build this bridge designed with the intent of achieving the maximum lightness possible.

64.3.3.2 Boulonnais Bridges

The three Boulonnais Viaducts are located on A16 highway which links Great Britain to the urban area of Paris, France, via the Channel Tunnel, and was completed in 1998. Their characteristics are as follows:

Name	Length, m	Span Distribution	Height above the Valley Floor, m
Quéhen	474	44.50 + 5 × 77.00 + 44.50	30
Herquelingue	259	52.50 + 2 × 77.00 + 52.50	25
Echinghen	1300	44.50 + 3 × 77.00 + 93.50	75
		5 × 110.00 + 93.50 + 3 × 77.00 + 44.50	

FIGURE 64.19 Boulonnais Bridges — pier transparency. (Courtesy of Jean Muller International.)

The foundations consist of diaphragm walls to a depth of 42 m. The typical pier is based on four diaphragm walls, whereas tallest piers are founded on eight diaphragm walls. These diaphragm walls were realized using drilling mud. Quantities are 3800 m of diaphragm walls, a third of which was excavated with a cutting bit; 10,000 m³ of concrete; 870 tons of reinforcing steel.

Each pier consists of two slender shafts, of diamond shape. These are linked on top by an aesthetically pleasing pier cap, on which the deck is supported (Figure 64.19).

The gap between the two pier shafts increases the bridge transparency created by the truss at deck level. The four tallest pier shafts are linked on their lower part by a transverse wall to increase the buckling stability.

The deck is a composite structure made of match-cast segments, assembled by cantilever method. The three bridges are formed by 524 segments. The deck structure consists of two prestressed concrete slabs, joined by four inclined V-shaped steel planes. Six inclined planes improve the transverse behavior of the deck near bridge supports.

The 23-cm-thick top slab is stiffened by four 70-cm-deep longitudinal ribs located in the diagonal planes. The top slab is prestressed transversely. The 27-cm-thick bottom slab is stiffened by longitudinal ribs and by two transverse beams per segment.

The deck is built by the cantilever method using a 132-m-long launching gantry weighing 500 tons. Segments, weighing 125 tons at the minimum, are put in place symmetrically in pairs. Imbalance between both cantilevers during erection never exceeds 20 tons.

FIGURE 64.20   Dole Bridge. (Courtesy of Campenon Bernard.)

The Echinghen Viaduct is located on a very windy site, a few kilometers from the Channel shore. Gusts of wind exceed 57 km/h 103 days a year, and 100 km/h 3 days a year. A project-specific calculation taking into account the turbulent wind was developed to study the bridge construction phases. This calculation led to imposition of very rigorous cantilever construction kinematics.

Moreover, a wind screen was designed for the windward side of the deck in prevailing wind to avoid very strict traffic limitations.

64.4. Long-Span Bridges

64.4.1 Girder Bridges

64.4.1.1 Dole Bridge

The Dole Bridge, completed in 1995, crossing the River Doubs in France, is 496 m long. It is a continuous seven-span box girder with variable depth. The typical span is 80 m long (Figure 64.20). The deck is erected by the balanced cantilever method using a traveling formwork.

The deck is a composite structure, 14.5 m wide, with two concrete slabs and two corrugated steel webs. The webs are welded to connection plates fixed to the top and bottom slabs by connection angles. Pier and abutment segments are strictly concrete segments.

The deck is longitudinally prestressed by three tendon families:

- Cantilever tendons, anchored on the top slab fillets: 12T15 tendons;
- Continuity tendons, located in the bottom slab in the central area of each span: 12T15 tendons;
- External prestressing, tensioned after completion of the deck, with a trapezoidal layout. The technology used allows removal and replacement of any tendon.

The Dole Bridge is the fourth bridge with corrugated steel webs erected in France.

FIGURE 64.21   Nantua Viaduct. (Courtesy of Campenon Bernard.)

64.4.1.2   Nantua Bridge

Nantua and Neyrolles Viaducts allow the A40 highway to link Geneva, Switzerland, to Macon, France. The Nantua viaduct is 1003 m long, divided in 10 spans. It was constructed in 1986. Its height above the ground varies from 10 to 86 m (Figure 64.21).

The western viaduct extremity is a 124-m-long span supported in a tunnel bored through the cliff. To balance this span, a concrete counterweight had to be constructed inside the cliff in a tunnel extension. The counterweight translates on sliding bearings of unusual size. The relatively large spans (approximately 100 m long) necessitated a variable-depth concrete box girder.

The construction principle for the deck is segments cast *in situ* symmetrically on mobile equipment. The 11.65-m-wide deck, for the first two-way roadway section of the highway, is longitudinally prestressed by cables located inside the concrete.

Various foundation methods were used, necessitated by differences in the soil bearing capacity.

64.4.2   Arch Bridges

64.4.2.1   Kirk Bridges

These concrete arch bridges were designed to provide a link between the Continent and the Isle of Kirk (former Yugoslavia). The two arches have spans of 244 and 390 m, respectively (Figure 64.22). The largest span represents a world record in its category. The box-girder arches are 8 m (width) × 4 m (height) and 13 m (width) × 6.50 (height), respectively.

The construction was carried out in two phases: In the first phase a box-girder arch, constituting the central part of the bridge, was made by using onshore precast segments. The assembly was performed by cantilevering from both banks by means of a mobile gantry (which was carried by the part of the arch already constructed) and of temporary stays. The use of precasting provided a better quality of concrete, a more precise tolerance of fabrication and reduced construction time. The keystone of the arch was likewise placed by means of a mobile gantry. The closure of the two

FIGURE 64.22 Kirk Bridges. (*Source:* Leonhardt, F., Ponts/Puentes — 1986 Presses Polytechniques Romandes. With permission.)

FIGURE 64.23 La Roche Bernard Bridge. (Courtesy of Campenon Bernard.)

semiarches was controlled by means of hydraulic jacks. The second phase of construction consisted of placing the lateral parts of the bridge, composed of large beams connected to the central arches. An *in situ* concreting of the joints between the precast segments and vertical and transversal prestressing ensure the monolithic integrity of the structure.

64.4.2.2 La Roche Bernard Bridge

La Roche Bernard Bridge, completed in 1996, is 376 m long and 20.80 m wide. It crosses the River Vilaine in Brittany, France, by an arch spanning 201 m and small approach spans (Figure 64.23).

The deck is a composite structure consisting of a steel box girder, 1.67 m deep with a trapezoidal shape, covered by a thin 23-cm-thick prestressed concrete slab. It is supported on four piers founded on the ground and six small piers fixed on the arch. The piers are spaced between 32 and 36 m. Like many other composite decks, the box girder is launched using a launching nose (20 m long); the slab is cast afterward. The concrete arch is 8 m wide with a height varying from 3.50 m at the springing to 2.90 m at the crown.

For the erection, the balanced cantilever process was applied using traveling formwork. Moreover, three temporary bents with 500 t jacks and two temporary pylons were successively used. The temporary bents were located below the segments S3 (the third), S5, or S15, and the temporary pylons were located on the riverbank or on the top of segment S15.

Except for segments S0 (springing segment) to S6 using the temporary, all other segments were erected by use of temporary pylons and temporary stays (11T15 and 13T15 units). The segments S7 to S13 were erected by means of stays fixed to the pylon on the riverbank and the temporary bent below S5.

The other segments S17 to 27 were erected by the use of stays fixed on the main pylon and by the use of bents below segments S5 and S15. The main pylon was placed on segment S15 and anchored in the previously erected segments.

While the number of stays fixed on the main pylon increased during erection, the number of stays on the other pylon decreased. Consequently, when the segment S20 was supported by the temporary stays, fixed to the main pylon, all stays on the other pylon had been removed.

64.4.2.3 Millau Bridge

To allow the highway A75, in France, to link two plateaus separated by the Tarn Valley five different crossings were designed. One of the proposals for traversing the 300-m-deep and 2500-m-wide valley was developed by JMI and consisted on a large arch and two approach viaducts. Two types of structures were designed for the deck: the basic scheme was based on a concrete box girder, while the alternative project was based on a steel–concrete composite box girder. Many features are common for the two designs, which is the reason only the basic project is described below:

The crossing is divided into three viaducts:

- The north approach viaduct: 486.50 m long, with four spans of between 66.50 and 168 m;
- The main viaduct: one arch spanning the 602 m over the river (Figure 64.24);
- The south approach viaduct: 1445.5 m long, with eight spans of 168 m and one shorter span of 101.50 m.

The 24 m wide roadway is carried by a 8-m-wide concrete box girder whose depth varies from 4 m at midspan to 10 m on pier, except at the central part of the arch where the depth is constant and equal to 4 m. Transversely, both 8-m-wide cantilevers are supported by struts, spaced 3.50 m. The box-girder webs are vertical and 500 mm thick. The bottom slab thickness decreases from 600 mm on pier to 300 mm at midspan.

For the approach viaducts and the first spans on the arch, the balanced cantilever method using traveling formwork is applied. Two families of PT are used: internal PT split in cantilever or continuity units and external PT for general continuity units.

Due to the great length of this bridge, an expansion joint is placed at midspan between P12 and P13, about 1500 m from the north abutment. This joint is equipped with two longitudinal steel girders simply supported on either side of the joint, which allow partial transfer of the bending moment and transfer of the shear force while reducing the deflections.

64.4.3. Truss Bridges

Bras de la Plaine Bridge

The future bridge, located on Isle of La Réunion, in the Indian Ocean (France), will span over the Bras de la Plaine valley which has highly inclined slopes (80°) and reaches a depth of 110 m.

The single-span prestressed composite truss deck, 270 m long, has an innovative static scheme: two cantilevers are restrained in counterweight abutments and linked at midspan by a hinge

FIGURE 64.24   Millau Bridge. (J. P. Houdry, Courtesy of Alain Spielmann.)

FIGURE 64.25   Bras de la Plaine Bridge. (Courtesy of Jean Muller International.)

(Figure 64.25). The deck structure, 17 m deep near the abutments and 4 m deep at midspan, comprises two concrete slabs linked by two inclined truss planes.

The upper 60 MPa (cylinder) concrete slab is 12 m wide. The lower 60 MPa (cylinder) concrete slab has a parabolic profile with variable thickness and width. Each truss panel consists of circular steel diagonals connected directly to the concrete slabs.

FIGURE 64.26 Theodor Heuss Bridge. (*Source*: Beyer, E., Bruckenbau, Beton Verlag, 1971. With permission.)

At midspan, four girders allow transmission of vertical and horizontal shear force and horizontal bending moment. The prestressing system is composed of internal tendons only located in the upper slab. Deck erection will begin by the end of 1999 using the standard cantilever erection method.

64.4.4. Cable-Stayed Bridges

64.4.4.1 Theodor-Heuss bridge

This bridge, also called "Northbridge", belongs to a family of three steel structures on the Rhine River in Düsseldorf, Germany. Northbridge is the first of the two, built in 1957, and belongs to the first generation of cable-stayed bridges.

This type of bridge was conceived to allow the crossing of large spans without intermediate ground support using cables to support the deck elastically in construction (Figure 64.26). The steel deck is 26.60 m wide and 476 m long divided into two approach spans of 108 m and the main span of 260 m. On the flooded riverbank, a five-span approach bridge extends the cable-stayed bridge. The four pylons are 41 m high, slender (1.90 long vs. 1.55 m wide) and spaced 17.60 m.

The main span is supported by four pairs of three cables fixed to the pylons. The three cables are parallel, set out like a harp in a single vertical plane and anchored in each edge of the deck with a spacing 36 m. Due to this spacing, the deck must be stiff, hence a depth of 3.14 m. This depth is extended further on to the approach bridge.

Regarding its erection, it was one of the first times that the balanced cantilever method was applied. The first cantilever segment of 36 m long was erected with the deck elastically supported with one pair of stays. The second segment and the others were erected at midspan.

64.4.4.2 Saint Nazaire Bridge

The bridge of St. Nazaire near the mouth of the Loire River in France, is approximately 3350 m long (Figure 64.27). It is composed of a central part, a 720-m cable-stayed steel bridge, and of two approach viaducts consisting, respectively, of 22 and 30 spans made up of precast concrete girders, each span being 50 m long.

The cable-stayed bridge has a central span 404 m long and two 158 m lateral spans. It is composed of steel box girders, 15 m wide. The construction of the cable stayed bridge, completed in 1975, was carried out in three phases.

FIGURE 64.27   Saint Nazaire Bridge. (Courtesy of Jean Muller International.)

1. The first phase consisted of the construction of the side spans. The steel box girders were assembled in the factory to pieces of 96 m. Then two segments each 96 m were assembled on site by welded joints and transported by two barges to be ultimately hoisted up to their final position.
2. In the second phase, the segments constituting the pylons were assembled on the bridge deck. Then the pylon was lifted by rotation to reach its definitive position.
3. The third phase consisted in erecting the central span as two cantilevers of 197.20 m of length with closure joint at midspan. The segments were lifted from barges with beam-and-winch system.

64.4.4.3 Brotonne Bridge

The Brotonne Bridge was designed to cross the River Seine downstream from Rouen, France (Figure 64.28). It was opened to traffic in 1977. It is composed of two approach viaducts and a cable-stayed structure with a 320-m-long central span. The deck consists of a prestressed concrete box girder 3.97 m deep and 19.20 m wide (Figure 64.29). The stays and the pylon (Figure 64.30) are placed in a single plane along the longitudinal axis of the bridge. The approaches and the main bridge were erected in the same way. In both cases a cantilever construction was used with success. The length of the segments was 3 m.

The segments were cast in place except for the webs which were precast and prestressed. The erection of the deck-girder consisted of extending the bottom slab form of the traveling formwork carried by the previous completed segment, then placing the precast webs that formed the basic shape and acted as a guide for the remaining traveling formwork. The webs were transported and lifted by a tower crane. Concerning the main bridge, the stays were tensioned in every two segments and were anchored in the top slab axis. For the segments, two inclined internal stiffeners were provided to transfer vertical loading generated by the stays. These stiffeners were prestressed.

FIGURE 64.28 Brotonne Bridge. (Courtesy of Campenon Bernard.)

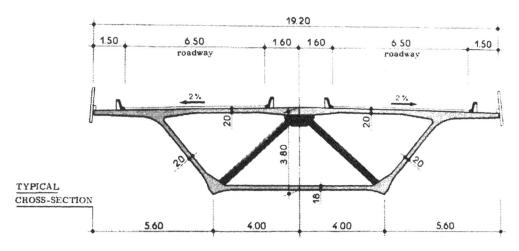

FIGURE 64.29 Brotonne Bridge — typical cross section. (Courtesy of Campenon Bernard.)

64.4.4.4 Normandie Bridge

Since 1994, the Normandie Bridge has allowed the A29 highway to pass over the River Seine near its mouth in northern France (see Figure 64.7). It is a cable-stayed bridge, 2141 m long with the following spans:

$$27.75 \text{ m} + 32.50 \text{ m} + 9 \times 43.50 \text{ m} + 96.0 \text{ m} + 856 \text{ m}$$
(longest cable stayed span in the world) $+ 96.00 \text{ m} + 14 \times 43.50 \text{ m} + 32.50 \text{ m}$

FIGURE 64.30   Brotonne Bridge — pylon base reinforcement. (Courtesy of Campenon Bernard.)

The central span is made of three parts: 116 m of prestressed concrete section, 624 m of steel section, and 116 m of prestressed concrete section.

The deck cross section is designed to reduce wind force on the bridge and to give a high torsional rigidity. At the same time its shape is adapted for both steel and concrete construction. It is 22.30 m wide and 3.0 m deep. The concrete deck is a three-cell box girder with two vertical webs and two inclined lateral webs. The steel deck is an orthotropic box girder constituted by an external envelope, stiffened by diaphragms and by trapezoidal stringers.

The A-shaped concrete pylon is extended by a vertical part where stays are anchored (Figure 64.31).

Three different construction methods were used for the Normandie Bridge erection:

1. The approach spans (southern approach 460 m, northern approach 650 m) were put in place by the incremental launching method from the embankment, using a launching nose.
2. On both sides of the 200-m-tall pylons, the superstructure was built by the cable-stayed balanced cantilever method with segments cast *in situ* in a traveling formwork. From the 90-m-long cantilevers, the 96-m side span was joined to the incrementally launched spans. Then the construction of the concrete deck was finalized with an additional 20 m of cast *in situ* cable-stayed cantilever on the river side.
3. The central part of the main span was erected by 19.65-m-long steel segments supplied by barge, lifted up by crane, and finally welded to the previous segment. A pair of cables was tensioned before moving the crane to lift the following segment.

The bridge foundations are the following:

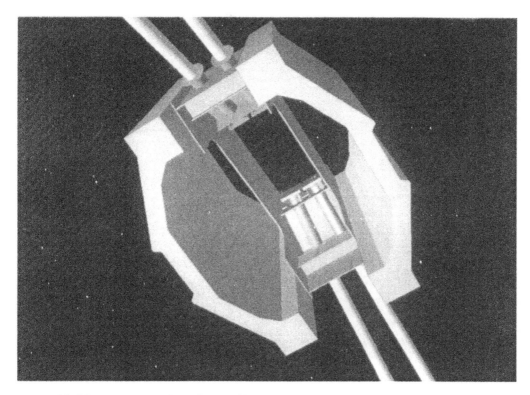

FIGURE 64.31 Normandie Bridge — cable stay anchorages. (Courtesy of Campenon BErnard.)

- Piers and abutments are founded on 1.50-m-diameter, 40-m-deep bored piles — four or five piles per pier.
- The towers are founded on 2.10-m-diameter, 50-m-deep bored piles — 14 piles for each pylon leg.

64.4.4.5 Bi-Stayed Bridge

The clear span of a conventional cable-stayed bridge is limited by the capacity of the deck to resist the axial compressive loads near the pylons created by the horizontal component of the stay forces. For the current materials (70 MPa high-strength concrete, for example), the limit span is between 1200 and 1500 m, depending upon the imagination and the boldness of the designer. Beyond this limit, only suspension bridges allow spanning very large crossings. This situation has now changed, thanks to the new so-called bi-stayed concept.

Deck construction still proceeds in the same fashion as for conventional cable-stayed bridges; starting from the pylons outwardly in a symmetrical sequence, the deck is suspended by successive stays. At a certain stage of construction [for a deck length equal to "a_1," Figure 64.32: (13a)] on either side of each pylon, for example), the deck axial load will have absorbed the full capacity of the materials (with provision for the future effect of live loads). No additional deck length may be added, without exceeding the allowable stresses.

At this stage, a second family of stays is installed [(Figure 64.32 (13b)], assigned to suspend the center portion of the main span. These additional stays are symmetrical with one another with regard to the main span centerline and no more with regard to the pylon. Furthermore, they are no longer anchored in the deck itself, but rather in outside earth abutments at both ends of the bridge, much in the same way as the main cables of a suspension bridge. The vertical load assigned to each stay is now balanced along a continuous tension chain, made up of the center portion of

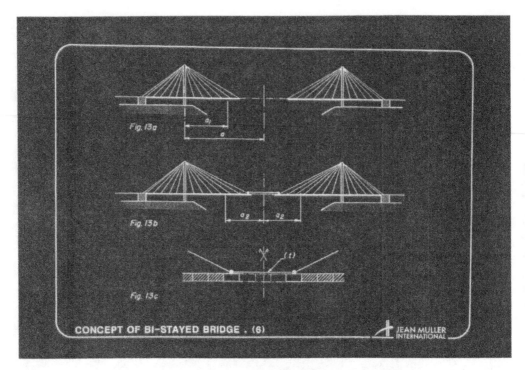

FIGURE 64.32   Bi-stayed bridge. (Courtesy of Jean Muller International.)

the deck (subjected to tension loads), associated with two symmetrical stays, deviated above the pylon heads, to be finally anchored outside the bridge deck.

Along the deck, an axial compression load appears in the vicinity of the pylons (created by the first family of stays), changed into a tension axial load at the centerline of the main span (created by the second family of stays).

In this first application of the new concept, one may increase the maximum clear span in the ratio $(a_1 + a_2/a_1)$, i.e., about 1.5.

In fact, it is possible to go much beyond that stage, while improving the quality of the structure, by using prestressing [Figure 64.32 (13c)]. On the deck length suspended by the second family of stays, prestressing tendons are installed to offset at least all axial tension forces due to dead and live loads. When no live load is applied, the deck is subjected to a compression load, which vanishes when the bridge is fully loaded.

With the usual proportions of dead to live loads, it is easily demonstrated that the maximum span length can be multiplied by 2.5. One can now consider with confidence the construction of a clear span of 3000 m.

A practical example of the new concept was prepared for an exceptional crossing in Southeast Asia with a 1 200 m clear main span. The deck carried six lanes of highway traffic, two train tracks, and two special lanes for emergency vehicles. The bridge was also subject to typhoons.

64.5. Large Projects

64.5.1. Second Severn Bridge

The second Severn Bridge provides a faster link between England and Wales. The structure, 5126 m long (Figure 64.33), consists of three parts: the eastern viaduct, 2103 m long; the main cable-stayed bridge, 946.6 m long; and the western viaduct, 2077 m long. From east to west the bridge span lengths are:

FIGURE 64.33   Second Severn Bridge at twilight time. (Courtesy of G.T.M.)

32 m, 58 m, 23 × 98.12 m, 456 m, 23 × 98.12 m, 65 m

1. The approach viaducts are founded on multicellular reinforced concrete caissons, one per pier, precast on land, with a weight varying from 1100 to 2000 according to the piers. These caissons are transported by barge from the precasting yard to the relevant site. The barge is equipped with a pair of crawler tractors of 1500-ton loading capacity.

Of the 47 concrete piers, 38 are precast, representing 338 concrete elements. Three to seven elements joined by wax-grouted vertical prestressing are necessary to build one pier. Two rectangular pier shafts are erected on each foundation caisson.

The approach viaduct deck consists of two parallel monocellular prestressed concrete box girders connected by the upper slab to provide a 33.20-m-wide platform on most of the bridge length. The deck depth varies from 7.0 m on pier to 3.95 m in the span central part.

The approach viaduct deck is divided into about 500-m-long sections. Expansion joints are located at midspan. The typical deck section consists of four spans and two cantilevers supported by five piers.

All spans are made of 3.643-m-long precast segments; these match-cast segments are put in place by the balanced cantilever method with epoxy joints. For this construction a 230-m-long launching gantry weighing 850 tons is used. All prestressing cables are external with all tendons individually protected in wax-grouted HDPE (high density polyethylene) sheaths. Four prestressing cable families can be distinguished:

- Cantilever tendons: 11 to 12 pairs per cantilever
- Continuity bottom tendons: 3 to 4 pairs per span
- Continuity top tendons: 1 to 2 pairs for span
- General continuity tendons: 5 to 6 pairs per span, spread over two spans

2. The bridge environment is particularly constraining: the Severn estuary is subject to the second strongest tides in the world which represents a differential capable of exceeding 14 m, with strong currents of 8 to 10 knots in certain places and occasionally strong winds. Furthermore, 80% of the foundations are exposed at low tide.

This means that the key to this challenge of the tides is a maximum use of prefabricated components. That explains choices made for the approach viaducts: precasting of foundation caissons, piers at sea, deck segments. That explains as well the main bridge pylon cross-beam and the anchorage block precasting, and precasting of the cable-stayed bridge deck elements.

3. The cable-stayed bridge is 946.60 m long. It is a symmetric work with the following span lengths: 49.06 m, 2 × 98.12 m, 456 m, 2 × 98.12 m, 49.08 m. The bridge towers are founded on precast multicellular caissons. Each 137-m-tall tower consists of two rectangular hollow concrete shafts: reinforced concrete for the typical section and prestressed concrete for the stay anchorage area.

Each pylon caisson is equipped with a 45 m³/h capacity ready-mix plant, and two metallic platforms to store reinforcement and formwork. This equipment allows one to give maximum autonomy to pylon teams. Pylon shafts are concreted *in situ* with a climbing formwork in 3.80-m-long sections. The cross-beams are precast on land and weigh 1300 and 900 tons, respectively. The lower cross-beam is lifted in place by a crane barge and then linked by concreting to the pylon legs. The upper tie beam is lifted and put down on the lower one, and then lifted to its definite position by jacking.

The first cross-beam is located at a 40-m height above the highest tide, the other forming a frame on the level of the stay-cable anchorage area. The main bridge deck is simply supported on the lower pylon cross-beams with transverse stops. It is supported on four secondary piers on both side spans with antiuplift bearings, and last simply supported on the access viaduct extremities.

The deck is a composite structure consisting of

- Two 2.50-m-deep I girders linked every 3.65 m by a truss beam. The distance between the two main girders is 25.2 m.
- A reinforced concrete slab about 35 m wide and 20 cm or 22.5 cm thick for a typical section.

The deck is assembled of 128 precast elements, 34.60 m wide and 7 m long. The steel structure is assembled by bolts at the precasting yard; then the concrete slab is poured except at the connection joint between two consecutive segments. Each standard segment weighing about 170 t is positioned by trailer and transported by barge to the site. The deck segments are lifted and positioned by a pair of mobile cranes located at the end of each cantilever, and bolted to the previous segment. Then two stays are tensioned and the joint with the previous segment is concreted.

The bridge deck is supported by four stay planes, each made up of 60 stays from 19 to 75 T15 strands with a length varying from 35 to 243 m.

4. The second Severn River crossing bridge provides three traffic lanes in each direction, emergency lanes, safety barriers, and lateral wind screens. The construction of this new bridge is financed by private sector. The existing and the new toll bridges are managed by a concessionary group which takes responsibility for design, construction, financing, operating, and maintenance of both bridges.

Over 2 years of study and 4 years of work on site, challenged by the extreme tides, were necessary to build this bridge, located 5 km downstream from the suspension bridge erected in 1966, 30 years earlier.

64.5.2. Great Belt Bridges

The construction of the fixed link across the Great Belt Strait is a bridge and a tunnel project of exceptional dimensions. The Great Belt fixed link consists of three major projects:

FIGURE 64.34   Great Belt Bridges — the railway tunnel. (Courtesy of Jean Muller International.)

1. The railway tunnel under the eastern channel between Zealand and the island of Sprogø, in the middle of the Belt. It is a bored tunnel comprising two single-track tubes each with an internal diameter of 7.7 m and an external diameter of 8.50 m (Figure 64.34). The total tunnel length is 8 km.

Four 220-m-long boring machines have worked down to 75 m below sea level. The twin tunnel tubes are lined with interlocking concrete rings made of precast concrete segments each of a width of 1.65 m in the direction of boring. Each ring consists of six circle segments and a smaller key segment. A total of 62,000 tunnel segments were manufactured.

The twin tunnel tubes are connected at 250-m intervals by cross-passages with an internal diameter of 4.5 m, lined with cast iron segments assembled as rings. Each cross-passage consists of 22 rings of each 18 elements.

The railway tunnel is the second longest underwater bored tunnel, the tunnel beneath the English Channel being the longest.

2. The highway bridge across the eastern channel is 6790 m long. It consists of a suspension bridge with a 1624-m-long main span (see Figure 64.8) and two 535-m-long side spans, and of 23 approach spans totaling 4096 m (14 spans for the eastern approach and 9 spans for the western approach).

The bridge towers are founded on concrete caissons weighing 32,000 tons, placed on the seabed. The two legs of the pylons are cast in climbing formwork from the base to the pylon top 254 m above sea level. Cross-beams interconnect the pylon legs at heights of 125 and 240 m.

The anchor blocks for the suspension cables are also founded on concrete caissons weighing 55,000 tons. The rest of the two anchor blocks, including the special distribution chambers in which the main cables are anchored, are cast *in situ* by a conventional method to the top height of 63.4 m above sea level.

Among the bridge piers, the most part, i.e., 18, are prefabricated. Each pier consists of three elements: a caisson, a lower pier shaft, and finally a top pier shaft. The bridge piers weigh 6000 t

FIGURE 64.35   Great Belt Bridges — 193-m-long approach span. (Courtesy of Cowi Consult.)

on average. Conventional floating cranes are used for assembly of both the caissons and the pier shafts.

The steel superstructure of the main span comprises a fully welded box girder 4 m deep and 31 m wide. After floating the 48-m-long segments to a position under the main cables, they are hoisted into place by winches, and then welded to the previous section.

The two main cables each have diameter of 85 cm and a length of approximately 3 km. Each main cable includes 148 cables, and each cable includes 126 wires with a diameter of 5.13 mm; 20,000 tons of the steel representing a length of 112,000 km constitute the suspension of the bridge.

The steel superstructure of the approach spans comprises a fully welded box girder with a constant girder depth of 6.7 m, a width of 26 m, and a typical span of 193 m (Figure 64.35). The cross section has the same wing shape as the main span girder. The steel girders, each weighing about 2300 tons, are hoisted from a barge by a large floating crane.

The steel panels for the road girders are manufactured in Italy and then shipped from Livorno to Sines, Portugal. Here they are processed into bridge sections, which are floated to Aalborg (Northern Denmark) and welded together into complete bridge spans.

3. The combined road and railway bridge crosses the western channel between Funen and Sprogø. This west bridge is a 6.6-km-long all-concrete bridge with separate decks for rail and highway traffic. The bridge consists of six continuous bridge sections of a length of about 1100 m; the individual bridge sections are linked by expansion joints and hydraulic dampers that transmit only instantaneous forces.

The box girder underneath the rail track is only 12.3 m wide, compared to the roadway girder width of 24.1 m. However, the railway girder is 1.36 m deeper than the roadway girder.

The piers of the west bridge are founded on precast caissons. Each caisson receives two pier shafts, one for roadway girder, the other for railway girder. Each of the 110.4-m-long girder elements is cast in fixed steel shuttering in five sections. These sections are progressively linked by prestressing at the precasting yard. A special vessel, *Svanen*, a self-propelled floating crane with a lift capacity of 7123 tons, was used to transport the foundation caissons to the relevant position, to lift the pier shafts into place and to place the 110 m long girders on the top on the pier shafts.

In addition to the bridge and tunnel sections, the Great Belt fixed link, opened to traffic in 1998, also includes new road and railway sections on land, connecting the existing highways and railways with the fixed link.

64.5.3. Tagus River Bridges

The Tagus Bridge, also named the Vasco da Gama Bridge, is a 17.2-km-long structure connecting the northern and southern banks of the Tagus estuary. This project will solve a great part of the traffic problems in Lisbon by creating a link between new highway systems in the north and in the south of the city, and makes the traffic flow more easily between the northern and the southern parts of Portugal (Figure 64.36).

The Vasco da Gama project is divided into seven distinct sections, five of which are bridges and viaducts, representing 12.3 km.

1. The northern viaduct, with a total length 488 m of 11 spans, crosses the northern railway line of the Portuguese Railway Company (C.P.) and several local junctions. The deck width is variable to accommodate connection to local roads by slip roads. Span lengths vary from 42 to 47 m. The deck, 3.50 m deep, is cast *in situ* span by span. The typical span is 29.3 m wide and made up of four T-shaped concrete beams.

2. The Exhibition viaduct, with a total length 672 m of 12 spans, is also situated on the northern bank of the Tagus. It crosses the area where the 1998 World Exhibition took place. The bridge span lengths are the following: 2×46.2 m, 3×52.3 m, 55.3 m, 6×61.3 m. The deck, 29.3 m wide, is made up of twin prestressed concrete box girders, connected by the upper slab. Each box girder consists of precast segments put in place by mobile cranes using the balanced cantilever method. After pouring the cantilever closure joints, external prestressing is tensioned inside the box girders to ensure deck continuity.

The deck is supported by concrete piers founded on 1700-mm-diameter piles through a 4.5-m-thick concrete pile cap.

3. The main bridge is a cable-stayed bridge, 829 m long with a 420-m-long main span. The H-shaped pylons are founded on 2.2-m-diameter piles; these 44 bored piles are 53 m long. A very stiff and robust pile cap allows the foundation to withstand impact from a 30,000 ton vessel traveling at 8 knots.

The pylons comprise two legs and reach a height of 150 m. These legs, with a cross section varying from 12×7.7 m to 5.5×4.7 m, are slip formed. They are linked by a 10-m-deep prestressed box girder at base level, and by a transverse cross beam 87 m above the base, poured *in situ* in four stages. The upper part of the pylon, above the cross-beam, consists of a composite steel–concrete structure in which stays are anchored.

The deck, 31.28 m wide, consists of two longitudinal 2.50-m-deep and 1.30-m-wide concrete girders, connected every 4.4 m by 2.0-m-deep steel cross-beams. This composite structure is completed by a 25-cm concrete top slab (Figure 64.37). The 8.83-m-long deck segments are cast *in situ* using traveling formworks by the balanced cantilever method. Two points should be noticed: during segment concreting the final stays are used as temporary stays and the traveling formwork is designed to pass beyond the rear piers.

Like all the other structures, the main bridge is designed to withstand violent earthquake effects without damage. Consequently, there is no fixed link between the deck and its supports. Dampers are installed, steel dampers transversely, and longitudinally steel dampers outfitted with hydraulic couplers. On top of that, damping guide deviators are placed at each end of the cable stays.

4. The central viaduct is 6531 m long of 80 spans and cross the Tagus estuary above sandbanks and two shipping channels. The deck for the most part of its length is less than 14 m above sea

FIGURE 64.36   Tagus Bridge — artist's view. (Courtesy of Campenon Bernard.)

level. The typical span is 78.62 m long, but over the two shipping channels the span length rises to 130 m and the height above sea level rises to over 30 m. The viaduct span lengths are the following: 79.62 m, 3 × 78.62 m/93.53 m, 130.00 m, 93.53 m/60 × 78.62 m/93.53 m, 130.00 m, 93.53 m/11 × 78.62 m.

The deck, 29.3 m wide, consists of two parallel prestressed concrete box girders with two webs of constant height of 3.95 m, connected by the upper slab. Over the shipping channel, the girder depth is variable from 3.95 m at midspan to 7.95 m on piers.

Every span is precast in eighths; these segments, with a unit weight about 240 tons, are assembled on a bench by prestressing after adjustment. Each 1800 ton to 2000 ton girder is lifted and stored by a gantry crane with 2200 tons capacity load.

To transport and place the girders at up to 50 m above sea level, a special catamaran is used, equipped with two cranes with a capacity of 1400 tons at a radius of 25 m, 82 m tall. The rhythm of transport and placing is at a standard rate of one beam every 2 days. Prestressing cables ensure

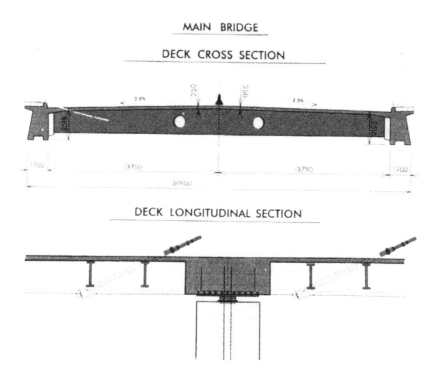

FIGURE 64.37 Tagus Bridge — main bridge deck cross section. (Courtesy of Campenon Bernard.)

longitudinal continuity of the deck. After concreting of continuity slab between the two parallel box girders, the transverse prestressing cables are tensioned. The deck is supported by concrete piers founded on 1700-mm-diameter deep piles.

5. The southern viaduct comprises 85 spans, each 45 m long, totaling 3825 m. As in the northern viaduct, the deck is composed by four T-shaped 3.50-m-deep concrete girders. The deck is cast *in situ* span by span with four mobile casting gantries working above the deck on two casting fronts. The deck is supported by concrete piers founded on bored piles for the land piers and on driven piles for the river piers.

The Vasco da Gama Bridge construction began in February 1995 and was finished in March 1998. It was privately financed and represents a cost of approximately $ 1 billion.

64.6 Future European Bridges

Future trends in bridge design can be classified in four categories:

1. Development of existing materials
2. Development of new materials
3. New structural association of materials
4. Structural control

1. The main materials used for bridges — concrete and steel — are still under development; their strength is always increasing. High-performance concrete (HPC) has been used for bridges for the first time in France and in the Scandinavian countries. Concrete with a compressive strength of 60 MPa (cylinder) at 28 days is becoming common for large bridges, especially for long spans and high piers. However, the advantage of HPC is not only strength, but durability, because this concrete is

much more compact and much less porous than ordinary concrete. A new type of concrete called reactive powder concrete (RPC) is being developed in France; its compressive strength at 28 days can reach 200 to 800 MPa. It is meant to be prestressed and does not include any passive reinforcement. High-strength steel with yield stress of 420 to 460 MPa has been used for bridges in Germany, Finland, France, Luxembourg, Norway, the Netherlands, and Sweden. It is used mostly for long-span bridges, and for parts of the bridge that are submitted to high concentrated forces.

2. New composite materials are being developed for bridges in Europe. The main ones are:

- Glass fiber–reinforced plastic (GFRP),
- Carbon fiber–reinforced plastic (CFRP),
- Aramide fiber–reinforced plastic (AFRP).

Their main advantages are high corrosion resistance and light weight, whereas they are still more expensive than steel.

GFRP bars and cables have been developed since 1980 in Germany, in Austria, and in France. At least five bridges have been built in Germany using GFRP prestressing cables. The Fidget Footbridge in England includes GFRP reinforcing bars. CFRP stays have been used in Germany. AFRP stays have been used for pedestrian bridges in Holland and in Norway.

New composite materials have also been used for the deck structure itself: Bends Mill movable bridge in England, Arnhem Footbridge in Holland. The Aberfeldy Footbridge, in Scotland, is the world's first all-composite bridge: deck, pylons, and stays.

FIGURE 64.38   Chavanon Bridge. (Courtesy of Jean Muller International.)

3. The association of steel I-girders with a concrete slab has become very common for medium-span bridges. We think that this association of steel and concrete will be developed for a large variety

of composite structures in the future: truss decks, arches, pylons, piers. A number of such innovative projects have been built in France and Switzerland, for example. The use of each material to its best capability will lead to more efficient and economical structures. A significant example of such a structure is implemented on the highway A89 which will link Clermond Ferrand to Bordeaux in southern France. To cross the deep valley of the River Chavanon respecting the natural environment, a suspension bridge is being built (Figure 64.38). The bridge deck is a steel concrete composite structure 22.4 m wide and 3.0 m deep. It is suspended by a single plane of suspension cables located at the cross section axis. The inverted V-shaped pylon straddles over the deck and leaves it free of any support. Its top is 52 m above the deck. This bridge, with a 300-m-long main span is an innovative, efficient, and very aesthetic projec4. With the development of high-strength materials, and possibly lightweight materials, bridges will become more and more slender and light, hence more sensitive to fatigue and dynamic problems, especially for long-span bridges. Consequently, it will become necessary to control vibrations due to traffic loads, wind, and earthquakes. This control can be achieved through passive devices, such as dampers, and active devices such as active pre-stressing tendons, active stays, active aerodynamic appendages. This will be the road toward "intel-ligent" bridges of the future...

An existing bridge could easily be equipped with an active device. Such a device was implemented in the Rogerville Viaduct, opened to traffic in 1996, located in northern France, on highway A29 not far from the Normandie Bridge (Figure 64.39). It is a continuous steel box girder, placed across the expansion joint, between two adjacent cantilever arms (Figure 64.40). It rests on two diaphragms on either side and may be adjusted before the bridge is opened to traffic to transfer shear force and bending moment, and consequently to compensate subsequent effects of steel relaxation and con-crete creep. At the moment, the continuity girder is a passive device

This connection could be equipped to transfer (long term under dead load, and short term under live load), shear force and moment in an active fashion at all times (Figure 64.41). In other words, the magnitude of shear load and moment across the joint may be monitored and adjusted at the designer's request to restore all the geometric and mechanical properties of a continuous deck across the expansion joint.

FIGURE 64.39   Rogerville Viaduct. (Courtesy of Jean Muller International.)

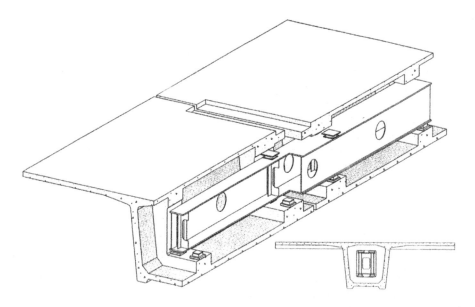

FIGURE 64.40   Rogerville Viaduct — expansion joint device. (Courtesy of Jean Muller International.)

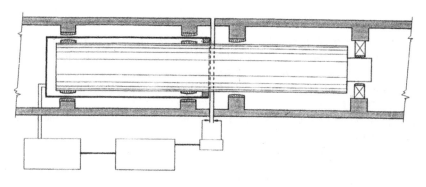

DISPLACEMENT MONITORING

THE ACTIVE CONNECTION

FIGURE 64.41   Active connection. (Courtesy of Jean Muller International.)

References

1. Eurocode 1. ENV 1991 Basis of design and actions on structures: Part 1 — Basis of design; Part 2 — Actions on structures; Part 3 — Traffic loads on bridges; Part 4 — Actions on silos and tanks; Part 5 — Actions due to cranes, traveling bridge cranes, and machinery.

    Experimental European norm XP ENV 1991-1 April 1996 Eurocode 1, *Basis of design and actions on structures.*

    Experimental European norm XP ENV 1991-2-1 October 1997 Eurocode 1, *Actions on structures: voluminal weights, self-weights, live loads.*

    Experimental European norm XP ENV 1991-2-2 December 1997 Eurocode 1, *Actions on structures exposed to fire.*

    European norm project P 06-102-2 March 1998 Eurocode 1. *Actions on structures exposed to fire.*

    Experimental European norm XP ENV 1991-2-3 October 1997 Eurocode 1. *Actions on structures: snow loads.*

    Experimental European norm XP ENV 1991-3 October 1997 Eurocode 1, *Traffic loads on bridges.*

    European norm project P 06-103 March 1998 Eurocode 1, *Traffic loads on bridges.*

2. Eurocode 2, ENV 1992 Concrete structure design; ENV 1992-1-1 General rules and rules for buildings; ENV 1992-1-2 Resistance to fire calculation; ENV 1992-1-3 Precast concrete elements and structures; ENV 1992-1-4 Lightweight concrete; ENV 1992-1-5 Structures prestressed by external or unbonded tendons; ENV 1992-1-6 Nonreinforced concrete structures; ENV 1992-2 Reinforced and prestressed concrete bridges; ENV 1992-3 Concrete foundations; ENV 1992-4 Retaining structures and tanks; ENV 1992-5 Marine and maritime structures; ENV 1992-6 Massive structures; ENV 1992-X Post-tension systems

    European norm project P 18-711 December 1992 Eurocode 2, *Concrete structure design.*

    Experimental European norm XP ENV 1992-1-3 May 1997 Eurocode 2, *General rules. Precast concrete elements and structures.*

    Experimental European norm XP ENV 1992-1-5 May 1997 Eurocode 2, *General rules. Structures prestressed by external or unbonded tendons.*

    European norm project P 18-712 March 1998 Eurocode 2, *General rules. Resistance to fire calculation.*

3. Eurocode 3, ENV 1993 Steel structure design; ENV 1993-1-1 General rules and rules for buildings; ENV 1993-1-2 Resistance to fire calculation; ENV 1993-1-3 Formed and cold-rolled element use; ENV 1993-1-4 Stainless steel use; ENV 1993-2 Plate-shaped bridges and structures; ENV 1993-3 Towers, masts, and chimneys; ENV 1993-4 Tanks, silos, and pipelines; ENV 1993-5 Piles and sheet piles; ENV 1993-6 Crane structures; ENV 1993-7 Marine and maritime structures; ENV 1993-8 Agricultural structures.

    Experimental European norm project P 22-311 December 1992 Eurocode 3, *Steel structure design.*

    Experimental European norm XP ENV 1993-1-2 December 1997 Eurocode 3, *General rules. Behavior under the action of fire.*

4. Eurocode 4, ENV 1994 — Steel–concrete composite structure design; ENV 1994-1-1 General rules for buildings; ENV 1994-1-2 Resistance to fire calculation; ENV 1994-2 Bridges.

    Experimental European norm project P 22-391 September 1994 Eurocode 4, *Steel–concrete composite structure design and dimensioning.*

    Experimental European norm XP ENV 1994-1-2 December 1997 Eurocode 4, *General rules. Behavior under the action of fire.*

    European norm project PROJECT P 22-392 March 1998 Eurocode 4, *General rules. Behavior under the action of fire.*

5. Eurocode 5, ENV 1995 Wooden work design; ENV 1995-1-1 General rules and rules for buildings; ENV 1995-1-2 Resistance to fire calculation; ENV 1995-2 Wooden bridges.

   Experimental European norm XP ENV 1995-1-1 August 1995 and AMDT.1 February 1998 Eurocode 5, *Wooden work design. General rules.*

   European norm project PROJECT P 21-712 March 1998 Eurocode 5, *General rules. Behavior under the action of fire.*

6. Eurocode 6, ENV 1996 Masonry structure design; ENV 1996-1-1 General rules and rules for reinforced or nonreinforced masonry; ENV 1996-1-2 Resistance to fire calculation; ENV 1996-1-X Cracking and deformation checking; ENV 1996-1-X Detailed rules for lateral loads; ENV 1996-1-X Complex-shaped section in masonry structures; ENV 1995-2 Guide for design, material choice, and construction of masonry structures; ENV 1995-3 Simple and simplified rules for masonry structures; ENV 1995-4 Masonry structures with low requirements.

   Experimental European norm XP ENV 1996-1-2 December 1997 Eurocode 6, *General rules. Behavior under the action of fire.*

   European norm project P 10-611B March 1998 Eurocode 6, *General rules. Rules for reinforced or nonreinforced masonry.*

   European norm project P 10-612 March 1998 Eurocode 6, *General rules. Behavior under the action of fire.*

7. Eurocode 7, ENV 1997 Geotechnical design; ENV 1997-1 General rules; ENV 1997-2 Laboratory test norms; ENV 1997-3 Sampling and test *in situ* norms; ENV 1997-4 Additional rules for special elements and structures.

   Experimental European norm XP ENV 1997-1 December 1996 Eurocode 7, *Geotechnical design. General rules.*

   European norm project PROJECT P 94-250-1 May 1997 Eurocode 7, *Geotechnical design. General rules.*

8. Eurocode 8, ENV 1998 Earthquake-resistant structure design and dimensioning; ENV 1998-1-1 General rules: seismic actions and general requirements for structures; ENV 1998-1-2 General rules: general rules for buildings; ENV 1998-1-3 General rules: special rules for various elements and materials; ENV 1998-1-4 General rules: Building strengthening and repairing; ENV 1998-2 Bridges; ENV 1998-3 Towers, masts, and chimneys; ENV 1998-4 Silos, tanks, and pipes; ENV 1998-5 Foundations, retaining structures, and geotechnical aspects.

   European norm project PROJECT P 06-031-1 March 1998 Eurocode 8, *Earthquake-resistant structure design and dimensioning.*

9. Eurocode 9, ENV 1999 Aluminum alloy structure design; ENV 1999-1-1 General rules and rules for buildings; ENV 1999-1-2 Additional rules for aluminum alloy structure design under the action of fire; ENV 1999-2 Rules concerning fatigue.

65

Design Practice in Japan

Masatsugu Nagai
Nagaoka University of Technology

Tetsuya Yabuki
University of Ryukyu

Shuichi Suzuki
Honshu-Shikoku Bridge Authority

65.1 Design

Tetsuya Yabuki

65.1.1 Design Philosophy

In the current Japanese bridge design practice [1], there are two design philosophies: ultimate strength design and working stress design.

1. Ultimate strength design considering structural nonlinearities compares the ultimate load-carrying capacity of a structure with the estimated load demands and maintains a suitable ratio between them. Generally, this kind of design philosophy is applied to the long-span bridge structures with spans of more than 200 m, i.e., arches, cable-stayed girder bridges, stiffened suspension bridges, etc.
2. Working stress design relies on an elastic linear analysis of the structures at normal working loads. The strength of the structural member is assessed by imposing a factor of safety between the maximum stress at working loads and the critical stress, such as the tension yield stress

0-8493-7434-0/00/$0.00+$.50
© 2000 by CRC Press LLC

TABLE 65.1   Loading Combinations and Their Multiplier Coefficients
for Allowable Stresses

No.	Loading Combination	Multiplier Coefficient for Allowable Stresses
1	P + PP + T	1.15
2	P + PP + W	1.25
3	P + PP + T + W	1.35
4	P + PP + BK	1.25
5	P + PP + CO	1.70 for steel members
		1.50 for reinforced concrete members
6	W	1.2
7	BK	1.2
8	P except L and I + EQ	1.5
9	ER	1.25

of material and the shear yield stress of or the compression buckling stress of material (see Section 65.1.4).

65.1.2   Load

The Japanese Association of Highways, the Standard Specification of Highway Bridges [1] (JAH-SSHB) defines all load systems in terms four load systems as follows:

1. Primary loads (*P*) — dead load (*D*), live load (*L*), impact load (*I*), prestressed forces (*PS*), creep (*CR*), shrinkage (*SH*), earth pressure (*E*), hydraulic pressure (*HP*), uplift force by buoyancy (*U*).
2. Secondary loads (*S*) — wind load (*W*), thermal force (*T*), effect of earthquakes (*EQ*).
3. Particular loads corresponding to the primary load (*PP*) — snow load (*SW*), effect of displacement of ground (*GD*), effect of displacement of support (*SD*), wave pressure (*WP*), centrifugal force (*CF*).
4. Particular loads (*PA*) — raking force (*BK*), tentative forces at the erection (*ER*), collision force(CO), etc.

The combinations of loads and forces to which a structure may be subjected and their multiplier coefficients for allowable stresses are specified as shown in Table 65.1. The most severe combination of loads and forces for a structure within combinations given in Table 65.1 is to be taken as the design load system. Details on the loads have not been given here and the reader should refer to the specification [1].

Limiting values of deflection are expressed as a ratio of spans for the individual superstructure types and span lengths.

65.1.3   Theory

In most cases, design calculations for both concrete and steel bridges are based on the assumptions of linear behavior (i.e., elastic stress–strain) and small deflection theory. It may be unreasonable, however, to apply linear analysis to a long-span structure causing the large displacements. The JAH-SSHB specifies that the ideal design procedure including nonlinear analyses at the ultimate loads should be used for the large deformed structure.

Bridges with flat stiffening decks raise some anxieties for the wind resistance. The designer needs to test to ensure the resistances for wind forces and/or the aerodynamic instabilities. In Japan, wind tunnel model testing including the full model and sectional model test is often applied for these verifications . The methods of model testing include full-model tests and sectional-model tests. The vibrations induced by vehicles, rain winds, and earthquakes are usually controlled by oil dampers, high damping rubbers, and/or vane dampers.

65.1.4 Stability Check

The JAH-SSHB specifies the strength criteria on stabilities for fundamental compression, shear plate, and arch/frame elements. The strength criteria for the stability of those elements are presented as follows:

1. Compressive strength for plate element — Fundamental plate material strength under uniform compression is mentioned here but details have not been given in all cases.

$$
\begin{aligned}
\frac{\sigma_{cl}}{\sigma_Y} &= 1.0 & \text{for } R \le 0.7 \\[2mm]
&= \frac{0.5}{R^2} & \text{for } 0.7 \le R
\end{aligned}
\tag{65.1}
$$

where σ_{cl} = plate strength under uniform compression, R = equivalent slenderness parameter defined as

$$
R = \frac{b}{t}\sqrt{\frac{\sigma_Y}{E}} \cdot \sqrt{\frac{12\left(1-\mu^2\right)}{\pi k^2}}
\tag{65.2}
$$

and b = width of plate, t = thickness of plate, μ = Poisson's ratio, k = coefficient applied in elastic plate buckling.

2. Compressive strength for axially loaded member — The column strength for overall instability is specified as

$$
\begin{aligned}
\frac{\sigma_{cg}}{\sigma_Y} &= 1.0 & \text{for } \bar{\lambda} \le 0.2 \\[2mm]
&= 1.109 - 0.545\bar{\lambda} & \text{for } 0.2 \le \bar{\lambda} \le 1.0 \\[2mm]
&= 1.0/\left(0.773 + \bar{\lambda}^2\right) & \text{for } 1.0 \le \bar{\lambda}
\end{aligned}
\tag{65.3}
$$

where σ_{cg} = column strength, σ_Y = yield-stress level of material, $\bar{\lambda}$ = slenderness ratio parameter defined as follows

$$
\bar{\lambda} = \frac{1}{\pi}\sqrt{\frac{\sigma_Y}{E}} \cdot \frac{\ell}{r}
\tag{65.4}
$$

and r = radius of gyration of column member and ℓ = effective column length. Thus, the ultimate stress of axially effective material, σ_c, is specified as

$$
\sigma_c = \frac{\sigma_{cg} \cdot \sigma_{cl}}{\sigma_Y}
\tag{65.5}
$$

3. Bending compressive strength — The ultimate strength for bending compression is specified, based on the lateral-torsional stability strength of beam under uniform bending moment as follows:

$$\frac{\sigma_{bg}}{\sigma_Y} = 1.0 \qquad \text{for } \alpha \le 0.2 \left.\right\}$$

$$= 1.0 - 0.412(\alpha - 0.2) \quad \text{for } 0.2 \le \alpha \qquad (65.6)$$

where σ_{bg} = lateral-torsional stability strength of beam under uniform moment, α = equivalent slenderness parameter defined as

$$\alpha = \frac{2}{\pi} K \sqrt{\frac{\sigma_Y}{E}} \cdot \frac{\ell}{b} \left.\right\}$$

$$K = 2 \qquad \text{for } A_w / A_c \le 2 \qquad (65.7)$$

$$= \sqrt{3 + 0.5 A_w / A_c} \quad \text{for } 2 \le A_w / A_c$$

and A_w = gross area of web plate, A_c = gross area of compression flange, ℓ = laterally unbraced length, b = width of compression flange. The effect of nonuniform bending is estimated by the multiplier coefficient, m, as follows

$$m = \frac{M}{M_{eq}} \qquad (65.8)$$

in which M = bending moment at a reference cross section, M_{eq} = equivalent conversion moment given as

$$M_{eq} = 0.6 M_1 + 0.4 M_2 \quad \text{or} \quad M_{eq} = 0.4 M_1 \quad \text{where} \quad M_1 \ge M_2 \qquad (65.9)$$

65.1.5 Fabrication and Erection

Fabrication and erection procedures depend on the structural system of the bridge, the site conditions, dimensions of the shop-fabricated bridge units, equipment, and other factors characteristic of a particular project. This includes methods of shop cutting and welding, the selection of lifting equipment and tackle, method of transporting materials and components, the control of field operation such as concrete placement, and alignment and completion of field joints in steel, and also the detailed design of special erection details such as those required at the junctions of an arch, a cantilever erection, and a cable-stayed erection. Therefore, for each structure, it is specified that the contractor should check

1. Whether each product has its specified quality or not.
2. Whether the appointed erection methods are used or not.

As a matter of course, the field connections of main members of the steel structure should be assembled in the shop.

Details on the inspections have not been given here and the reader should refer to the specifications [1].

FIGURE 65.1 Tennyo-bashi.

65.2 Stone Bridges

Tetsuya Yabuti

It is possible that stone bridges were built in very ancient times but that through lack of careful maintenance and/or lack of utility they were destroyed so that no trace remains. Since stone masonry is generally suited to compressive stresses, it is usually used for arch spans. Therefore, most stone bridges that have survived to the present are arch bridges. Generally, stone arch bridges are classified into two types: the European voussoirs are built of bricks and the Chinese type where each voussoir in arch is curved and behaves as a rib element.

Figure 65.1 shows Tennyo-bashi (span length, 9.5 m) located in Okinawa prefecture. This is the only area in Japan that has the Chinese type. This bridge is the oldest Chinese-type stone arch bridge in Japan that has survived to the present time; it was originally constructed in 1502. Figure 65.2 shows Tsujyun Bridge located in Kumamoto prefecture (length of span, 75.6 m; raise of arch, 20.2 m; width of bridge, 6.3 m). This bridge is typical of aqueduct stone arch bridges that have survived to the present in Japan and was originally constructed in 1852. Figure 65.3 shows Torii-bashi Bridge located in Oita prefecture which is one of the multispanned stone arch bridges constructed early in the 20th century. This bridge is a five-span arch bridge (length of bridge, 55.15 m; width of bridge; 4.35 m; height of bridge, 14.05 m) constructed in 1916.

Separate stones sometimes have enough tensile strength to permit their being used for beams and slabs as seen in Hojyo-bashi which is the clapper bridge shown in Figure 65.4. This bridge located in Okinawa prefecture has a span of 5.5 m and was originally constructed in 1498.

65.3 Timber Bridges [2,3]

Masatsugu Nagai

Since 1990, the number of timber bridges constructed has increased. Most of them use glue-laminated members and many are pedestrian bridges. To date, about 10 bridges have been constructed to carry 14 or 20 tf trucks. All were constructed on a forest road. We have no design code for timber bridges. However, there is a manual for designing and constructing timber bridges.

The following is an introduction to timber arch, cable-stayed, and suspension bridges in Japan.

FIGURE 65.2   Tsujyun Bridge.

FIGURE 65.3   Torii-bashi.

1. Arch Bridges — Table 65.2 shows nine arch bridges. Figure 65.5 shows Hiraoka bridge, which, of timber arch bridges, has the longest span in Japan. Figure 65.6 shows Kaminomori bridge [4]. It is the first arch bridge, which carries a 20 tf truck load.
2. Cable-Stayed Bridges — Table 65.3 shows three cable-stayed bridges. Figure 65.7 shows Yokura bridge. It has a world record span length of 77.0 m, and has a concrete tower.
3. Suspension Bridges — Table 65.4 shows two suspension bridges. Figure 65.8 shows Momosuke bridge. Momosuke bridge is an oldest timber suspension bridge in Japan. It was constructed in 1922, and reconstructed in 1993.

FIGURE 65.4 Hojyo-bashi.

TABLE 65.2 Arch Bridges

Name	Span and Width (m)	Construction Year	Remarks
Yunomata	13.0 6.0	1990	Tied arch bridge for 14 tf truck loading
Kisoohashi	33.0 6.0	1991	Fixed arch pedestrian bridge
Deai	39.0 2.0	1992	Tied arch pedestrian bridge
Hiroaka	45.0 3.0	1993	Three hinged arch pedestrian bridge
Yasuraka	30.0 1.5	1993	Nielsen Lohse type pedestrian bridge
Chuo	21 5.0	1993	Lohse type pedestrian bridge
Kaminomori	23 5.0	1994	Two hinged arch bridge for 20 tf truck loading
Awaiido	24.0 8.0	1994	Lohse type bridge for 20 tf truck loading
Meoto	20 1.5	1994	Two hinged arch pedestrian bridge

65.4 Steel Bridges

Tetsuya Yabuti

In Japan, metal as a structural material began with cast and/or wrought iron used on bridges after the 1870s. Through lack of these utilities in urban areas, however, almost all of those bridges have broken down. Since 1895, steel has replaced wrought iron as the principal metallic bridge material.

After the great Kanto earthquake disaster in 1923, high tensile steels have been positively adopted for bridge structural uses and Kiyosu Bridge (length of bridge, 183 m; width of bridge, 22 m) shown in Figure 65.9 is a typical example. This eyebar-chain-bridge over the Sumida river in Tokyo is a self-anchored suspension bridge and a masterpiece among riveted bridges. It was completed in 1928.

FIGURE 65.5   Hiraoka Bridge.

FIGURE 65.6   Kaminomori Bridge.

Figure 65.10 shows one of the curved tubular girder bridges located in the metropolitan express-way. Curved girder bridges have become an essential feature of highway interchanges and urban expressways now common in Japan.

TABLE 65.3 Cable-Stayed Bridges

Name	Span and Width (m)	Construction Year	Remarks
Midori Kakehashi	27.5 2.0	1991	Two-span continuous pedestrian bridge
Yokura	77.0 5.0	1992	Three-span continuous pedestrian bridge
Himehana	21.5 1.5	1995	Three-span continuous pedestrian bridge

FIGURE 65.7 Yokura Bridge.

TABLE 65.4 Suspension Bridges

Name	Span and Width (m)	Construction Year	Remarks
Momosuke	104.5 2.3	1993	Four-span continuous pedestrian bridge
Fujikura	32.0 1.8	1994	Single-span pedestrian bridge

Figure 65.11 shows Katashinagawa Bridge (length of span; 1033.8 m = 116.9 + 168.9 + 116.9; width of bridge, 18 m) located in Gunma prefecture. This bridge is the longest curved-continuous truss bridge in Japan and was completed in 1985. Figure 65.12 shows Tatsumi Bridge (length of bridge, 544 m; width of bridge, 8 m) located in Tokyo. This viaduct bridge in the metropolitan expressway is a typical example of rigid frame bridges in an urban area and was completed in 1977. Figure 65.13 shows a typical π-shaped rigid frame bridge. This structural type is used as a viaduct over a highway or a highway bridge in mountain areas and is common in Japan.

FIGURE 65.8   Momosuke Bridge.

FIGURE 65.9   Kiyosu Bridge.

FIGURE 65.10 A curved tubular girder bridge in a metropolitan express highway.

FIGURE 65.11 Katashinagawa Bridge.

FIGURE 65.12   Tatsumi Bridge.

FIGURE 65.13   A typical π-shaped rigid frame bridge.

FIGURE 65.14 Saikai Bridge.

Saikai Bridge located in Nagasaki prefecture (length of span, 216 m; width of bridge, 7.5 m) shown in Figure 65.14 was completed in 1955. Construction of bridges in Japan after the World War II began in earnest with the Saikai Bridge. This bridge is a fixed arch bridge and the stress condition was improved by prestressing when the main arch was finally completed.

Figure 65.15 shows Ooyano Bridge located in Kumamoto prefecture. This bridge is a typical through-type arch bridge (length of bridge, 156 m; width of bridge, 6.5 m) and was constructed in 1966.

Figure 65.16 shows Ikuura Bridge located in Mie prefecture and completed in 1973. The main span of this bridge (length of span, 197 m; width of bridge, 8.3 m) is a tied arch with inclined hangers of the Nielsen system that is one of the most favored bridge types in Japan, along with the cable-stayed bridge.

Figure 65.17 shows Tsurumi-Tsubasa Bridge (length of bridge, 1021 m = center span of 510 m + two side spans of 255 m; height of towers, 136.7 m) located in Yokohama Bay, Kanagawa prefecture. This bridge is a single-plane, cable-stayed bridge with continuous three spans. It is the longest bridge of this type including those under design all over the world and was completed in 1994.

Figure 65.18 shows Iwagurojima Bridge (length of bridge, 790 m = center span of 420 m + two side spans of 185 m; height of towers, 148.1 m; width of bridge, 22.5 m) located Kagawa prefecture. This bridge completed in 1988 is a double-plane cable-stayed bridge with continuous three spans and is a combined bridge with highway and railway traffic. It has four express railways. The cable-stayed bridge with four express railways is unprecedented, including those under design worldwide.

Figure 65.19 shows Kanmon Bridge (length of bridge, 1068 m = center span of 420 m + two side spans of 185 m) completed in 1973. This bridge spans over the Kanmon channel and links Moji in Fukuoka prefecture, Kyushu Island, and Shimonoseki in Yamaguchi prefecture, main island. It is the first bridge in Japan spanning a channel.

The Japan Association of Steel Bridge Construction has contributed to the preparation of some photographs in this section.

FIGURE 65.15   Ooyano Tsubasa Bridge.

FIGURE 65.16   Ikuura Bridge.

FIGURE 65.17 Tsurumu-Tsubasa Bridge.

FIGURE 65.18 Iwakuro Bridge.

FIGURE 65.19   Kanmon Bridge.

65.5   Concrete Bridges

Tetsuya Yabuki

Construction of reinforced concrete bridges began in the 1900s in Japan but has gradually become useless because of change of the utility conditions in urban areas. Since the 1950s, the use of prestressing spread to nearly every type of simple structural element and spans of concrete bridges became much longer. Probably the most significant observable feature of prestressed concrete is its crack-free surface under service loads. Especially, when the structure is exposed to weather conditions, elimination of cracks prevents corrosion. Many reinforced concrete bridges constructed previously are being replaced by prestressed concrete ones in Japan.

Figure 65.20 shows a typical reinforced concrete bridge damaged by corrosion. Most of these kinds of bridges have been replaced by prestressed concrete structures. Figure 65.21 shows Chousei Bridge (length of bridge, 10.8 m = three continuous spans of 3.6 m) located in Ishikawa prefecture. This bridge is a pretensioned simple composite slab bridge. It was completed in 1952 and is the first prestressed concrete bridge in Japan. Figure 65.22 shows Ranzan Bridge (length of bridge, 75 m = main span of 51.2 + two side spans of 11.9 m) located in Kanagawa prefecture. This bridge is a rigid frame bridge composed by three spans with a hinge. It was completed in 1959 and is the first bridge in Japan that was constructed by the cantilever erection.

Figure 65.23 shows International Expo No. 9 Bridge (length of bridge, 27 m; width of bridge, 5.5 m; thickness of slab, 0.1 m) located in Osaka. This bridge is a pedestrian bridge. It was completed in 1969 and is the first suspended slab bridge in Japan. Figure 65.24 shows Takashimadaira Bridge (length of bridge, 230 m = 75 + 75 + 80; width of bridge, 18 m) located in Tokyo. This viaduct in the metropolitan expressway is composed by linking three bridges and completed in 1973. Each one is a continuous three span-bridge (length of spans, 25 m + 25 + 25 m).

FIGURE 65.20   A typical reinforced concrete bridge damaged by corrosion.

FIGURE 65.21   Chosei-bashi.

FIGURE 65.22   Ranzan Bridge.

FIGURE 65.23   International Expo No. 9 Bridge.

FIGURE 65.24 Takashimadaira Bridge.

Figure 65.25 shows Akayagawa Bridge (length of bridge, 298 m; arch span, 116 m) located in Gunma prefecture. This rib arch bridge is the longest concrete arch railway bridge in Japan and was completed in 1979. Its arch rib is composed of a plate with thickness of 0.8 m and mainly receives compressive stress. Figure 65.26 shows Omotogawa Bridge located in Iwate prefecture. This bridge was completed in 1979 and is the first prestressed concrete stayed railway bridge in the world.

The Japan Prestressed Concrete Association has contributed to preparation of some of the photographs in this section.

65.6 Hybrid Bridges

Masatsugu Nagai

Hybrid bridges consist of composite and compound bridges. Composite bridges have a cross section of steel and concrete connected by shear connectors. Compound bridges consist of different materials, such as steel and concrete. In Japan, many composite girder bridges have been constructed. However, since 1980, the number of composite girder bridges has decreased. One of main reasons is the damage of concrete decks due to overloading by heavy trucks. In recent years, for economic reasons, the choice of composite girder construction has been reconsidered. In bridge systems, prestressed precast concrete slabs are used to attain higher durability. The following is an introduction of the practices and plans of the hybrid bridges of Japan Highway Public Corporation.

Figure 65.27 shows Hontani Bridge (total span length, 197 m = 44 + 97 + 56 m; width, 11.4 m) constructed in Gifu prefecture in 1998. It has a box section with corrugated steel plate used as a web between upper and lower concrete slabs. To reduce the total weight of the concrete box girder, instead of concrete webs, a steel web was used. This kind of structural system was first employed in the Cognac Bridge in France. However, for the connection between concrete and steel plate, a simple system with reinforcing bars attached to the corrugated plate and without steel flange was used.

FIGURE 65.25 Akayagawa Bridge.

FIGURE 65.26 Omotogawa Bridge.

FIGURE 65.27 Hontani Bridge. (Courtesy of Japan Highway Public Corporation.)

FIGURE 65.28 Tomoegawa Bridge. (Courtesy of Japan Highway Public Corporation.)

Figure 65.28 shows Tomoegawa Bridge (total span length, 475 m = 59 + 3 × 119 + 59 m; width, 16.5 m) which will be constructed on the route of the New Tomei Expressway discussed in Section 65.8.2. It has a box section, and a steel truss is used as a web between the upper and lower concrete slabs, whose main purpose is to reduce the total weight of the superstructures. When the span length becomes long, if a steel plate is used as the web, a horizontal connection in the bridge

FIGURE 65.29   Ibigawa Bridge. (Courtesy of Japan Highway Public Corporation.)

longitudinal direction is necessary for transportation from the shop to the site. To avoid the problem, a truss member is planned to be used. This kind of bridge system was first used in Arbois Bridge in France. However, in this bridge, a new connecting system between truss member and concrete slab will be used.

Figure 65.29 shows Ibigawa Bridge (total span length, 1397 m = 154 + 4 × 271.5 + 157 m; width 33 m) which will be constructed on the route of the New Meishin Expressway. It crosses the Ibi River near Nagoya City. Kisogawa Bridge, which crosses the Kiso River running parallel to the Ibi River, has a structural form similar to that of Ibigawa Bridge. The bridge consists of concrete and steel girders. The steel box girders are used for the middle part with a length of 100 m in the central four spans, and are connected to the concrete girder. The concrete girder is suspended by diagonal stays from concrete towers. The height of the tower from the deck level is lower than that used in conventional cable-stayed bridges. Since the span length is 271.5 m, if the concrete bridges are used, considerably greater depth is inevitable. By suspending the concrete girder and using a steel girder with lighter weight in the central portion of the bridge, reduction of depth is attained.

Figure 65.30 shows Shinkawa Bridge (total span length, 278 m = 2 × 40 + 113 + 2 × 40 m; width 21.4 m) which will be constructed on the highway in Shikoku Island. It consists of the concrete and steel box girders. Since the side span length is planned to be short, a concrete box girder is used and connected to the steel girder in the side span adjacent to the main span. The purpose of this countermeasure is to balance the weight.

Figure 65.31 shows Kitachikumagawa Bridge (total span length, 346.7 m = 84.35 + 2 × 89 + 84.35 m; width, 10.4 m) which is located near Nagano City and was constructed in 1997. In this bridge, reinforced concrete piers were connected to the steel box girders. By employing a rigid frame structure, the bearings are avoided and good performance against earthquake is attained.

Shigehara Bridge (total span length; 166.8 m = 47.4 + 72 + 47.4 m; width, 10.4 m) was constructed on the highway in Kyushu Island in 1995. It has composite piers, in which steel pipes are encased instead of reinforcing bars as shown in Figure 65.32. This system is developed to reduce the volume of the reinforcing bars, and resulted in the reduced construction work.

FIGURE 65.30 Shinkawa Bridge. (Courtesy of Japan Highway Public Corporation.)

FIGURE 65.31 Kitachikuma Bridge. (Courtesy of Japan Highway Public Corporation.)

FIGURE 65.32 Piers of Shigehara Bridge. (Courtesy of Japan Highway Public Corporation.)

65.7 Long-Span Bridges (Honshu–Shikoku Bridge Project)

Masatsugu Nagai and Shuichi Suzuki

The Honshu–Shikoku Bridge Project is a national project to link the Honshu and Shikoku Islands. Construction of the long-span bridge started in 1975 and was completed in 1999. Figure 65.33 shows three routes, in which many long-span cable-supported bridges are constructed. The following is an introduction of the super- and substructures of these cable-supported bridges.

65.7.1 Kobe–Naruto Route

This route is 89 km long, and two suspension bridges are arranged. Figure 65.34 shows the Akashi Kaikyo Bridge [5] (total span length, 3911 m = 960 + 1991 + 960 m) which is a three-span, two-hinged suspension bridge, and has a world-record span length of 1991 m. The distance between two cables is 35.5 m. This bridge was opened to traffic in 1998. The original plan was to carry both rail and road traffic. In 1985, this plan was changed so that the bridge carries highway traffic only. It is known, in the design of long-span suspension bridges, that ensuring safety against static and dynamic instabilities under wind load is an important issue. Aerodynamic stability was investigated through boundary layer

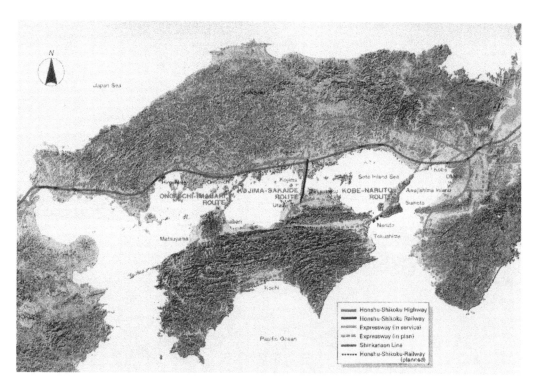

FIGURE 65.33 Honshu–Shikoku Bridge connecting route.

wind tunnel test with a ¹⁄₁₀₀ full model, and the test was conducted as a cooperative study between the Honshu–Shikoku Bridge Authority and the Ministry of Construction. Through the test, it was confirmed that flutter occurred at a wind velocity exceeding 78 m/s, a value was well above the required wind velocity. To avoid two main cables on each side, wires with a tensile strength of 1760 MPa (higher than that used in other suspension bridges) was used. Further, since the ratio of live load to dead load is small, the factor of safety of 2.2 against tensile strength was utilized.

The laying-down caisson method was adopted for the main tower foundations. The method is to install a prefabricated steel caisson shown in Figure 65.35 on a dredged seabed and subsequently complete a rigid foundation structure by filling concrete inside the caisson compartments. Desegregation concrete was developed for this purpose. The anchorage foundation was constructed with an underground slurry wall method on the Kobe side and the spread foundation method on the Awajishima side. Highly workable concrete was developed for the body of anchorages shown in Figure 65.36.

Figure 65.37 shows the Ohnaruto Bridge completed in 1985. The center and side spans are 876 and 330 m, respectively. The distance between two cables are 34 m. To achieve aerodynamic stability, vertical stabilizing plates were installed under the median strip of the deck to change the wind flow patterns. This kind of stabilizing countermeasure was also adopted in the stiffening truss of the Akashi Kaikyo Bridge.

The multicolumn method, aiming to avoid disrupting the famous Naruto Whirlpools in the Naruto Straits, was used for the main and side tower foundations.

65.7.2 Kojima–Sakaide Route

This route is 37 km long, and three suspension bridges and two cable-stayed bridges were constructed. Since the bridges carry both roadway and railway traffic, a truss girder was selected; its upper deck is used for the roadway and the lower deck for the railway.

FIGURE 65.34   Akashi Kaikyo Bridge.

The laying-down caisson method was utilized for 11 underwater foundations in this route. Prepacked concrete, in which coarse aggregate was at first packed inside the steel caisson and then mortar was injected into voids of the aggregate, was developed for the underwater concrete of the foundations.

Figure 65.38 shows the Shimotsui-Seto Bridge completed in 1988, which is a single-span bridge with 940 m in span length. The distance between two cables is 35 m, a value that is common to all suspension bridges on this route. The main cables of most long-span suspension bridges in Japan have been erected by the prefabricated strand (PS) method. However, this bridge employed the air spinning (AS) method. Since the cable is anchored to the rock directly (a tunnel anchor), the AS method enabled making the anchoring system small.

Figure 65.39 shows the Kita Bisan-Seto Bridge and the Minamai Bisan-Seto Bridge. These bridges have a three-span continuous truss girder. The center spans of theses two bridges are 990 and 1100 m, respectively, and their side spans are each 274 m. These bridges were constructed in 1988. The side view is similar to that of the San Francisco–Oakland Bay Bridge in the United States. The cables of two bridges are anchored to opposite sides of one common anchorage. Hence, the anchorage is subjected only to the difference between the horizontal components of tension in the cables of the two bridges. At the end the bridge, an expansion joint allowing 1.5 m movement was installed, and, also, a transition girder system was used to absorb large amounts of changes in inclination in the track for ensuring running stability of the train. The laying-down caisson method was used for the six underwater foundations. The Sikoku side anchorage foundation is the largest one in this route, reaching 50 m below sea level.

Figure 65.40 shows the Iwakurojima Bridge and the Hitsuishijima Bridge. Both bridges have center spans of 420 m and side spans of 185 m. These bridges were constructed in 1988. First, these bridges were designed as a Gerber-type truss girder. After carrying out a study on the

FIGURE 65.35 Steel caisson.

possibility of cable-stayed bridges, the bridges were changed to that type. Because the weight of the girder required to carry road and rail traffic is substantial, two parallel cables were anchored to an upper chord on one side of the truss girder. Due to the narrow distance between the two cables, wake galloping in the leeward cables was observed. To suppress oscillations, a damping device connecting two cables was used. At each end of the girders, elastic springs in the bridge longitudinal direction were installed. This elastic support adjusts the natural period of the bridge, resulting in a reduction of inertia force due to earthquake. The laying-down caisson method was used for five underwater foundations.

FIGURE 65.36 Body of anchorage.

65.7.3 Onomichi–Imabara Route

This route is 60 km long. In this route, five suspension bridges and four cable-stayed bridges were opened to traffic.

Figure 65.41 shows the Ohsima Bridge completed in 1988, which is a single-span bridge with a 560-m span length. The distance between two cables is 22.5 m. For the stiffening girder, a trapezoidal box section with a depth of 2.2 m was used because of its economical efficiency and maintenance. This is the first application of the box girder to the stiffening girder of a suspension bridge in Japan. A spread foundation was adopted for all the anchorages and the main towers which were constructed on land.

Figure 65.42 shows the Innoshima Bridge whose center and side spans are 770 and 250 m, respectively. The distance between two cables is 26 m. This bridge was constructed at an early stage of the grander project. One strand consisting of 127 wires, which are larger than the more commonly used size of 61 or 91, were developed. A reduced construction period was attained by using this strand. The same type of foundations as the Ohsima Bridge was adopted for this bridge.

Figure 65.43 shows the Kurishima Kaikyo Bridges, which are three consecutive suspension bridges. These bridges were opened to traffic in 1999. The center spans of the three bridges are 600, 1020, and 1030 m. The distance between two cables is 27 m. The girder has a streamlined box section with a depth of 4.3 m. The block of the girder is transported by a self-sailing barge, lifted up with the lifting machine, and finally fixed to hanger ropes. For connection of the block of some parts of the tower, tension bolts instead of friction bolts were employed. Two common anchorages that have a similar cable strand anchoring system as the Kita and Minami Bisan-Seto

FIGURE 65.37   Ohnaruto Bridge.

Bridges were constructed in these three bridges. The laying-down caisson method was adopted for six underwater foundations. A concrete caisson was used for the two relatively small caissons, while the other caissons were made of steel.

Figure 65.44 shows the Tarata Bridge [6,7], which has a world-record span of 890 m and side spans of 320 and 270 m. This bridge was opened to traffic in 1999. Since the side span is short, prestressed concrete girders with a span of 110 and 62.5 m from both abutments are used, and connected to steel girder. These prestressed concrete girders work to counter the large uplifting forces and contribute to an increased in-plane flexural rigidity of the bridge. The depth of the girder is 2.7 m and the ratio of the center span length to girder depth is around 330. Buckling instability of the girder was investigated through analytical and experimental studies. In the experimental study, a ⅟₅₀ full model was used. Aerodynamic stability was investigated by wind tunnel tests with a ⅟₇₀ full model and a ⅟₂₀₀ full model. In the latter model, the influence of the surrounding topography on aerodynamic behavior was also investigated. The laying-down caisson method was used for two of the tower foundations.

Figure 65.45 shows the Ikuchi Bridge completed in 1989. The center span is 490 m and the side span is 150 m. Again, since the side span length is short, prestressed concrete girders were used in both side spans and were connected to the steel girder. Various connecting methods were investigated, through which a combination of bearing plate and shear studs was used. Each cable consists of galvanized steel wires with a diameter of 7 mm, and is coated with polyethylene (PE) tube. A pile foundation was adopted for all piers since the ground was composed of weathered rock.

FIGURE 65.38   Shimotsui-Seto Bridge.

65.8   New Bridge Technology Relating to Special Bridge Projects

Masatsugu Nagai

65.8.1   New Material in the Tokyo Wan Aqua-Line Bridge [8]

The Tokyo Wan Aqua-Line with a 15.1 km length crosses the Tokyo Bay, and links Kanagawa and Chiba prefectures. It was opened to traffic in December 1997. It has a marine section of 14.3 km, and consists of an approximately 9.5-km-long tunnel and 4.4-km-long bridge structures. Here, an outline of the bridge and new material for corrosion protection of steel piers are introduced.

Figure 65.46 shows a general view of the bridges, which consists of 3-, 10-, 11-, 10-, and 9-span continuous steel box girder bridges. Since the maximum span length of 240 m is needed for a navigation channel and the strength of soil foundation is weak, steel box girders were designed and constructed. For erection of most of the bridges, floating cranes and deck barges were used. After completion of the superstructure, vortex-induced oscillation with an amplitude over 0.5 m was observed. To suppress it, 16 tuned mass dampers, as shown in Figure 65.47, were installed. Part of the steel deck was cut out and stiffened. After installing the tuned mass dampers, the removed plates were welded to the original position.

To attain higher durability of steel piers, titanium-clad steel is attached to the steel piers. Figure 65.48 shows the pier with the titanium-clad steel. It was used in the region of the steel piers affected by tidal movement and saltwater spray. By employing this system, a maintenance-free system more than 100 years is expected.

FIGURE 65.39   Kita-Bisan Seto and Minami-Bisan Seto Bridges.

65.8.2 New Bridge System in the New Tomei Meishin Expressway

Japan Highway Public Corporation started constructing the New Tomei (between Tokyo and Nagoya City) Meishin (between Nagoya and Kobe Cities) Expressway which is around 600 km long and links the big cities of Tokyo, Nagoya, Osaka, and Kobe.

Figure 65.49 shows a plate I-girder bridge with a small number of main girders, which is a new bridge system employed for Ohbu Viaduct near Nagoya City. This viaduct was constructed in 1998. Conventionally, the distance of the I-girder, which corresponds to the span of concrete slabs, has been designed to be less than 3 m. However, in this project, using prestressed, precast concrete slabs,

FIGURE 65.40 Iwakurojima and Hitsuishijima Bridges.

the span of the slab extended to 6 m. Hence, for a three-lane bridge with a width of around 15 m, the number of the I-girders is reduced from 6 to 3. Further, the I-girders are connected by simple beams arranged at a distance of around 10 m only. Conventionally, I-girders have been stiffened by cross-beams or bracing, which are installed at a distance less than 6 m, and a lateral bracing member, which is installed at a lower level of the girder. This simple bridge system can reduce the construction cost and also the painting area. Further, this system leads to easy inspection.

65.8.3 Superconducting Magnetic Leviation Vehicle System [9]

Japan Railway Corporation has a plan of constructing a new line, the Chou–Shinkansen line. Using a high-speed train, it will run through the central part of Japan from Tokyo to Osaka. Now, in a test line with a 18.4 km length constructed in Yamanashi prefecture, the running stability, etc. of the high-speed train is being tested.

Figure 65.50 shows the Maglev car running through Ogatayama Bridge which will be explained later. It levitates 100 mm from the ground and runs at a speed of 500 km/h. A repulsive force and an attractive force induced between the superconducting magnets on the vehicle and the coils on the side walls (propulsion coils and levitation coils) are used for propelling and levitating the car. The following are the major technical issues involved in designing the structures.

1. The deflection of the structures should be small to ensure running stability and riding comfort of the train.
2. High accuracy should be attained when positioning the coil on the side walls.
3. Magnetic drag force due to magnetic phenomenon produced between the magnet and steel should be small.

FIGURE 65.41 Ohshima Bridge.

The countermeasure of the third issue is to use a low-magnetic metal such as austenitic high-manganese steel when the metal is positioned within 1.5 m of the superconducting magnet.

Ogatayama Bridge completed in 1995, a Nielsen Lohse bridge, is shown in Figure 65.50. The span and arch rise are 136.5 and 23 m, respectively. The width between the center of the arch chords is 15 m at springing and 9.6 m at arch crown. In this bridge, the steel structures such as the arch chord and floor system are designed to be positioned 1.5 m apart from the magnet. However, for the reinforcing bars encased in the guide way, high-manganese low-magnetic steel is used.

65.8.4 Menshin Bridge on the Hanshin Expressway

Early in the morning on 17 January 1995, a huge earthquake shook the densely populated southern part of Hyogo prefecture. Many steel and concrete bridges fell due to the collapse of reinforced concrete piers. Many steel piers suffered buckling damage and two collapsed. Immediately after the earthquake, investigation began to identify the causes of damage and repair work, such as encasing the concrete piers and increasing longitudinal ribs in the steel piers. The concrete deck changed to the steel deck, and the collapsed prestressed concrete girders changed to steel bridges. This is to reduce the weight of the superstructures. In addition, metal bearings were also replaced by rubber supports for greater damping.

Figure 65.51 shows the typical rubber support. Almost all metal bearings were changed to this type. Figure 65.52 shows the rigid-frame-type piers. At the foot of a column, rubber bearings were installed to isolate earthquake acceleration. Bridges with these isolating system are called Menshin bridges.

FIGURE 65.42   Imnoshima Bridge.

FIGURE 65.43   Kurushima Kaikyo Bridges.

FIGURE 65.44 Tatara Bridge.

FIGURE 65.45 Ikuchi Bridge.

FIGURE 65.46   Steel bridges on Tokyo Wan Aqua-Line. (Courtesy of Trans-Tokyo Bay Highway Corporation.)

65.8.5 Movable Floating Bridge in Osaka City [10]

The world's first swing and floating bridge is under construction in the Port of Osaka. It will be completed in 2000. The main role of the bridge is to connect two reclaimed islands (the names are Maishima and Yumenoshima). The width of the waterway between the two islands is around 400 m. In case of the occurrence of unforeseen accidents in the main waterway of the Port of Osaka nearby, this warterway (subwaterway) will provide an alternative entrance. On such occasions, large-sized vessels will pass the subwaterway. In addition, the soil foundation is not strong enough to resist the loads of a conventional bridge. Hence, a movable floating bridge with two pontoon foundations and a swing type has been conceived. The bridge has a total length of 940 m and a width of 38.4 m. The floating part has a length of 410 m and a main span length of 280 m with a double-arch rib rigidly connected to two steel pontoons as shown in Figure 65.53. Safety against dynamic responses subjected to waves, winds, earthquake, and heavy track loading have been investigated through numerical and experimental studies.

65.9 Summary

Tetsuya Yabuki

A chronological table of the major revisions of the standard specification of highway bridges and the concrete standard specification in Japan during the latter half of the 20th century is shown in

FIGURE 65.47 Tuned mass damper. (Courtesy of Trans-Tokyo Bay Highway Corporation.)

TABLE 65.5 Chronological Table on the Revisions of the Standard Specifications and the Major Earthquake Disasters in Japan

Epoch	Occurrences
1948	Fukui earthquake disaster
1956	Revision of the Standard Specification of Highway Bridges
1964	Niigata earthquake disaster
1968	Tokachi offshore-earthquake disaster
1971	Establishment of self-editing on the seismic design specification in Standard Specification of Highway Bridges
1973	Revision of the Standard Specification of Highway Bridges
1978	Miyagi Prefecture offshore-earthquake disaster
1980	Revision of the Standard Specification of Highway Bridges
1986	Revision of the Concrete Standard Specification
1995	Great Hanshin-Awaji earthquake disaster
1996	Revision of the Standard Specification of Highway Bridges
	Revision of the Concrete Standard Specification

Table 65.5. The major earthquake disasters in Japan are also shown in Table 65.5 for reference purposes. The design specifications of bridges have been revised mainly whenever strong earthquakes have occurred as shown in Table 65.5. We may be able to make the statement that the most important influence on the evolution of bridges in Japan has been earthquakes. This shows that disasters have controlled bridge engineering. The evaluation of conditions to produce an efficient bridge for seismic motion is distinctly an engineering problem. Therefore, bridge engineers have to escalate their efforts from now on so that bridge engineering can control bridge disasters. The evolution of bridges will be achieved by bridge engineers' efforts to develop optimum structural performance with materials that will be in shorter supply, be longer in durability, give higher strength, and bring less dynamic inertia force.

FIGURE 65.48   Steel pier with titanium-clad steel. (Courtesy of Nippon Steel.)

FIGURE 65.49 Plate I-girder with simple cross-beams.

FIGURE 65.50 Maglev car and Ogatayama Bridge. (Courtesy of Japan Railway Corporation.)

FIGURE 65.51   Rubber bearings.

FIGURE 65.52 Rigid frame pier of Benten section. (Courtesy of Hanshin Expressway Public Corporation.)

FIGURE 65.53   Yumeshima-Maishima Bridge. (Courtesy of Osaka Municipal Government.)

References

1. The Japanese Association of Highway, Standard Specification of Highway Bridges, rev. 1996.
2. S. Usuki and K. Komatsu, Two timber road bridges, *J. Struct. Eng. Int.*, IABSE, 23–24, 1998.
3. S. Usuki, Y. Horie, and K. Hasebe, Timber arch bridge on city roadway, in *Proc. of the 1990 Int. Timber Engineering Conference*, Vol. 2, Japan, 1990, 431–438.
4. S. Usuki, K. Komatsu, I. Kagiwada, and H. Abe, Drift pin connections with II-shaped internal steel plates for an arch bridge, in *Proc. of Pacific Timber Engineering Conference*, Australia, 1994, 51–58.
5. S. Saeki, Y. Fujino, K. Tada, M. Kitagawa, and T. Kanazaki, Technological aspect of the Akashi Kikyo Bridge, ASCE Structures Congress, Boston, 1995, 237–252.
6. T. Endo, T. Iijima, A. Okukawa, and M. Ito, The technical challenge of a long cable-stayed bridge — Tatara bridge, in *Cable-Stayed Bridges — Recent Development and Their Future*, Elsevier, New York, 417–436, 1991.
7. K. Tada, H. Sato, T. Kanazaki, and H. Katsuchi, Full model wind tunnel test of the Tatara bridge, in *Proc. of Bridges into the 21st Century*, Hong Kong, 1995, 671–678.
8. H. Ikeda, The Trans-Tokyo Bay Highway project, in *Proc. of the 6th East Asia–Pacific Conference on Structural Engineering & Construction*, Taiwan, 1998, 41–48.
9. H. Wakui and N. Matsumoto, Dynamic study on Twin-Beam-Guidway for JR Maglev, in *Proc. of Int. Conference on Speedup Technology for Railway and Maglev Vehicles*, Vol. 1, 1993, 126–131.
10. T. Maruyama, E. Watanabe, T. Utsunomiya, and H. Tanaka, A new movable floating bridge in Osaka Harbor, in *Proc. of the 6th East Asia–Pacific Conference on Structural Engineering & Construction*, Taiwan, 1998, 429–434.

66

Design Practice in Russia

Simon A. Blank
California Department of Transporation

Oleg A. Popov
Joint Stock Company Giprotransmost (Tramos), Russia

Vadim A. Seliverstov
Joint Stock Company Giprotransmost (Tramos), Russia

66.1 Introduction

Bridge design and construction practice in former USSR, especially Russia, is not much known by foreign engineers. Many advanced structural theories and construction practices have been established. In view of the global economy, the opportunities to apply such advanced theories to practice became available with the collapse of the iron curtain.

In 1931, Franklin D. Roosevelt said, "There can be little doubt that in many ways the story of bridge building is the story of civilization. By it, we can readily measure a progress in each particular country." The development of bridge engineering is based on previous experiences and historical aspects. Certainly, the Russian experience in bridge engineering has it own specifics.

0-8493-7434-0/00/$0.00+$.50
© 2000 by CRC Press LLC

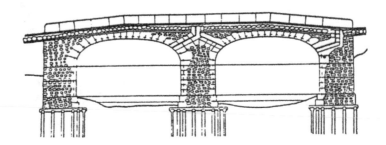

FIGURE 66.1   Typical masonry bridge (1786).

66.2   Historical Evolution

66.2.1   Masonry and Timber Bridges

The most widespread types of bridges in the old time were timber and masonry bridges. Because there were plenty of natural wood resources in ancient Russia, timber bridges were solely built up to the end of the 15th century.

For centuries masonry bridges (Figure 66.1) have been built on territories of such former republics of the USSR as Georgia and Armenia. From different sources it is known that the oldest masonry bridges in Armenia and Georgia were built in about the 4th to 6th centuries. One of the remaining old masonry bridges in Armenia is the Sanainsky Bridge over the River Debeda-chai built in 1234. The Red Bridge over the River Chram in Georgia was constructed in the 11th century. Probably the first masonry bridges built in Russia were in Moscow. The oldest constructed in 1516 was the Troitsky arch masonry bridge near the Troitsky Gate of the Kremlin. The largest masonry bridge over the Moskva River, named Bolshoi Kamenny, was designed by Yacobson and Kristler. The construction started in 1643 but after 2 years the construction was halted because of Kristler's death. Only in 1672 was this construction continued by an unknown Russian master and the bridge was mostly completed in 1689. Finally, the Russian Czar Peter I completed the construction of this bridge. Bolshoi Kamenny Bridge is a seven-span structure with total length of 140 m and width of 22 m [1]. This bridge was rebuilt twice (in 1857 and in 1939). In general, a masonry bridge cannot compete with the bridges of other materials, due to cost and duration of construction.

Ivan Kulibin (Russian mechanics engineer, 1735 to 1818) designed a timber arch bridge over the Neva River having a span of about 300 m, illustrating one of the attempts in searching for efficient structural form [2]. He tested a 1/10 scale model bridge to investigate the adequacy of members and found a large strength in the new structural system. However, for unknown reasons, the bridge was not built.

66.2.2   Iron and Steel Bridges

The extensive progress in bridge construction in the beginning of the 19th century was influenced by overall industrial development. A number of cast-iron arch bridges for roadway and railway traffic were built. At about the same time construction of steel suspension bridges was started in Petersburg. In 1824 the Panteleimonovsky Bridge over the Fontanka River having a span of about 40 m was built. In 1825 the pedestrian Potchtamsky Bridge over the Moika Bankovsky was built, and the Lion's and Egyptian Bridges over the Fontanka River having a span of 38.4 m were constructed in 1827.

The largest suspension bridges built in the 19th century were the chain suspension bridge over the Dnepr River in Kiev (Ukraine), with a total length 710 m including six spans (66.3 and 134.1 m) and

FIGURE 66.2   Bridge over the Luga River (1857).

approaches, constructed in 1853. In 1851 to 1853 two similar suspension bridges were constructed over the Velikaya River in Ostrov City with spans of 93.2 m. The development in suspension bridge systems was based on an invention of wire cables. In Russia, one of the first suspension bridges using wire cables was built in 1836 near the Brest-Litovsk fortress. This bridge crossed the River West Bug, having a span of about 89 m. However, suspension bridges of the first half of the 19th century, due to the lack of structural performance understanding, had inadequate stiffness in both vertical and horizontal directions. This appeared to be the main cause of a series of catastrophes with suspension bridges in different countries. Any occurrence of catastrophes in Russia was not noted. To reduce the flexibility of suspension bridges, at first timber and then steel stiffening trusses were applied. However, this innovation improved the performance only partly, and the development of beam bridges became inevitable in the second half of the 19th century.

The first I-beam bridge in Russia was the Semenovsky Bridge over the Fontanka River in Petersburg constructed in 1857 and the bridge over the Neman River in Kovno on the railway of Petersburg–Warsaw constructed in 1861. However, I-beams for large spans proved to be very heavy, and this dictated a wider application of truss systems.

The construction of the railway line in 1847 to 1851 between Petersburg and Moscow required a large number of bridges. Zhuravsky modified the structural system of timber truss implemented by Howe in 1840 to include continuous systems. These bridges over the Rivers Volga, Volhov, and some others had relatively large spans; e.g., the bridge over the Msta River had a 61.2 m span. This structural system (known in Russia as the Howe–Zhuravsky system) was further widely used for bridge construction up to the mid-20th century.

The steel truss bridges at first structurally repeated the types of timber bridges such as plank trusses or lattice trusses. A distinguished double-track railway bridge with deck truss system, designed by Kerbedz, was constructed in 1857 over the Luga River on the Petersburg–Warsaw railway line (Figure 66.2). The bridge consists of two continuous spans (each 55.3 m.). The P-shape cross sections were used for chord members, and angles for diagonals. Each track was carried by a separate superstructure which includes two planes of trusses spaced at 2.25 m. The other remarkable truss bridge built in 1861 was the Borodinsky Bridge over the Moskva River in Moscow, with span length of 42.7 m. The cornerstone of the 1860s was an introduction of a caisson foundation for bridge substructures.

Up to the 1880s, steel superstructures were fabricated of wrought steel. Cast-steel bridge superstructures appeared in Russia in 1883. And after the 1890s, wrought steel was no longer used for superstructures. In 1884, Belelubsky established the first standard designs of steel superstructures covering a span range from 54.87 to 109.25 m. For spans exceeding 87.78 m, polygonal trusses were designed. A typical superstructure having a 87.78-m span is shown in Figure 66.3. Developments in structural theory and technology advances in the steel industry expanded the capabilities of shop fabrication of steel structures and formed a basis for further simplification of truss systems and an increase of panel sizes. This improvement resulted in application of a triangular type of trusses. By the end of the 19th century, a tendency to transition from lattice truss to triangular truss was outlined, and Proskuryakov initiated using a riveted triangular truss system in bridge superstructures. The first riveted triangular truss bridge in Russia was constructed in 1887 on the Romny–Krmenchug railway line.

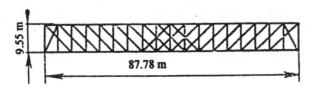

FIGURE 66.3   Standard truss superstructure.

Many beam bridges built in middle of the 19th century were of a continuous span type. Continuous bridge systems have economic advantages, but they are sensitive to pier settlement and have bigger movement due to temperature change. To take these aspects into account was a complicated task at that time. In order to transfer a continuous system to a statically determined cantilever system, hinges were arranged within spans. This was a new direction in bridge construction. The first cantilever steel railway bridge over the Sula River designed by Proskuryakov was built in 1888. The first steel cantilever highway bridge over the Dnepr River in Smolensk was constructed. The steel bridge of cantilever system over the Dnestr River, a combined railway and highway bridge, having a span of 102 m, was designed by Boguslavsky and built in 1894. In 1908, the steel bridge over the River Dnepr near Kichkas, carrying railway and highway traffic and with a record span of 190 m, was constructed. Development of new techniques for construction of deep foundations and an increase of live loads on bridges made the use of continuous systems more feasible compared with the other systems.

In the middle of the 19th century arch bridges were normally constructed of cast iron, but from the 1880s, steel arch bridges started to dominate the cast iron bridges. The first steel arch bridges were designed as fixed arches. Hinged arch bridges appeared later and became more widely used. The need to apply arches in plain areas led to the creation of depressed through-arch bridges.

66.3   Modern Development

The 20th century has been remarkable in the rapid spread of new materials (reinforced concrete, prestressed concrete, and high-strength steel), new structural forms (cable-stayed bridges), and new construction techniques (segmental construction) in bridge engineering. The steel depressed through arch truss railway bridge over the Moskva River built in 1904 is shown in Figure 66.4. To reduce a thrust, arches of cantilever system were used. For example, the Kirovsky Bridge over the Neva River having spans of 97 m was built in Petersburg in 1902. Building new railway lines required construction of many long, multispan bridges. In 1932, two distinguished arch steel bridges designed by Streletsky were constructed over the Old and New Dnepr River (Ukraine) having main spans of 224 m and 140 m, respectively (Figure 66.5).

The first reinforced concrete structures in Russian bridge construction practice were culverts at the Moscow–Kazan railway line (1892). In the early 20th century, the use of reinforced concrete was limited to small bridges having spans of up to 6 m. In 1903, the road ribbed arch bridge over the Kaslagach River was built (Figure 66.6). This bridge had a total length of 30.73 m and a length of arch span of 17 m. In 1904, the road bridge over the Kazarmen River was constructed. The bridge had a total length of 298.2 m and comprised 13 reinforced concrete arch spans of 21.3 m, having ribs of box section.

The existing transportation infrastructure of Russia is less developed compared with other European countries. The average density of railway and highway mileage is about five times less than that of the United States. For the past two decades, railway and highway construction activities have slowed down, but bridge design and construction have dramatically increased.

FIGURE 66.4   Steel depressed arch railway bridge over the Moskva River.

FIGURE 66.5   Steel arch bridge with a span of 224 m.

66.3.1   Standardization of Superstructures

An overview of the number and scale of bridges constructed in Russia shows that about 70% of railway bridges have a span length less than 33.6 m, and 80% of highway bridges have less than 42 m. Medium and small bridges are, therefore, predominant in construction practice and standard structural solutions have been developed and used efficiently, using standardized design features for modern bridges.

The current existing standardization covers the design of superstructures for certain bridge types. For railway bridges, standard designs are applicable to spans from 69 to 132.0 m. These are reinforced concrete superstructures of slab, stringer types, box girder; steel superstructures of slab, stringer types; composite superstructures; steel superstructures of through plate girder, deck truss, through-truss types. For highway bridges, standard designs cover the span range from 12 to 147 m. These are reinforced concrete superstructures of voided slab, stringer, channel (P-shaped) girder, solid web girder, box girder types, composite superstructures of steel web girder types, and steel super-structures of web and box girder types.

Modern highway bridges having spans up to 33 m and railway bridges up to 27.6 m are normally constructed with precast concrete simple beams. For highway bridges, continuous superstructures of solid web girder types of precast concrete segments are normally used for spans of 42 and 63 m, and box girder type of precast box segments are used for spans of 63 and 84 m. The weight of

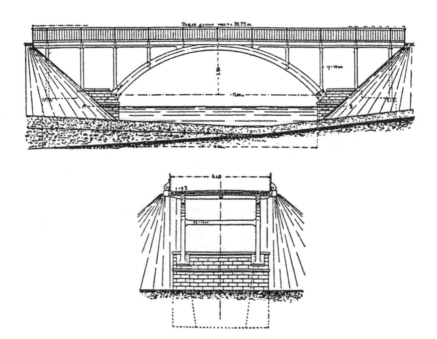

FIGURE 66.6   Reinforced concrete ribbed arch bridge over the Kashlagash River (1903).

precast segments does not exceed 60 tons and meets the requirements of railway and highway transportation clearances. For railway bridges, steel box superstructures of full span (33.6 m) shop-fabricated segments have become the most widespread in current practice.

The present situation in Russia is characterized by a relative increase in a scale of application of steel superstructures for bridges. After the 1990s, a large number of highway steel bridges were constructed. Their construction was primarily based on modularization of superstructure elements: shop-fabricated segments having a length of up to 21 m. Many of them were built in the city of Moscow, and on and over the Moscow Ring Road, as well as the bridges over the Oka River in Nizhni Novgorod, over the Belaya River in the Ufa city, and many others. A number of steel and composite highway bridges having spans ranging from 60 to 150 m are currently under design or construction. In construction of bridges over the Moskva, Dnepr, Oka, Volga, Irtish, Ob Rivers, on the peripheral highway around Ankara (Turkey), on the Moscow Ring Road, and some others, continuous steel superstructures of web and box girder types permanent structure depth are typically used. The superstructures are assembled with modularized, shop-fabricated elements, which are welded at the shop or construction site to form a complete cross section configuration. Erection is normally accomplished by incremental launching or cantilever segmental construction methods.

The extensive use of steel bridges in Russia is based on optimum structural solutions, which account for interaction of fabrication technology and erection techniques. The efficiency is proved by high-quality welded connections, which allow erection of large prefabricated segments, and a reduction in quantity of works and in the construction period. A low maintenance cost that can be predicted with sufficient accuracy is also an advantage, while superstructures of prestressed reinforced concrete in some cases require essential and frequently unpredictable expenses to ensure their capacity and durability.

66.3.2 Features of Substructure

Construction of bridge piers in Russia was mainly oriented on the use of precast concrete segments in combination with cast-in-place concrete. The usual practice is to construct piers with columns of uniform rectangular or circular sections fixed at the bottom of the foundations. For highway bridges with spans up to 33 m full-height precast rectangular columns of 50 × 80 cm are of standard design. For longer-span bridges, the use of precast contour segments forming an outer shape of piers and cast-in-place methods has become more widespread.

A typical practice is use of driven precast concrete piles for the pile foundation. Standard types of precast concrete piles are of square section 0.35 × 0.35, 0.40 × 0.40, 0.45 × 0.45 m and of circular hollow section of 0.60 m diameter. Also, in the last two decades, CIDH (cast-in-place drilled holes) piles of 0.80 to 1.7 m in diameter have been widely used for foundations.

An increasing tendency is to use pile shafts especially in urban areas, when superstructures are borne directly by piles extended above the ground level (as columns). The efficiency was reached by implementing piles of square sections 35 × 35, 40 × 40 cm, and of circular Section 80 cm in diameter. Bored and cast-in-place piles of large diameter ranging between 1.6 to 3.0 m drilled to a depth of up to 50 m and steel casings are widely employed in bridge foundations. In foundations for bridges over the rivers and reservoirs, bored piles using a nonwithdrawable steel casing within the zone of change in water level and scour depth are normally used. These foundation types were applied for construction of bridges over the Oka River on the peripheral road around Nizhni Novgorod, over the Volga River in Kineshma City, over the Ob River in Barnaul City, over the Volga River in Ulyanovsk City, and some others.

Open abutments are most commonly used for highway bridges. Typical shapes are bank seats, bank seats on piles, and buried skeleton (spill-through). Wall abutments and bank seat on piles are the types of substructure mainly used for railroad bridges. Wingwalls are typically constructed back from the abutment structure and parallel to the road.

66.4 Design Theory and Methods

66.4.1 Design Codes and Specifications

The Russian Bridge Code SNIP 2.05.03-84 [5] was first published in 1984, amended in 1991, and reissued in 1996. In Russia the new system of construction codes was adopted in 1995. In accordance with this system the bridge design must satisfy the requirements of the bridge code, local codes, and industry standards. The Standards introduce new requirements resolving inconsistencies found in the bridge code. The bridge code covers design of new and the rehabilitation of existing bridges and culverts for highways, railways, tramways, metro lines, and combined highway–railway bridges. The requirements specified are for the location of the structures in all climatic conditions in the former USSR, and for seismic regions of magnitude up to 9 on the Richter scale. The bridge code has seven main sections: (1) general provisions, (2) loads, (3) concrete and reinforced concrete structures, (4) steel structures, (5) composite structures, (6) timber structures, and (7) foundations.

In 1995, the Moscow City Department of Transportation developed and adapted "Additional Requirements for Design and Construction of Bridges and Overpasses on the Moscow Ring Highway" to supplement the bridge code for design of the highway widening and rehabilitation of the 50 bridge structures on the Moscow Ring Highway. The live load is increased by 27% in the "Additional Requirements." In 1998 the Moscow City Department issued the draft standard TSN 32 "Regional Building Norms for Design of Town Bridge Structures in Moscow" [6] on the basis of the "Additional Requirements." The new standard specifies an increased live load and abnormal loading and reflects the necessity to improve the reliability and durability of bridge structures. The final TSN 32 was issued in 1998.

66.4.2 Design Concepts and Philosophy

In the former Soviet Union, the ultimate strength design method (strength method) was adopted for design of bridges and culverts in 1962. Three limit states — (1) the strength at ultimate load,

(2) deformation at service load, and (3) cracks width at service load — were specified in the bridge design standard, a predecessor of the current bridge code. Later, the limit states 2 and 3 were combined in one group. The State Standard: GOST 27751-88 "Reliability of Constructions and Foundations" [4] specifies two limit states: strength and serviceability. The first limit state is related to the structural failure such as loss of stability of the structure or its parts, structural collapse of any character (ductile, brittle, fatigue) and development of the mechanism in a structural system due to material yielding or shear at connections. The second limit is related to the cracking (crack width), deflections of the structure and foundations, and vibration of the structure.

The main principles for design of bridges are specified in the Building Codes and Regulations — "Bridges and Culverts" SNIP 2.05.03-84 [5]. The ultimate strength is obtained from specified material strengths (e.g., the concrete at maximum strength and usually the steel yielding). In general, bridge structures should satisfy ultimate strength limit in the following format:

$$\gamma_d S_d + \gamma_l (1+\mu) S_l \eta \le F\left(m_1, m_2, \gamma_n, \gamma_m, R_n, A\right) \tag{66.1}$$

where S_d and S_l are force effects due to dead load and live load, respectively; γ_d and γ_l are overload coefficients; μ is dynamic factor; η is load combination factor; F is function determining limit state of structure; m_1 is general working condition factor accounting for possible deviations of constructed structure from design dimensions and geometrical form; m_2 is coefficient characterizing uncertainties of structure behavior under load and inaccuracy of calculations; γ_n is coefficient of material homogeneity; γ_m is working condition factor of material; R_n is nominal resistance of material; A is geometric characteristic of structure element.

The serviceability limit state requirement is

$$f \le \Delta \tag{66.2}$$

where f is design deformation or displacement and Δ is ultimate allowable deformation or displacement.

The analysis of bridge superstructure is normally implemented using three-dimensional analysis models. Simplified two-dimensional models considering interaction between the elements are also used.

66.4.3 Concrete Structure Design

Concrete structures are designed for both limit states. Load effects of statically indeterminate concrete bridge structures are usually obtained with consideration for inelastic deformation and cracking in concrete. A proper consideration is given to redistribution of effects due to creep and shrinkage of concrete, forces adjustment (if any), cracking, and prestressing which are applied using coefficients of reliability for loads equal to 1.1 or 0.9.

The analyses to the strength limit state include calculations for strength and stability at the conditions of operation, prestressing, transportation, storing, and erection. The fatigue analysis of bridge structures is made for operation conditions.

The analyses to the serviceability limit state comprise calculations for the same conditions as indicated above for the strength limit state. The bridge code stipulates five categories of requirements to crack resistance: no cracks; allowing a small probability of crack formation (width opening up to 0.015 mm) due to live-load action on condition that closing cracks perpendicular to longitudinal axis of element under the dead load is assured; allowing the opening of cracks after passing of live load over the bridge within the limitations of crack width opening 0.15, 0.20, 0.30 mm, respectively.

The bridge code also specifies that the ultimate elastic deflections of superstructures are not to exceed, for railroad bridges, $L/600$ and, for highway bridges, $L/400$. The new standard TSN 32 provides a more strict limit of $L/600$ to the deflections of highway bridges in Moscow City.

66.4.4   Steel Structure Design

Steel members are analyzed for both groups of limit states. Load effects in elements of steel bridge structures are determined usually using elastic small deformation theory. Geometric nonlinearity is required to be accounted for in the calculation of systems in which such an account causes a change in effects and displacements more than 5%. The strength limit states for steel members are limited to member strength, fatigue, general stability, and local buckling. The calculations for fatigue are obligatory for railroad and highway bridges.

For steel superstructures in calculations for strength and stability to the strength limit state the code requires consideration of physical nonlinearity in the elastoplastic stage. Maximum residual tensile strain is assumed as 0.0006, and shear strain is equal to 0.00105.

The net sections are used for strength design of high-strength bolt (friction) connections and the gross sections for fatigue, stability, and stiffness design. A development of limited plastic deformations of steel is allowed for flexural members in the strength limit state. The principles for design of steel bridges with consideration for plastic deformations are reviewed in a monograph [9]. Stability design checks include the global flexure and torsion buckling as well as flange and web local buckling.

A composite bridge superstructure is normally based on a hypothesis of plane sections. Elastic deformations are considered in calculations of effects that occur in elements of statically indeterminate systems as well as in calculations of strength and stability, fatigue, crack control, and ordinates of camber.

66.4.5   Stability Design

Piers and superstructures of bridges are required to be checked for stability with respect to over-turning and sliding under the action of load combinations. The sliding stability is checked with reference to a horizontal plane. The working condition factors of more than unity should also be applied for overturning and driving and less than unity for resisting forces.

66.4.6   Temporary Structures

The current design criteria for temporary structures used for bridge construction are set forth in the Guideline BSN (Department Building Norms) 136-78 [10]. This departmental standard was developed mainly as an addition to the bridge code and also some other codes related to bridge construction. The BSN 136-78 is a single volume first published in 1978 and amended in 1984. These guidelines cover the design of various types of temporary structures (sheet piling, cofferdams, temporary piers, falsework, etc.) and devices required for construction of permanent bridge structures. It specifies loads and overload coefficients, working condition factors to be used in the design, and requirements for design of concrete and reinforced concrete, steel, and timber temporary structures. Also, it provides special requirements for devices and units of general purpose, construction of foundations, forms of cast-in-place structures, and erection of steel and composite superstructures.

The temporary structures are designed to the two limit states similar to the principles established for permanent bridge structures. Meanwhile, the overload coefficients and working condition factors have a lower value compared to that of permanent structures. The recent study has shown that the Guideline BSN 136-78 requires revision, and therefore initial recommendations to improve the specified requirements and some other aspects have been reviewed in Reference [11].

The BSN 136-78 specifies a 10-year frequency flood. Also on the basis of technical-economical justification up to 2-year return period may be taken in the design, but in this case special measures for high-water discharge and passing of ice are required. The methods of providing technical hydrologic justification for reliable functioning of temporary structures are reviewed in the Guideline [12]. To widen the existing structures, the range of the design flood return period for temporary structures in the direction of lower and higher probabilities of exceedance is also recommended [12].

66.5   Inspection and Test Techniques

The techniques discussed herein reflect current requirements set forth in SNIP 3.06.07-86, the rules of inspection and testing for bridges and culverts [13]. This standard covers inspection and test procedures of constructed (new) or rehabilitated (old) bridges. Also, this standard is applicable for inspection and tests of structures currently under operation or for bridges designed for special loads such as pipelines, canals, and others. Inspection and testing of bridges are implemented to determine conditions and to investigate the behavior of structures. These works are implemented by special test organizations and contractor or operation agencies.

All newly constructed bridges are to be inspected before opening to traffic. The main intention of inspection is to verify that a bridge meets the design requirements and the requirements on quality of works specified by SNIP 3.06.04-91 [14]. Inspection is carried out by means of technical check up, control measurements, and instrumentation of bridge structural parts. If so required, inspection may additionally comprise nondestructive tests, laboratory tests, and setting up long-term instrumentation, etc. Results obtained during inspection are then compared to allowable tolerances for fabrication and erection specified in SNIP 3.06.04-91. If tolerances or other standard requirements are breached, the influence of these noted deviations on bridge load-carrying capacity and the service state are estimated.

Prior to structural testing, various details should be precisely defined on the basis of inspection results. Load tests require elaborate safety procedures to protect both the structure and human life. Maximum load, taking into account the design criteria and existing structural deviations, needs to be established. The position of structure (before testing is started) for future identification of changes resulting from load tests needs to be recorded (marked up). For the purpose of dynamic load tests, conditions of load passing over the bridge are evaluated.

66.5.1   Static Load Tests

During static load testing, displacements and deformations of the structure and its parts, stresses in typical sections of elements, local deformations (crack opening, displacements at connections, etc.) are measured. Moreover, depending on the type of structure, field conditions, and the testing purpose, measurements of angle strain and load effects in stays or struts may be executed.

For static load tests a bridge is loaded by locomotives, rolling stock of railways, metro or tramway trains, trucks as live loads. In cases where separate bridge elements are tested or the stiffness of the structure is determined, jacks, winches, or other individual loads may be needed. Load effects in members obtained from tests should not exceed the effects of live loads considered in the design accounting for an overload factor of unit and the value of dynamic factor taken in the design. At the same time, load effects in members obtained from the test are not to be less than 70% of that due to design live loading. Weight characteristics of transportation units used for tests should be measured with an accuracy of at least 5%. When testing railroad, metro, tramway bridges, or bridges for heavy trucks, load effects in a member normally should not be less than those due to the heaviest live loads passing over a given bridge.

Quantities of static load tests depend on bridge length and complexity of structures. Superstructures of longer span are usually tested in detail. In multispan bridges having similar equal spans, only one superstructure is tested in detail; other superstructures are tested on the basis of a reduced program, and thus only deflections are measured.

During testing, live loads are positioned on deck in such a manner that maximum load effects (within the limits outlined above) occur in a member tested. Time for test load carrying at each position is to be determined by a stabilization of readings at measuring devices. Observed deformation increments within a period of 5 min should not exceed 5%. In order to improve the accuracy of measurements, time of loading, unloading, and taking readings is to be minimized as much as possible. Residual deformations in the structure are to be determined on the basis of the first test loading results. Loading of structures by test load should normally be repeated. The number of repeated loading is established considering the results that are obtained from the first loading.

66.5.2 Dynamic Load Tests

Dynamic load tests are performed in order to evaluate the dynamic influence (impact factor) of actual moving vehicles and to determine the main dynamic characteristics of the structure (free oscillation frequency and oscillation form, dynamic stiffness, and characteristics of damping oscillation). During dynamic load testing, overall structure displacements and deformations (e.g., midspan deflections, displacements of superstructure end installed over movable bearings), as well as, in special cases, displacements and stresses in individual members of the structure are measured.

The heavy vehicles that may really pass over the bridge are to be used in dynamic load tests to determine dynamic characteristics of structures and moving impact. Vibrating, wind, and other loads may be used for dynamic tests. To investigate oscillations excited by moving vehicles, the trucks are required to pass over the bridge at different speeds, starting from a speed of 10 to 15 km/h. This allows us to determine the behavior of the structure within a range of typical speeds. It is recommended to conduct at least 10 heats of trucks at various speeds and repeat those heats, when increased dynamic impact is noted. In some cases when motorway bridges are tested, to increase the influence of moving vehicles (e.g., to ascertain dynamic characteristics of the structure), a special measure may be applied. This measure is to imitate the deck surface roughness e.g., by laying planks (e.g., 4 cm thick) perpendicular to the roadway spaced at the same distance as the distance between the truck wheels.

When testing pedestrian bridges, excitation of structure free oscillations is made by throwing down a load or by a single pedestrian or a group of pedestrians walking or running over bridges. Throwing down a load (e.g., castings having a weight of 0.3 to 2 t) creates the impact load on a roadway surface typically from 0.5 to 2.0 m height. The location of test load application should coincide with the section where maximum deflections have occurred (midspan, cantilever end). To protect the roadway surface a sand layer or protective decking is placed. The load is dropped down several times and each time the height is adjusted. The results of these tests give diagrams of free oscillations of superstructures. Load effects in structural members during the test execution should not exceed those calculated in the design as stipulated in the section above.

66.5.3 Running in of Bridge under Load

To reveal an adequate behavior of structure under the heaviest operational loads, running in of bridges is conducted. Running in of railroad and metro bridges is implemented under heavy trains, the bridges designed for AB highway loading run-in by heavy trailers. Visual observations of structure behavior under load are performed. Also, midspan deflection may be measured by simple means such as leveling. A number of at least 12 load passes (shuttle-type) with different speeds are recommended in the running procedures of railroad and metro bridges. The first two or three passes are performed at a low speed of 5 to 10 km/h; if deflections measurements are required, the trains are stopped. Positioning trailers over marginal lane having 10 m spacing between the back and front wheels of adjacent units is recommended in running in bridges designed for AB highway loading of two or more lanes. It is recommended that single trailers pass over the free lane at a speed of 10 to 40 km/h, and a number of passes are normally taken, at least five. When visual observations are completed, trailers are moved to another marginal lane and single trailers pass over the lane, which is set free. For running in of single-lane bridges, the passes of single trailers only are used.

FIGURE 66.7 Typical through-truss bridge of 44 m span on the Baikal-Amour Railway Bridge Line.

66.6 Steel and Composite Bridges

In recent decades, there has been further development in design and construction of steel bridges in Russia. New systems provide a higher level of standardization and reduce construction cost.

66.6.1 Superstructures for Railway Bridges

Most railroad bridges are steel composite bridges with single-track superstructure. Standard super-structures [15] are applicable to spans from 18.2 to 154 m (steel girder spans 18.2 to 33.6 m; composite girder spans 18.2 to 55 m; deck truss spans 44 to 66 m; and through-truss 33 to 154 m). Figures 66.7 and 66.8 show typical railroad bridges. Steel box girders as shown in Figure 66.9 have a span of 33.6 m. Similar box girders can be assembled by connecting two prefabricated units. These units are shop-welded and field-bolted with high-strength bolts of friction type. More details are given in References [3] and [15]. For truss systems, the height of trusses is from 8.5 to 24 m with panels of 5.5 and 11 m. Two types of bridge deck — ballasted deck with roadbed of reinforced concrete or corrosion-resistant steel and ballastless deck with track over wooden ties or reinforced concrete slabs — are usually used. For longer than 154 m or double-track bridge, special design is required.

66.6.2 Superstructures for Highway Bridges

Standard steel superstructures covering spans from 42 to 147 m are based on modularization of elements: 10.5 m length box segment, 21 m double-T segment, and 10.5 m orthotropic deck segment. Typical standard cross sections are shown in Figure 66.10. Main technological features of shop production of steel bridge structures have been maintained and refined. Automatic double arc-welding machines are used in the 90% shop fabrication.

66.6.3 Construction Techniques

Steel and composite superstructures for railway bridges are usually erected by cantilever cranes (Figure 66.11) having a capacity up to 130 tons and boom cranes with a larger capacity. The superstructures of 55-m-span bridges may be erected by incremental launching method using a

FIGURE 66.8 Deck girder composite bridge of 55 m span over the Mulmuta River (Siberia).

FIGURE 66.9 Steel box superstructure of 33.6 m span for railway bridges.

nose. When the cantilever crane is applied to erection of the superstructure of 55 m span, a temporary pier is required. For truss superstructures, cantilever and semicantilever erection methods are widely implemented. Figure 66.12 shows the cantilever erection of bridge over the Lena River (span of 110 + 132 + 110 m) using derrick cranes.

For highway bridges, the incremental launching method has been the main erection method for plate girder and box girder bridges since 1970, although this method has been known in Russia for a long time. Cantilever and semicantilever methods are also used for girder bridges. Floating-in is an effective method to erect superstructures over waterways when a large quantity of assembled superstructure segments are required. The application of standard pontoons simplified the assembly of the erection floating system. Equipping of floating temporary piers by an air leveling system and other special equipment made this erection method more reliable and technological.

66.6.4 Typical Girder Bridges

Pavelesky Railroad Overhead

Design of overpasses in Moscow was a really challenge to engineers. It requires low construction depth and minimum interruption of traffic flow. Figure 66.13 shows the Pavelesky Overhead built

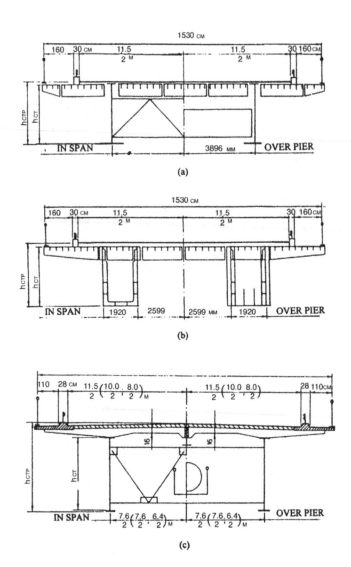

FIGURE 66.10 Typical highway superstructure cross sections. (a) Steel plate sirder; (b) box girder; (c) composite girder.

in 1996 over the widened Moscow Ring Road. The bridge has a horizontal curve of $R = 800$ m and carries a triple-track line. The steel superstructure (Figure 66.14) was designed of low construction depth to meet the specified 5.5 m highway clearance and to maintain the existing track level. This new four-spans (11 + 30 + 30 + 11 m) skewed structure was designed as simple steel double-T girders with a depth of 1.75 m. The orthotropic plate deck of a thickness of 20 mm with inverted T-ribs was used. An innovation of this bridge is the combination of a cross beam (on which longitudinal ribs of orthotropic deck are borne) and a vertical stiffener of the main girder connected to the bottom flange, thus forming a diaphragm spaced at 3 m. A deck cover sheet in contact with ballast was protected by metallized varnish coating.

Moskva River Bridge

Figure 66.15 shows the Moskva River Bridge in the Moscow region, built in 1983. The superstructure is a continuous three-span (51.2 + 96 + 51.2 m) twin box girders (depth of 2.53 m) with orthotropic deck. The bridge carries two lanes of traffic in each direction and sidewalks of 3 m. In transverse

FIGURE 66.11   Erection of steel box superstructure of 33.6 m by boom cranes.

FIGURE 66.12   Cantilever erection of typical truss bridge using derrick cranes.

FIGURE 66.13   Paveletsky Railroad Overhead.

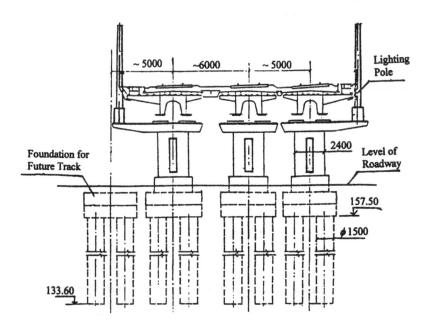

FIGURE 66.14   Typical cross-section of Paveletsky Railroad Overhead (all dimensions in mm).

direction, boxes are braced, connected by cross frames at 9 m. Piers are of reinforced concrete Y-shaped frames. Foundation piles are 40 × 40 cm section and driven to a depth of 16 m. The box girders were launched using temporary piers (Figure 66.16) from the right bank of the Moskva River. Four sliding devices were installed of each pier. The speed of launching reached 2.7 m/h.

Oka River Bridge

Oka River Bridge, near Nizhny Novgorod City, consisting of twin box section (Figure 66.17) formed with two L-shaped elements with orthotropic deck was open to traffic in 1993. A steel bridge of 988 m length is over the Oka River with a span arrangement of 2 × 85 + 5 × 126 + 2 × 84 m.

FIGURE 66.15   Moskva River Bridge in Krilatskoe.

FIGURE 66.16   Superstructure launching for Moskva River Bridge in Krilatskoe.

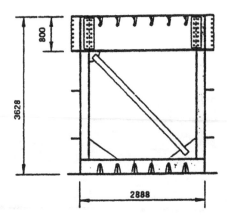

FIGURE 66.17   Cross section of main box girder formed of two L-shaped elements (all dimensions in mm).

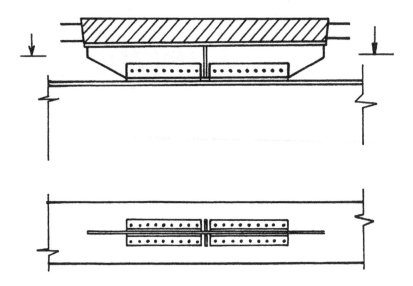

FIGURE 66.18   Connection details of precast reinforced concrete deck and steel girder.

Ural River Bridge

The Ural River Bridge near Uralsk City was completed in 1998. This five-span (84 + 3 × 105 + 84 m) continuous composite girder bridge (depth of 3.6 m) on a gradient of 2.6 m carries a single lane of highway with overall width of 14.8 m. Precast concrete deck segments are connected with steel girders by high-strength friction bolts (Figure 66.18). Connection details are presented in Reference [3].

Chusovaya River (Perm-Beresniki) Highway Bridge

The Chusovaya River Bridges, with a length of 1504 m and on convex vertical curves in radius of 8000 and 25,000 m was completed in 1997. The superstructure comprises two continuous steel composite girders (4 × 84 + 84 + 126 + 5 × 147 + 126 + 84 m). One main problem of bridge construction is the construction of pier foundations within the river under complex geologic and hydrologic conditions. All river piers were constructed under protection of sheet pilling with the use of scows. Piers were constructed using floating cranes. The superstructure segments having lengths of 94.8, 84, and 99.7 m were assembled on the right bank, and then slipped to the river and placed over the floating pier.

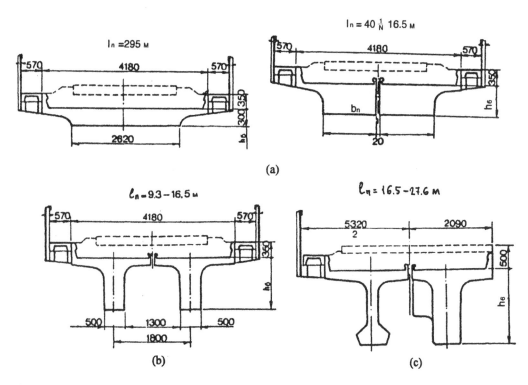

FIGURE 66.19 Typical cross sections of concrete superstructures for railway bridges (all other dimensions in mm). (a) Nonprestressed reinforced concrete slab; (b) nonprestressed reinforced concrete I-beam; (c) prestressed concrete T-beam.

66.7 Concrete Bridges

About 90% of modern bridges are concrete bridges. The recent decades have been characterized by intensive development of standardized precast concrete bridge elements. Main operational requirements and various conditions of construction are taken into account by the existing standardization. This allows a flexible approach in solving architectural and planning tasks and construction of bridges of various span lengths and clearances. Precast concrete bridges, both railway and highway, have been built in many environments, ranging from urban to rural areas. General characteristics of the common types of concrete bridge structures are provided in the following subsections.

66.7.1 Superstructures for Railway Bridges

For railway bridges, standardized shapes for girders and slabs have been widely used for span lengths of 2.95 to 27.0 m. The simplest type of bridge superstructure is the deck slab, which may be solid or voided. The standard structures are designed to carry live load S14 (single-track) and may be located on curved sections of alignment having a radius of 300 m and more. Typical nonprestressed concrete slabs with structural depth from 0.65 to 1.35 m are applicable for spans ranging 2.95 to 16.5 m. Nonprestressed precast T-beams with depth from 1.25 to 1.75 m are used for spans ranging from 9.3 to 16.5 m. Prestressed concrete T-beams with depths from 1.75 to 2.6 m are applicable for spans from 16.5 to 27.6 m. Single-track bridges consist of two precast full segments connected at diaphragms by welding of steel joint straps and then pouring concrete.

All three types of precast superstructure segments are normally fabricated at the shop and transported to a construction site. The waterproofing system is shop-applied. Longitudinal gaps between the segments are covered by steel plates. All superstructure segments starting from 1 to 5 m

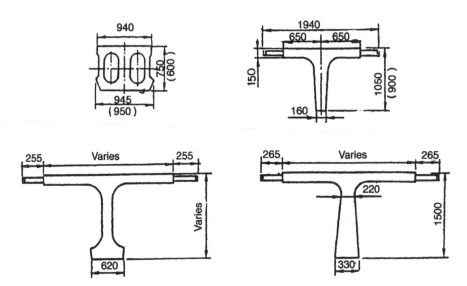

FIGURE 66.20 Standard precast beams for highway bridges (all dimensions in mm).

FIGURE 66.21 Typical section of overhead on the Moscow Ring Road.

length are placed on steel bearings. Superstructures may be connected into a partially continuous system, thus allowing adjustment of horizontal forces transferred to piers.

66.7.2 Superstructures for Highway Bridges

More than 80% of bridges on federal highway networks have span lengths not exceeding 33 m. Figure 66.20 shows typical standard cross sections for highway bridges. Nonprestressed concrete void slabs with structural depth from 0.6 to 0.75 m are applicable for spans ranging 12 to 18 m. Nonprestressed precast T-beams with depth from 0.9 to 1.05 m are used for spans ranging from 11.1 to 17.8 m. Prestressed concrete T-beams with depths from 1.2 to 1.7 m are applicable for spans from 32.3 to 41.5 m. Figure 66.21 shows a construction site of precast concrete beams on the Moscow Ring Road.

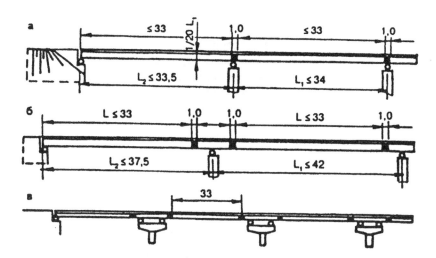

FIGURE 66.22 Typical continuous bridge schemes composed of standard beams (all dimensions in mm).

Fabrication of pretensioned standard beams are conducted at a number of specialized shops. However, by experience some technological difficulties have been noted. These are operations related to installation of reinforcement (space units) into molds of a complex configuration, e.g., bulbous bottom of the girder section, placing of concrete into and taking off the beam from these molds. To improve the fabrication procedure, a special shape of the beam web in the form of a drop has been developed. The standard design of a beam having a 27 m length has been elaborated. Precast prestressed T-beams which are the most widespread in the current construction practice have structural depth from 1.2 to 1.7 m with typical top width of 1.8 m and weight of one beam from 32.3 to 59 tons.

Where special transport facilities are not available or transportation limitations exist or manufacturing facilities such as stressing strands are expensive, precast-in-segment post-tensioned beams are considered to be beneficial. The designs of standard T-beams have been worked out considering both prestressing systems: post-tensioning and pretensioning. The standard T-beams have been designed precast in segments with subsequent post-tensioning for the span lengths of 24, 33, and 42 m.

Due to transportation limitations or restrictions or other reasons, a rational alternative for the post-tensioned beams is beams with transverse joints. In this case the beam consists of segments of limited length and weight (up to 11.8 tons) which may be transported by the usual means. Such segments may be precast either on site or in short lengths at the factory with subsequent post-tensioning at the site. The joints between segments may be implemented by filling the joint gap of 20 to 30 mm thickness with concrete (thin joint); by placing concrete of a minimum thickness of 70 mm (thick joint); or epoxy glued having a 5 mm thickness.

To reduce expansion joints and improve road conditions, the partially continuous system — simple beams at the erection stage and continuous beams at service stage — has been widely used for superstructure; spans up to 33 m. The girders are connected by casting the deck slab over support locations slabs are connected by welding of steel straps on the top of deck. Figure 66.22 shows a typical continuous bridge scheme composed of standard precast slabs or T-beams.

Figure 66.23 shows a standard design for continuous superstructures with double-T beams. The overall width of the precast segment may reach 20 m, but its length is limited to 3 m due to transportation constraints. Ducts for prestressing cables are placed in the web only.

In 1990, another standard design for box-girder continuous systems was initiated. A typical box section is shown in Figure 66.24. Each segment has 1.4 m at bent and 2.2 m within spans with weight up to 62 tons. Using standard precast segments, superstructures can have spans of 63, 71.8, 84, and 92.8 m. Due to financial difficulties the design work on standard box segmental superstructures has not been completed. However, the general idea has been implemented at a number of bridges.

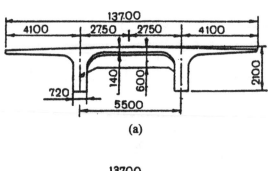

(a)

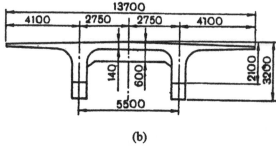

(b)

FIGURE 66.23 Typical cross sections of standard solid web girders. (a) For span arrangement of 33 + *n*(42) + 33 m; (b) for span arrangement of 42 + *n*(63) + 42 m (all dimensions in mm).

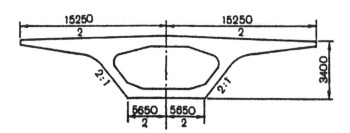

FIGURE 66.24 Typical cross section of precast box segment (all dimensions in mm).

66.7.3 Construction Techniques

The precast full-span beam segments are fabricated at specialized shops on the basis of standard designs. The erections of solid girder segments up to 33 m are conducted by mobile boom cranes of various capacities, gantry cranes, and launching gantries. Special scaffoldings have been designed to erect the solid girder of 24 to 63 m (Figure 66.25).

When large-span bridges are constructed, different cranes may be used simultaneously. At a low-water area of rivers, gantry cranes may be applied, but at deep water, cranes SPK-65 or others may be used. A typical erection scheme of precast cantilever bridge over the Volga River built in 1970 is shown in Figure 66.26.

Although concrete bridge superstructures are constructed mainly by precast segments, in recent years cast-in-place superstructure construction has been reviewed on a new technological level. By this method, construction of superstructure is organized in the area behind the abutment on the approaches to a bridge. The successive portion of the superstructure is cast against the preceding segment and prestressed to it before proceeding to erection by the method of incremental launching.

Cast-in-place concrete main girders and modified precast concrete decks are usually used. Traditionally, when precast deck slabs were used for bridge construction, these precast elements were fabricated with provision of holes for shear connectors that were later filled with concrete. This

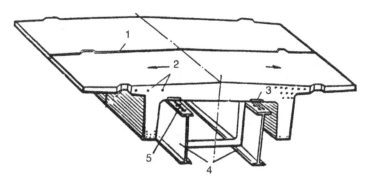

FIGURE 66.25 Position of precast segments of solid web girders over special scaffoldings. 1–glued joints; 2–ducts; 3–embeds; 4–movable special scaffoldings; 5–rails for segment moving.

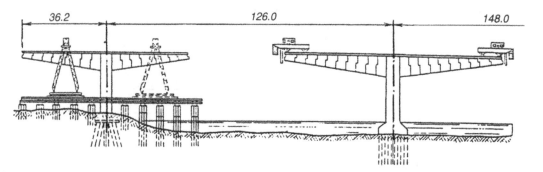

FIGURE 66.26 Typical erection scheme of the bridge over the Volga River (all dimensions in mm).

solution has several disadvantages. To improve the practice, new types of joint between structural members were introduced. The use of steel embeds in precast slabs allows connection of main girders by means of angles and high-strength bolts [3].

66.7.4 Typical Bridges

Komarovka Bridge

A railway bridge over the Komarovka River has recently been built in Ussuryisk (far east of Russia). The bridge of 106.85 m length has a span arrangement of 6 × 16.5 m (Figure 66.27). The superstructure is of reinforced concrete beams of standard design. The intermediate piers are of cast-in-place reinforced concrete. Abutments are spill-through type of precast elements and cast-in-place concrete. Foundations are on bored 1.5-m-diameter piles. Steel casings of 1.35 m diameter were placed in the top portion of the piles. Separate connections of the superstructure into the two systems reduced the temperature forces in the girt. Structurally, the girt (Figure 66.28) consists of two angles of 125 × 125 × 10 mm, which are jointed to the bearing nodes by means of gussets, installed between the bottom flange of superstructure and the sole plate of the fixed bearings. The girt is attached to each gusset plate by high-strength bolts. To reduce the temperature stresses, the girts were fixed to the superstructure and embeds over the abutments at 0° C.

Kashira Oka River Bridge

In 1995, a bridge of total length of 1.96 km crossing over the Oka River near Kashira was constructed. The main spans over the navigation channel are 44.1 + 5 × 85.5 + 42.12 m. The bridge carries three traffic lanes in each direction. Seven-span continuous superstructures are precast concrete box

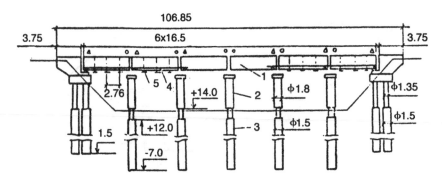

FIGURE 66.27 General arrangement of bridge over the Komarovka River. 1–RC super structure; 2–cast-in-place pier; 3–bored pile; 4–special girt; 5–key to limit sideward movement (all dimensions in mm).

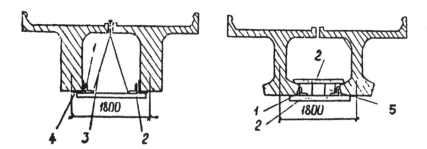

FIGURE 66.28 Special girt details. (Left) Nonprestressed and (right) prestressed girder. 1–girt; 2–Crosstie; 3–anchor element; 4–noftlen pad; 5–jointing plank (all dimensions in mm).

segments with depth of 3.4 m and width of 16 m constructed using a cantilever method with further locking in the middle of the spans. A typical cross section is shown in Figure 66.29. The precast segments vary from 1.5 to 1.98 m and are governed by a capacity of erection equipment limited to 60 tons. Piers are cast-in-place, slip-formed. Foundations are on bored piles of 1.5 and 1.7 m diameter.

Frame Bridge with Slender Legs

Development of structural forms and erection techniques for prestressed concrete led to a construction of fixed rigid frames for bridges. Compared to continuous-span frame bridge systems it is less commonly used. To form a frame system, precast superstructure and pier elements are concreted at overpier section (1 m along the bridge) and at the deck section (0.36 m wide in a transverse direction). Figure 66.30 shows a typical frame bridge with slender legs. The design foundations need an individual approach depending on the site geologic conditions. Simple forms of precast elements, low mass, clear erection scheme, and aesthetic appearance are the main advantages of such bridge systems.

Buisky Perevoz Vyatka River Bridge

This highway bridge (a nonconventional structure) as shown in Figure 66.31 was open to traffic in 1985. The bridge is a cantilever frame system with a suspended span of 32.3 m. The superstructure is a single box rectangular box girder with depth of 3.75 m and width of 8.66 m. Overall deck width is 10 m. The river piers are of cast-in-place concrete and the foundations are on bored 1.5 m piles penetrated to a depth of 16 m. One of the piers is founded on a caisson placed at a depth of 8 m. The 32-mm-diameter bars of pile reinforcement were stressed with 50 kN force per bar for better crack resistance.

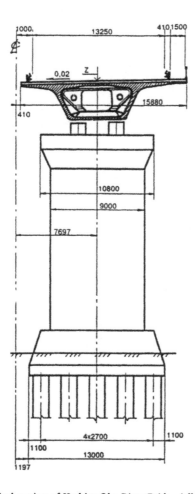

FIGURE 66.29 Typical section of Kashira Oka River Bridge (all dimensions in mm).

Penza Sura River Bridge

This highway bridge comprises a precast prestressed concrete frame of a two-hinge system (Figure 66.32) built in 1975. The superstructure is three boxes with variable sections (Figure 66.33) with a total of 66 prestressing strands. The inclined legs of frames have a box section of 25 × 1.5 m at the top and a solid section 1.45 × 0.7 m at the bottom. The legs of each frame are reinforced by 12 prestressing strands. Each strand diameter of 5 mm consists of 48 wires. Piers are cast-in-place concrete. Foundations are on driven hollow precast concrete piles of 0.6 m. Each frame structure was erected of 60% precast segments. These segments are 5.6 m wide and 2.7 to 3.3 m long. Each frame system superstructure was erected with 12 segments in a strict sequence, starting from the center of the span to the piers, The span segments are placed into the design position, glued, and prestressed. Leg segments are connected with span segments by cast-in-place concrete. A special sequence, as shown in Figure 66.34, for tensioning the strands was established for this bridge.

Moscow Moskva River Arch Bridge

The cantilever arch bridge (Figure 66.35) over the Moskva River on the Moscow Ring Road was built in 1962. The bridge is a three-span structure with 48.65 + 98 + 48.64 m. The overall road width is 21 m and sidewalks are 1.5 m on each side. Half arches are connected by a tie of 10 prestressing strands at the level of the roadway. The erection of the superstructure was implemented on steel scaffoldings. Arches are erected of precast elements weighing from 10 to 20 tons.

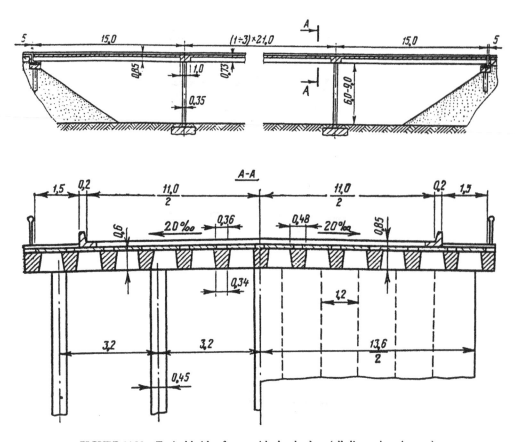

FIGURE 66.30 Typical bridge frame with slender legs (all dimensions in mm).

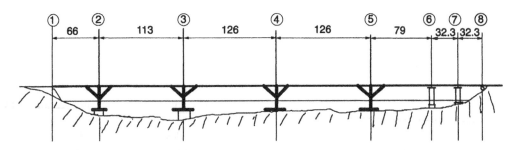

FIGURE 66.31 General scheme of Buisky Perevoz Vyatka River Bridge (all dimensions in mm).

66.8 Cable-Stayed Bridges

The first cable-stayed bridges were constructed in the former USSR during the period 1932–1936. The cable-stayed highway bridge with a span of 80 m designed by Kriltsov over the Magna River (former Georgian SSR) was constructed in 1932. The bridges over the Surhob River, having a span of 120.2 m, and over the Narin River, having a span of 132 m, were constructed in 1934 and 1936, respectively. Figure 66.36 shows a general view of the Narin River Bridge. A stiffening girder of steel truss system was adopted in these bridges.

FIGURE 66.32 Penza Sura River Bridge.

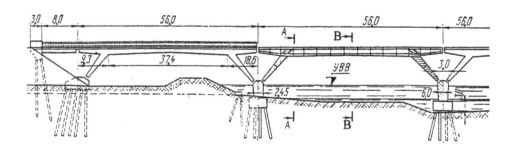

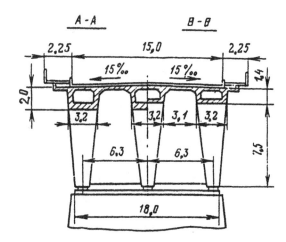

FIGURE 66.33 Typical cross section of Penza Sura River Bridge (all dimensions in mm).

The modern period of cable-stayed bridge construction may be characterized by the following projects: the Dnepr River Bridge in Kiev (1962), the Moscow Dnepr River Bridge in Kiev (1976), Cherepovets Scheksna River Bridge (1980), Riga Daugava River Bridge (1981), the Dnepr River South Bridge in Kiev (1991). Two cable-stayed bridges over the Volga River, Uiyanovsk, and over the Ob River near Surgut are currently under construction.

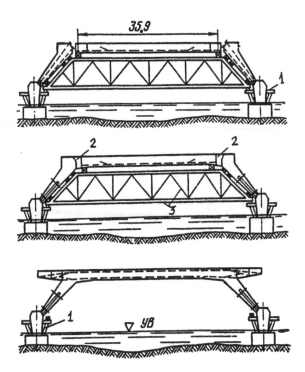

FIGURE 66.34   General sequence of frame structure erection (all dimensions in mm).

FIGURE 66.35   Moscow Moskva River arch bridge.

Kiev Dnepr River Bridge

The first concrete cable-stayed bridge (Figure 66.37) crossing over the harbor of the Dnepr River in Kiev was constructed in 1962. The three-span cable-stayed system has spans of 65.85 + 144 + 65.85 m. The bridge carries highway traffic having a width of roadway of 7 m and sidewalks of 1.5 m on each side. The superstructure comprises two main II-shaped prestressed concrete beams of 1.5 m deep, 1.4 m wide, and spaced at 9.6 m. Cable arrangement is of radiating shape. The stays are composed of strands of 73 and 55 mm in diameter. Towers are cast-in-place reinforced concrete structure.

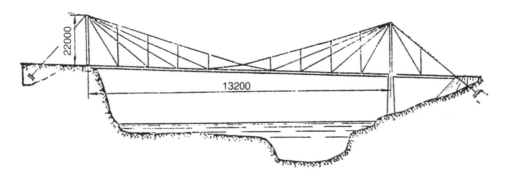

FIGURE 66.36   Narin River Bridge (all dimensions in mm).

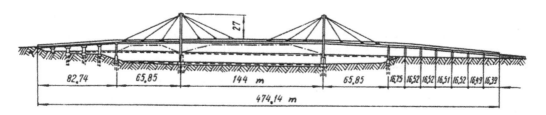

FIGURE 66.37   Kiev Dnepr River Bridge.

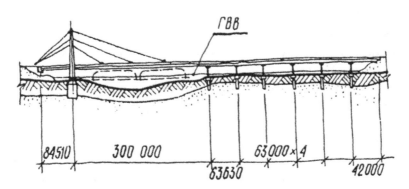

FIGURE 66.38   Moscow Dnepr River Bridge (all dimensions in mm).

Moscow Dnepr River Bridge

In 1976 the Moscow bridge (Figure 66.38) with a cable-stayed system was constructed. The bridge carries six lanes of traffic and five large-diameter pipes below the deck, and has an overall width of 31 m. The three-span continuous structure has a span arrangement of 84.5 + 300 + 63 m. The stiffening girder comprises twin steel box beams with orthotropic deck fabricated of 10 XCND low-alloyed steel grade. To meet the transportation clearances, the depth of the girder was limited to 3.6 m. In a cross section the main beams are 5.5 m wide with a distance between inner webs of adjacent girders equal to 20.2 m and diaphragms spaced at 12.5 m. The stiffening girder has a fixed connection to the abutment and movable connections at the pylon and intermediate pier. A shaped single reinforced concrete pylon 125 m high has a box section of its legs. Each stay is formed from 91 parallel galvanized wires (diameter 5 mm). The stays have a hexagonal section of 55 × 48 cm and are installed in two inclined planes.

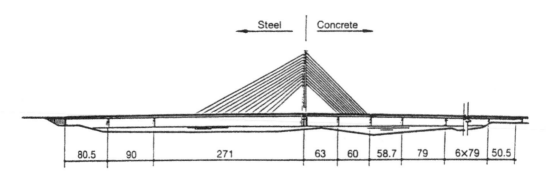

FIGURE 66.39 Dnepr River South Bridge (all dimensions in mm).

Dnepr River South Bridge

The south bridge crossing over the Dnepr River in Kiev (Ukraine) was opened to traffic in 1993. This bridge crossing includes a cable-stayed bridge of a length of 564.5 m and a concrete viaduct of a length of 662 m. A general scheme of the bridge is shown in Figure 66.39. The bridge allows traffic of four lanes and two rail tracks. The bridge also carries four large-diameter water pipes. The design and construction features of this bridge have been presented in Reference [19].

The superstructure comprises steel and concrete portions. The steel portion is a three-span continuous box girder formed of vertical I-beams with orthotropic deck. The concrete portion provides the required counterweight and was constructed of segmental prestressed concrete box sections. The pylon is a two-column cast-in-place concrete frame structure having cross struts between the columns. Cables stays are positioned in two planes, thus torsional rigidity of the bridge is effectively provided.

Ulyanovsk Volga River Bridge

The bridge crossing over the Volga River near Ulyanovsk City is more than 5 km long and includes the cable-stayed bridge. This bridge is currently under construction and will carry combined traffic: highway traffic of four lanes and two tracks for streetcar lines. A span arrangement of this cable-stayed bridge is based on spans of 220 + 2 × 407 + 220 m with a single pylon. The scheme of the cable-stayed bridge is shown in Figure 66.40.

The stiffening girder is of trapezoidal configuration comprising two planes of truss interacting with an orthotropic top and bottom deck plate. The steel orthotropic deck for highway traffic is located on top of the truss superstructure. Truss members are hermetically sealed; therefore their inner surfaces are not required to be painted. A peculiarity of structural detailing for the joints is that the node in a form of hermetically sealed welded box was fabricated at the shop. Connections of flanges and diagonals are moved from the node center. Longitudinal ribs of the orthotropic deck system are of box section; transverse beams are spaced at 5.5 m. The stiffening girder is supported by two planes of stays. Each stay includes cables of parallel wires with diameter of 7 mm. The number of wires in the cables varies from 127 to 271. The weight of stiffening girder of cable-stayed bridge is 20,115 t, and stays, 1900 t. The superstructure is erected by the cantilever method.

The approach steel superstructure is also of truss system with continuous spans of 2 × 220 m. This steel superstructure differs from that of the main span by a transverse rectangular shape only. The structural details are similar to the main-span superstructure. The weight of 2 × 20 m superstructure is 7640 t. The superstructure is assembled on the bank and then erected by 220 m spans using the floating-in method.

The cast-in-place pylon of 204 m height has an inverted Y-shape in the direction along the bridge and a frame of 2 h (H configuration) with cross struts in the transverse direction. For a general

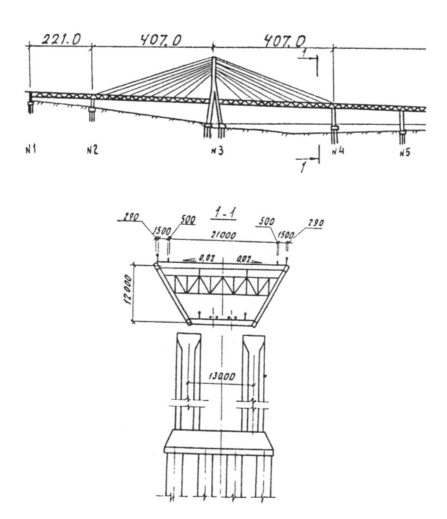

FIGURE 66.40 Ulyanovsk Volga River Bridge (all dimensions in mm).

view of the design alternative for the pylon see Figure 66.41. Towers have a hollow box section with thickness of walls equal to 0.8 to 1.0 m. This type of the pylon structure was influenced by symmetrical side spans of 407 m and absence of permanent guys. The pylon interacts with the superstructure under the unsymmetrical loading by means of its bending stiffness. Foundations are on bored piles of 1.7 m diameter having a bell shape of 3.5 m in diameter at the end.

Surgut Ob River Bridge

Recently, the construction of new bridge crossing the River Ob near Surgut City has started. The overall length of this bridge crossing is a little more than 2 km. A general scheme of the bridge is shown in Figure 66.42. The bridge has an overall width of 15.2 m and will allow traffic of two lanes. It is located in profile on a convex curve having a radius of 120,000 m. The superstructure is a single steel box girder with orthotropic deck. A single pylon of 146 m high is to be constructed in the bottom portion of precast segments forming an outer shape and cast-in-place concrete core, and in the upper portion of two parallel steel towers (transverse section) with struts creating a frame. Intermediate piers are constructed of precast segments with cast-in-place concrete core and with foundation on bored piles with steel casing of 1420 mm in diameter. Abutment are of cast-in-place concrete with foundation on reinforced concrete piles of hollow section 0.6 m in diameter and filled by concrete.

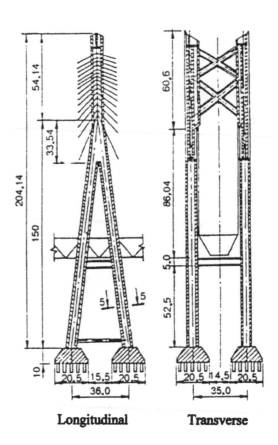

Longitudinal   **Transverse**

FIGURE 66.41   Ulyanovsk Volga River Bridge pylon (design alternative) (all dimensions in mm).

66.9 Prospects

In recent years, fewer new bridges have been designed and constructed in Russia. The demands for rehabilitation and strengthening of bridge structures are increasing every year. The future directions of bridge design practice are

- Revision and modification of national standards considering Eurocode and standards of other leading countries;
- Considerations for interactions of structural solutions with technological processes considering aesthetic, ecological, and operational requirements;
- Development of new structure forms such as precast or cast-in-place reinforced concrete and prestressing concrete, steel and composite structures for piers and superstructures to improve reliability and durability;
- Redesign of standard structures considering practical experience in engineering, fabrication, and erection practices;
- Unification of shop-fabricated steel elements ready for erection to a maximum dimension fitting the transportation requirements, improvement of precast concrete decks, improvement of corrosion protection systems to a life span up to 12 years; and
- Development of relevant mobile equipment and practical considerations of cast-in-place concrete bridges.

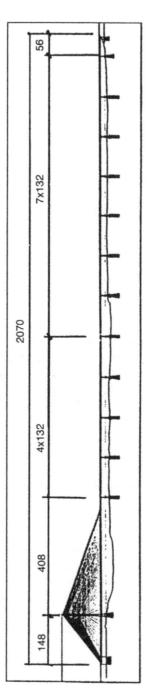

FIGURE 66.42 Surgut Ob River Bridge (all dimensions in mm).

References

1. Belyaev, A.V., *Moskvoretsky's Bridges*, Academy of Science of the USSR, Moscow, Lenningrad, 1945, chap. 1 [in Russian].
2. Evgrafov, G. K. and Bogdanov, N. N., *Design of Bridges*, Transport, Moscow, 1996, chap. 1, [in Russian].
3. Monov, B. and Seliverstov, V., Erection of composite bridges with precast deck slabs, in *Proceedings of Composite Construction Conventional and Innovative*, IABSE, Zurich, 1997, 531.
4. GOST, 27751-88 (State Standard), *Reliability of Constructions and Foundations, Principal Rules of Calculations*, Grosstroy of USSR, Moscow, 1989 [in Russian].
5. SNIP 2.05.03-84 (Building Norms and Regulations), *Bridges and Culverts*, Ministry of Russia, Moscow, 1996 [in Russian].
6. TSN 32, Regional Building Norms for Design of Town Bridge Structures in Moscow, 1st Draft, Giprotransmost, Moscow, 1997 [in Russian].
7. GOST, 9238-83 (State Standard), *Construction and Rolling Stock Clearance Diagrams for the USSR Railways of 1520 mm Gauge*, Grosstroy of USSR, Moscow, 1988 [in Russian].
8. GOST, 26775-97 (State Standard), *Clearances of Navigable Bridge Spans in the Inland Waterways, Norms and Technical Requirements*, Grosstroy, Moscow, 1997 [in Russian].
9. Potapkin, A. A., *Design of Steel Bridges with Consideration for Plastic Deformation*, Transport, Moscow, 1984 [in Russian].
10. BSN 136-78 (Departmental Building Norms), *Guidelines to Design of Temporary Structures and Devices for Construction of Bridges*, Minstransstory, Moscow, 1978, amended 1984 [in Russian].
11. Seliverstov, V. A., Specified requirements to determine forces from hydrologic and meteorological factors for design of temporary structures, *Imformavtodor*, 8, Moscow, 1997 [in Russian].
12. Perevoznikov, B. F., Ivanova, E. N., and Seliverstov, V. A., Specified requirements and recommendations to improve the process of design of temporary structures and methods of hydrologic justification of their functioning, *Imformavtodor*, 7, Moscow, 1997 [in Russian].
13. SNIP 3.06.07-86 (Building Norms and Regulations), *Bridges and Culverts, Rules of Inspection and Testing*, Grosstroy of USSR, Moscow, 1988 [in Russian].
14. SNIP 3.06.04-91 (Building Norms and Regulations), *Bridges and Culverts*, Grosstroy of USSR, Moscow, 1992 [in Russian].
15. Popov, O. A., Monov, B., Kornoukhov, G., and Seliverstov, V., Standard structural solutions in steel bridge design, in *Proceedings of 2nd World Conference on Steel in Construction*, May 11–13, 1998, The Steel Construction Institute, Elsevier, 1998.
16. GOST 6713-91 (State Standard), *Low Alloyed Structural Rolled Stock for Bridge Building*, Grosstroy, Moscow, 1992 [in Russian].
17. Zhuravov, L. N., Chemerinsky, O. I., and Seliverstov, V. A., Launching steel bridges in Russia, *Struct. Eng. Int.*, 6(3), IABSE, Zurich, 1996.
18. Popov, O. A., Chemerinsky, O. I., and Seliverstov, V. A., Launching construction of bridges: the Russian experience, in *Proceedings of International Conference on New Technologies in Structural Engineering*, Lisbon, Portugal, July 2–5, Vol. 2, LNEC/IABSE, Zurich, 1997.
19. Korniyiv M. H. and Fuks, G. B., The South Bridge: Kiev, Ukraine, *Struct. Eng. Int.*, 4(4), IABSE, Zurich, 1994.
20. Kriltsov, E. I., Popov, O. A., and Fainstein, I. S., *Modern Reinforced Concrete Bridges*, Transport, Moscow, 1974 [in Russian].

67

The Evolution of Bridges in the United States

Norman F. Root
California Department of Transportation

0-8493-7434-0/00/$0.00+$.50
© 2000 by CRC Press LLC

67.1   Introduction

American civilization with its bridges is relatively recent compared with the ancient civilizations of Asia, Europe, and even South America. The Americas are the last continents to have become heavily populated and industrialized.

The evolution of bridges in the United States is probably not much different from anywhere else in the world. Civilizations have borrowed their bridging ideas from each other for centuries. Fallen logs across streams served as primitive bridges that led to the concept of girder spans in use today. Suspension spans across deep chasms is a primitive idea used throughout the world. The stone arch introduced by the ancient Romans is a naturally occurring, efficient, and pleasing structural shape that has been used with various evolving materials.

FIGURE 67.1   The aqueduct bridge at La Purisima Mission, Santa Barbara County, California, is an example of a primitive bridge, a short-span stone slab. Built in 1813, it is the oldest bridge in California. (Courtesy of California Department of Transportation.)

Bridge practice evolves as user needs, traffic, and vehicles change, technology progresses, and new materials are developed. But span length is still the primary determining factor for bridge type selection.

67.2   Early U.S. Bridges

The first recorded bridge in the United States was built at James Towne Island, Virginia in 1611. This is the site of one of the earliest European colonies. It was a timber structure, actually a wharf accessing ships anchored in deeper water (Figure 67.1).

67.3 The Canal Era

By water was an early method of heavy transport as the United States began to expand inland from the Eastern Seaboard. Canal builders in the late 1700s and early 1800s were the first to construct U.S. bridges of any consequence. The concept of stone arches, borrowed from Roman aqueducts, was common during this era. Besides, the stone arch readily adapts to the loads imposed (Figure 67.2).

FIGURE 67.2 Scholarie Creek Aqueduct is the Erie Canal over Scholarie Creek at Fort Hunter, New York. It was built by John Jervis in 1841. Canals were the first major users of bridges in the United States. (Courtesy of American Society of Civil Engineers.)

Turnpikes

Private toll roads during the colonial period, 1600s and 1700s, often built timber structures. Logs are natural beams and their ready availability made them natural materials for early bridges.

Timber Bridges

Timber is easy to work and build with. But timber bridges require constant maintenance; joints loosen as the wood shrinks and vibrates from traffic, and wood must be protected from the elements (Figure 67.3).

FIGURE 67.3 Dolan Creek Bridge on the Monterey Coast in California was built in 1932. This is one of only two three-pin timber arch bridges ever built on the California State Highway system. It lasted only a few years, and has since been replaced with a concrete bridge in 1961. (Courtesy of California Department of Transportation.)

Covered Timber Bridges

Many timber bridges of the 19th century were covered to protect the wood from the elements and in northern climates to keep snow off the decks (Figures 67.4 and 67.5).

FIGURE 67.4   The Bridgeport Covered Bridge in California may be the longest single-span, 70.1 m, covered bridge in the world. The superstructure is a Burr arch superimposed on a Howe truss. It was a toll bridge built by David Wood in 1862, and was later purchased by the Virginia Turnpike Company. (Courtesy of California Department of Transportation.)

FIGURE 67.5 The Cornish–Windsor Covered Bridge is a two-span town-lattice truss crossing the Connecticut River between Cornish, New Hampshire and Windsor, Vermont. Built in 1866 it is the longest covered bridge, 140.2 m, in the United States. It has been designated a National Civil Engineering Landmark by the American Society of Civil Engineers. (Courtesy of American Society of Civil Envineers.)

Iron Bridges

Cast-iron bridge members were first considered due to the proximity of several foundries near the National Road. The material turned out to be quite strong and very durable. Cast iron is resistant to normal corrosion associated with ferrous metals (Figures 67.6 and 67.7).

FIGURE 67.6 Dunlap's Creek Bridge, built in 1839, is the first iron bridge in the United States. It was built for the National Cumberland Road, at Brownsville, Pennsylvania, by Captain Richard Delafield of the Army Corps of Engineers. The bridge is still in service today. (Courtesy of Federal Highway Administration.)

FIGURE 67.7 Bow Bridge in Central Park, New York, is the oldest surviving wrought-iron bridge in the United States, built in 1862. It has the longest span, 26.5 m, of five ornately decorated bridges in the park, all designed by Calvert Vaux and Jacob Wrey Mould. (Courtesy of American Society of Civil Engineers.)

67.4 The Railroad Era

The age of steam ushered in an era where bridge building in the United States came of age. Railroads became the dominant mode of transportation for both passengers and freight. Easy grades required for railroads, in turn, required lots of bridges. Canals were all but forgotten and wagon roads went into a 50-year period of neglect (Figure 67.8).

FIGURE 67.8 Starrucca Viaduct, built in the form of the ancient Roman aqueducts, was designed by James Kirkwood for the New York and Erie Rail Road in 1848. It is located over the Starrucca Creek plain at Lanesboro, Pennsylvania. This was the first bridge to use a concrete foundation. This bridge is still in service. (Courtesy of American Society of Civil Engineers.)

Trusses

Squire Whipple and Herman Haupt, two American railroad bridge engineers, are credited with being the first to calculate methods for determining stresses in truss members and were thereby able to determine their appropriate sizes. Each worked independently of the other, in the mid-19th century, using ancient knowledge of mathematics, physics, and strength of materials.

The knowledge to engineer trusses made their construction popular. They provided strength with considerable savings in materials and weight. The concepts of rational principles are equally applicable to both timber and metal trusses. Many other engineers quickly embraced the concepts and patented various truss diagonal configurations for their own use. Many of their names are familiar today: Pratt, Parker, Howe, Burr, Fink, and Warren, to name a few.

Railroad Trestles

See Figures 67.9 through 67.11.

FIGURE 67.9 Theodore Judah took advantage of timber to build trestles quickly and move on, while racing to build the Central Pacific Railroad, the California end of the Transcontinental Railroad. He solved the long-term maintenance problem by later filling in the trestle with cut and tunnel spoil, forming an embankment which would remain long after the timber had rotted away. This is the Secrettown Trestle in the California Sierras, built in 1865, being buried in earth fill. (Courtesy of California State Library.)

FIGURE 67.10 The Devil's Gate High Bridge at Georgetown, Colorado, appears too spindly to support a railroad. But clever use of tension counters distributes the reversing loads throughout the towers. The bridge was prefabricated by Clark Reeves and Company of Phoenixville, Pennsylvania, for the Colorado Central Railroad in 1884. The trestle was in continuous use until torn down in 1939. A replica rebuilt in 1984 is now in use by the Georgetown Loop Mining and Railroad Park. (Courtesy of Missouri Historical Society.)

FIGURE 67.11 Keddie Wye is a unique steel tower trestle built by the Union Pacific Railroad in California's rugged Feather River Canyon in 1912. The wye trestle emerges from a tunnel in the south wall of the canyon splitting rail traffic over the river; one leg heads north to meet with the Burlington Northern Railroad and the other is the main line heading east toward Chicago. (Courtesy of the Feather River Rail Society.)

Steel Arch Bridges

See Figures 67.12 through 67.14.

FIGURE 67.12 Eads' Bridge over the Mississippi River at Washington Street in Saint Louis shattered engineering precedents of the time. It was the first extensive use of steel for bridge construction. The three 175+ m arch spans are each four 464-mm steel truss-stiffened wrought iron tubes. The spandrels are extensive steel truss and lattice work. Built by James B. Eads in 1874. Eads' Bridge is pictured on the two dollar denomination United States postage stamp series commemorating the Trans-Mississippi Exposition of 1896. (Courtesy of U.S. Bureau of Engraving and Printing.)

FIGURE 67.13 Navajo Bridge at Marble Canyon, near Lee's Ferry, Arizona, is the classic example of an arch sprung between canyon walls. This is also an example of a deck truss, an evolution for automobiles, beyond the through truss. When built, in 1929, it was the highest bridge in the world, 162.5 m, from deck to water. It was designed by Ralph Hoffman of the Arizona Highway Department. A parallel twin designed by Cannon Associates has since been constructed, in 1996. (Courtesy of American Society of Civil Engineers.)

FIGURE 67.14 The Cold Springs Canyon steel plate girder arch, in Santa Barbara County, California, is the longest arch span at 213.4 m, and a rise of 121.9 m. The bridge has won a Lincoln Foundation welding award, American Institute of Steel Construction beauty award, and the Governor's Design Award. Built in 1963, it was designed by the California Division of Highways, Marv Shulman, design engineer. (Courtesy of California Department of Transportation.)

Kit Bridges

During the late 19th and early 20th centuries, several bridge companies sold "American Standard," prefabricated wrought iron bridge pieces (bridge in a box), of given span lengths that could be erected on site. All one had to do was order a bridge from a catalog, build abutments for the appropriate span length, and assemble the pieces erector-set-style. Kit bridges are readily adaptable to disassembly, transport, and reuse elsewhere, as has been the case for many of these bridges still in use (Figures 67.15 and 67.16).

FIGURE 67.15 Laughery Creek Bridge, near Aurora, Indiana, was built by the Wrought Iron Bridge Company of Canton, Ohio, in 1878. Its 92-m span was unprecedented. This bridge appeared on the cover of the company's catalog in 1893. (Courtesy of American Society of Civil Engineers.)

FIGURE 67.16 This detail at Haupt Creek, in Sonoma County, California, shows a typical pin connection of a kit bridge and a "Phoenix Column," a patented cast-iron member built exclusively by the Phoenix Iron Works of Pennsylvania. This bridge was built in 1880. (Courtesy of California Department of Transportation.)

67.5   The Motor Car Era

Almost instantaneously, at the turn of the 20th century, the nation was swept up into the automotive age. Long-neglected wagon roads became important once again. State Highway Departments sprang up and road and bridge building, under the "Good Roads Movement," took on a new fervor. Railroad engineering became almost stagnant. Most new highway bridge engineers were former railroad bridge engineers, so many of the early highway bridges looked just like railroad through-truss bridges.

Steel Truss Bridges

See Figures 67.17 through 67.19.

FIGURE 67.17   The Carquinez Straits Bridge in California, built by the American Toll Bridge Company as a private toll bridge in 1927, is an example of a cantilevered truss with eye bar tension members. A parallel twin using welded hybrid high-strength steels was designed and built in 1954 by the California Division of Highways, Roger Sunbury, engineer. Steel truss bridges are considered by many to be ugly. Carquinez is not one of the worst examples, but when a candidate bridge architect interviewing for the California Department was shown a picture of the twin spans and asked for comments, he answered, "Why make the same mistake twice?" He got the job as Chief Bridge Architect. (Courtesy of California Department of Transportation.)

FIGURE 67.18 Coos Bay Bridge on the Oregon Coast Highway is one of several landmark bridges designed by Conde B. McCullough of the Oregon Highway Department. The 225.2 m main span is a classic example of a cantilever truss. Built in 1936, it is the largest of McCullough's coastal gems. The concrete arch end spans and spires are a McCullough trademark. The bridge is now named the McCullough Memorial Bridge in honor of the engineer. (Courtesy of American Society of Civil Engineers.)

FIGURE 67.19 The San Francisco–Oakland Bay Bridge east, is part of the longer 13.3 km crossing composed of the west suspension span, a tunnel through Yerba Buena Island, and this cantilever truss east span. The seismic retrofitting solution at this site is to replace the bridge. There is local controversy over the type of span to be used. There are cost concerns, fear by San Francisco that an east side signature span could overshadow their west suspension span, and aspirations by Oakland that their city is also deserving of a signature span on their side of the Bay. (Courtesy of California Department of Transportation.)

Reinforced Concrete

About the same time as the motor car era began, the turn of the 20th century, the concept of reinforced concrete was introduced. It was generally unaccepted until the San Francisco earthquake of 1906. The few reinforced concrete buildings were the only structures to survive. From that time on, reinforced concrete has been widely used (Figure 67.20).

FIGURE 67.20 Alvord Lake Bridge is the first reinforced concrete bridge, built by Ernest Ransome, the developer of reinforced concrete, in 1888. This bridge is still in service carrying State Route 1 over Golden Gate Park in San Francisco. The facia is hammered to resemble familiar stone arch work. The bridge is a National Historic Civil Engineering Landmark. (Courtesy of California Department of Transportation.)

Concrete Arches

Reinforced concrete arches were popular during the early part of the 20th century. Reinforced concrete was the modern material, and arches were a comfortable, tried, and true shape. Thousands of reinforced concrete arches were built until the 1950s (Figures 67.21 through 67.27).

FIGURE 67.21 The Colorado Street Bridge over the Arroyo Seco in Pasadena, California, is the highest scoring bridge for historical significance in the state. The main span is 46.6 m with a height of 45.7 m. The structure is highly adorned with Beaux Art ornamentation. It was designed in 1912 by John Waddell, the "Dean" of American bridge engineering. The bridge served the famed Route 66 for many years. Seismic retrofitting was a challenge in trying to maintain the bridge's historic aesthetic features. (Courtesy of California Department of Transportation.)

FIGURE 67.22 Fern Bridge near Ferndale, California, is a remarkable structure that has withstood the test of time. Six major floods since it was built have washed out other bridges on the lower Eel River, but Fernbridge still stands. It is composed of seven 61-m rubble-filled closed spandrel concrete arches, each on 250 timber piles. it was designed by John B. Leonard in 1911 for Humboldt County. It is now part of the California State Highway system. (Courtesy of California Department of Transportation.)

FIGURE 67.23 Harlan D. Miller (Dog Creek) Bridge is an example of state-of-the-art bridge development by the State of California under Bridge Engineer Harlan D. Miller in 1926. The State Legislature named the bridge in his honor for the great strides he accomplished with state bridges. Miller died only a week after receiving the honor, so the bridge became the Harlan D. Miller Memorial Bridge. (Courtesy of California Department of Transportation.)

FIGURE 67.24 Bixby Creek Bridge on the Monterey Coast in California is one of the most picturesque and photographed bridges in California. This Monterey Coast Highway was the first designated Scenic Highway in California, in 1961. The route is also the first to be designated an All American Road. Built in 1932, it has a main span of 109.7 m and is 79.2 m above the streambed. Construction required 26 stories of falsework. It was designed by Harvey Stover of the California Division of Highways. Seismic retrofitting is complicated due to aesthetic restrictions established by historical preservation codes. (Courtesy of California Department of Transportation.)

FIGURE 67.25 Conde McCullough, of the Oregon State Highway Department, designer of the two bridges shown in Figures 67.25 and 67.26, gained fame as the designer of several landmark bridges on the Oregon Coast Highway. The Rogue River Bridge at Gold Beach, Oregon, is a typical open spandrel concrete arch. The monumental spires at the abutment piers are a McCullough trademark. Both of these bridges were built in 1932. (Courtesy of Oregon Department of Transportation.)

FIGURE 67.26 The double-tiered concrete arch end spans at Cape Creek, on the Oregon Coast, are reminiscent of Roman aqueducts. The north-bound highway at this point emerges from a tunnel providing a picturesque view of the Heceta Head Lighthouse, as the traveler glides out over Cape Creek. (Courtesy of American Society of Civil Engineers.)

FIGURE 67.27 The Lilac Road arch gracefully frames the southern entrance to the fertile San Luis Rey Valley, of Southern California. It was built over Interstate Route 15 in San Diego County in 1978. The designer was Fred Michaels of the California Department of Transportation. (Courtesy of California Department of Transportation.)

Concrete Girders

See Figures 67.28 and 67.29.

FIGURE 67.28 Rockcut Bridge, owned by Stevens and Ferry Counties, earned a Portland Cement Association design award in 1997. The designer was Nicholls Engineering. (Courtesy of Portland Cement Association.)

FIGURE 67.29 The North Santiam (Gates) Bridge, in Marion County, Oregon, earned a Portland Cement Association design award in 1997. It was designed by the Oregon Department of Transportation. (Courtesy of Portland Cement Association.)

Canticrete

See Figure 67.30.

FIGURE 67.30 The landmark Alsea Bay Bridge at Waldport, Oregon, is the only Conde McCullough bridge that has required replacement due to deterioration. The new structure is reminiscent of McCullough's style and utilizes the best of the two most popular building materials, concrete and steel. The concrete-covered steel members are called canticrete. This new span, built in 1992, was designed by Howard Needles Tamman Bergendoff. It won awards from both the American Institute of Steel Construction and the Portland Cement Association. (Courtesy of Howard Needles Tamman Bergendoff.)

Suspension Bridges

Suspension bridges are one of the oldest concepts in the world. The first recorded suspension bridge in the United States was a chain-link catenary over Jacobs Creek in 1801 at Uniontown, Pennsylvania. Suspension bridges have continued to be a favored type into modern times. They are graceful and especially practical for long spans (Figures 67.31 through 67.35).

FIGURE 67.31   The Brooklyn Bridge is probably the best known of the classic U.S. bridges. It is one of the early uses of wire rope, being a combination suspension and cable-stayed span. Designed and built in 1883 by John and Washington Roebling for the City of New York. (Courtesy of American Society of Civil Engineers.)

FIGURE 67.32   The west span of the San Francisco–Oakland Bay Bridge is really two suspension bridges end to end with a central anchorage between the two. It is the only double-suspension bridge in the world. Opened in 1936, it is owned by the California Department of Transportation and designed by its predecessor, the California Division of Highways, Charles Andrew, Chief Bridge Engineer. (Courtesy of California Department of Transportation.)

FIGURE 67.33 The Golden Gate Bridge is one of the best-known landmarks in the United States. It spans the entrance to San Francisco Bay. It held the longest span, 1280 m, record for 27 years. Designed and built in 1937 by Charles B. Strauss. It is owned by the Golden Gate Bridge, Highway and Transportation District. (Courtesy of California Department of Transportation.)

FIGURE 67.34 The Mackinac Straits Bridge was the winner of the 1958 American Institute of Steel Construction's Artistic Bridge Award and several gold medals. Its design provided a level of aerodynamic stability never before attained in a suspension bridge. It has a main span of 1158 m. It was designed by David Steinman and is owned by the Mackinac Bridge Authority in northern Michigan. (Courtesy of David Steinman.)

FIGURE 67.35 The Verrazano Narrows Bridge, in New York City, has the longest span, 1298 m, of any bridge in America. Designed by Amman and Whitney, it was opened in 1964. (Courtesy of Metropolitan Transportation Authority, New York.)

Movable Bridges

See Figures 67.36 and 67.37.

FIGURE 67.36 The Tower Bridge in Sacramento earned an AISC design award in 1936, the year in which it was built. This unique lift span is clad in steel plate to cover the moving parts. It was designed by Leonard Hollister of the California Division of Highways. (Courtesy of California Department of Transportation.)

FIGURE 67.37 The double-swing span George P. Coleman Bridge over the York River in Yorktown, Virginia, was recently widened. To minimize impacts on heavy traffic flows, new truss sections were built in dry dock and floated into place as the old were floated out. The replacement designed by Parsons, Brinckerhoff, Quade and Douglas won the 1997 George P. Richardson Medal for outstanding achievement. (Courtesy of Parsons Brinckerhoff.)

Floating Bridge

See Figure 67.38.

FIGURE 67.38   The 2377-m-long Lacey V. Murrow Floating Bridge across Lake Washington near Seattle is composed of hollow concrete pontoons. The depth of water, 45.7 m, precludes piers, but there are some bridge spans over shallow water near the shore that can pass small vessels. It was designed by Charles Andrew and Clark Elkridge in 1940. The bridge is listed on the National Register of Historic places. (Courtesy of American Society of Civil Engineers.)

67.6   The Interstate Era

The Federal System of Interstate and Defense Highways following World War II gave another boost to highway and bridge building. The system designed to be nonstop, separated, and controlled access requires many bridges in order to function as planned. Old-time bridge engineers had a difficult time trying to adapt. Their experience up until then had been to bridge the low spot in valleys crossing over waterways. Now, bridge engineers found themselves building bridges over dry land, at ridges, and over the highways themselves. Several new innovations were spawned during this prolific period. Composite steel, concrete box girders, and prestressed concrete became routine.

Concrete Box Girders

This superstructure type, developed by Jim Jurkovich of the California Division of Highways, has good torsional stability and provides exceptional wheel load distribution across the girders. Concrete structures evolved into the preferred types, starting in California. California has an abundant source of aggregates and cement. Contractors learned to build them at costs competitive with steel (Figures 67.39 and 67.40).

FIGURE 67.39   The Four Level Interchange in downtown Los Angeles, built in 1950, is the first multilevel interchange of two freeways. It is a reinforced concrete box girder, a type developed by, and to become the hallmark of, the California Division of Highways. (Courtesy of California Department of Transportation.)

FIGURE 67.39 Mission Valley Viaduct sweeps Interstate Route 805 over the San Diego River floodplain in southern California. (Courtesy of California Department of Transportation.)

Prestressed Concrete

Prestressed concrete is a natural evolution of concrete girders. It makes the best use of the compressive qualities of concrete and the tensile properties of steel. Prestressing allows shallower structure depth, and a tremendous savings in approach roadway earthwork for interstate separations. Prestressed concrete can be either pretensioned or post-tensioned, precast or cast in place. All of these options have their place under different situations. The California Division of Highways pioneered this system in the 1940s and has since made extensive use of cast-in-place post-tensioned concrete box girders. The type has become so prevalent that construction contractors are able to build them for the same or less cost than normal reinforced concrete structures (Figures 67.41 and 67.42).

FIGURE 67.41   The Interstate Routes 105/110 Interchange in Los Angeles, California, is a massive forest of concrete columns supporting intertwined roadways. The *Smithsonian Magazine* highlighted the edifice as an artistic concrete creation in their January 1994 cover story. It was designed by Elweed Pomeroy of the California Department of Transportation. (Courtesy of Smithsonian Institution.)

FIGURE 67.42 Kellogg Central Business District Viaduct in Wichita, Kansas, designed by Howard Needles Tamman Bergendoff, won a Portland Cement Association award in 1996. (Courtesy of Howard Needles Tamman Bergendoff.)

Composite Steel

Composite steel girders, where a concrete deck is attached to the top flange of a steel girder through mechanical connectors, utilizes the best advantages of the compressive properties of concrete and the tensile properties of steel. While concrete was dominating the California and western bridge scene, steel remained the primary building material in the eastern and midwestern states (Figures 67.43 and 67.44).

FIGURE 67.43   The South Fork of the Eel River Bridge in Northern California exemplifies the virtues of composite steel structures. This 1958 bridge won an AISC award that year. (Courtesy of California Department of Transportation.)

FIGURE 67.44 The Cuyahoga River Valley Bridge built in 1980 for the Ohio Turnpike Authority earned an AISC award that year. It was designed by Howard Needles Tamman Bergendoff. (Courtesy of American Institute of Steel Construction.)

A Resurgence of Steel

As the Interstate Highway program began to utilize more and more concrete structures, during the 1960s, the steel and welding industry struggled to maintain its share of the bridge market. Many innovations were introduced for the use of steel through this campaign, by the development of exotic steels, distribution of design aids and examples, and conducting of design contests.

Steel Girders

See Figures 67.45 and 67.46.

FIGURE 67.45 The Eugene A. Doran Memorial (San Mateo Creek–Crystal Springs Reservoir) Bridge is a prize-winning bridge in a park setting. Sloping exterior facia plates provide web stiffening and aesthetic treatment. This welded plate steel girder bridge, built in 1970, was designed by Bob Cassano of the California Division of Highways. (Courtesy of California Department of Transportation.)

FIGURE 67.46 The Sacramento River Bridge at Elkhorn is a steel girder utilizing high-strength steel. Built in 1970 for Interstate Route 5, the bridge earned an AISC award that year. It was designed by Bert Bezzone of the California Division of Highways. (Courtesy of California Department of Transportation.)

Steel Box Girders and Orthotropic Steel Decks

See Figures 67.47 through 67.49.

FIGURE 67.47 The Klamath River crossing at Orleans in Northern California is a picturesque setting on a back road. There have been seven structures at this site, one burned and five have been washed away during major floods. The current steel box girder suspension span has lasted longer than any of its predecessors. Built in 1967, it was designed by Bert Bezzone of the California Division of Highways. (Courtesy of California Department of Transportation.)

FIGURE 67.48 San Mateo–Hayward Bridge over the San Francisco Bay, not only has composite steel approach spans, but the main span has an orthotropic steel deck. Listed among distinctive bridges, it won an American Institute of Steel Construction prize in 1968. It was designed by the California Division of Bay Toll Crossings. (Courtesy of California Department of Transportation.)

FIGURE 67.49 The Coronado Island Bridge over San Diego Bay in Southern California is a steel box girder. It earned an AISC award in 1970. It was designed by the California Division of Bay Toll Crossings. (Courtesy of California Department of Transportation.)

67.7 Era of the Signature Bridge

With the energetic vision of the great Interstate Era virtually complete, and the ensuing rush of the Seismic Retrofit Age winding down, bridge engineers turned their imaginative minds toward the building of great monuments.

Segmental Prestressed Bridges

Advanced technology of high-strength concrete and prestressing allows the cantilevering of structures out over deep valleys and bodies of water (Figures 67.50 and 67.51).

FIGURE 67.50 The California Department of Transportation experimented with and built its only segmental bridge, on Interstate Route 8 over Pine Valley in San Diego County in 1974. Bert Bezzone was the design engineer. The bridge received an American Society of Civil Engineers award in 1974. (Courtesy of California Department of Transportation.)

FIGURE 67.51   This graceful arch by Figg Engineering, carries the historic Natchez Trace over the park in Tennessee. It is the first and longest, 317 m, precast segmental arch. It received a design award of excellence in 1996. (Courtesy of Figg Engineering.)

Cable-Stayed Bridges

See Figures 67.52 and 67.53.

FIGURE 67.52 The new cable-stayed Sunshine Skyway, built in 1987, clearly a signature bridge, makes a bright statement over the entrance to Tampa Bay, Florida. This structure by Figg Engineering replaced the former Sunshine Skyway truss brought down by an errant barge that weighed more than that bridge. The new piers are protected by caissons as big and heavy as that barge. (Courtesy of Figg Engineering.)

FIGURE 67.53 The Cheasapeake and Delaware Canal Bridge owned by Delaware Department of Transportation was designed by Figg Engineering. It received an Excellence in Design Award in 1996. (Courtesy of Figg Engineering.)

Composites

The new definition of composite bridges has nothing to do with steel or concrete. Composites in modern usage refer to groups of organic chemical polymers commonly known as plastics. These are still experimental materials as far as bridges are concerned, but have been used successfully in other industries for some time now. Composites are now being used with fiber-wrap bridge columns as a seismic retrofit technique. The California Department of Transportation is currently designing an experimental span which will be concrete-filled composite tube girders with a composite deck.

FIGURE 67.54 Laurel Lick Bridge is the second all-composite bridge to be completed It is owned by the West Virginia Department of Highways and was built experimentally in conjunction with West Virginia University, in 1997. (Courtesy of West Virginia Department of Highways.)

67.8 Epilogue

All superstructure types are seen in combination, and with many variations. Even though seemingly prevalent during evolving eras, type periods greatly overlap, with type selections being more dependent upon crossing length and foundation conditions. In fact, every superstructure type is still being built today in response to various needs.

Let us all admire and learn from those Americans who have contributed, pioneered, and those who have consistently created award-winning structures of which all in the bridge-building profession can be proud. These include Squire Whipple, James Eads, Theodore Cooper, Gustav Lindenthal, Othmar Amman, David Steinman, Ralph Modjeski, Leon Moisseiff, John and Washington Roebling, Joseph Strauss, John Waddell, Conde McCullough, T.Y. Lin, Eugene Figg, the California Department of Transportation, and Howard Needles Tamman Bergendoff.

Index

U